Annual Reports in Organic Synthesis—1987

Annual Reports in Organic Synthesis

ANNUAL REPORTS IN ORGANIC SYNTHESIS—1970
John McMurry and R. Bryan Miller, Eds.

ANNUAL REPORTS IN ORGANIC SYNTHESIS—1971
John McMurry and R. Bryan Miller, Eds.

ANNUAL REPORTS IN ORGANIC SYNTHESIS—1972
John McMurry and R. Bryan Miller, Eds.

ANNUAL REPORTS IN ORGANIC SYNTHESIS—1973
R. Bryan Miller and Louis S. Hegedus, Eds.
John McMurry, Series Editor

ANNUAL REPORTS IN ORGANIC SYNTHESIS—1974
Louis S. Hegedus and Stephen R. Wilson, Eds.
R. Bryan Miller, Series Editor

ANNUAL REPORTS IN ORGANIC SYNTHESIS—1975
R. Bryan Miller and L. G. Wade, Jr., Eds.

ANNUAL REPORTS IN ORGANIC SYNTHESIS—1976
R. Bryan Miller and L. G. Wade, Jr., Eds.

ANNUAL REPORTS IN ORGANIC SYNTHESIS—1977
R. Bryan Miller and L. G. Wade, Jr., Eds.

ANNUAL REPORTS IN ORGANIC SYNTHESIS—1978
L. G. Wade, Jr., and Martin J. O'Donnell, Eds.

ANNUAL REPORTS IN ORGANIC SYNTHESIS—1979
L. G. Wade, Jr., and Martin J. O'Donnell, Eds.

ANNUAL REPORTS IN ORGANIC SYNTHESIS—1980
L. G. Wade, Jr., and Martin J. O'Donnell, Eds.

ANNUAL REPORTS IN ORGANIC SYNTHESIS—1981
L. G. Wade, Jr., and Martin J. O'Donnell, Eds.

ANNUAL REPORTS IN ORGANIC SYNTHESIS—1982
L. G. Wade, Jr., and Martin J. O'Donnell, Eds.

ANNUAL REPORTS IN ORGANIC SYNTHESIS—1983
Martin J. O'Donnell and Louis Weiss, Eds.

ANNUAL REPORTS IN ORGANIC SYNTHESIS—1984
Martin J. O'Donnell and Louis Weiss, Eds.

ANNUAL REPORTS IN ORGANIC SYNTHESIS—1985
Martin J. O'Donnell and Eric F. V. Scriven, Eds.

ANNUAL REPORTS IN ORGANIC SYNTHESIS—1986
Eric F. V. Scriven and Kenneth Turnbull, Eds.

Annual Reports in Organic Synthesis—1987

edited by

Eric F. V. Scriven
Reilly Tar and Chemical Corporation
Indianapolis, Indiana

Kenneth Turnbull
Wright State University
Dayton, Ohio

ACADEMIC PRESS, INC.
Harcourt Brace Jovanovich, Publishers
SAN DIEGO NEW YORK BERKELEY
BOSTON LONDON SYDNEY TOKYO TORONTO

ACADEMIC PRESS, INC.
1250 Sixth Avenue
San Diego, California 92101

United Kingdom Edition published by
ACADEMIC PRESS INC. (LONDON) LTD.
24-28 Oval Road, London NW1 7DX

LIBRARY OF CONGRESS CATALOG CARD NUMBER: 71-167779

ISBN 0-12-040818-X (alk. paper)

PRINTED IN THE UNITED STATES OF AMERICA
88 89 90 91 9 8 7 6 5 4 3 2 1

CONTENTS

PREFACE

One of the most difficult problems facing chemists today is that of "keeping up with the literature." For several reasons, the problem is particularly severe for the synthetic organic chemist. Bits of information of potential use are scattered throughout common chemistry journals and can be found in any paper, not just those dealing strictly with synthesis. Thus, synthetic chemists must read a large number of journals and must organize and index what they read to make the information available for future reference. All synthetic chemists do this, but the task is becoming more difficult each year as the flow of information increases.

The problem, however, is shared to some extent by all. Most organic chemists are at some time faced with the problem of synthesizing a desired material, and for many the problems are formidable. Nonspecialists faced with the synthetic problem are not likely to have kept pace with the developments in synthetic chemistry that may well solve their problems, and they will not have the necessary information in their files.

Thus, we felt that an organized annual review of synthetically useful information would prove beneficial to nearly all organic chemists, both specialist and nonspecialist in synthesis. It should help relieve some of the information-storage burden of the specialist and should enable the nonspecialist who is seeking help with a specific problem to become rapidly aware of recent synthetic advances. Ideally also, it should appear as promptly as possible after the close of the abstracting period. As in past years, we have placed particular emphasis on keeping the abstracts as concise as possible, while indicating the generality of the reactions involved. We have tried to combine similar publications into inclusive abstracts, particularly in Chapter 1. This practice has allowed us to include a larger number of references without a substantial increase in the book's length.

In producing *Annual Reports in Organic Synthesis—1987*, we have abstracted 49 primary chemistry journals, selecting useful synthetic advances. We have tried to present the information in an organized manner, emphasizing rapid visual retrieval. Only the common journals received by our libraries have been abstracted. Any journal received after March 1, 1988, will be covered in the next volume. We have also exercised selectivity in choosing which papers to abstract. Our general guidelines have been to include all reactions and methods that are new, synthetically useful, and reasonably general. Each entry is composed primarily of structures, accompanied by very few comments. The purpose of this emphasis is to aid the reader in scanning the book. The mind is capable of absorbing a whole picture in an instant, but is considerably slowed by having to read sentences. If the pictures presented catch the reader's interest, he or she should then seek details from the original paper.

We have included an author index based on the name of the senior author or sometimes the first author. No subject index is included because to do so would greatly increase both the cost of the book and the lead time for publication. Instead, we have chosen to use an extensive table of contents. Chapters I–III are organized by reaction type and constitute a major part of the book. The organization of these sections is self-explanatory; thus, there should be no difficulty in locating a new method of oxidation or a new cyclopropanation procedure. Chapter IV deals with methods of synthesizing heterocyclic systems, and Chapter V covers the use of new protecting groups. Chapter VI is divided into three main parts and covers those synthetically useful transformations that do not fit easily into the first three chapters. Chapter VII has been divided into sections in order to help the reader find quickly a review on a specific topic.

Any undertaking of this type involves a series of compromises. We have chosen to emphasize reasonable cost, rapid publication, and rapid visual retrieval of information at the admitted expense of detail and beauty.

The arduous task of drawing the multitude of structures appearing in this review was carried out by Andrea Ernest and Ursula Scriven. We thank them very much for their efforts.

Eric F. V. Scriven
Kenneth Turnbull

JOURNALS ABSTRACTED

Accounts of Chemical Research
Acta Chemica Scandinavica
Aldrichimica Acta
Angewandte Chemie International Edition in English
Australian Journal of Chemistry
Bulletin of the Chemical Society of Japan
Bulletin de Sociétés Chimiques Belges
Bulletin de la Société Chimique de France
Canadian Journal of Chemistry
Chemical Communications
Chemical and Pharmaceutical Bulletin
Chemical Reviews
Chemical Society Reviews
Chemische Berichte
Chemistry of Heterocyclic Compounds
Chemistry and Industry
Chemistry Letters
Collection of Czechoslovakian Chemical Communications
Gazzetta Chimica Italiana
Helvetica Chimica Acta
Heterocycles
Indian Journal of Chemistry
Journal of the American Chemical Society
Journal of Chemical Research
Journal of the Chemical Society (Perkin I)
Journal of the Chemical Society (Perkin II)
Journal of General Chemistry (USSR)
Journal of Heterocyclic Chemistry
Journal of Medicinal Chemistry
Journal of Organic Chemistry
Journal of Organic Chemistry (USSR)
Journal of Organometallic Chemistry
Journal für Praktische Chemie
Liebigs Annalen der Chemie
Monatschefte für Chemie
New Journal of Chemistry
Organic Preparations and Procedures International
Organic Syntheses
Organometallics
Pure and Applied Chemistry
Recueil des Travaux Chimiques des Pays-bas
Russian Chemical Reviews
Synthesis
Synthetic Communications
Tetrahedron
Tetrahedron Letters
Topics in Current Chemistry
Zeitschrift für Chemie
Zeitschrift für Naturforschung, Teil B

GLOSSARY OF ABBREVIATIONS

Ac acetyl
acac acetonylacetone
AIBN azobisisobutyronitrile
Am amyl
A-26 Amberlyst [R] A-26 resin
AMSO *N*-acetyl-(S)-methionine(R,S)sulfoxide
AOCOBT allyl-1-benzotriazolyl carbonate
9-BBN 9-borabicyclo[3.3.1]nonane
BINAL-H binaphthol modified aluminum hydride
BOC (*t*-Boc) *t*-butyloxycarbonyl
bpy bipyridyl
BSA *N,O*-bis silylacetamide
Bu butyl
Bn benzyl
Bz benzoyl
CAN ceric ammonium nitrate
Cbz benzyloxycarbonyl
clayfen clay-supported ferric nitrate
claycop clay-supported cupric nitrate
COD 1,5-cyclooctadiene
COT 1,3,5,7-cyclooctatetraene
Cp cyclopentadienyl
CRA complex reducing agents
CSA camphorsulfonic acid
CTAN ceric triethyl ammonium nitrate
DABCO 1,4-diazabicyclo[2.2.2]octane
DAM diaminomaleonitrile
DAST diethylaminosulfur trifluoride
dba dibenzylidene acetone
DBN 1,5-diazabicyclo[4.3.0]non-3-ene
DBTCE dibromotetrachloroethane
DBU 1,5-diazabicyclo[5.4.0]undec-5-ene
DCC dicyclohexylcarbodiimide
DCE dichloroethane
DCCI dicyclohexylcarbodiimide
DDQ 2,3-dichloro-5,6-dicyanobenzoquinone
de diastereomeric excess
DEAD diethyl azodicarboxylate
DEAEA 2-diethylaminoethylamine
DIAD diisopropyl azodicarboxylate
DIBAH (DIBAL) diisobutylaluminum hydride
diphos 1,2-bis(diphenylphosphino)ethane
DIOP 2,3-*O*-isopropylidene-2,3-dihydroxy-1,4-bis(diphenylphosphino)butane
DIPT diisopropyl tartrate
DMAD dimethyl acetylenedicarboxylate
DMAP 4-*N,N*-dimethylaminopyridine
DME dimethoxyethane
DMF dimethylformamide
DMSO dimethylsulfoxide
DPMS diphenylmethylsilyl
DPPE, dppe diphenylphosphinoethane
E^- general electrophile
EDAC-HCl 1-ethyl-3-(3-dimethylaminopropyl)carbodiimide hydrochloride
ee enantiomeric excess
Et ethyl
Fp η^5-$C_5H_5Fe(Co)_2$
FVT flash vacuum thermolysis
Hex hexyl
HMDS 1,1,1,3,3,3-hexamethyldisilazane
HMPA, HMPT hexamethyl phosphoramide
hν irradiation with light
KAPA potassium 3-aminopropylamide
K9-O-DIPGF-9-BBNH potassium 9-*O*-(1,2:5,6-di-*O*-isopropylidene-α-D-glucofuranosyl)-9-borabicyclo[3.3.1]nonane
L triphenylphosphine ligand
LAH lithium aluminum hydride
LDA lithium diisopropylamide
LICA lithium isopropylcyclohexylamide
Ln lanthanide metal
LTA lead tetraacetate
MCPBA m-chloroperbenzoic acid
Me methyl
MEM β-methoxyethoxymethyl
MOM methoxymethyl
MoOPH oxodiperoxymolybdenum(pyridine)hexamethylphosphoramide
Ms methanesulfonyl
MSA methanesulfonic acid
MTM methylthiomethyl
NBS *N*-bromosuccinimide
NCS *N*-chlorosuccinimide
NIS *N*-iodosuccinimide
Ni(R) Raney nickel
NMIM *N*-methylimidazole
NPSP *N*-phenylselenophthalimide
[O] general oxidation
Ⓟ polymeric backbone
PCC pyridinium chlorochromate
PCy_3 tricyclohexylphosphine

PDC pyridinium dichromate
PEG polyethylene glycol
Ph phenyl
(Phen) 1,10-phenanthroline
Phth phthaloyl
PLE pig liver esterase
Pn pentyl
PPA polyphosphoric acid
PPE polyphosphate ester
PPL porcine pancreatic lipase
PPSE polyphosphoric acid trimethylsilyl ester
Pr propyl
Py, pyr pyridine
PTC phase-transfer catalysis
PTSA *p*-toluene sulfonic acid
PPTS pyridinium *p*-toluenesulfonate
Q^+ quaternary ammonium
Reillex™425 polyvinyl pyridine 25% crosslinked with divinylbenzene
RT room temperature
SDS sodium dodecyl sulfonate
SEM β-trimethylsilylethoxymethyl
TBAB tetrabutylammonium bromide
TBDMS t-butyldimethylsilyl
TCNQ 7,7,8,8-tetracyanoquinodimethane
TDA-1 tris[(methoxyethoxy)ethyl]amine
Teoc 2-(trimethylsilyl)ethoxycarbonyl
Tf trifluoromethane sulfonate
TFA trifluoroacetic acid
TFFA trifluoroacetic anhydride
TFSA trifluoromethane sulfonic acid
Th 2-thienyl
THF tetrahydrofuran
THP tetrahydropyranyl
TMEDA, tmed tetramethylethylenediamine
TMP 2,2,6,6-tetramethylpiperidine
TMS trimethylsilyl
Tol tolyl
TPSCl 2,4,6-triisopropylbenzenesulfonyl chloride
TPPC tetraphenylporphyrin cobalt (II)
Tr trityl
Ts, Tos *p*-toluenesulfonyl
TSA toluenesulfonic acid
Z benzyloxycarbonyl
ziram zinc *N*, *N*-dimethylthiocarbamate
Δ heat
ϕ phenyl
18-C-6 18-crown-6
(((ultrasonication

I.A. Carbon - Carbon Single Bonds

(See also: I.E., I.F., I.G., I.H.)

I.A.1. Alkylations of Aldehydes, Ketones and Their Derivatives

I.A.1-1 D.L. Hughes and U.-H. Dolling et al., J. Org. Chem., 52, 4745 (1987); A.J. Poss and R.K. Belter, ibid., 52, 4810 (1987); B. Zwanenburg et al., Tetrahedron Lett., 28, 3027 (1987); C.S. Swindell and B.P. Patel, ibid., 28, 5275 (1987); E.S. Rao and J.S. Yadav, Ind. J. Chem., 25B, 1174 (1987); C. Iwata et al., Chem. Commun., 1802 (1987); M. Lissel et al., Leibigs Ann. Chem., 263 (1987); O. Hoshino, B. Umezawa et al., Chem. Pharm. Bull., 35, 2734 (1987); T.V. Lee and J. Toczek, J. Chem. Soc., Perkin Trans. 1 759 (1987); S.V. Ley et al., ibid, 121 and 131 (1987); M.L. Christiansen et al., Acta Chem. Scand., B41, 536 (1987).

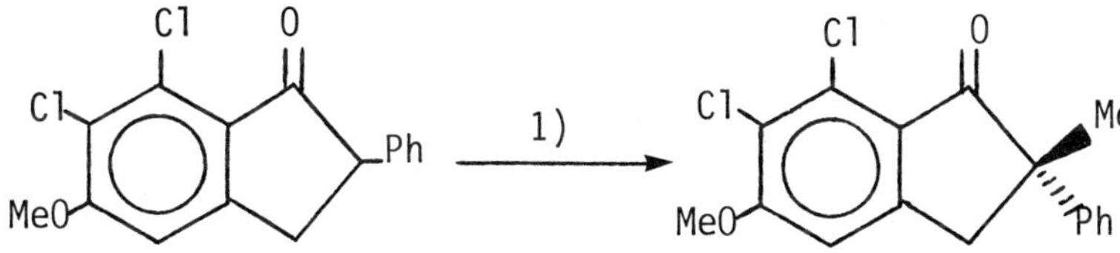

98%, 94% ee

1) MeCl, 50% NaOH, toluene, N-benzylcinchoninium halides (other papers use different bases, boron or zinc enolates or PTC)

I.A.1-2 S.G. Davies and M. Wills, J. Organomet. Chem., 328, C29 (1987); G. Bashiardes and S.G. Davies, Tetrahedron Lett., 28, 5563 (1987); S.G. Davies et al., J. Chem. Soc., Perkin Trans. 1, 489 (1987).

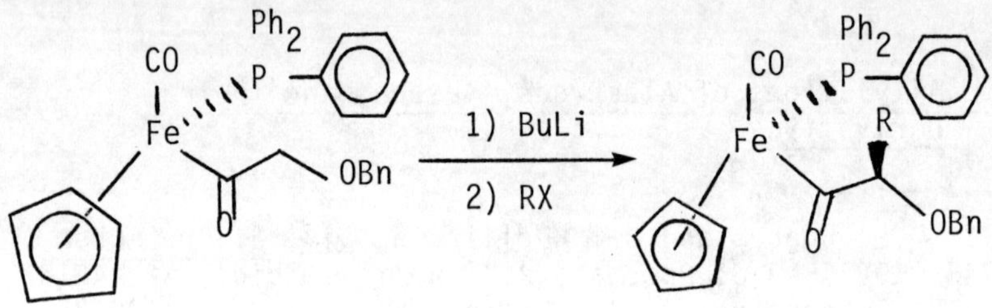

58-80%

>60 : 1 to >100 : 1 selectivity

I.A.1-3 S.E. Denmark et al., J. Org. Chem., 52, 165 (1987); M. Hayashi and T. Mukaiyama, Chem. Lett., 1283 (1987); J.H. Rigby and C. Senanayake, J. Am. Chem. Soc., 109, 3147 (1987).

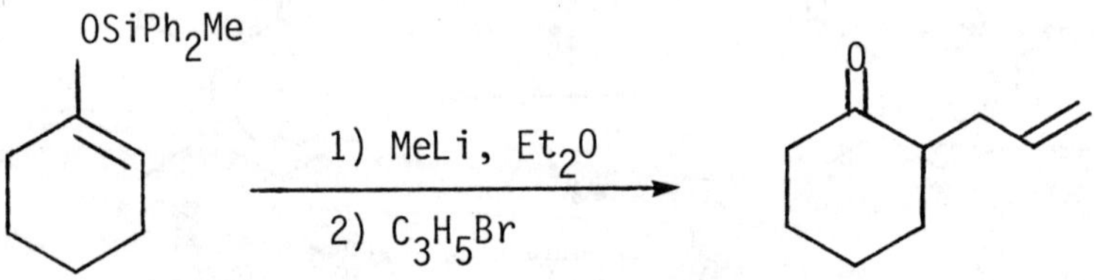

72%

DPMS - new protective group - stability intermediate between TMS and TBDMS

I.A.1-4 W.S. Johnson et al., J. Org. Chem., 52, 180 (1987); J.M. Cook et al., ibid., 52, 4110 (1987); A. Oku et al., Tetrahedron Lett., 28, 4847 (1987); I. Matsuda et al., ibid, 28, 6657 (1987); S. Torii et al., Bull. Chem. Soc. Jpn., 60, 2173 (1987); T. Mukaiyama et al., Chem. Lett., 1051 (1987); S. Kobayashi and T. Mukaiyama, ibid., 1183 (1987); M.T. Reetz and M. Sauerwald, J. Organomet. Chem., 328, 155 (1987); R.A. Cormier and T.E. Morris, Synth. Commun., 17, 1673 (1987); E. Nakamura and I. Kuwajima, Org. Synth., 65, 17 (1987).

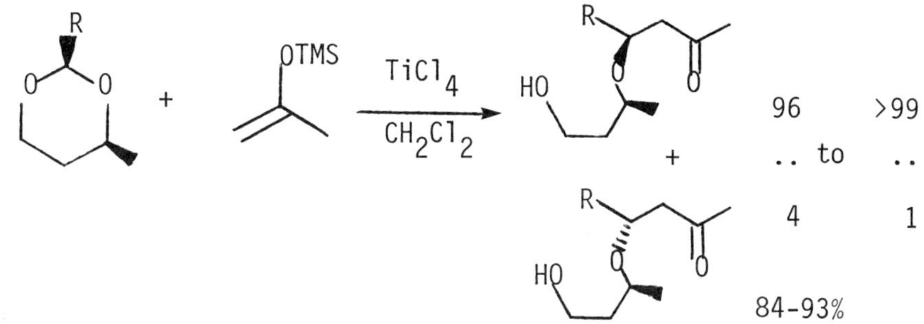

I.A.1-5 Y. Nagao et al., Chem. Lett., 133 (1987); H. Hagiwara and H. Uda, Chem. Commun., 1351 (1987); S.F. Martin et al., Org. Synth., 65, 119 (1987); T. Tsuda, T. Saegusa et al., J. Org. Chem., 52, 439 (1987).

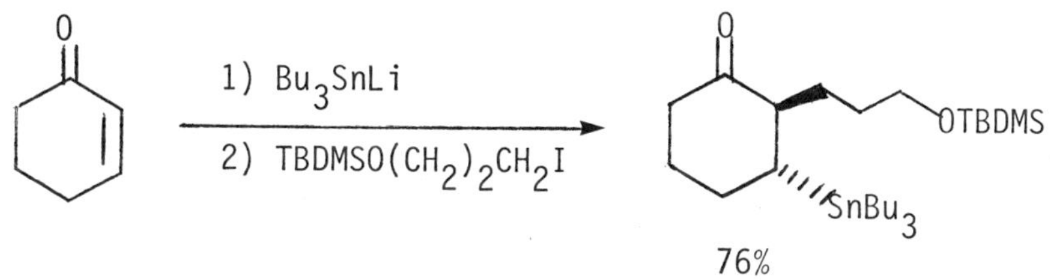

I.A.1-6 B.M. Foxman, M. Rosenblum et al., Organometallics, 6, 2394 (1987); A.J. Pearson and M.K. O'Brien, Chem. Commun., 1445 (1987).

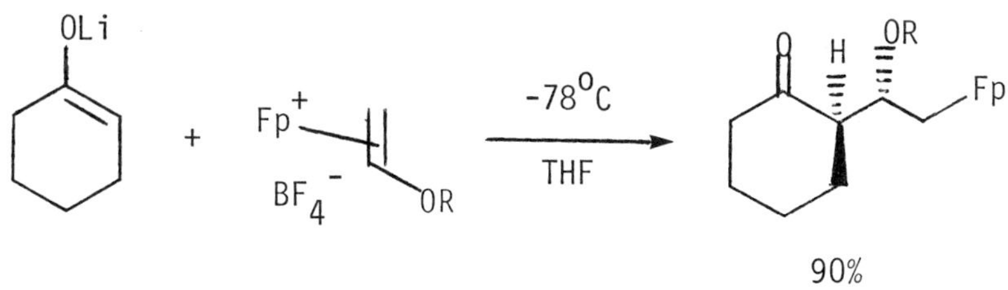

I.A.1-7 T. Tokoroyama et al., Chem. Commun., 358 (1987); P. Welzel et al., Tetrahedron Lett., 28, 3091 and 3095 (1987); D. Enders et al., ibid., 28, 3795 (1987); idem, Org. Synth., 65, 183 (1987); E. Brown and J. Lebreton, Tetrahedron, 43, 5827 (1987); I.R. Trehan et al., Ind. J. Chem., 25B, 1243 (1986).

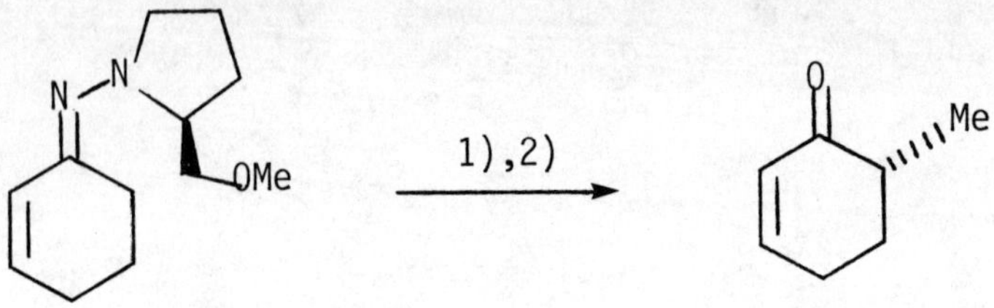

1) LDA then TsOMe 2) MeI then 2M HCl

I.A.1-8 T. Nakai et al., Tetrahedron Lett., 28, 5879 (1987); A. Hosomi et al., Chem. Commun., 570 (1987).

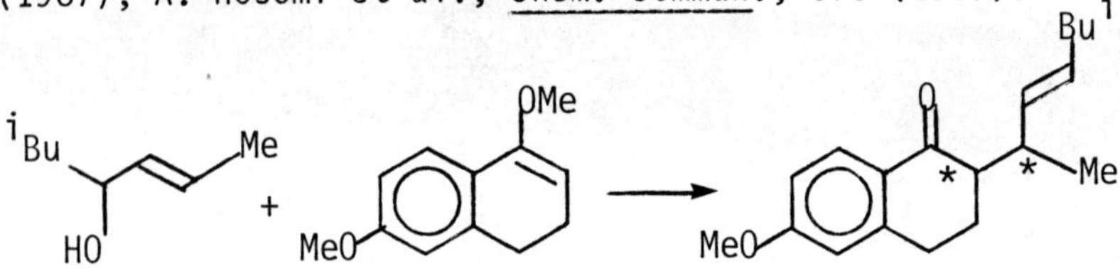

DMP 100°C, 10h anti:syn >98:<2, >95%

$PdCl_2(PhCN)_2$ rt, <1h anti:syn >2:>98, >95%

I.A.1-9 T. Mukaiyama et al., Chem. Lett., 1691 and 1695 (1987).

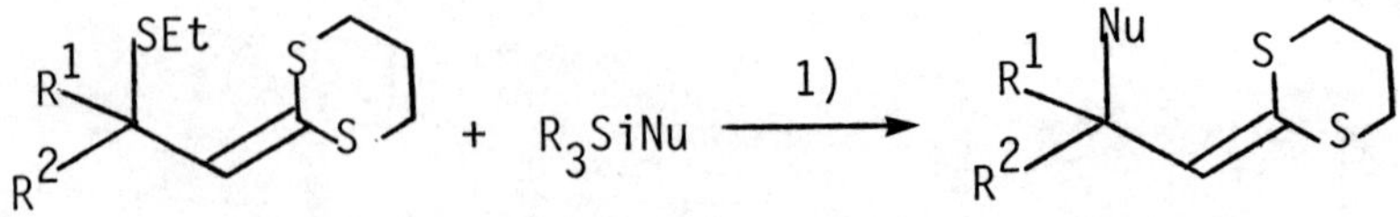

54-98%

1) Ph_3CBF_4

R_3SiNu = silyl enol ethers

I.A.1-10 J.H. Rigby and A.S. Kotnis, Tetrahedron Lett., 28, 4943 (1987); J. Nakayama et al., Chem. Lett., 939 (1987); A.T. Hewson et al., J. Chem. Soc., Perkin Trans. 1, 137 (1987); J. Drouin and M.A. Boaventura, Tetrahedron Lett., 28, 3923 (1987).

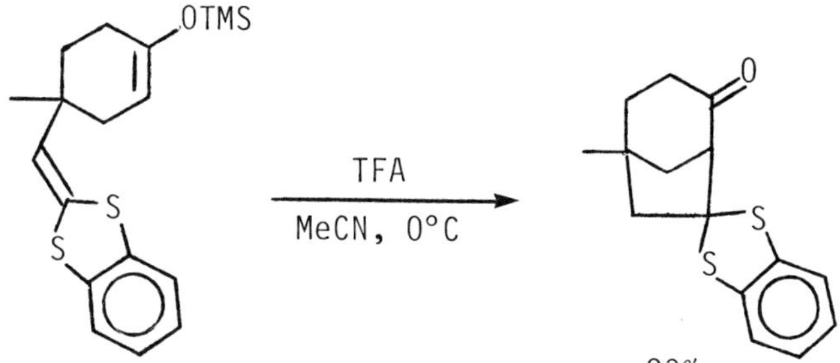

83%

similar results with $HgCl_2$, HMDS followed by NaI, HCl

I.A.1-11 T. Umemoto and Y. Gotoh, Bull. Chem. Soc. Jpn., 60, 3823 (1987); R.M. Moriarty et al., J. Chem. Soc., Perkin Trans. 1, 559 (1987)

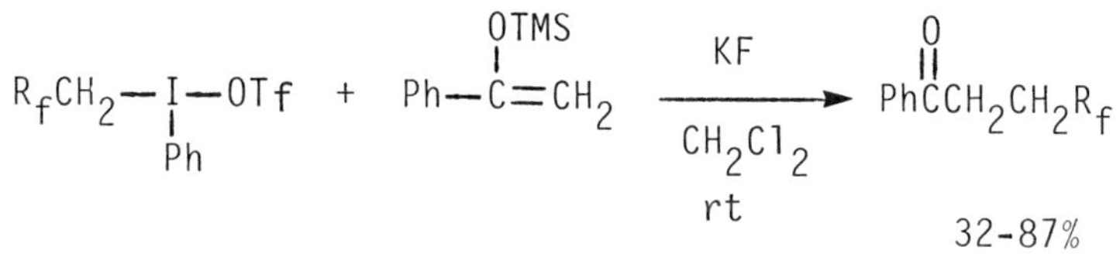

32-87%

R_f = CF_3, C_2F_5, i-C_3F_7, n-C_7F_{15}

I.A.1-12 Y. Hayashi and T. Mukaiyama, Chem. Lett., 1811 (1987); R.M. Moriarty et al., Tetrahedron Lett., 28, 873 (1987); E. Baciocchi, G. Civitarese and R. Ruzziconi, ibid., 28, 5357 (1987).

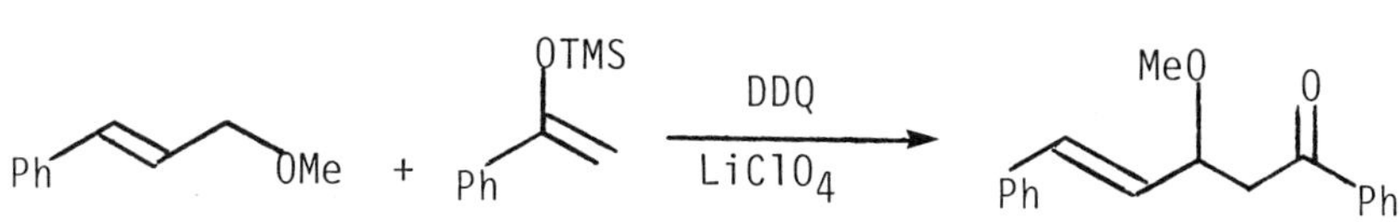

84%

other nucleophiles also used

38% without $LiClO_4$

LTA and CAN used similarly

I.A.1-13 J.L. Belletire and E.G. Speltzer, Synth. Commun., 17, 1701 (1987); J.L. Belletire and D.F. Spry, J. Org. Chem., 52, 2549 (1987); P. Camps et al., Tetrahedron Lett., 28, 1831 (1987); T.F. Titova and A.P. Krysin, J. Org. Chem. (USSR), 23, 540 (1987).

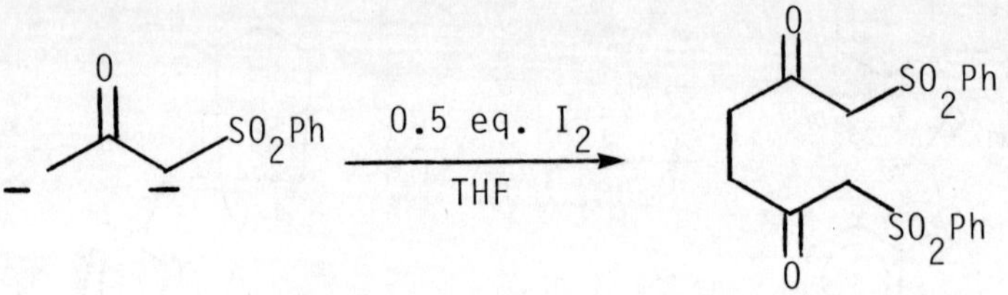

>64%

I.A.1-14 J. Tsuji et al., J. Org. Chem., 52, 2988 (1987); idem, Synthesis, 992 (1987); idem, Tetrahedron Lett., 28, 2397 (1987).

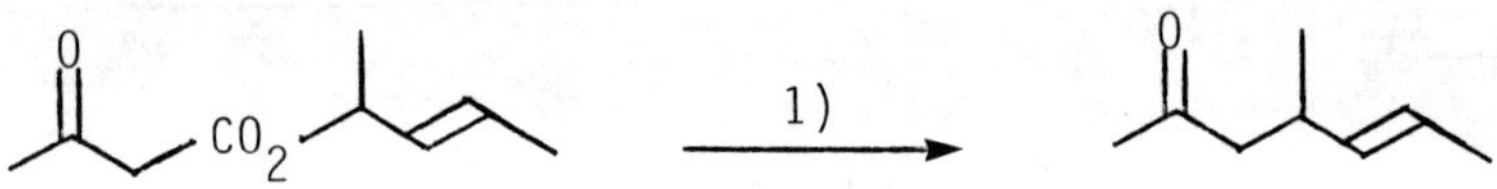

74%

1) $Pd(OAc)_2$- PPh_3, 90°C, 1h

I.A.1-15 J.D. Elliot et al., Tetrahedron Lett., 28, 6557 (1987).

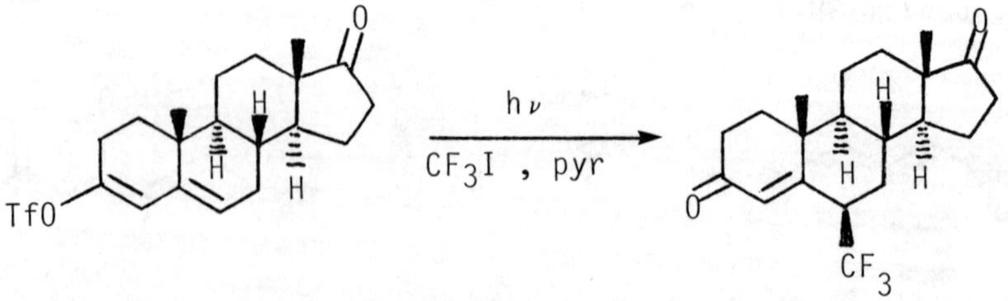

80%

I.A.2. Alkylations of Nitriles, Acids and Acid Derivatives

I.A.2-1 K. Narasaka et al., Bull. Chem. Soc. Jpn., 60, 1457 (1987); D. Seebach and H. Estermann, Tetrahedron Lett., 28, 3103 (1987); G. Lannoye and J.M. Cook, ibid., 28, 4821 (1987); P.J. Garrat et al., ibid., 28, 589 (1987); J. Sterling et al., ibid., 28, 1685 (1987); B.D. Gray and P.W. Jeffs, Chem. Commun., 1329 (1987); A. Zapata et al., Synth. Commun., 17, 1165 (1987); M. Vaultier et al., Tetrahedron, 43, 1811 (1987); T. Kitahara, M. Mori and K. Mori, ibid., 43, 2689 (1987).

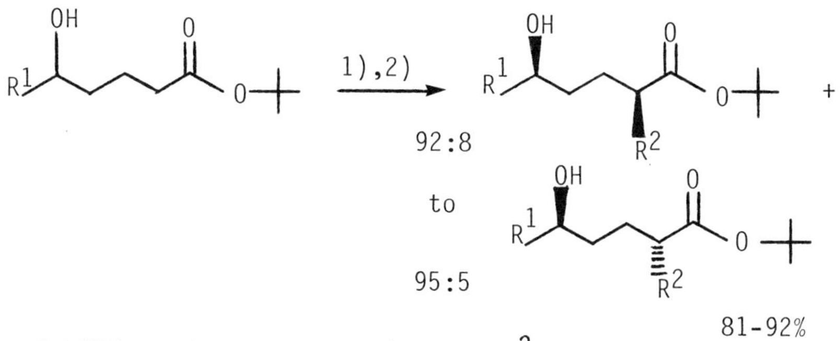

1) 3 $LiNMe_2$, THF-HMPA, -100°C 2) R^2X

I.A.2-2 C. Agami and F. Couty, Tetrahedron Lett., 28, 5659 (1987); C. Agami et al., Synth. Commun., 17, 241 (1987); D. Seebach et al., Helv. Chim. Acta, 70, 1194 (1987); G. Antoni and B. Langstrom, Acta Chem. Scand., B41, 511 (1987); D. Kim and H.S. Kim, J. Org. Chem., 52, 4634 (1987); M. Larcheveque and S. Henrot, Tetrahedron, 43, 2303 (1987); I. Ojima and X. Qiu, J. Am. Chem. Soc., 109, 6537 (1987).

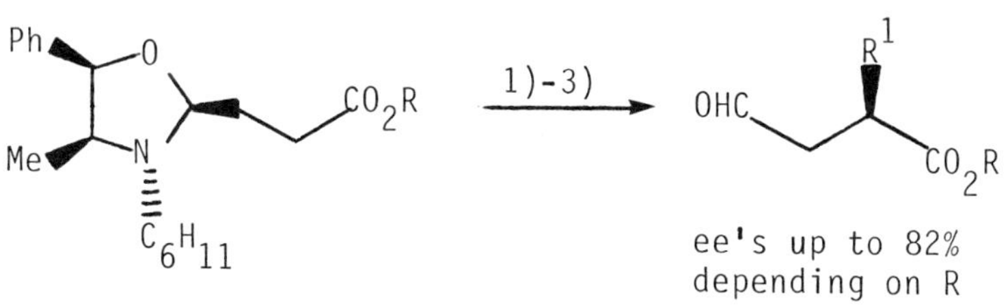

1) LDA, THF 2) R^1X 3) SiO_2, H_2O

I.A.2-3 T. Oesterle and G. Simchen, Liebigs Ann. Chem., 693 (1987); T. Mukaiyama et al., Chem. Lett., 1243 (1987); E. Akgun et al., Monatsh. Chem., 118, 363 (1987); T. Frejd et al., Tetrahedron Lett., 28, 2753 (1987); Y. Kita et al., Chem. Pharm. Bull., 35, 562 (1987).

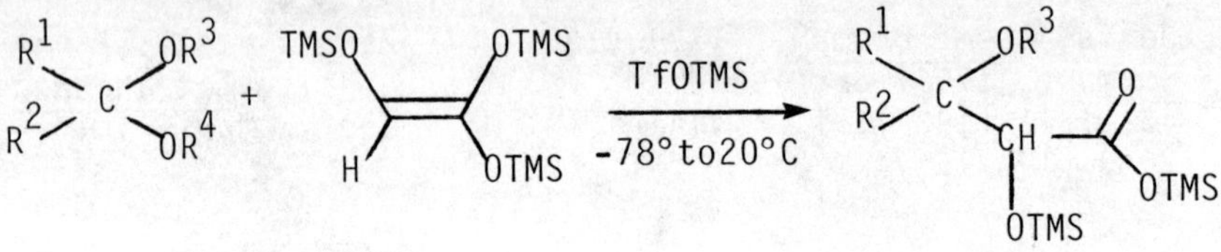

42-100%

I.A.2-4 Y. Watanabe et al., J. Org. Chem., 52, 1695 (1987); Y. Ito, T. Hayashi et al., Tetrahedron Lett., 28, 4849 (1987); S.A. Lebedev et al., J. Org. Chem. (USSR), 22, 1410 (1986).

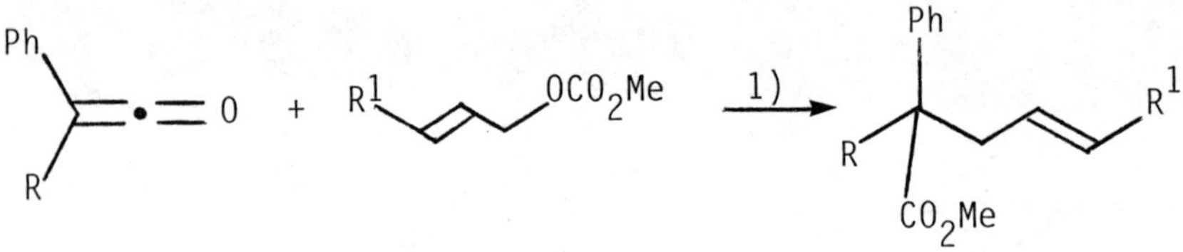

R = Ph, Et

1) $Pd(PPh_3)_4$, DMF, 0°C

41-90%

I.A.2-5 P. Duhamel et al., Tetrahedron Lett., 28, 3801 (1987).

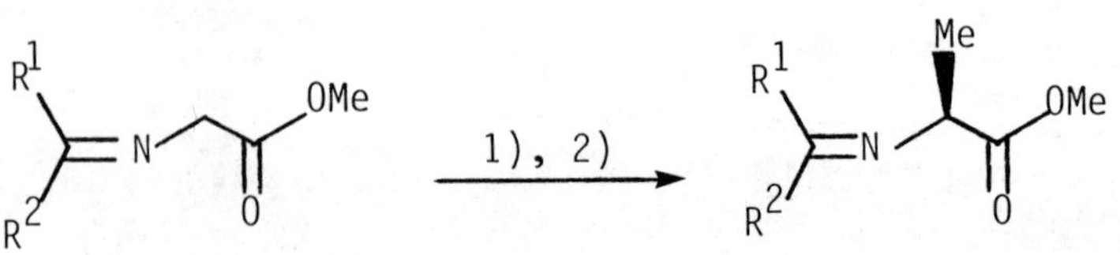

57-60% , 33-40% ee

1) LDA 2) A^*OSO_2OMe, THF-HMPA

A^* = sugar moieties

I.A.2-6 J.H. Balder and S.J. Sarussi, J. Org. Chem., 52, 3462 (1987); J.-Y Nedelec et al., J. Chem. Res. (S), 315 (1987).

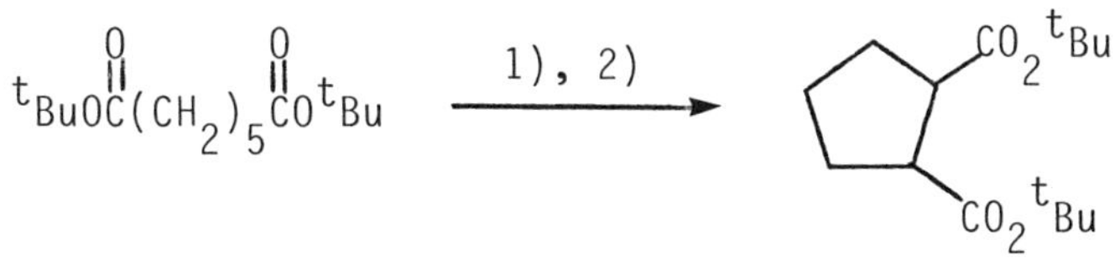

1) 2LDA 2) $CuCl_2$ >50%

I.A.2-7 Z. Vit and M. Hajek, Coll. Czech. Chem. Commun., 52, 1280 (1987).

80%, 95 : 5

I.A.2-8 A. Ando and T. Shiori, Chem. Commun., 656 (1987).

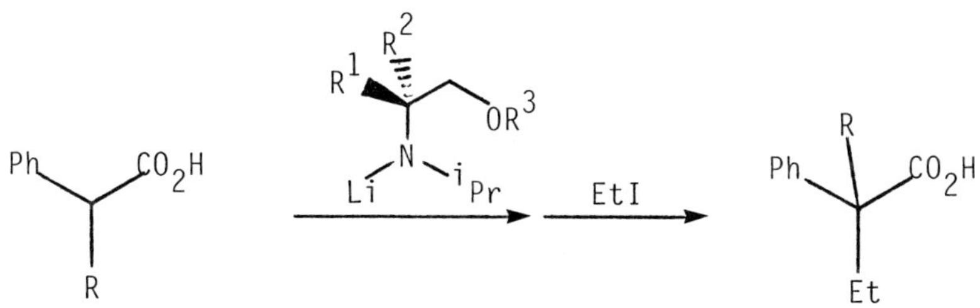

R = H, Me 38-82% , 4-20% ee

I.A.2-9 D. Seebach et al., Helv. Chim. Acta, 70, 1676 (1987).

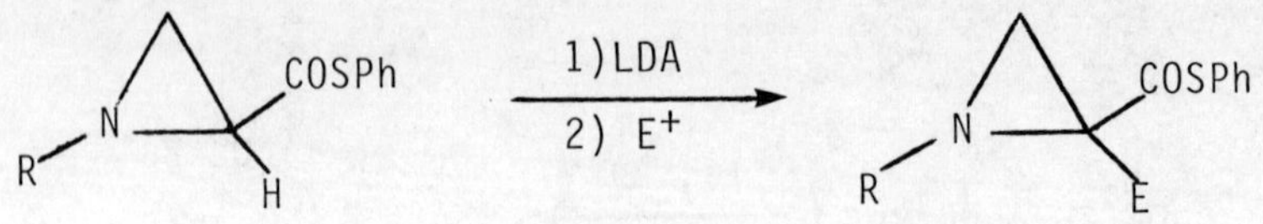

61-87%

88:12 to 72:18 ds

I.A.2-10 A. Amrollah-Madjdabadi and L. Stella, Bull. Soc. Chim. Fr., 350 (1987); E. Piers and J.S.M. Wai, Chem. Commun., 1342 (1987); R.K. Venkateswaran et al., Tetrahedron Lett., 28, 4329 (1987); H. Barrios et al., Org. Prep. Proced. Int., 19, 427 (1987); P. Assithianakis and H. Stamm, Chem. Ber., 120, 855 and 1239 (1987).

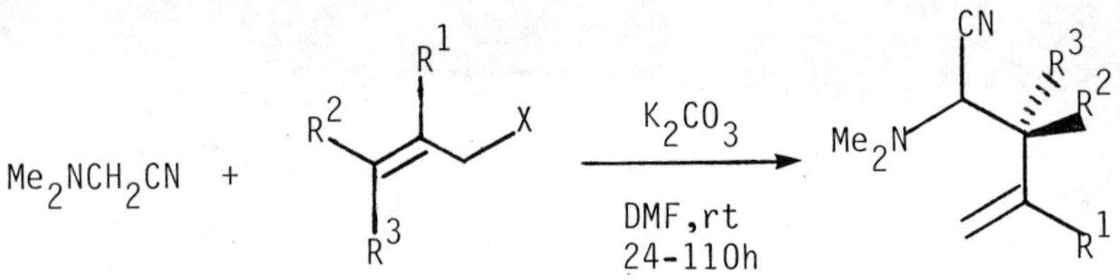

85-95%

X = Br, Cl

I.A.2-11 H.J. Bestmann and M. Schmidt, Angew. Chem., Int. Ed. Engl., 26, 79 (1987); idem, Tetrahedron Lett., 28, 2111 (1987); J.E. Baldwin et al., Chem. Commun., 153 (1987).

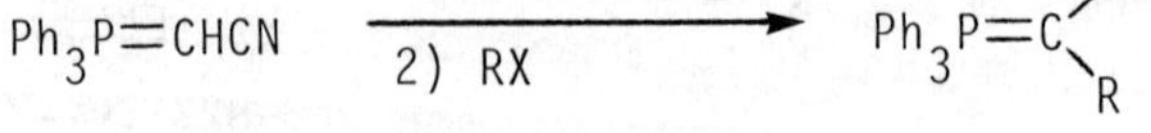

51-93%

I.A.2-12 A.I. Meyers and B.A. Lefker, Tetrahedron, 43, 5663 (1987); S.L. Schreiber et al., J. Am. Chem. Soc., 109, 5749 (1987); A.T. Russell and G. Procter, Tetrahedron Lett., 28, 2041 and 2045 (1987).

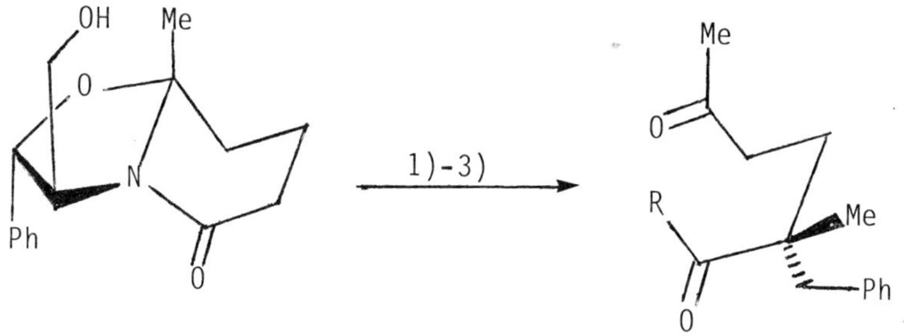

R = Me, Bu ~ 40-45%

1) LDA, MeI ; LDA, BnBr 2) RLi 3) H_2PO_4-H_2O

I.A.3. Alkylation of β-Dicarbonyl, β-Cyanocarbonyl Systems and Other Active Methylene Compounds

I.A.3-1 P. Deslongschamps et al., Can. J. Chem., 65, 43 and 1298 (1987); T. Fuchigami and Y. Nakagawa, J. Org. Chem., 52, 5276 (1987); M. Okahara et al., ibid., 52, 3192 (1987); S.C. Welch et al., ibid., 52, 1440 (1987); H. Saito and T. Hirata, Tetrahedron Lett., 28, 4065 (1987); E. Ghera and R. Maurya, ibid., 28, 709 (1987); L. Toke et al., Synthesis, 565 (1987); K. Mullen et al., Angew Chem., Int. Ed. Engl., 26, 1291 (1987); T. Osa et al., Chem. Lett., 2235 (1987).

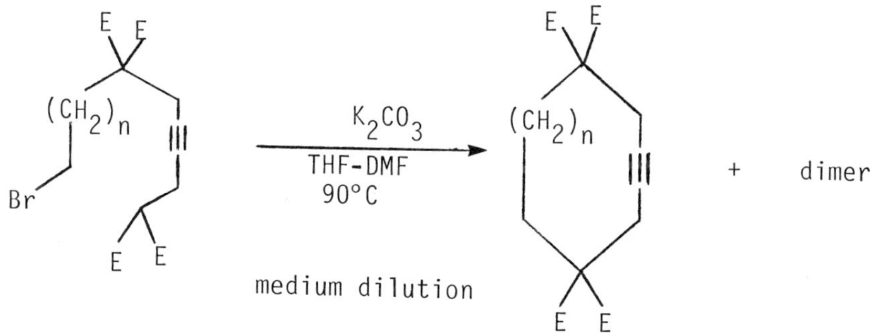

E = CO_2Me n = 4,5 37-42% 18-19%

other papers use different substrates and bases, electrogenerated bases or PTC

I.A.3-2 P. Tundo et al., J. Chem. Soc., Perkin Trans. 1, 2159 (1987).

Alkylation Reactions of Ethyl Malonate, Ethyl Acetoacetate, and Acetylacetone by Gas-Liquid Phase-Transfer Catalysis (G.L. - P.T.C.)

I.A.3-3 B.M. Trost et al., J. Am. Chem. Soc., 109, 2176 (1987); Y. Xu and B. Zhou, J. Org. Chem., 52, 974 (1987); T. Cuvigny and M. Julia, J. Organomet. Chem., 331, 121 (1987); J. Gore et al., Tetrahedron., 43, 3441 and 3453 (1987); Y. Masuyama et al., Bull. Chem. Soc. Jpn., 60, 1525 and 1527 (1987).

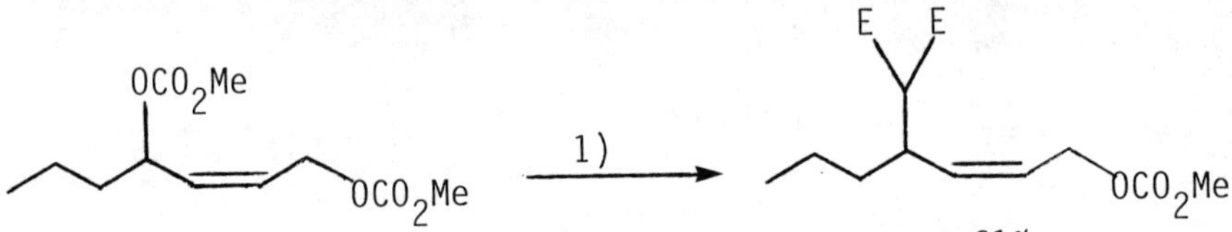

61%

$E = CO_2Me$

displacement of terminal carbonate with Pd(0)

1) $NaCHE_2$, $(MeCN)_3W(CO)_3$, bpy

$Bu_4N[Fe(CO)_3NO]$, TBA-$CuBr_2$, $Pd(dba)_2/6PPh_3$

and bpy, $Mo(CO)_6$ used similarly

I.A.3-4 J.-E. Backvall and S.K. Juntunen, J. Am. Chem. Soc., 109, 6396 (1987); B.M. Trost and M. Lautens, ibid., 109, 1469 (1987); J. Zhu and X. Lu, Tetrahedron Lett., 28, 1897 (1987); F.E. Ziegler and A. Kneisley, ibid., 28, 1725 (1987); K. Burgess, J. Org. Chem., 52, 2046 (1987).

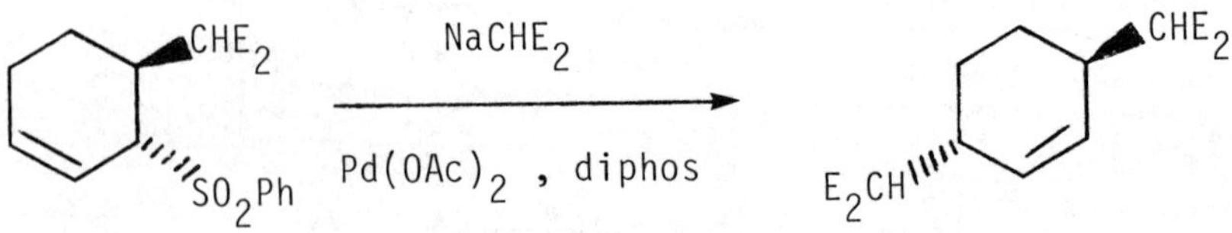

$E = CO_2Me$

86%, 87 : 13 trans : cis

other catalysts and leaving groups utilized similarly

I.A.3-5 T. Hayashi, A. Yamamoto and Y. Ito, Chem. Lett., 177 (1987); B.M. Trost and M. Lautens, Tetrahedron., 43, 4817 (1987); N. Ono et al., J. Chem. Soc., Perkin Trans. 1, 1439 (1987); B. Akermark et al., J. Organomet. Chem., 335, 133 (1987); Y. Masuyama et al., Tetrahedron Lett., 28, 443 (1987); idem, Bull. Chem. Soc. Jpn., 60, 3431 (1987).

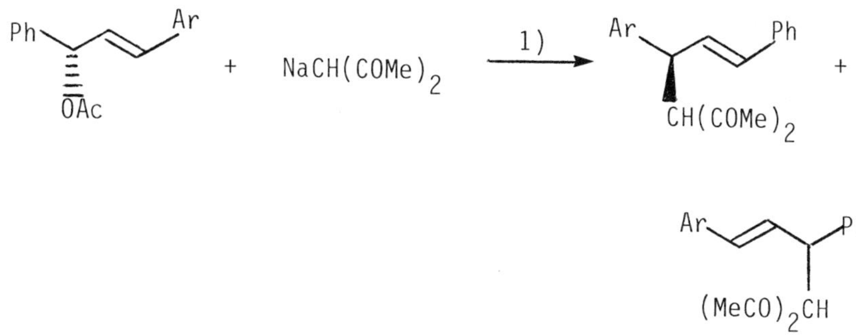

92-97%, 44 : 56 to 100 : 0

1) 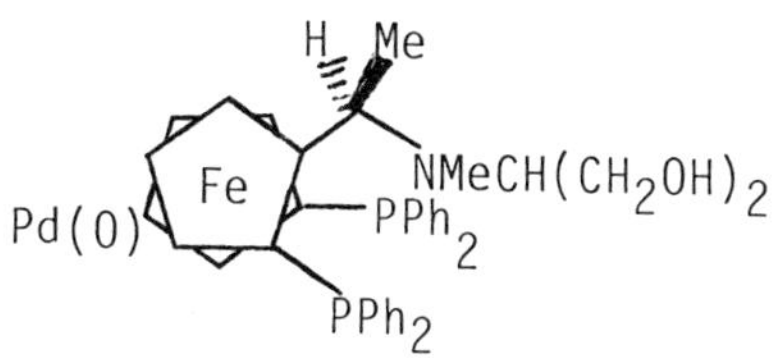

, THF, 40°C

non chiral catalysts such as $Mo(CO)_4bpy$, $Mo(CO)_6$, $Pd(PPh_3)_4$ also used

---

I.A.3-6 S. Takahashi et al., Chem. Commun., 785 (1987); M. Ahmar et al., Tetrahedron, 43, 513 (1987).

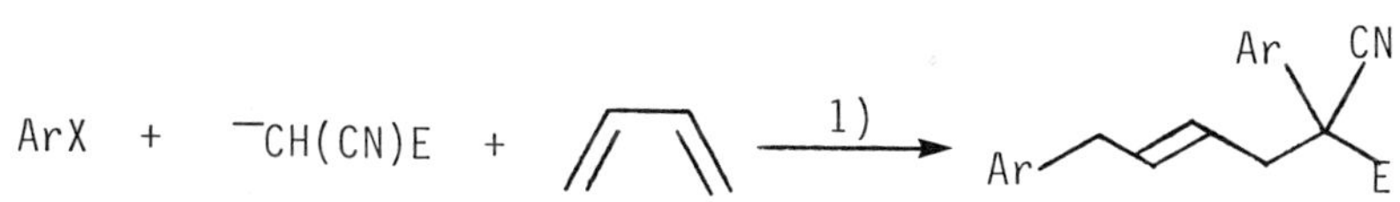

E = CO_2Me 36-62%

1) $(Ph_3P)_2PdCl_2$, THF, Δ

I.A.3-7 A.R. Katritzky and J.M. Aurrecoechea, <u>Synthesis</u>, 342 (1987).

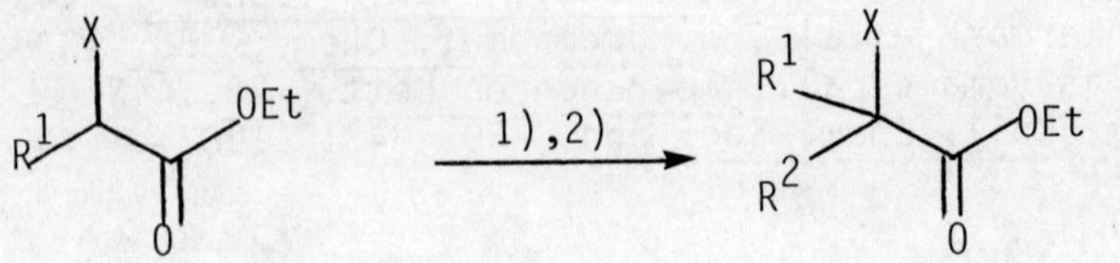

X = CO_2Et, COMe, CN

1) NaH, toluene or DMSO

2) 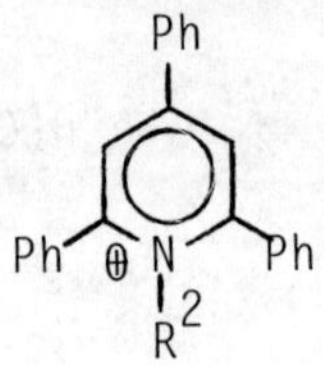 , 60-110°C

I.A.3-8 T. Sakai et al., <u>Bull. Chem. Soc. Jpn.</u>, <u>60</u>, 1945 (1987).

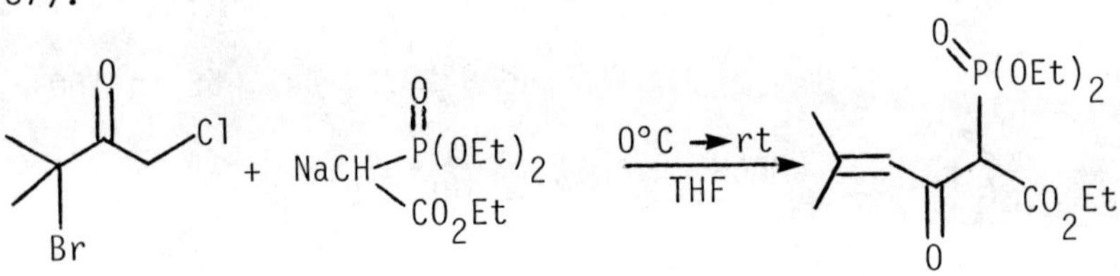

79%

I.A.3-9 E.J. Corey and A.K. Ghosh, <u>Tetrahedron Lett.</u>, <u>28</u>, 175 (1987); B.B. Snider et al., <u>ibid.</u>, <u>28</u>, 841 and 845 (1987); B.B. Snider and M.A. Dombroski, <u>J. Org. Chem.</u>, <u>52</u>, 5487 (1987).

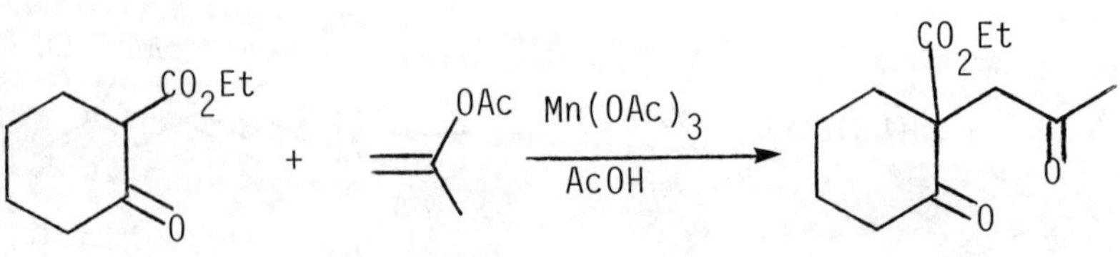

71%

I.A.3-10 J. Marquet et al., Tetrahedron Lett., 28, 3715 (1987); M. Moreno-Manas et al., Chem. Commun., 644 (1987).

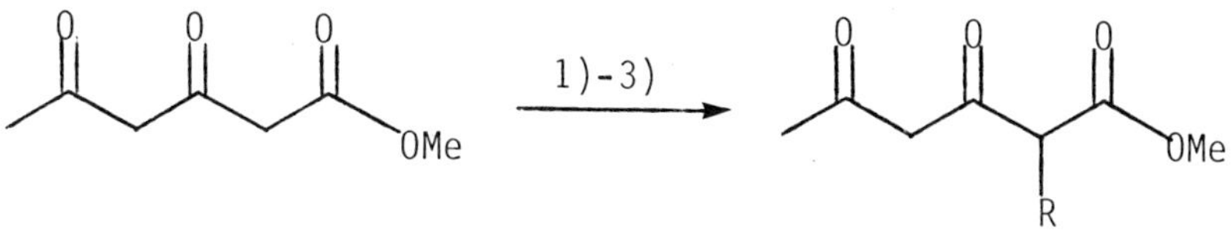

60-94%

1) copper acetate, H_2O-MeOH 2) NaH, THF, reflux 3) RX

I.A.3-11 L. Zhiyu and Y. Junjie, Synth. Commun., 17, 1617 (1987); R.K. Haynes and A.G. Katsifis, Chem. Commun., 340 (1987); M.-A. Boaventura and J. Drouin, Synth. Commun., 17, 975 (1987).

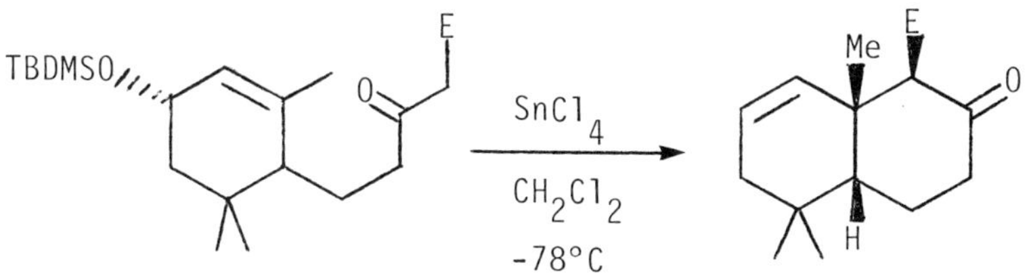

E = CO_2Me 79%

I.A.4. Alkylation of N-, P-, S-, Se- and Similar Stabilized Carbanions

I.A.4-1 H. Shirahama et al., Tetrahedron Lett., 28, 5665 (1987); R. Gleiter et al., J. Org. Chem., 52, 3979 (1987).

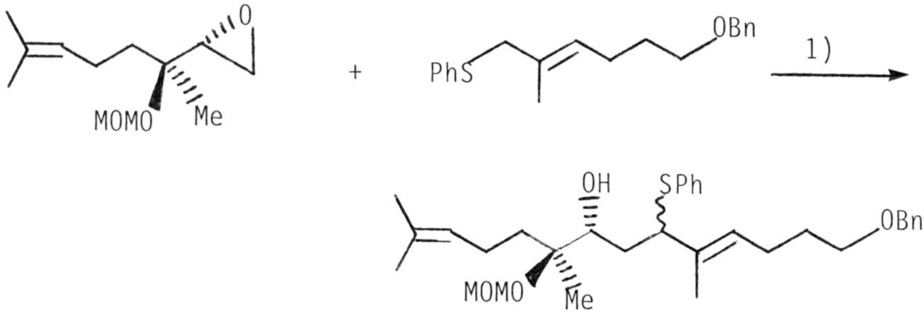

78%

1) BuLi, DABCO, THF, -78°C, 3h

I.A.4-2 R. Tanikaga et al., *Tetrahedron Lett.*, **28**, 3705 (1987); T. Fujisawa et al., *ibid.*, **28**, 5677 (1987); P. Bravo et al., *J. Chem. Soc., Perkin Trans. 1*, 1405 (1987); L. Guang-jian et al., *Synth. Commun.*, **17**, 1839 (1987).

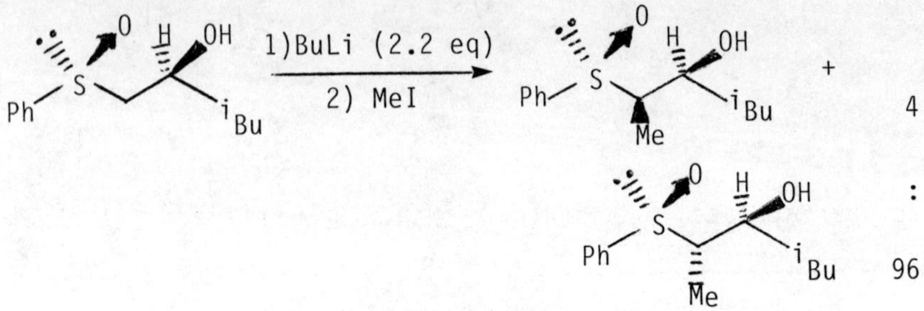

96%

I.A.4-3 Y. Kishi et al., *J. Am. Chem. Soc.*, **109**, 6205 (1987); L.A. Paquette et al., *ibid.*, **109**, 3730 (1987); T. Chou et al., *J. Org. Chem.*, **52**, 244 and 5082 (1987); P.L. Fuchs et al., *Synth. Commun.*, **17**, 55 (1987); N.S. Simpkins, *Tetrahedron Lett.*, **28**, 989 (1987).

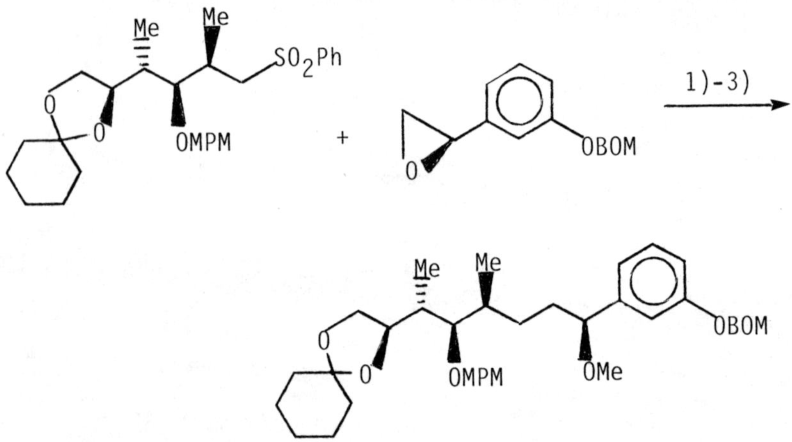

54%

BOM = benzyloxymethyl MPM = p-methoxybenzyl

1) 2 BuLi, rt 2) Na-Hg, Na_2HPO_4, MeOH, rt

3) MeI, KOH, DMSO, rt

I.A.4-4 A.J. Pearson et al., Tetrahedron Lett., 28, 2459 (1987).

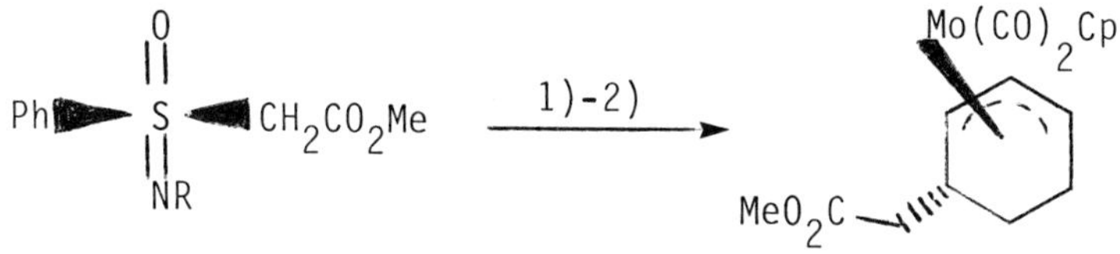

45-70%, 11-89% ee

1) NaH then 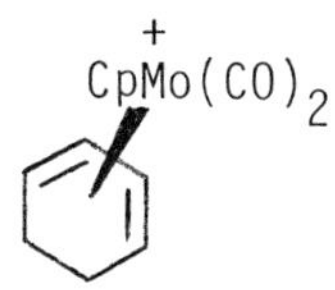 , 0°C 2) Na/Hg or Al/Hg

---

I.A.4-5 M.P. Crozet et al., Tetrahedron Lett., 28, 5531 (1987); N. Kornblum et al., J. Org. Chem., 52, 196 and 5301 (1987); K.E. Kim and H.G. Adolph, Synthesis, 1029 (1987).

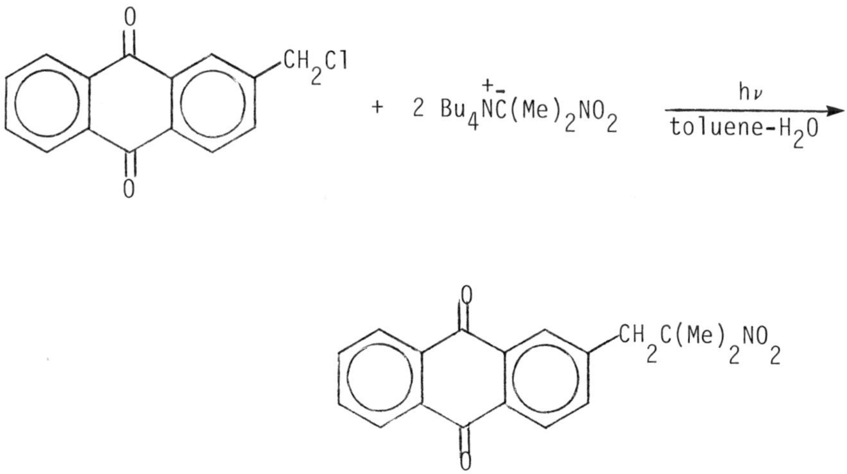

77%

I.A.4-6 U. Schollkopf and R. Schutze, Liebigs Ann. Chem., 45 (1987).

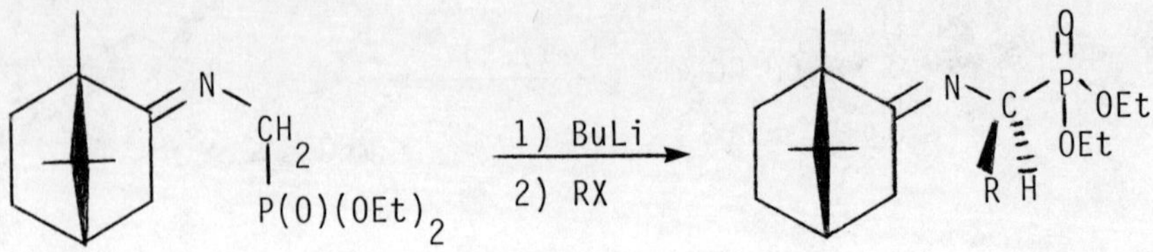

from (+) camphor 54-88%, 11 to ≥ 95% de

I.A.5. Alkylations of Organometallic Reagents

(see also: I.F., I.G.)

I.A.5-1 T. Katsuki et al., Tetrahedron Lett., 28, 6191 and 6195 (1987); J.M. Chong et al., ibid., 28, 5009 (1987); M. Larcheveque et al., ibid., 28, 1781 and 1993 (1987); M.J. Kurth and M.A. Abreo, ibid., 28, 5631 (1987); R.D. Tung and D.H. Rich, ibid., 28, 1139 (1987); D. Tanner and P. Somfai, Tetrahedron, 43, 4395 (1987).

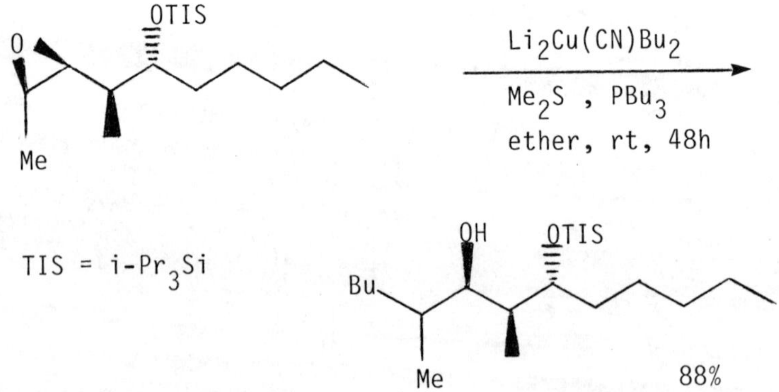

$BF_3 \cdot OEt_2$ also used on occasion

I.A.5-2 J.P. Marino et al., J. Org. Chem., 52, 4898 (1987); J.A. Marshall and V.H. Audia, ibid., 52, 1106 (1987); J.A. Marshall and J.D. Trometer, Tetrahedron Lett., 28, 4985 (1987); P. Mangeney et al., ibid., 28, 2363 (1987).

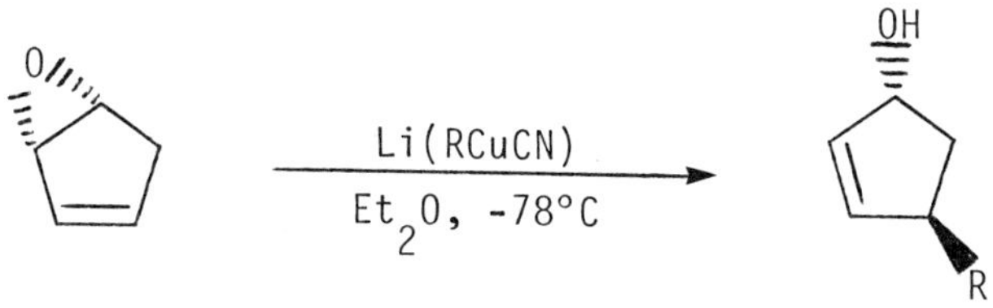

R = Me, Bu, ^{t}Bu 78-95%

towards prostaglandins

I.A.5-3 H. Stamm et al., Tetrahedron, 43, 565 (1987).

Reactions with Aziridines - 39. Ring Opening of Activated 2,2-Dimethylaziridines by Grignard Reagents. A Mechanistic Study.

I.A.5-4 S.H. Bertz et al., Tetrahedron Lett., 28, 4251 (1987).

The Preparation of Lithium Organocuprates from Various Cu(I) Salts.

I.A.5-5 E.C. Ashby and D. Coleman, J. Org. Chem., 52, 4554 (1987).

Evidence for Single Electron Transfer in the Reactions of Lithium Dimethylcuprate with Alkyl Halides.

I.A.5-6 J. Soulie et al., Tetrahedron, 43, 2701 (1987); G.L. Larson et al., Synth. Commun., 17, 385 (1987); A. Hosomi et al., J. Org. Chem., 52, 2947 (1987); Z. Yoshida et al., ibid., 52, 4418 (1987); N.K. Hao et al, J. Org. Chem. (USSR), 23, 449 (1987); R.H. Schlessinger and E.J. Iwanowicz, Tetrahedron Lett., 28, 2083 (1987); H. Amri and J. Villieras, ibid., 28, 5521 (1987); E. Nakamura, K. Sekiya and I. Kuwajima, ibid., 28, 337 (1987).

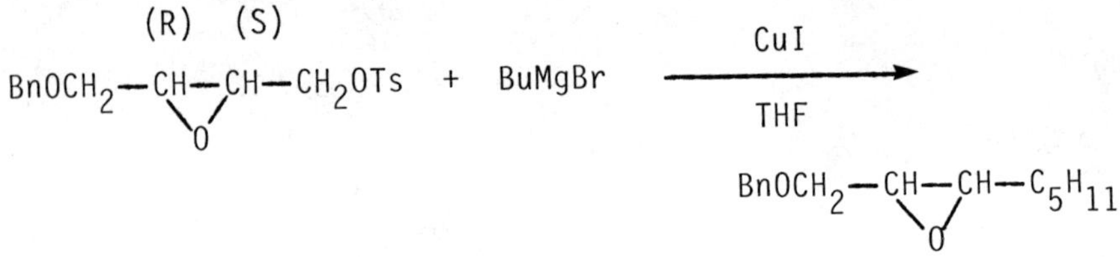

60%

alkyl zincs and other leaving groups also employed

I.A.5-7 T. Ibuka, Y. Yamamoto et al., Chem. Commun., 1596 (1987); J.-E. Backvall and M. Sellen, ibid., 827 (1987); P. Yates et al., Can. J. Chem., 65, 69 (1987); W.H. Okamura et al., Tetrahedron Lett., 28, 2917 and 2921 (1987).

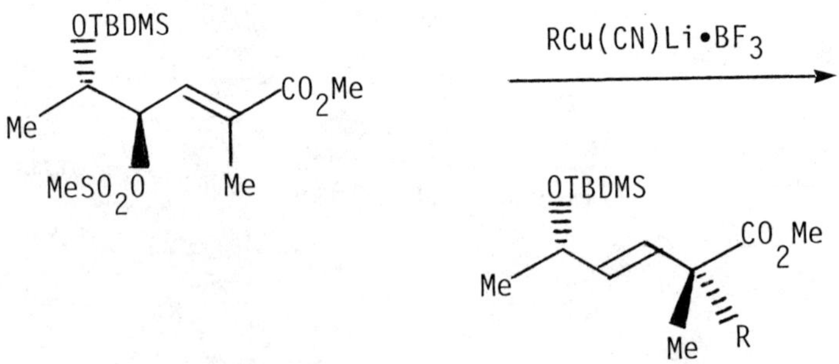

94-99%

>99 : 1 (R)

I.A.5-8 J. Otera et al., Tetrahedron Lett., 28, 6299 (1987); W.S. Johnson et al., ibid., 28, 6535 (1987); S. Kano et al., Chem. Lett., 1531 (1987); T. Takeda et al., ibid., 1345 (1987); T. Mukaiyama et al., ibid., 1167 (1987); Y. Naruta and K. Maruyama, ibid., 963 (1987); Y. Yamamoto and J. Yamada, Chem. Commun., 1218 (1987); D. Seebach et al., Helv. Chim. Acta, 70, 448 (1987); R.D. Walkup and N.U. Obeyesekere, Synthesis, 607 (1987).

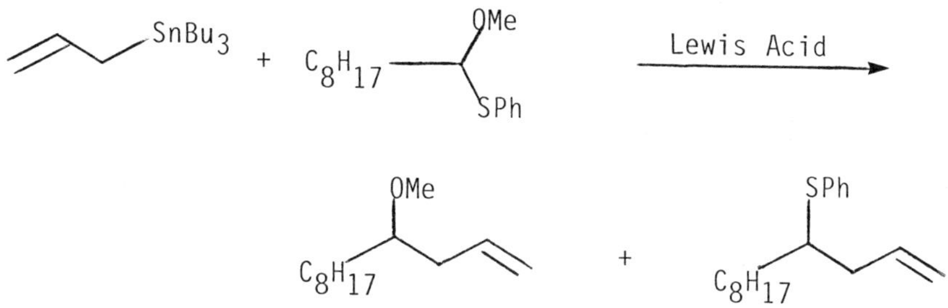

Lewis Acid = BF_3•OEt_2 72%, 100 : 0
= $TiCl_4$ 79% 6 : 94

allyl silanes, acetals or $TrClO_4$ also employed

I.A.5-9 P. Bauer et al., Can. J. Chem., 65, 2428 (1987); L. Miginiac et al., ibid., 65, 859 (1987); P. Munster and W. Steglich, Synthesis, 223 (1987); E. Negishi et al., Tetrahedron Lett., 28, 5793 (1987); M. Nojima et al., J. Am. Chem. Soc., 109, 3391 (1987); J. Barluenga et al., Chem. Commun., 915 (1987); M.E. Jung and J.A. Hagenah, J. Org. Chem., 52, 1889 (1987); E.C. Ashby and T.N. Pham, ibid., 52, 1291 (1987); J.T. Gupton and W.J. Layman, ibid, 52, 3683 (1987).

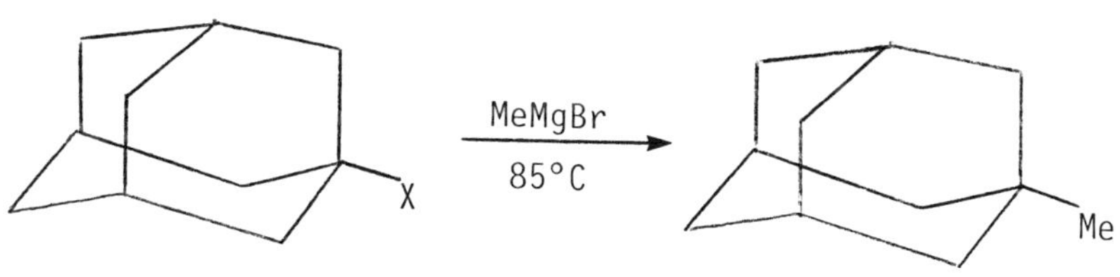

X = Br, Cl >99%

alkyl lithio nucleophiles and different leaving groups also used

I.A.5-10 A. Padwa et al., J. Org. Chem., 52, 3278 (1987); G. Picotin and Ph. Miginiac, J. Organomet. Chem., 328, 249 (1987); M.T. Reetz and T. Seitz, Angew. Chem., Int. Ed. Engl., 26, 1028 (1987).

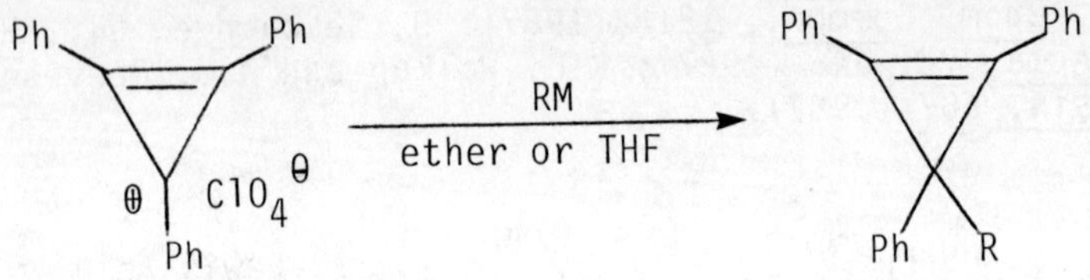

76-96%

I.A.5-11 M. Ohno, Y. Kobayashi et al., Chem. Pharm. Bull., 35, 647 (1987); T. Takeda et al., Chem. Commun., 1963 (1987); P. Helquist et al., Organometallics, 6, 988 (1987).

$$CF_3(CF_2)_nI \quad + \quad CH_2{=}CH(CH_2)_8OAc \xrightarrow{1)-3)} CF_3(CF_2)_n(CH_2)_{10}OH$$

37-71%

1) $Fe_3(CO)_{12}$, EtOH, pyr 2) Zn, AcOH 3) KOH, MeOH

I.A.5-12 B. Bogdanovic et al., Angew. Chem., Int. Ed. Engl., 26, 1025 (1987).

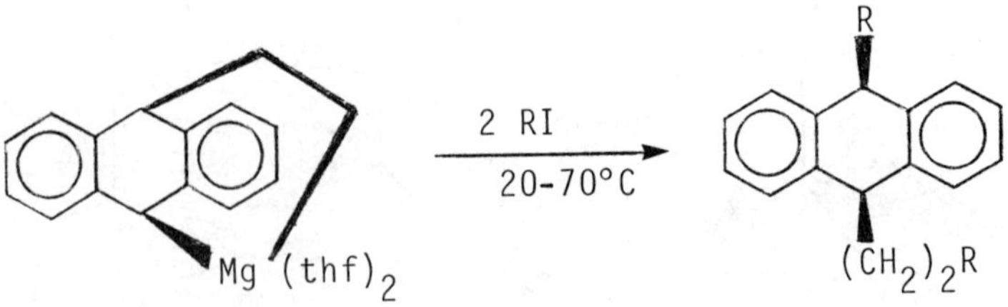

no yields given

I.A.5-13 J.C. Vederas et al., J. Am. Chem. Soc., 109, 4649 (1987).

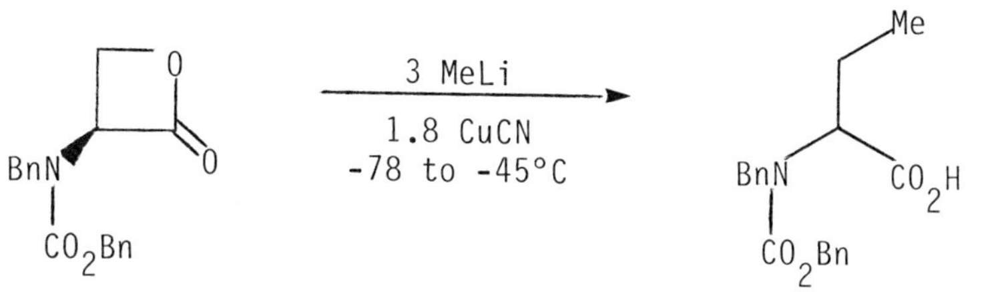

92%, 1% decrease in ee

I.A.5-14 A. Alberola et al., J. Chem. Soc., Perkin Trans. 1, 2125 (1987); S. Matsubara, S. Kozima et al., J. Org. Chem., 52, 4416 (1987).

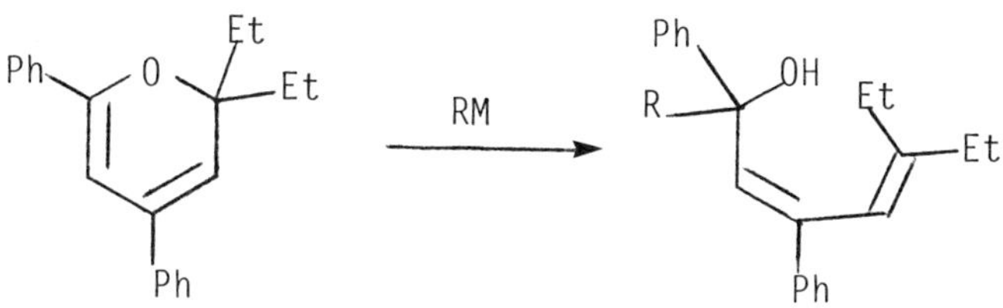

48-83%

I.A.5-15 T. Hayashi, Y. Matsumoto and Y. Ito, Organometallics, 6, 884 (1987); N. Ono et al., Chem. Commun., 947 (1987); M. Uemura et al., Tetrahedron Lett., 28, 641 (1987).

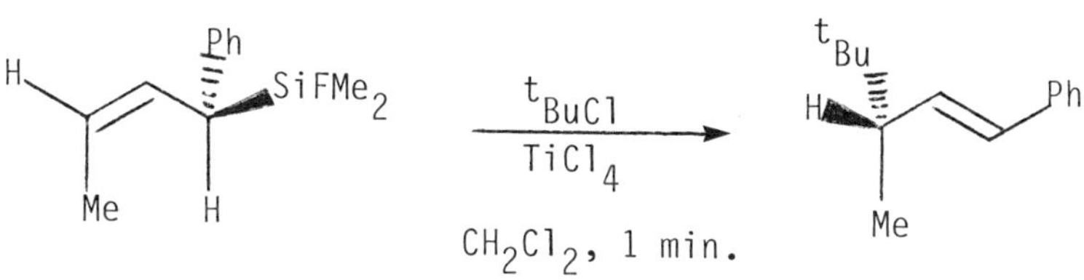

26% ee

73%, 29% ee

other Lewis Acids and leaving groups also used

I.A.5-16 V.V. Shchepin et al., J. Org. Chem. (USSR), 23, 836, 840 and 844 (1987).

$$Cl_3CCH{=}C(Me)CHO \xrightarrow{1),2)} Cl_2C{=}CHC(Me)(CH_2OR)CHO$$

10-38%

1) Zn 2) $ClCH_2OR$

I.A.5-17 N. Auner, J. Organomet. Chem., 336, 59 (1987); M.P. Cooke, Jr. and R.K. Widener, J. Am. Chem. Soc., 109, 931 (1987); A. Krief and P. Barbeaux, Chem. Commun., 1214 (1987).

$$Me_3SiCH{=}CH_2 \xrightarrow[2)\ Me_3SiCH{=}CH_2]{1)\ {}^tBuLi} Me_3SiCH(CH{=}CHSiMe_3)CH_2{}^tBu$$

80%

I.A.5-18 S. Torii et al., Bull. Chem. Soc. Jpn., 60, 1951 (1987); F. Vogtle et al., J. Org. Chem., 52, 5560 (1987).

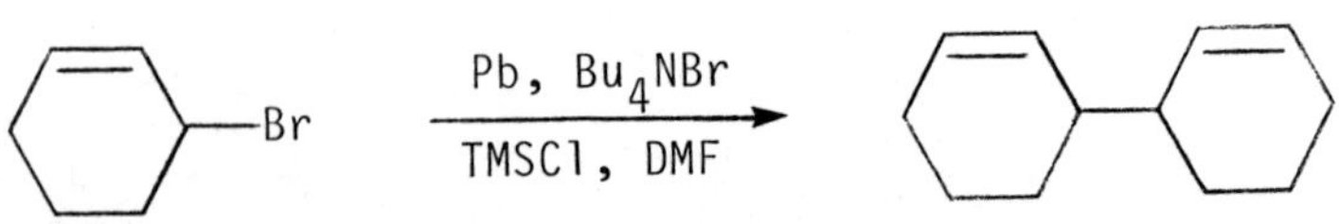

100%

Fe in DMF also used

I.A.5-19 H.-J. Kreuzfeld et al., J. Organomet. Chem., 336, 287 (1987); F. Naso et al., Synthesis, 1034 (1987); K. Oshima et al., Chem. Lett., 2203 (1987).

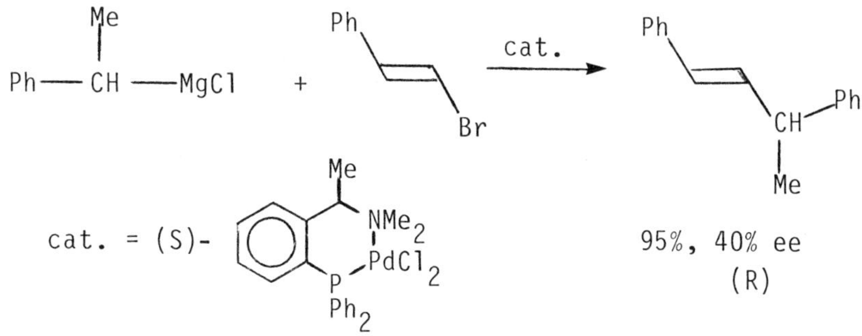

95%, 40% ee (R)

$NiCl_2$ (dppp) and Li_2MnCl_4 (catalysts) and OTf (leaving group) also employed

---

I.A.5-20 F. Orsini and F. Pelizzoni, Synth. Commun., 17, 1389 (1987); G.W. Klumpp et al., Tetrahedron Lett., 28, 5929 (1987).

$$BrZnCH_2CO_2{}^tBu \quad + \quad ROTf \xrightarrow{(Ph_3P)_2PdCl_2} RCH_2CO_2{}^tBu$$

48-77%

intramolecular cyclic ether cleavage also examined

I.A.5-21 S. Matsubara et al., Tetrahedron Lett., 28, 5857 (1987); M. Kosugi, T. Migita et al., Chem. Lett., 1237 (1987).

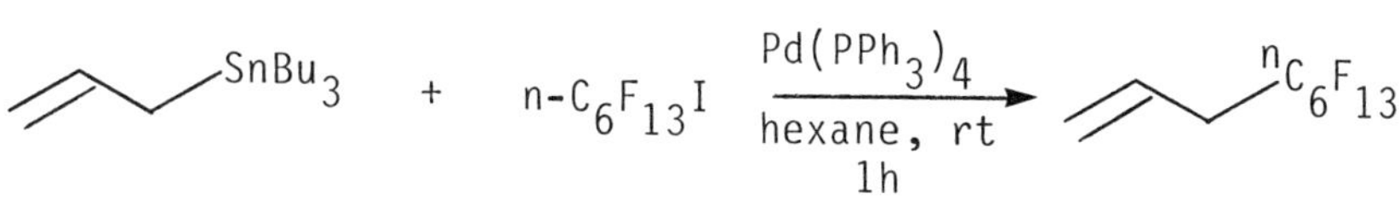

100%

I.A.5-22 M. Vandewalle et al., Tetrahedron Lett., 28, 4195 (1987); L. Brandsma et al., J. Org. Chem., 52, 5261 (1987); J.N. Marx and Q.-R. Bih, ibid., 52, 336 (1987); K. Tamao, E. Nakajo and Y. Ito, Synth. Commun., 17, 1637 (1987); C. Szantay et al., ibid., 17, 173 (1987); M. Nojima, S. Nagase et al., J. Chem. Soc., Perkin Trans. 2, 673 (1987); P. Beak et al., J. Am. Chem. Soc., 109, 5403 (1987).

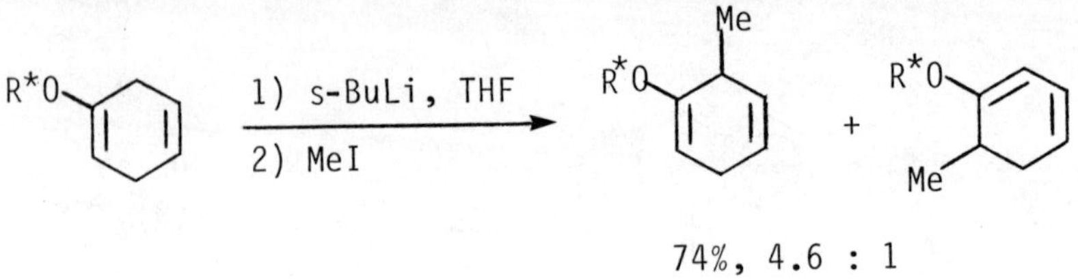

R* = per-O-methyl-β-D-glucopyranosyl

I.A.5-23 M.F. Semmelhack and E.J. Fewkes, Tetrahedron Lett., 28, 1497 (1987); Y.-C. Xu and W.D. Wulff, J. Org. Chem., 52, 3263 (1987).

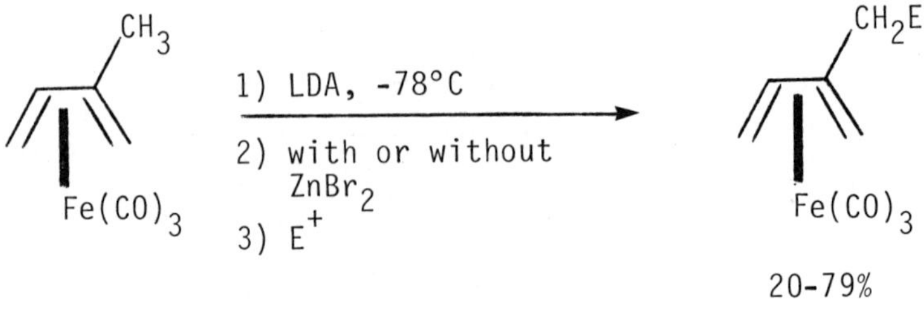

I.A.5-24 P.E. Eaton et al., J. Am. Chem. Soc., 109, 948 (1987).

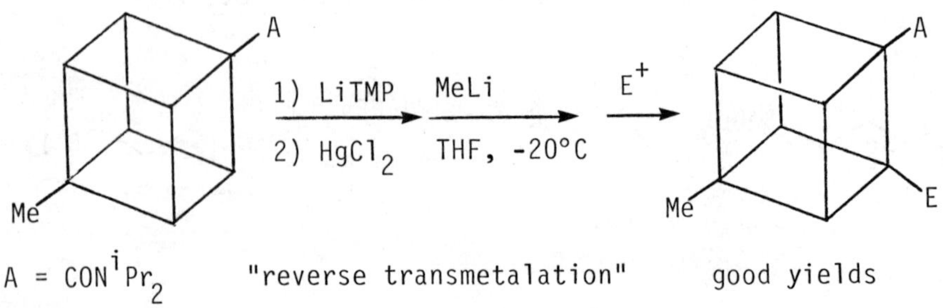

A = CON^iPr_2 "reverse transmetalation" good yields

If LiTMP alone is used only ~3% ortho-lithio formed

I.A.6. Other Alkylation Procedures

I.A.6-1 A.R. Katritzky et al., Synthesis, 415 (1987).

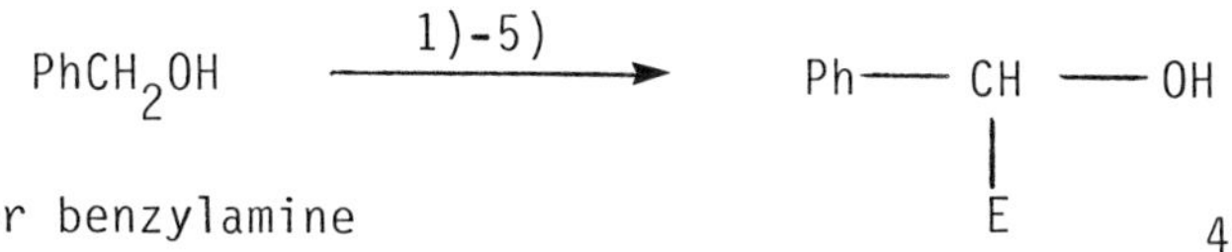

also for benzylamine

45-68%

1) BuLi 2) CO_2 3) $^{t}BuLi$ 4) E^+ 5) aq. HCl

I.A.6-2 D. Seebach et al., Helv. Chim. Acta, 70, 1507 (1987).

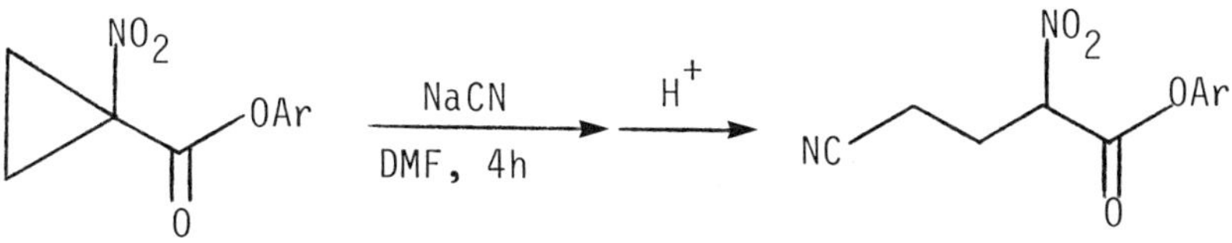

81%

Ar = 2,6-di-t-butyl-4-methoxyphenyl

I.A.6-3 S. Murahashi and T. Shiota, Tetrahedron Lett., 28, 6469 (1987).

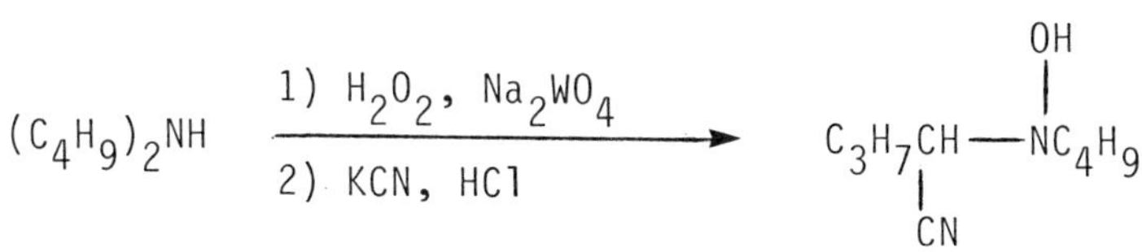

76%

I.A.6-4 S.L. Buchwald and S.J. LaMaire, Tetrahedron Lett., 28, 295 (1987).

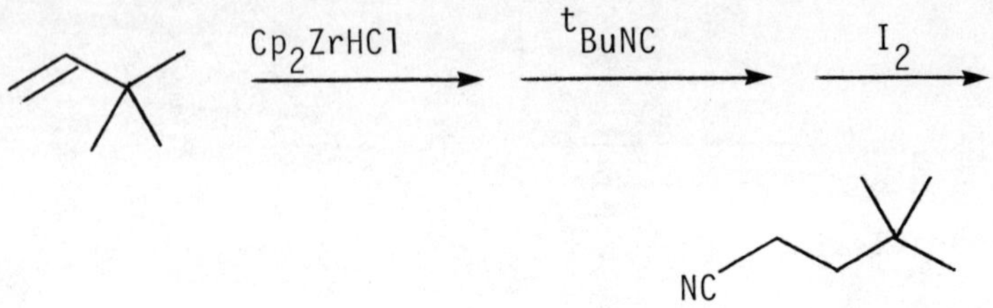

84%

I.A.6-5 Y. Otsuji et al., Chem. Lett., 2103 (1987).

$$R^1COCH{=}CHR^2 \xrightarrow{1)-4)} R^1COCH_2CH(R^2)CH_2C{\equiv}CH$$

50-83%

1) Me_3SiI 2) $Bu_4NFe(CO)_3NO$ 3) $P(OPh)_3$ 4) 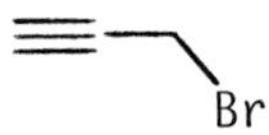

I.A.7. Nucleophilic Addition to Electron Deficient Carbon

I.A.7.a.1a Intermolecular Aldol-Type 1,2- Additions

I.A.7.a.1a-1 M. Braun, Angew. Chem., Int. Ed. Engl., 26, 24 (1987).

Review: "Stereoselective Aldol Reactions with α-Unsubstituted Chiral Enolates".

I.A.7.a.1a-2 D.A. Evans et al., Tetrahedron Lett., 28, 39 (1987); S. Schreiber and M.T. Goulet, ibid., 28, 6001 (1987); I. Paterson and C.K. McClure, ibid., 28, 1229 (1987); C.-N. Hsiao, L. Liu and M.J. Miller, J. Org. Chem., 52, 2201 (1987); S. Thaisrivongs et al., J. Med. Chem., 30, 976 (1987).

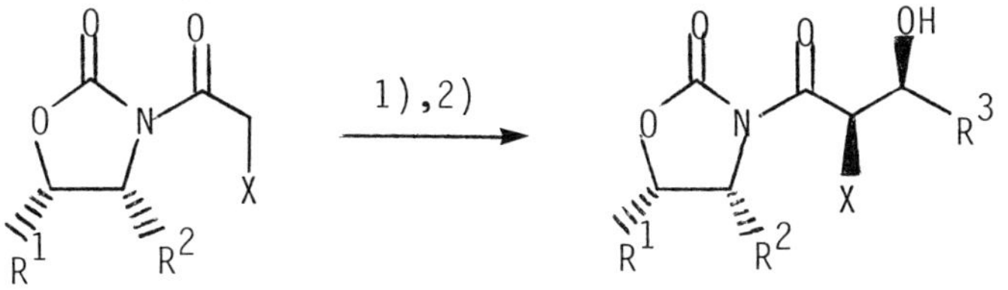

63-94%, 94 : 6 to 98 : 2

X = Cl, Br

1) Bu_2BOTf , NEt_3 2) R^3CHO

I.A.7.a.1a-3 G.P. Boldrini et al., J. Organomet. Chem., 336, 23 (1987); R.P. Short and S. Masamune, Tetrahedron Lett., 28, 2841 (1987); R.W. Hoffmann et al., Liebigs Ann. Chem., 977 (1987); J. Hooz et al., J. Org. Chem., 52, 1347 (1987).

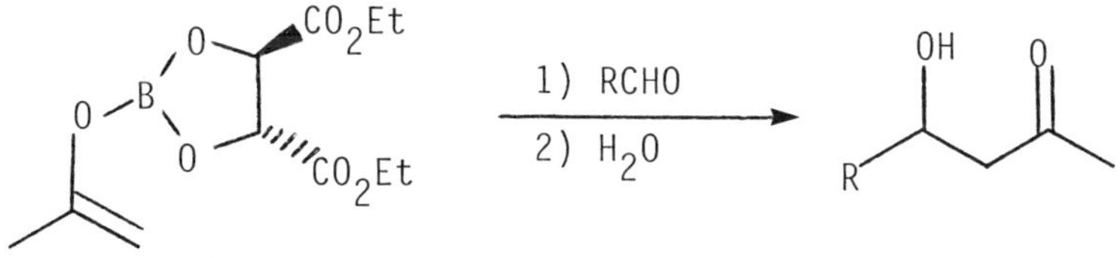

55-77%
41-65% ee

I.A.7.a.1a-4 T. Mukaiyama and K. Narasaka, Org. Synth., 65, 6 and 12 (1987); A.E. Vougioukas and H.B. Kagan, Tetrahedron Lett., 28, 5513 (1987); M.T. Reetz and A.E. Vougioukas, ibid., 28, 793 (1987); C. Scolastico et al., J. Org. Chem., 52, 888 (1987); idem, Tetrahedron, 43, 2317 (1987); C. Gennari et al., J. Org. Chem., 52, 2754 (1987); U. Widmer et al., Helv. Chim. Acta, 70, 1412 (1987); I. Mori, P.A. Bartlett and C.H. Heathcock, J. Am. Chem. Soc., 109, 7199 (1987); M. Onaka et al., Bull. Chem. Soc. Jpn., 60, 2689 (1987).

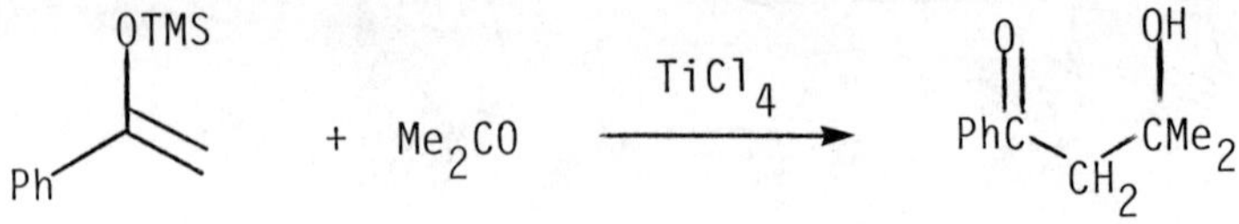

70-74%

$LnCl_3$ (Ln = La, Ce, Sm), $MgBr_2$,

Al-Montmorillonite and

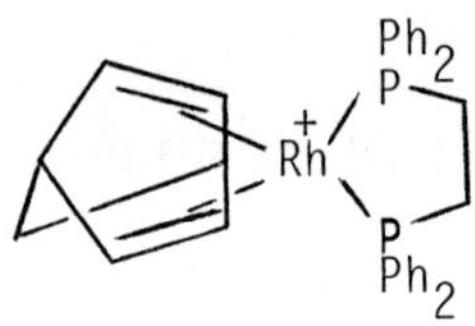

also used as catalysts

I.A.7.a.1a-5 R.A. Pilli and D. Russowsky, Chem. Commun., 1053 (1987); G. Guanti et al., Tetrahedron Lett., 28, 4331 and 4335 (1987); I. Ojima et al., ibid., 28, 613 (1987); Y. Kita et al., ibid., 28, 1431 (1987); J.W. Scheeren et al., J. Org. Chem., 52, 365 (1987); L.Yu. Kryukova et al., J. Org. Chem. (USSR), 23, 205 (1987).

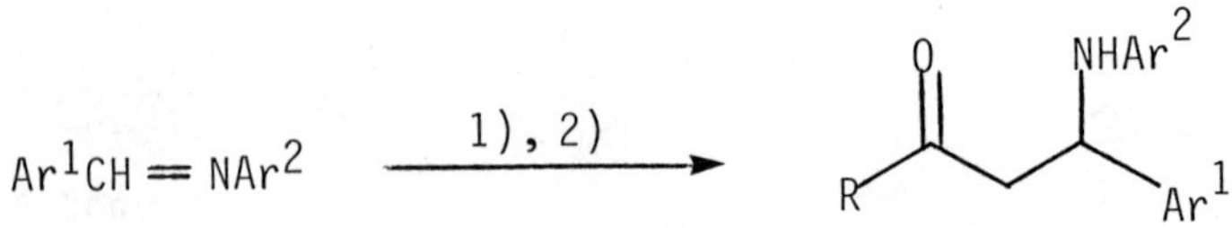

50-98%

1) TMSOTf 2) CH_2=C(R)OTMS

I.A.7.a.1a-6 T. Mukaiyama et al., Chem. Lett., 293 and 479 (1987); D.A. Evans and A.E. Weber, J. Am. Chem. Soc., 109, 7151 (1987); Y. Yamamoto et al., Chem. Commun., 561 (1987).

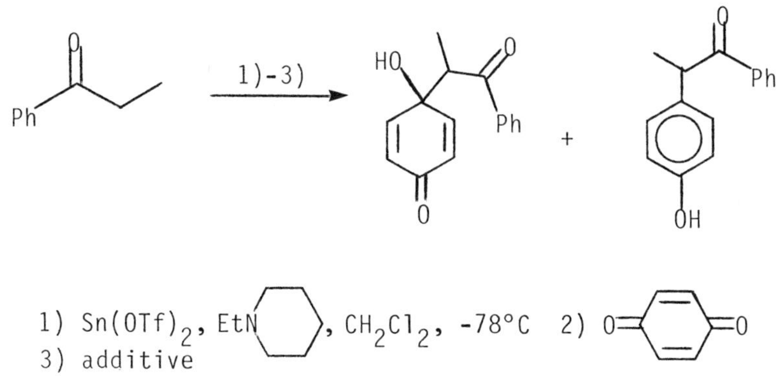

with no additive	68%	0%
with 2 TMSCl	0%	63%

I.A.7.a.1a-7 J.E. Lynch et al., Tetrahedron Lett., 28, 1385 (1987); J. Barluenga et al., Synthesis, 318 (1987); G. Bianchi et al., J. Chem. Res. (S), 302 (1987)

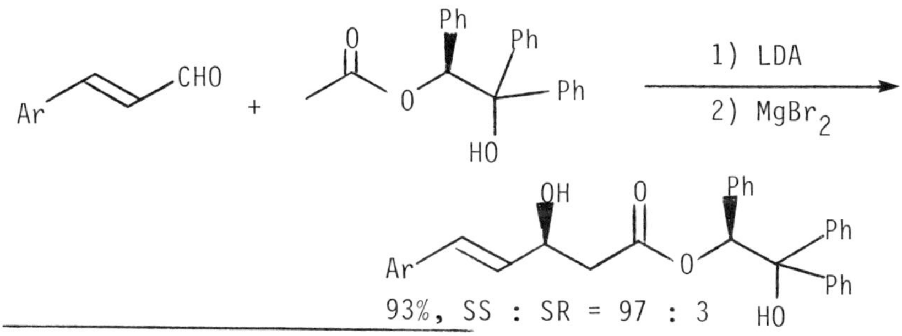

93%, SS : SR = 97 : 3

I.A.7.a.1a-8 T. Sakai et al., Tetrahedron Lett., 28, 3817 (1987); S. Fukuzawa et al., J. Chem. Soc., Perkin Trans. 1, 1473 (1987).

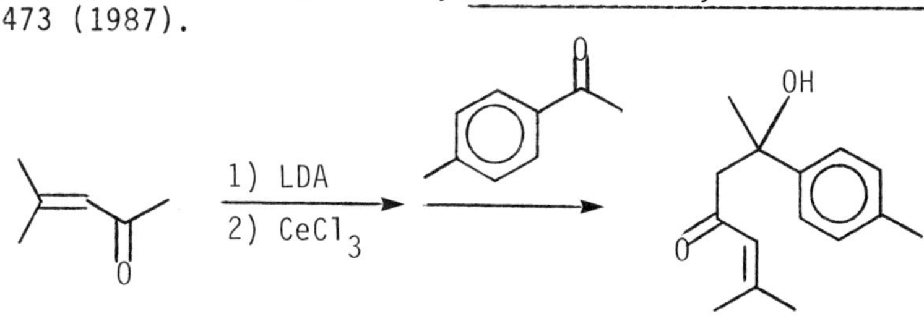

54%

alcohol reduced to alkane with TMSCl-NaI-MeCN

I.A.7.a.1a-9 Y. Ito and S. Terashima, Tetrahedron Lett., 28, 6629 (1987); A.J. Fry et al., J. Org. Chem., 52, 2496 (1987).

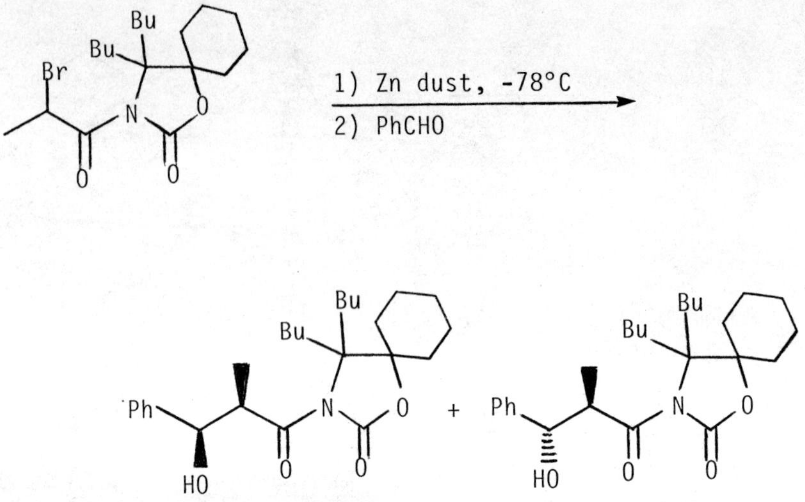

98%, 98 : 2

electrolysis/$LaBr_3$ also employed

I.A.7.a.1a-10 G. Procter et al., Tetrahedron Lett., 28, 2037 (1987).

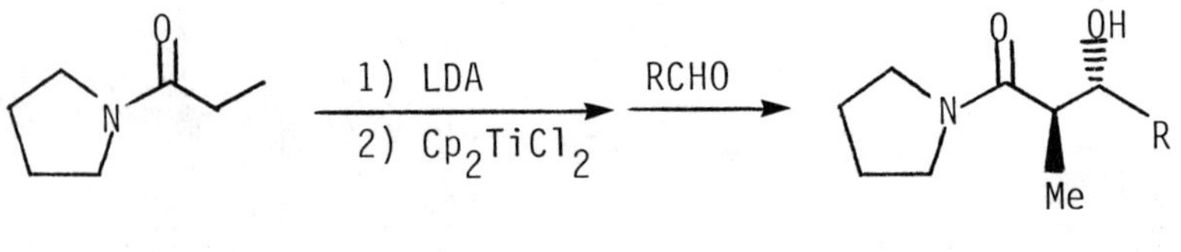

64-77%

anti : syn 79 : 21 to 98 : 2

in contrast to high syn stereoselectivity with Zr enolates

I.A.7.a.1a-11 G. Iwasaki and M. Shibasaki, Tetrahedron Lett., 28, 3257 (1987).

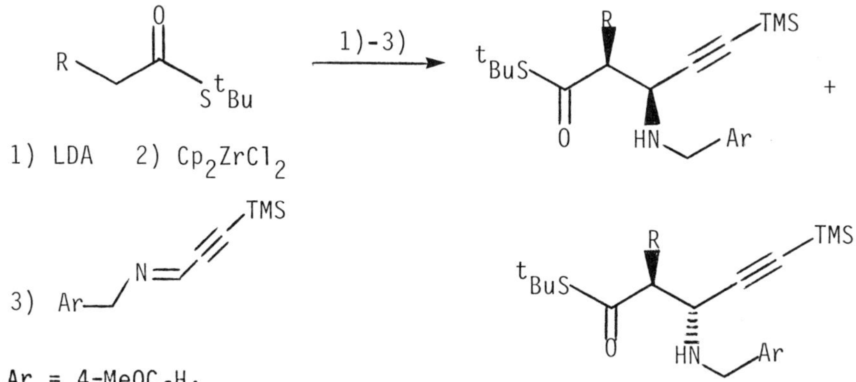

1) LDA 2) Cp_2ZrCl_2

3) Ar–CH_2–N=CH–C≡C–TMS

Ar = 4-$MeOC_6H_4$

43-58%, 4 : 1 to 21 : 1

I.A.7.a.1a-12 C.R. Harrison, Tetrahedron Lett., 28, 4135 (1987).

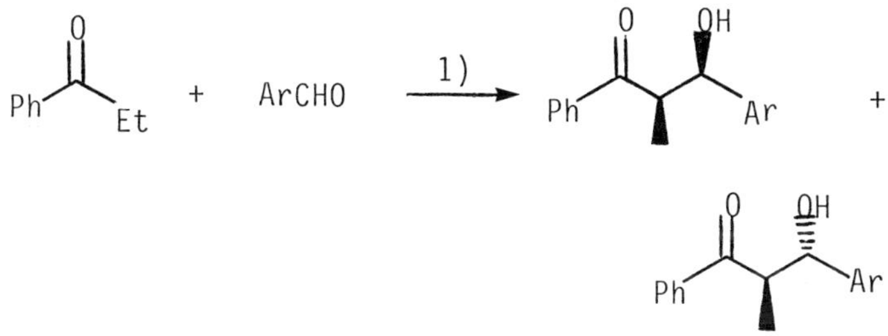

95 : 5, 91%

1) $TiCl_4$, NEt_3, CH_2Cl_2, 0°C, 0.5h

I.A.7.a.1a-13 R.G. Bergman, C.H. Heathcock et al., J. Am. Chem. Soc., 109, 2022 (1987).

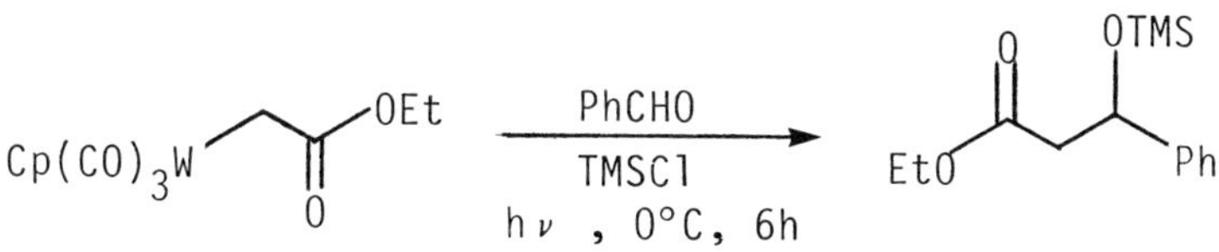

47% (84% NMR)

I.A.7.a.1a-14 A. Battaglia et al., Tetrahedron Lett., 28, 2995 (1987).

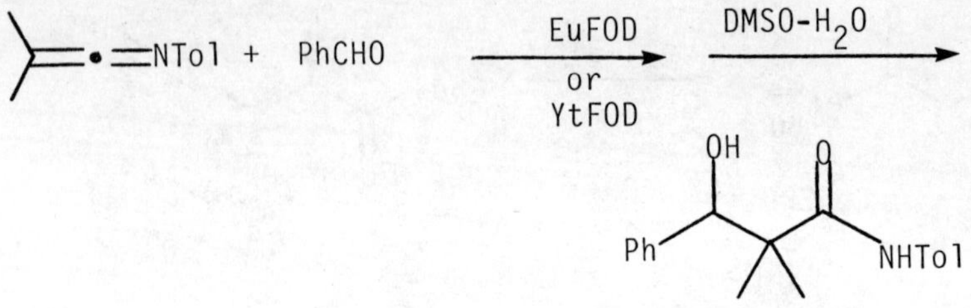

56%

I.A.7.a.1a-15 Sh.T. Akhmedov et al., J. Org. Chem. (USSR), 22, 2226 (1986); A. Takeda et al., Synthesis, 292 (1987).

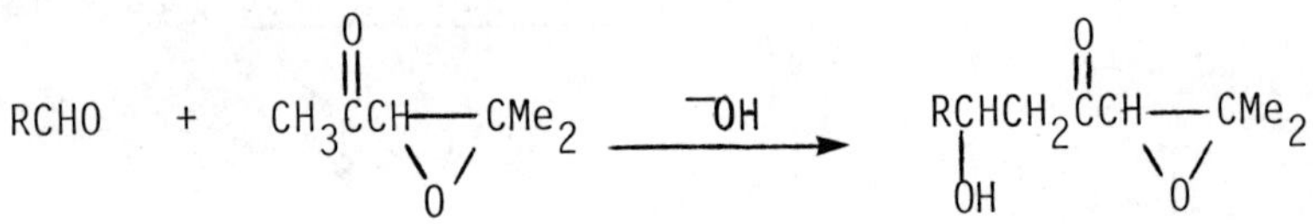

45-50%

I.A.7.a.1a-16 A. Ando and T. Shioiri, Chem. Commun., 1620 (1987).

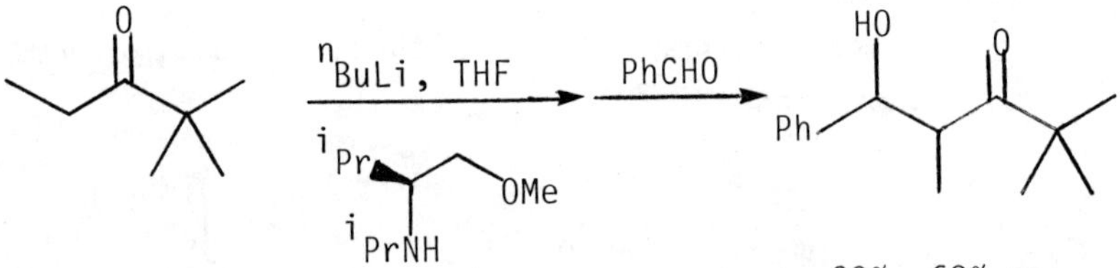

93%, 68% ee

I.A.7.a.1a-17 T. Oshima et al., Chem. Lett., 1745 (1987); B.M. Trost et al., J. Am. Chem. Soc., 109, 613 (1987); E.P. Lodge and C.H. Heathcock, ibid., 109, 3353 (1987); M.M. Campbell et al., Tetrahedron Lett., 28, 985 (1987); R. Carrie et al., ibid., 28, 1761 (1987); R. Mestres et al., ibid, 28, 3853 (1987); J. Barluenga et al., J. Chem. Res. (S), 402 (1987).

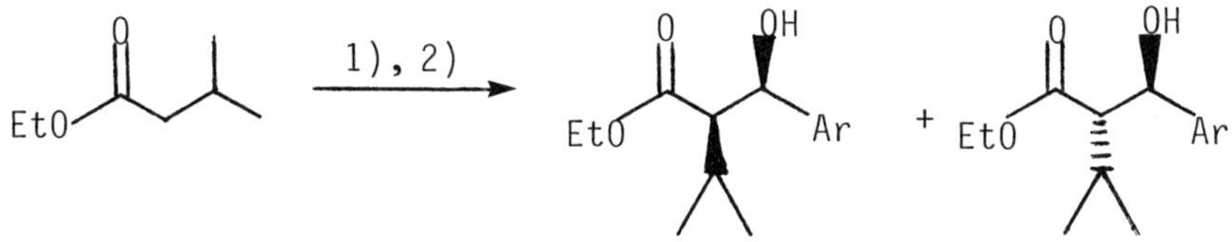

additive = none, 88%, 6 : 1

= 2.3 eq. HMPT, ~75%, 1 : 4

1) LDA, THF, additive 2) veratraldehyde

I.A.7.a.1a-18 J.F. Wolfe et al., Synth. Commun., 17, 13 (1987).

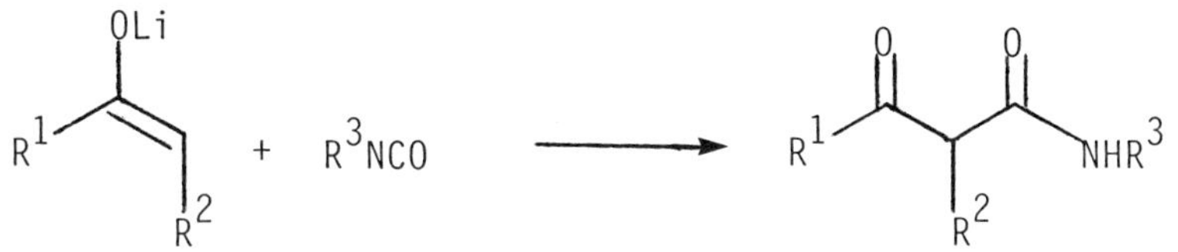

76-97%

I.A.7.a.1a-19 N.K. Kochetkov et al., Tetrahedron Lett., 28, 3835 and 3839 (1987); U. Schmidt and W. Siegel, ibid., 28, 2849 (1987); H. Schick et al., J. Prakt. Chem., 329, 385 (1987); T. Shono and N. Kise, Chem. Lett., 697 (1987).

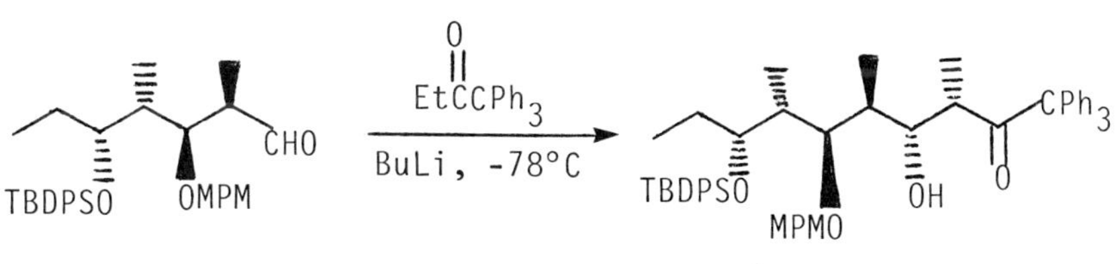

MPM = p-methoxybenzyl 84%

similar reaction with electrogenerated anions

I.A.7.a.1a-20 S.G. Davies et al., Bull. Soc. Chim. Fr., 608 (1987).

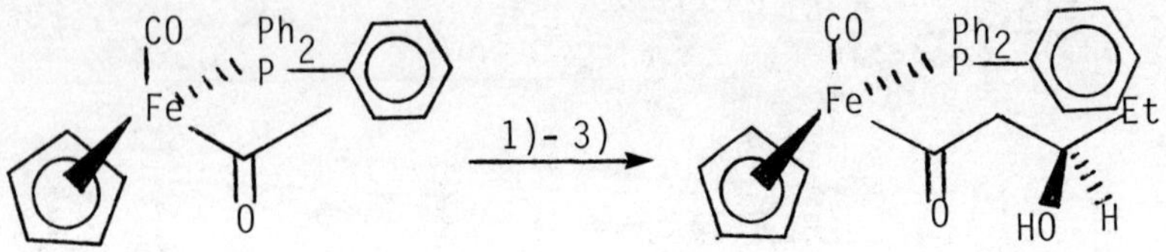

>100 : 1

1) BuLi 2) Et_2AlCl 3) EtCHO

I.A.7.a.1a-21 D. Seebach et al., Helv. Chim. Acta, 70, 237, 1930 and 1944 (1987); S.E. Drewes et al., Synth. Commun., 17, 795 (1987); W.H. Pearson and M.-C. Cheng, J. Org. Chem., 52, 3176 (1987); R.A. Holton et al., J. Am. Chem. Soc., 109, 1597 (1987); M. Hatanaka and H. Nitta, Tetrahedron Lett., 28, 69 and 83 (1987).

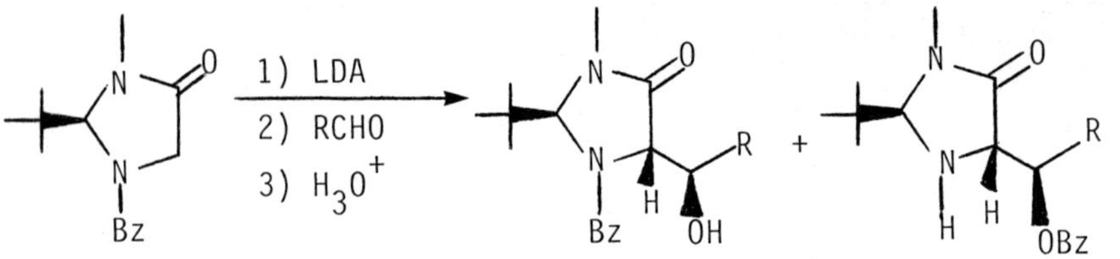

41-85%, 63-95% ds

masked glycines

I.A.7.a.1a-22 T. Ishihara et al., Tetrahedron Lett., 28, 3501 (1987).

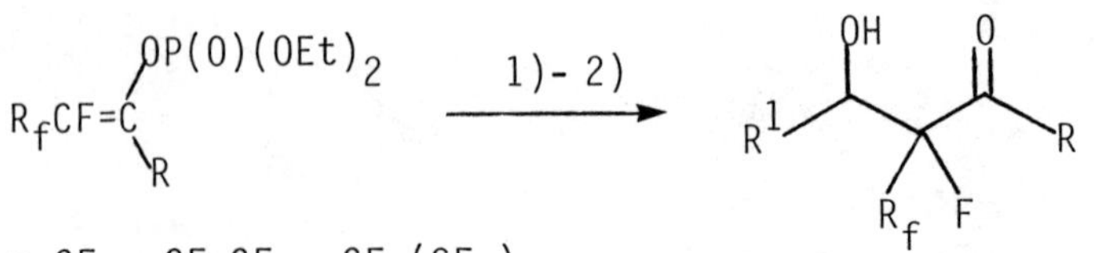

$R_f = CF_3$, CF_3CF_2, $CF_3(CF_2)_5$

1) $CuBr_2$, LAH, -30°C 2) R^1CHO 34-84%

I.A.7.a.1a-23 C.H. Heathcock et al., J. Org. Chem., 52, 1686 (1987).

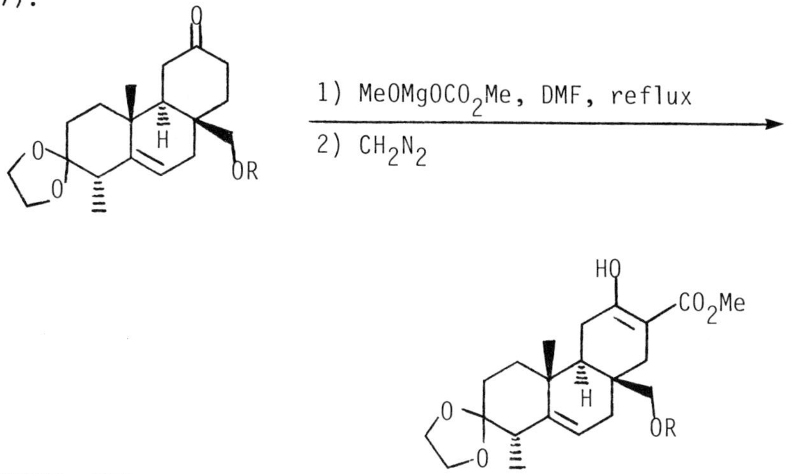

R = TBDMS, THP

85-95%

I.A.7.a.1a-24 F. Effenberger and A. Straub, Tetrahedron Lett., 28, 1641 (1987).

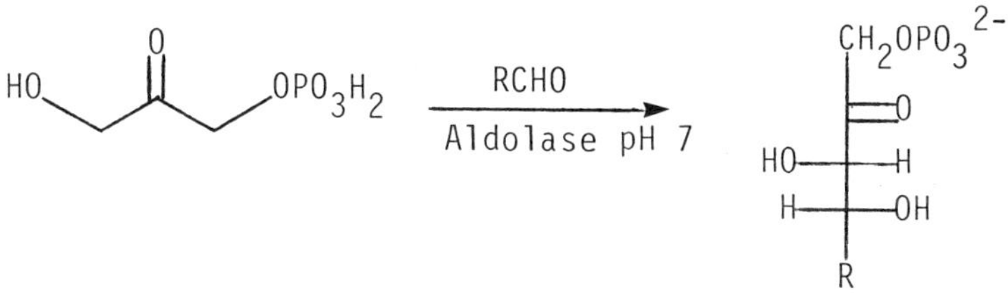

65-100%

I.A.7.a.1a-25 A. Revis and T.K. Hilty, Tetrahedron Lett., 28, 4809 (1987).

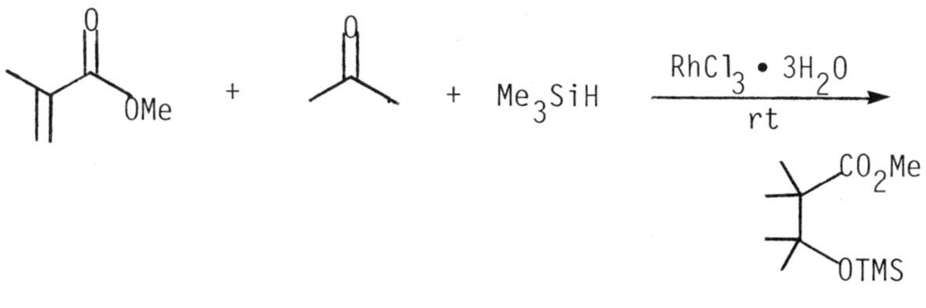

95% (major)

I.A.7.a.1a-26 B. Zwanenburg et al., Tetrahedron Lett., 28, 1329 (1987).

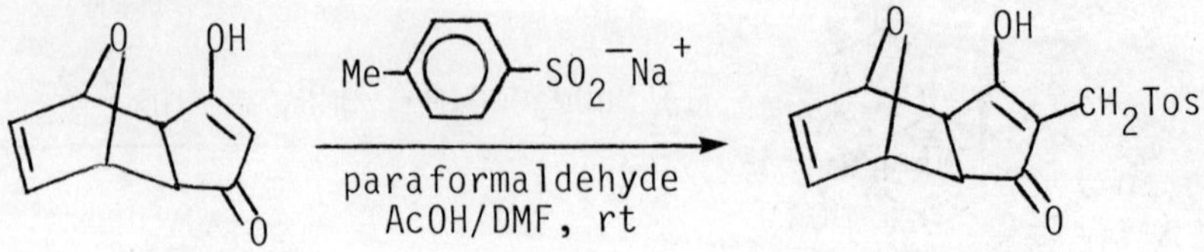

85%

I.A.7.a.1a-27 M.M. Joullie et al., Tetrahedron Lett., 28, 2837 (1987); E.J. Thomas et al., Chem. Commun., 1228 (1987); M.E. Jung et al., J. Org. Chem., 52, 4570 (1987); N. Ono et al., Synthesis, 532 (1987); S. Brandange and B. Rodriguez, Acta Chem. Scand., B41, 740 (1987).

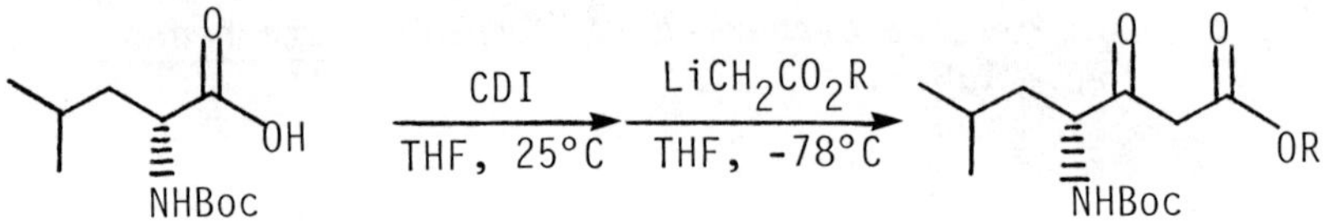

CDI = carbonyldiimidazole

82-88%

I.A.7.a.1a-28 R.K. Olsen et al., J. Org. Chem., 52, 4531 (1987); Y. Nagao et al., Chem. Lett., 1861 (1987); K.E. Henegar and J.D. Winkler, Tetrahedron Lett., 28, 1051 (1987); H. Griengl et al., Monatsh. Chem., 118, 415 (1987); M.A. Metwally et al., J. Prakt. Chem., 329, 563 and 732 (1987); T.S. Mansour, Synth. Commun., 17, 1315 (1987).

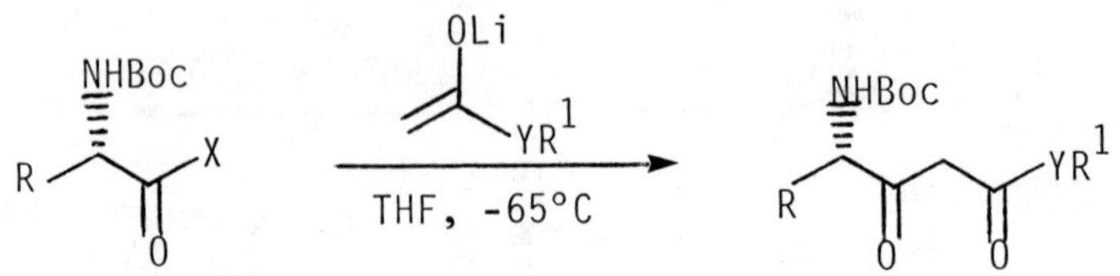

X = Cl, $OCO_2{}^iBu$ YR^1 = OEt

40-78%

I.A.7.a.1b. Intramolecular Aldol-Type 1,2-Additions

I.A.7.a.1b-1 C. Agami et al., Bull. Soc. Chim. Fr., 358 (1987); idem, Tetrahedron, 43, 1091 (1987); E.R. Koft, ibid., 43, 5775 (1987); W. Kreiser et al., Helv. Chim. Acta, 70, 445 (1987).

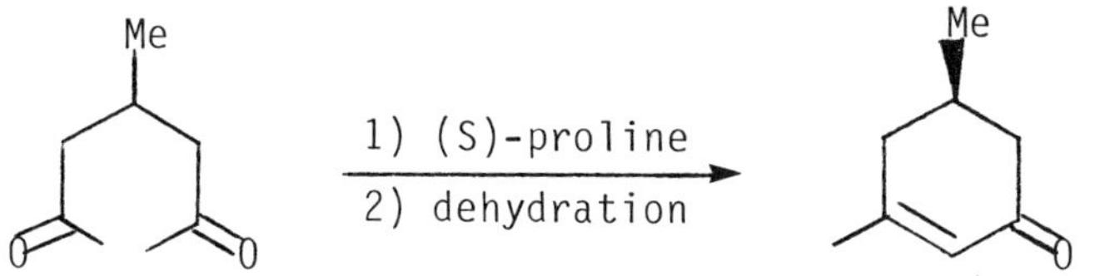

(-) 43% ee

I.A.7.a.1b-2 J.W. ApSimon and R.F. Lawuyi, Synth. Commun., 17, 1773 (1987).

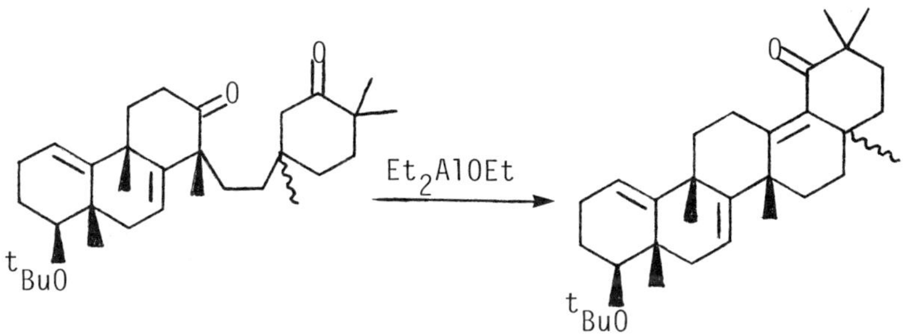

other reagents ineffective 78%

I.A.7.a.1b-3 K. Tadano, T. Suami et al., J. Org. Chem., 52, 1946 (1987); A. Takahashi and M. Shibasaki, Tetrahedron Lett., 28, 1893 (1987).

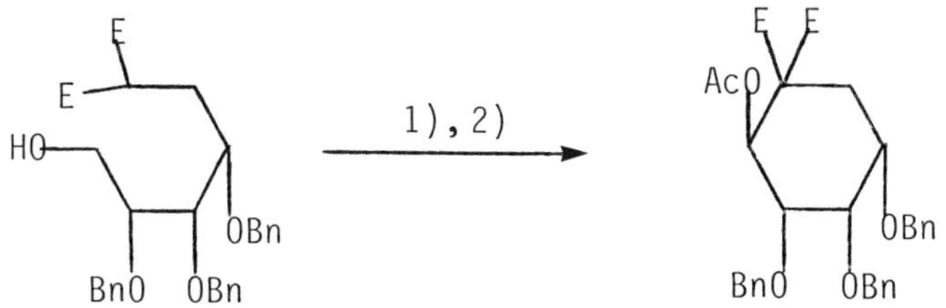

69%

1) PCC, CH_2Cl_2, 4A sieves 2) Ac_2O, pyr

I.A.7.a.1b-4 J.M. Harrison et al., J. Chem. Soc., Perkin Trans. 1, 21 (1987).

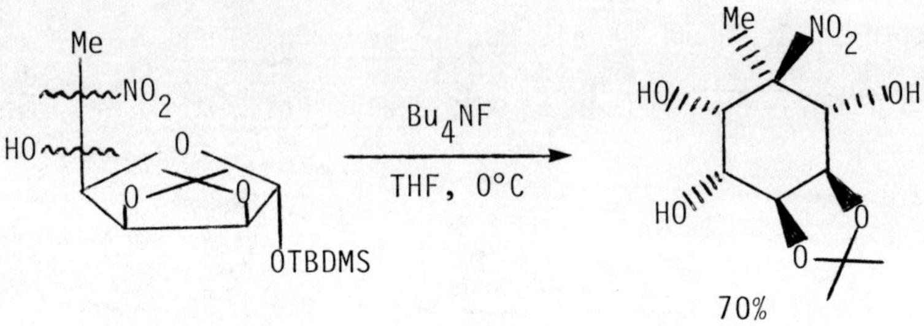

I.A.7.a.2. 1,2-Additions of N-, P-, S-, or Similar Stabilized Carbanions

I.A.7.a.2-1 A. Manfredi et al., J. Organomet. Chem., 330, 357 (1987).

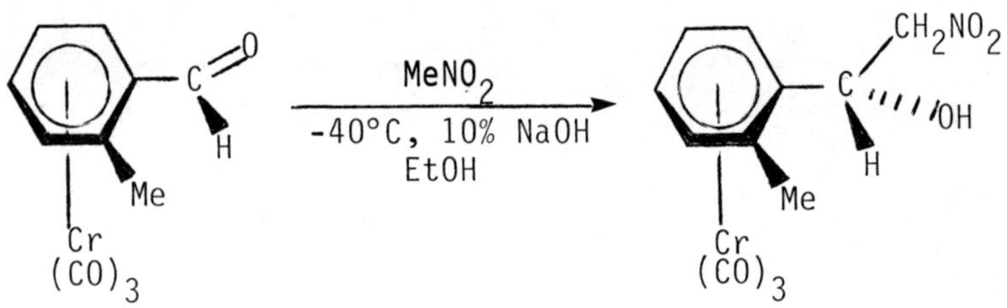

towards ephedrine 90%, 97 : 3 diastereomeric mixture

I.A.7.a.2-2 C.H. Heathcock et al., J. Med. Chem., 30, 1858 (1987); R.A. Hill et al., J. Chem. Soc., Perkin Trans. 1, 2209 (1987); G.M. Blackburn et al., ibid., 181 (1987).

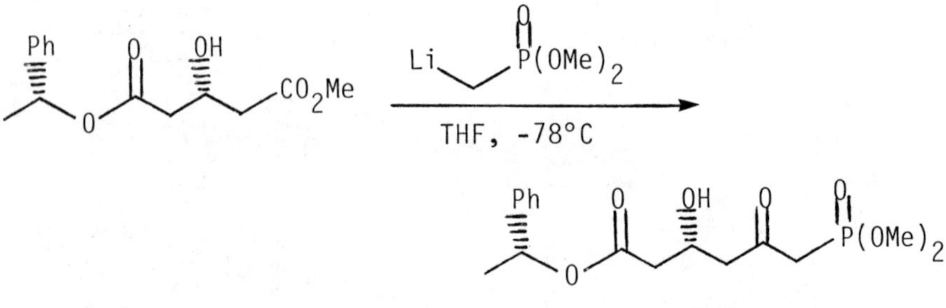

towards compactin and analogues 43%

I.A.7.a.2-3 H. Rehman and J.M. Rao, Tetrahedron, 43, 5335 (1987).

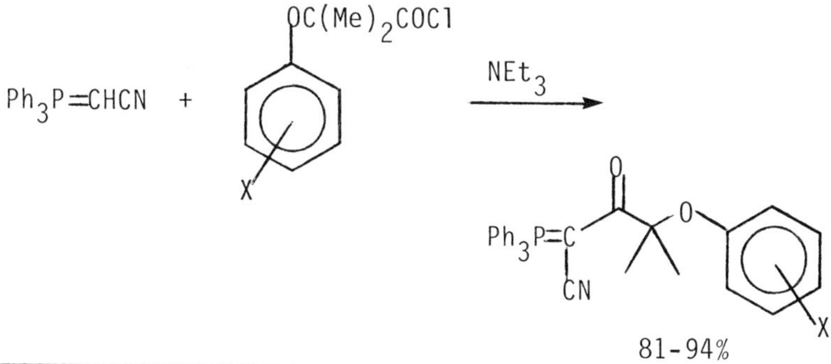

I.A.7.a.2-4 J.-M. Fang et al., J. Org. Chem., 52, 855 (1987).

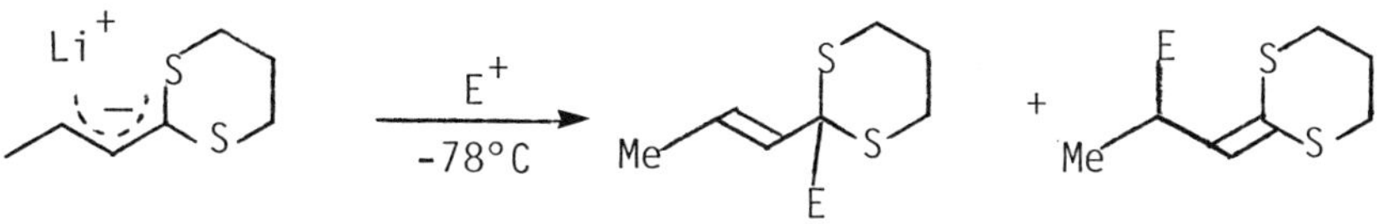

E^+	= aldehydes		major
	= MeI	exclusive	
	= medium size ketones	major	
	= bulky ketones		major

I.A.7.a.2-5 Ae. de Groot et al., Rec. Trav. Chim., 106, 489 and 505 (1987).

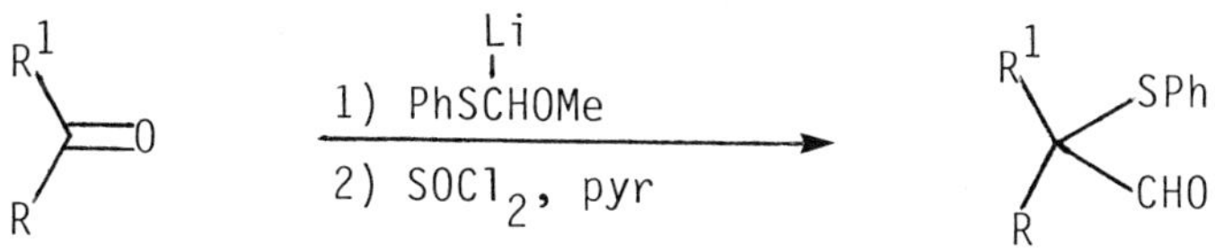

~ 70-85%

I.A.7.a.2-6 E.G. Baggiolini et al., Tetrahedron Lett., 28, 2095 (1987); M. Delmas et al., Synth. Commun., 17, 1593 (1987); J.V. Sinisterra, M. Delmas et al., Tetrahedron, 43, 4141 (1987).

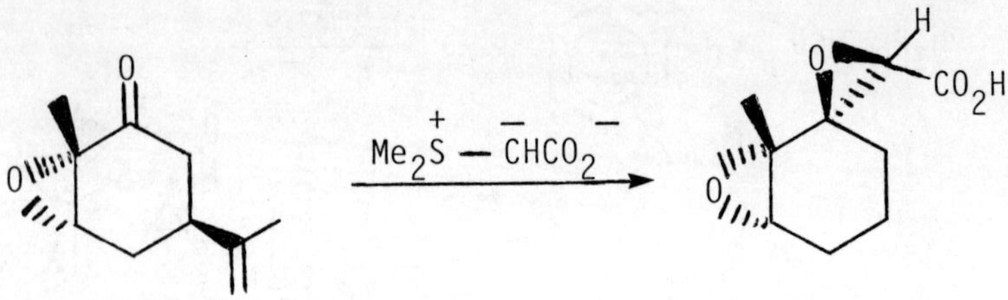

88%

I.A.7.a.2-7 R.C. Gadwood et al., J. Org. Chem., 52, 774 (1987).

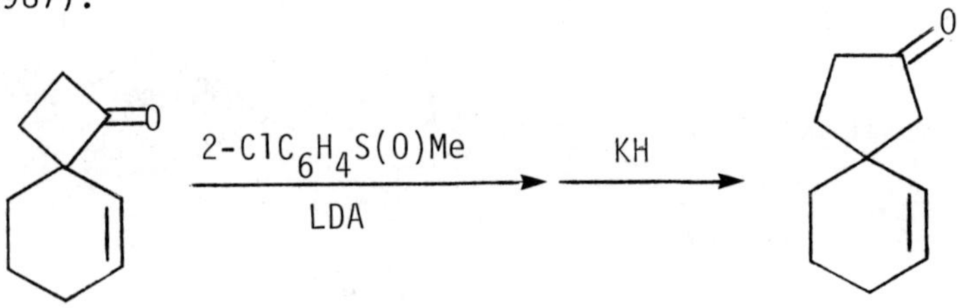

94%

I.A.7.a.2-8 K. Hiroi et al., J. Am. Chem. Soc., 109, 1249 (1987); P.F. Schuda and W.A. Price, J. Org. Chem., 52, 1972 (1987); R. Tanikaga et al., J. Chem. Soc., Perkin Trans. 1, 825 (1987); P.R. Jenkins et al., Chem. Commun., 1844 (1987).

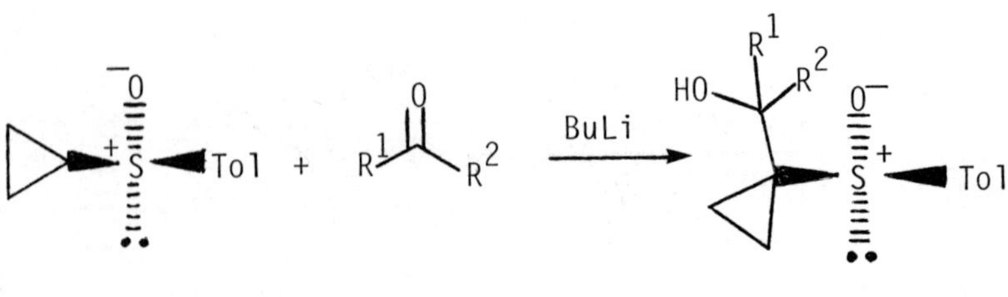

78%, 3 : 2

I.A.7.a.2-9 G. Solladie et al., Tetrahedron Lett., 28, 61 (1987); C. Iwata et al., ibid., 28, 3131 and 3135 (1987); P. Bravo et al., Tetrahedron, 43, 4635 (1987); R.K. Haynes et al., Aust. J. Chem., 40, 1331 (1987); C.R. Johnson and D.S. Dhanoa, J. Org. Chem., 52, 1885 (1987).

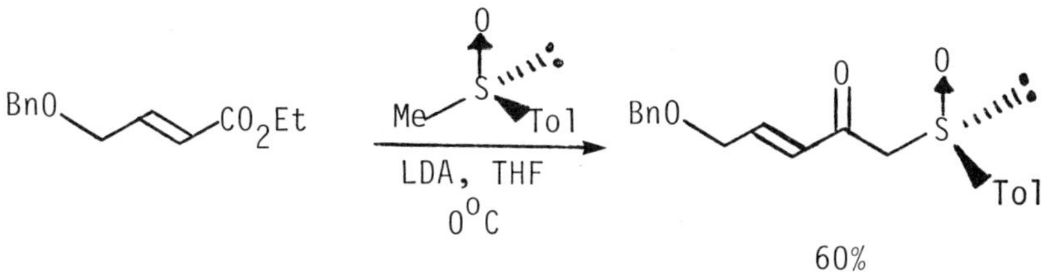

significant Michael addition occurs at -78^oC

I.A.7.a.2-10 R. Tanikaga et al., Chem. Lett., 829 (1987); E. Brown et al., Bull. Soc. Chim. Fr., 473 (1987); K. Fuji et al., Chem. Commun., 449 (1987).

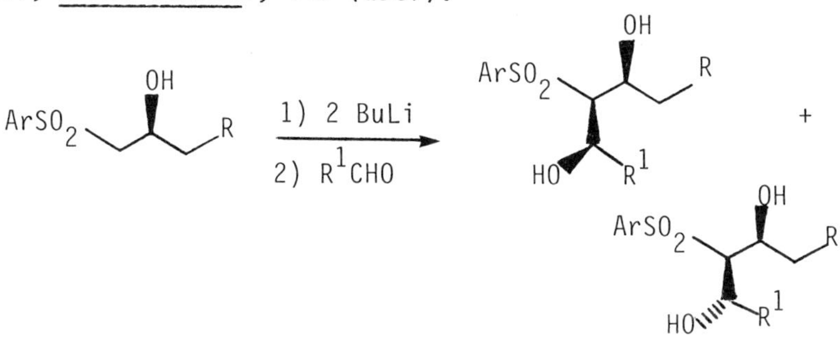

68-84%, 80:20 to 64:36

I.A.7.a.2-11 A.M. van Leusen et al., Rec. Trav. Chim., 106, 159 (1987); J. Moskal and A.M. van Leusen, ibid., 106, 137 (1987); A.M. van Leusen et al., Tetrahedron, 43, 5073 (1987); K.-J. Hwang, E.W. Logusch et al., J. Org. Chem., 52, 3435 (1987).

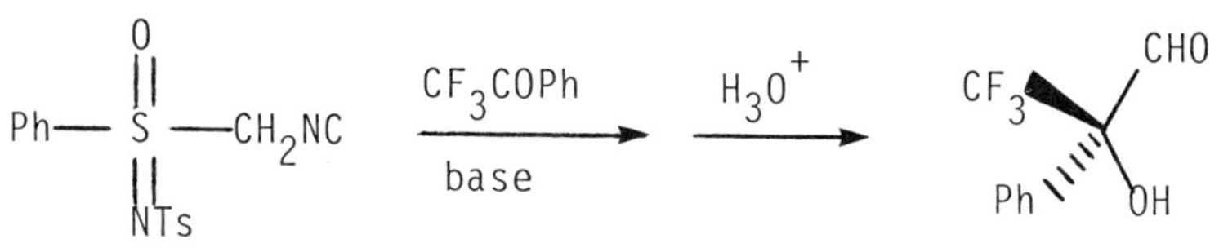

58%, 27% ee

I.A.7.a.3. 1,2-Additions of Grignard-Type Carbanions

I.A.7.a.3-1 Y. Yamamoto, Acc. Chem. Res., 20, 243 (1987).

Review: "Acyclic Stereocontrol via Allylic Organometallic Compounds."

I.A.7.a.3-2 H.C. Brown et al., J. Org. Chem., 52, 319 and 3701 (1987); E.J. Thomas et al., Chem. Commun., 1826 (1987).

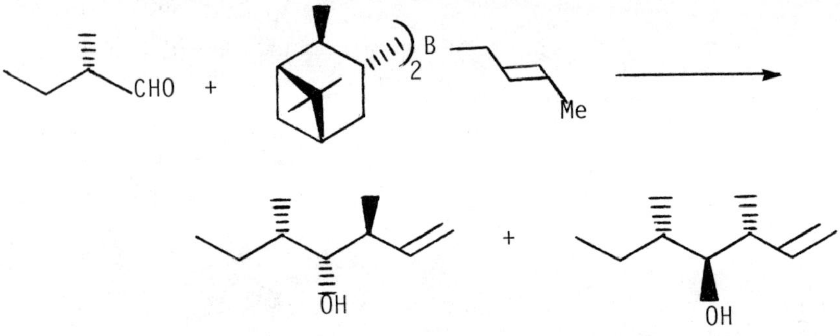

75%, 96 : 4

I.A.7.a.3-3 W.R. Roush et al., J. Org. Chem., 52, 316 (1987); W.R. Roush and A.D. Palkowitz, J. Am. Chem. Soc., 109, 953 (1987); S.J. Danishefsky et al., ibid., 109, 8117 (1987); R.W. Hoffmann et al., Leibigs Ann. Chem., 215 and 881 (1987); R.W. Hoffmann and S. Dresely, Tetrahedron Lett., 28, 5303 (1987); S. Masamune et al., J. Org. Chem., 52, 4831 (1987).

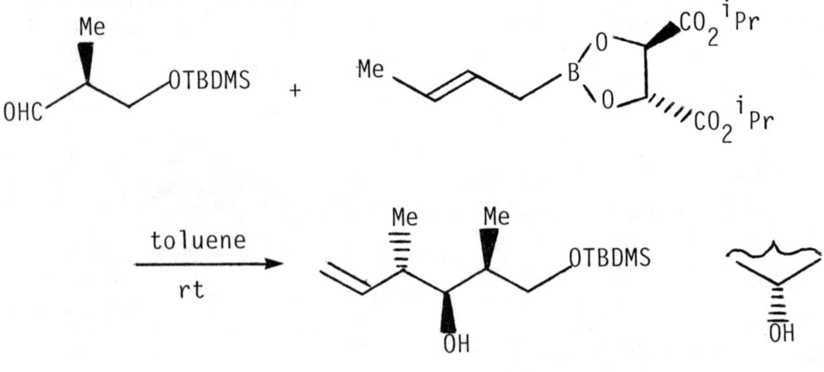

80%, 97 : 3

I.A.7.a.3-4 G.E. Keck et al., Tetrahedron Lett., 28, 139 (1987); E.J. Thomas et al., ibid., 28, 5343 (1987); M. Koreeda and Y. Tanaka, ibid., 28, 143 (1987); R.H. Schlessinger and D.D. Graves, ibid., 28, 4381 (1987); J.A. Marshall et al., ibid., 28, 527 (1987); A. Takuwa et al., Chem. Lett., 1091 (1987).

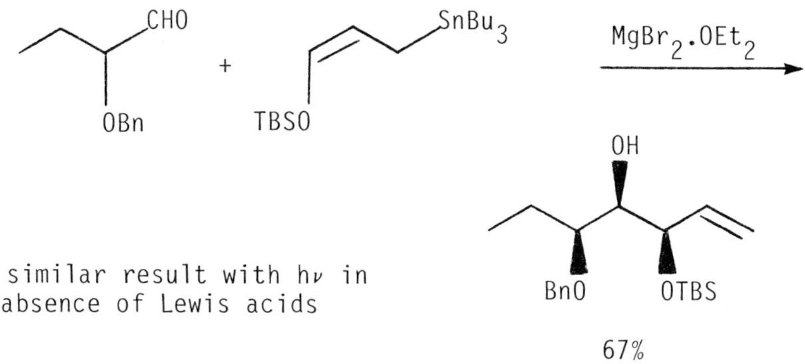

similar result with hν in absence of Lewis acids

I.A.7.a.3-5 C. Trombini, A. Umani-Ronchi et al., J. Org. Chem., 52, 5447 (1987); N. Minowa and T. Mukaiyama, Bull. Chem. Soc. Jpn., 60, 3697 (1987).

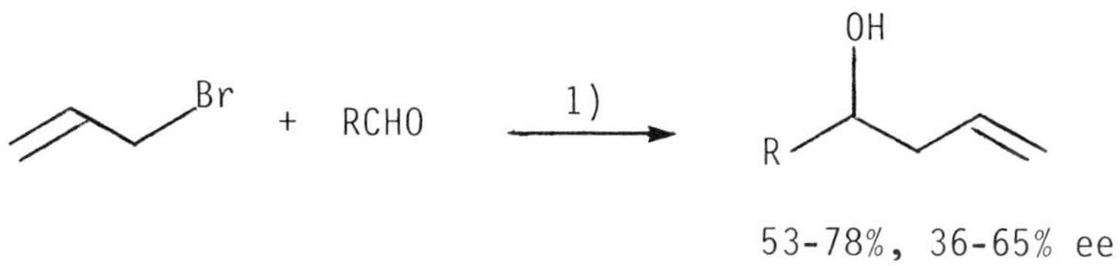

1) $SnCl_2$, (+)-diethyl tartrate, 18-22h

similar results with $Sn(OTf)_2$, a chiral amine and allyl diisobutyl aluminum

---

I.A.7.a.3-6 E. Erdik, Tetrahedron, 43, 2203 (1987).

Review: "Use of Activation Methods for Organozinc Reagents."

I.A.7.a.3-7 Y. Kobayashi et al., Tetrahedron Lett., 28, 659 (1987); K.T. Mead, ibid., 28, 1019 (1987); M. Kuroboshi and T. Ishihara, ibid., 28, 6481 (1987); G. Picotin and P. Miginiac, ibid., 28, 4551 (1987); idem, J. Org. Chem., 52, 4796 (1987); K. Takai et al., ibid., 52, 4410 (1987); A. Furstner, J. Organomet. Chem., 336, C33 (1987); C. Wakselman et al., Chem. Commun., 642 (1987); T. Ishihara and M. Kuroboshi, Chem. Lett., 1145 (1987); T. Nakai et al., ibid., 1971 (1987).

$$R^1C{\equiv}CCF_2Br \;+\; R^2CHO \xrightarrow{1)} R^1C{\equiv}CCF_2CH(OH)R^2$$

35-82%

1) Zn, activated with $HgCl_2$

I.A.7.a.3-8 J. Brocard et al., J. Organomet. Chem., 336, C47 (1987).

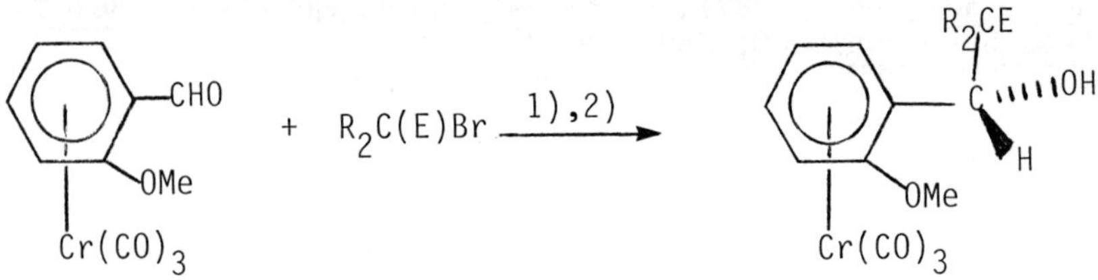

E = CO_2R^1, CN

70-80%

1) activated Zn, THF, Δ 2) NH_4Cl

I.A.7.a.3-9 Y. Tamaru, Z. Yoshida et al., Angew. Chem., Int. Ed. Engl., 26, 1157 (1987); Y. Masuyama et al., J. Org. Chem., 52, 3702 (1987).

$$RCO(CH_2)_nZnI \;+\; R^1COCl \xrightarrow{Pd(PPh_3)_4} RCO(CH_2)_nCOR^1$$

53-91%

I.A.7.a.3-10 E.J. Corey and F.J. Hannon, *Tetrahedron Lett.*, *28*, 5233 and 5237 (1987); G. Buono et al., *ibid.*, *28*, 6163 (1987); P.A. Chaloner and S.A.R. Perera, *ibid.*, *28*, 3013 (1987); K. Soai et al., *ibid.*, *28*, 4841 (1987); *idem*, *Chem. Commun.*, 467 and 1690 (1987); *idem*, *J. Am. Chem. Soc.*, *109*, 7111 (1987); S. Itsuno and J.M.J. Fréchet, *J. Org. Chem.*, *52*, 4140 (1987); Ab.A. Smaardijk and H. Wynberg, *ibid.*, *52*, 135 (1987).

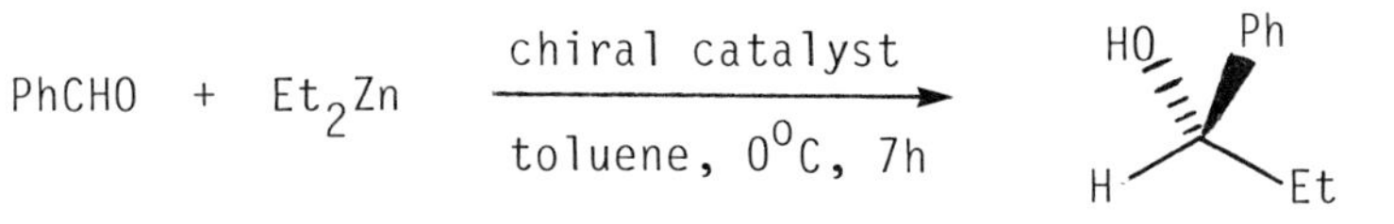

68%, 95% ee

chiral catalyst = Li complex of

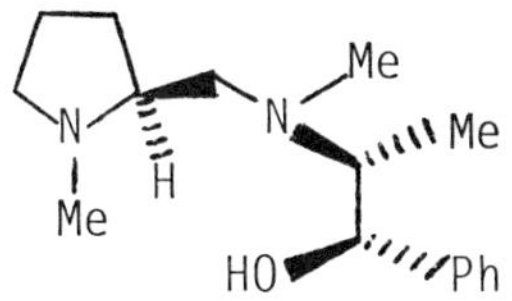

various other chiral catalysts also employed

I.A.7.a.3-11 A.A. Solov'yanov and I.P. Beletskaya, *Russ. Chem. Rev.*, *56*, 465 (1987).

Review: "Reactions of Carbonyl Compounds with Organic Derivatives of Alkali Metals and Hydride Reductants: The Effects of the Medium and the Cation."

I.A.7.a.3-12 R.W. Hoffmann, D. Hoppe et al., *Angew. Chem., Int. Ed. Engl.*, *26*, 1145 (1987).

Estimation of the Configurational Stability of Chiral Organometallic Reagents.

I.A.7.a.3-13 S.V. Frye et al., J. Am. Chem. Soc., 109, 1862 (1987).

Rapid-Injection Nuclear Magnetic Resonance Investigation of the Reactivity of α- and β-Alkoxy Ketones with Dimethylmagnesium : Kinetic Evidence for Chelation.

I.A.7.a.3-14 M. Okubo et al., Bull. Chem. Soc. Jpn., 60, 2085 (1987).

Correlation of Oxidation and Reduction Potentials of Reactants with Product Distribution in the Reactions of Various Organomagnesium Reagents with Aromatic Carbonyl and Nitro Compounds.

I.A.7.a.3-15 T. Kramer and D. Hoppe, Tetrahedron Lett., 28, 5149 (1987); J.A. Marshall and B.S. DeHoff, Tetrahedron, 43, 4849 (1987); D.J. Kempf et al., J. Med. Chem., 30, 1978 (1987); M. Hosomi et al., Chem. Pharm. Bull., 35, 3058 (1987).

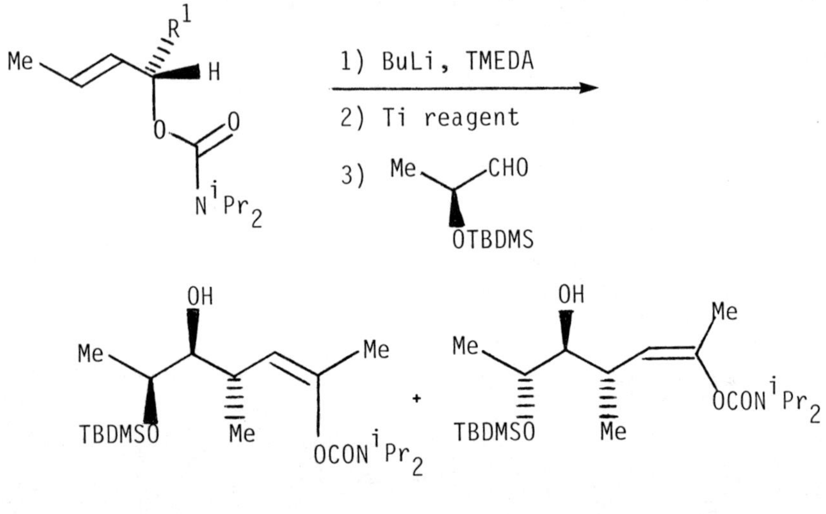

Ti reagent = $Ti(O^iPr)_4$ 61%, 93 : 7

= $ClTi(NEt_2)_3$ 60%, 2 : 98

I.A.7.a.3-16 X.-C. He and E.L. Eliel, Tetrahedron, 43, 4979 (1987); E. Buschmann et al., Liebigs Ann. Chem., 349 (1987); M.T. Reetz et al., Angew. Chem., Int. Ed. Engl., 26, 1141 (1987); Y. Tamura et al., Chem. Pharm. Bull., 35, 2305 (1987); S.D. Burke et al., Tetrahedron Lett., 28, 3905 (1987); G. Procter et al., ibid., 28, 2629 (1987); R. Gree et al., ibid., 28, 2959 (1987); G. Stork and S.D. Rychnovsky, J. Am. Chem. Soc., 109, 1564 and 1565 (1987); S.J. Danishefsky et al., J. Am. Chem. Soc., 109, 1572 (1987); D. Bernard et al., Synth. Commun., 17, 1807 (1987).

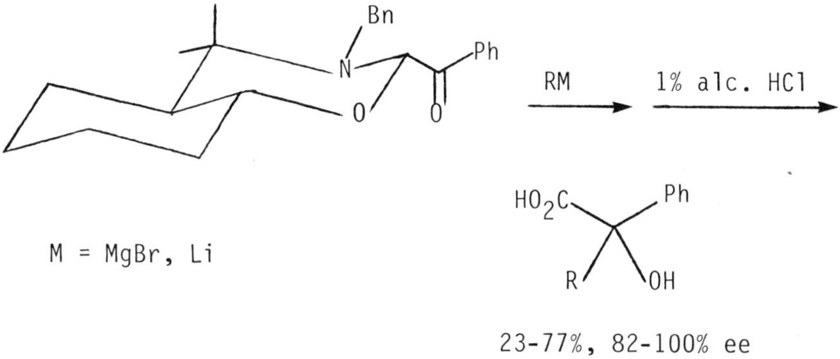

I.A.7.a.3-17 P. Canonne and M. Bernatchez, J. Org. Chem., 52, 4025 (1987); P. Canonne et al., Tetrahedron Lett., 28, 4997 (1987); idem, Tetrahedron, 43, 663 (1987); J. Barluenga et al., Synthesis, 819 (1987); idem, J. Chem. Res. (S), 400 (1987); S. Chandrasekhar and T. Murthy, ibid., 414 (1987).

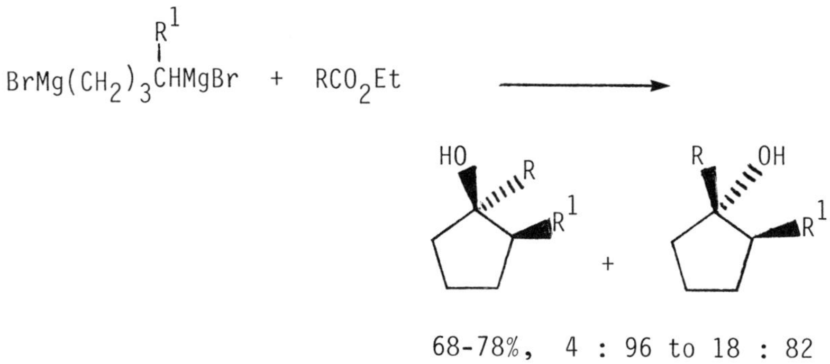

I.A.7.a.3-18 K. Tomioka et al., Chem. Lett., 65 (1987); idem, Tetrahedron Lett., 28, 1291 (1987).

$$R^1CHO + R^2MgBr \xrightarrow[-100^oC]{1)} R^1R^2CHOH$$

68-92%

40-71% ee (S)

1) 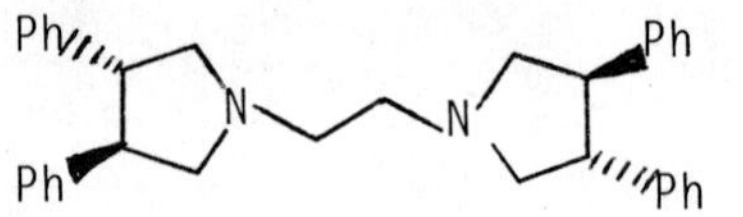

I.A.7.a.3-19 J.M. Frechet et al., J. Org. Chem., 52, 4644 (1987).

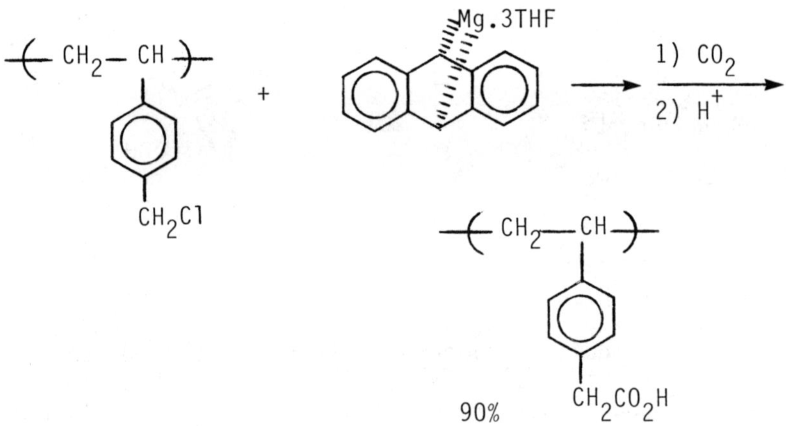

I.A.7.a.3-20 Y. Zhang et al., Tetrahedron Lett., 28, 3815 (1987).

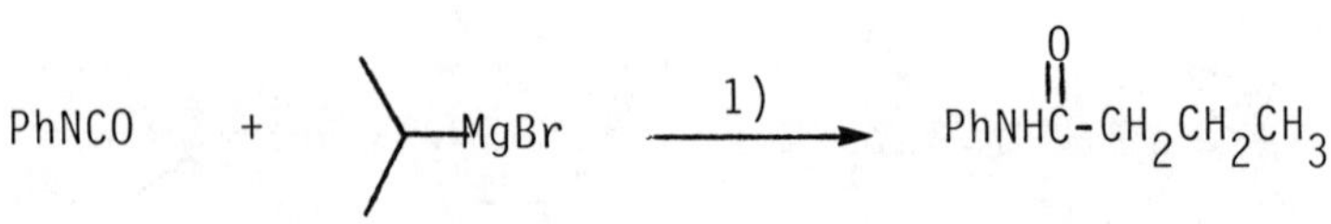

68%

1) 10 mol % Cp_2TiCl_2

I.A.7.a.3-21 P.G. Gassman and N.J. O'Reilly, J. Org. Chem., 52, 2481 (1987); L.A. Paquette et al., Tetrahedron, 43, 5567 (1987); W.R. Roush et al., J. Am. Chem. Soc., 109, 7575 (1987); W.F. Bailey et al., ibid, 109, 2442 (1987); S.V. Ley et al., Tetrahedron Lett., 28, 5755, 5759 and 5763 (1987); C.J. Kowalski and G.S. Lal, ibid., 28, 2463 (1987); T. Kolasa, S. Sharma and M.J. Miller, ibid., 28, 4973 (1987); A. Basha and D.W. Brooks, Chem. Commun., 305 (1987); A. Pelter et al., ibid., 297 (1987).

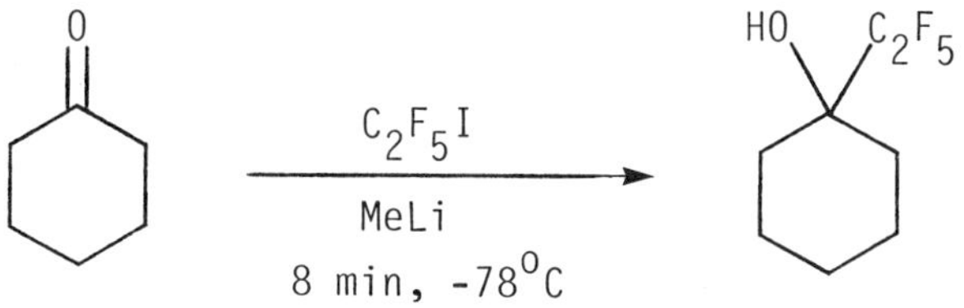

100%

no observable addition of MeLi

I.A.7.a.3-22 G. Tsuchihashi et al., Tetrahedron Lett., 28, 6335 and 6339 (1987); R. Bloch and L. Gilbert, ibid., 28, 423 (1987); B. Milenkov and M. Hesse, Helv. Chim. Acta, 70, 308 (1987); D. Hoppe et al., Angew. Chem., Int. Ed. Engl., 26, 1034 (1987); M.T. Reetz et al., ibid., 26, 477 (1987); M.T. Reetz and S. Maus, Tetrahedron, 43, 101 and 109 (1987); M.T. Reetz and S.-H. Kyung, Chem. Ber., 120, 123 (1987); N. Krause and D. Seebach, ibid., 120, 1845 (1987); A.N. Kasatkin et al., J. Org. Chem. (USSR), 23, 693 (1987).

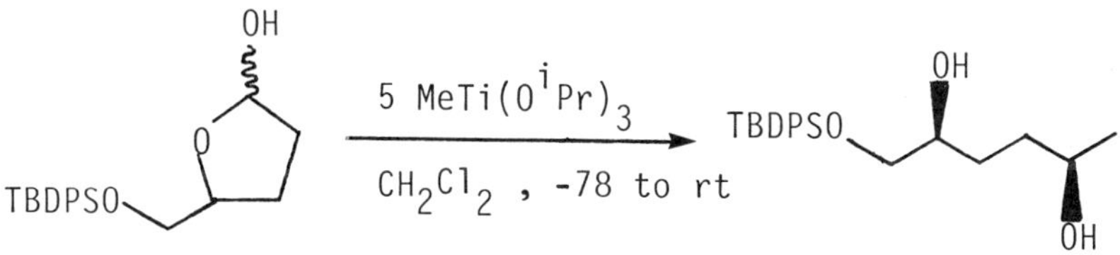

87%, 12 : 1

I.A.7.a.3-23 D. Seebach et al., Helv. Chim. Acta, 70, 954 (1987); H. Takahashi et al., Chem. Pharm. Bull., 35, 1604 (1987).

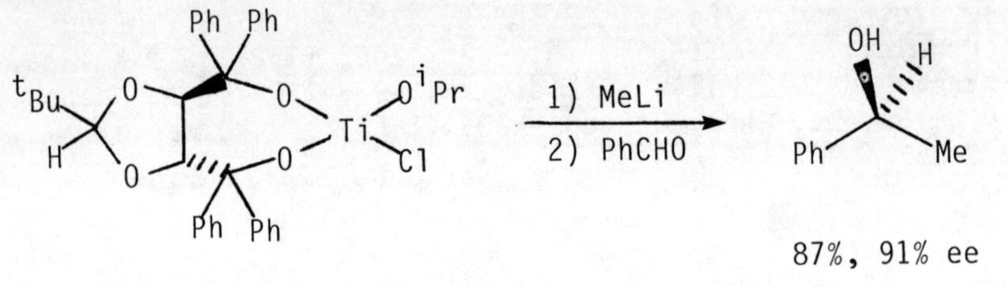

I.A.7.a.3-24 E.A. Mash, J. Org. Chem., 52, 4142 (1987); C.R. Johnson and B.D. Tait, ibid., 52, 281 (1987); Y. Tamura, Y. Kita et al., Tetrahedron Lett., 28, 4583 (1987); B.A. Narayanan and W.H. Bunnelle, ibid., 28, 6261 (1987); T. Cohen et al., J. Am. Chem. Soc., 109, 4710 (1987); S.E. Denmark et al., ibid., 109, 2224 (1987); M.B. Anderson and P.L. Fuchs, Synth. Commun., 17, 621 (1987).

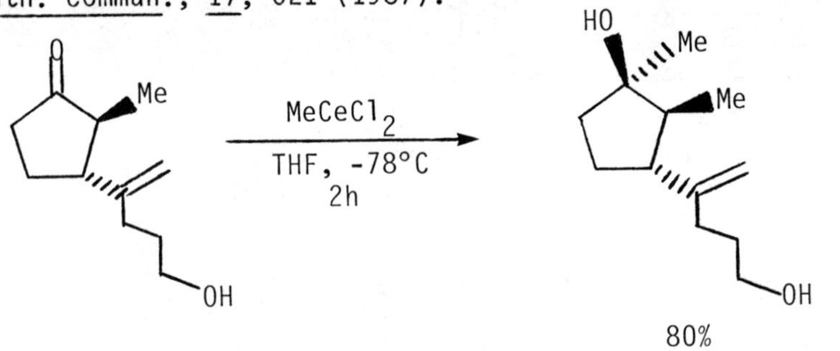

I.A.7.a.3-25 S. Torii et al., Synth. Commun., 17, 789 (1987); idem, Chem. Lett., 673 (1987); M. Wada, H. Ohki and K. Akiba, Chem. Commun., 708 (1987); Y. Butsugan et al., Tetrahedron Lett., 28, 3707 (1987); S. Sibille et al., ibid., 28, 55 (1987).

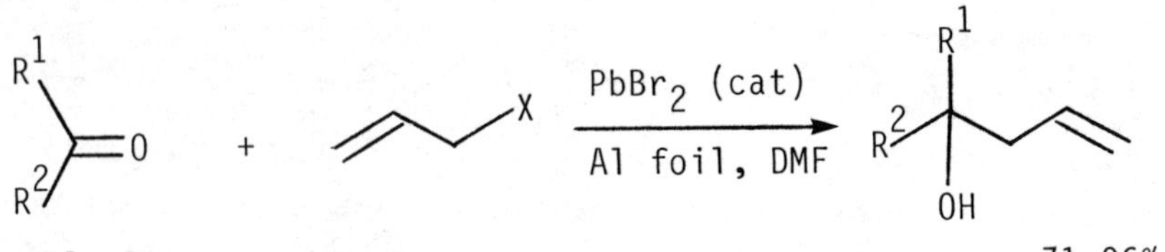

X = Cl, Br

similar results with $BiCl_3$-Al, Sb or $NiBr_2$bipy, e^-

I.A.7.a.3-26 A. Gorgues et al., Tetrahedron, 43, 5841 (1987); K. Takai et al., J. Am. Chem. Soc., 109, 951 (1987); K.C. Nicolaou et al., Angew. Chem., Int. Ed. Engl., 26, 1019 (1987).

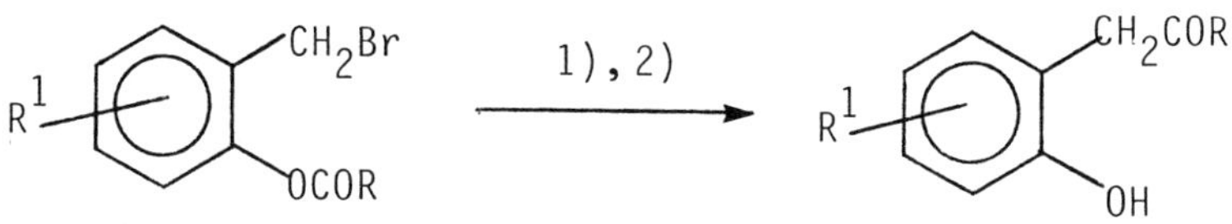

1) $CrCl_2$, THF-HMPT 2) H_2O, NH_4Cl 20-50%

very slow addition required

I.A.7.a.3-27 Y. Fujiwara et al., J. Org. Chem., 52, 3524 (1987).

$$\mathrm{RYbI} + \mathrm{R^1CH{=}CHCOR^2} \xrightarrow{\text{rt}} \mathrm{R^1CH{=}CHC(OH)R^2R}$$

37-66%

I.A.7.a.3-28 N. Kambe, N. Sonoda et al., Angew. Chem., Int. Ed. Engl., 26, 1187 (1987).

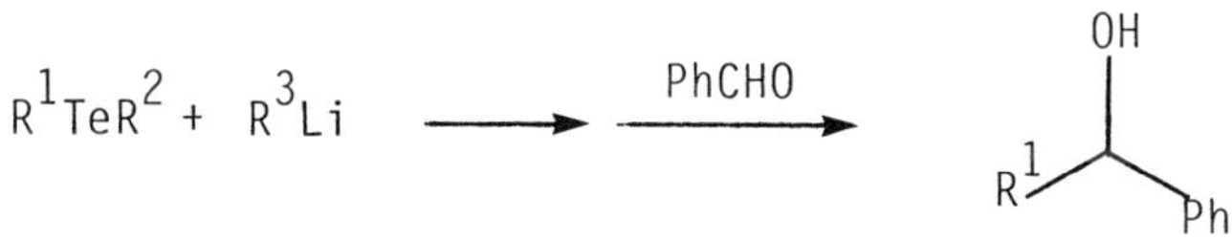

66-89%

R^1 = various, R^2 = ^{n}Bu, ^{n}Pr, Ph

R^3 = ^{n}Bu, ^{s}Bu, ^{t}Bu

I.A.7.a.3-29 H. Yasuda, A. Nakamura et al., J. Organomet. Chem., 327, 193 (1987).

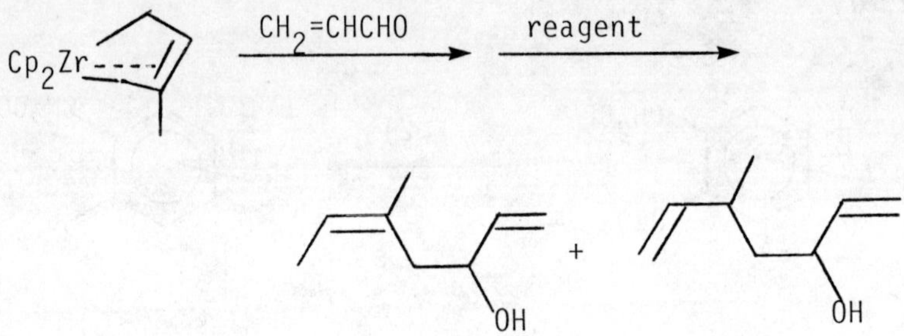

reagent = AcOH 96%, 6 : 94
= Et_2NH 97%, 86 : 14

I.A.7.a.3-30 S. Kiyooka et al., Chem. Lett., 495 (1987).

$$\text{RCHO} + \text{Ph}_3\text{GeCH}_2\text{COGePh}_3 \xrightarrow[\substack{\text{CH}_2\text{Cl}_2 \\ -78^\circ\text{C to rt}}]{\text{BF}_3 \cdot \text{OEt}_2} \text{RCH(OH)CH}_2\text{COGePh}_3$$

66-81%

no reaction with ketones or acetals

I.A.7.a.3-31 Y. Yamamoto and K. Matsuoka, Chem. Commun., 923 (1987); F.J. Weiberth and S.S. Hall, J. Org. Chem., 52, 3901 (1987); D.M. Ryckman and R.V. Stevens, ibid., 52, 4274 (1987); H.D. Verkruijsse and L. Brandsma, J. Organomet. Chem., 332, 95 (1987).

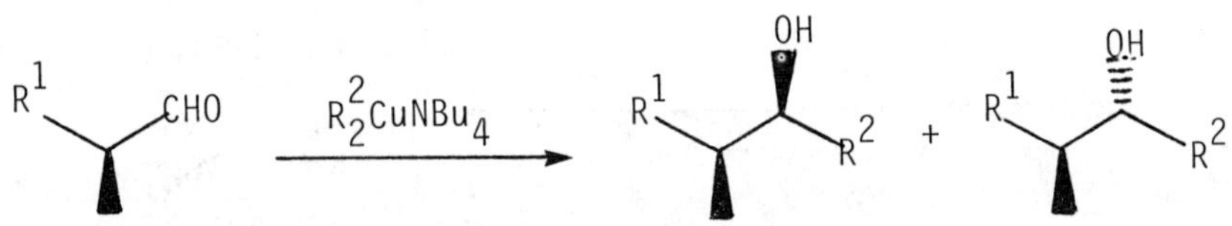

62-100%, 3.4 : 1 to 1 : 1.9

I.A.7.a.3-32 D.R. Williams and F.D. Klingler, Tetrahedron Lett., 28, 869 (1987); K. Nishitani and K. Yamakawa, ibid., 28, 655 (1987); I. Fleming et al., ibid., 28, 5921 (1987); M. Taddei et al., ibid., 28, 969 (1987); S.E. Denmark et al., J. Am. Chem. Soc., 109, 2512 (1987); K. Soai et al., Chem. Lett., 341 (1987); T. Mukaiyama et al., ibid., 1121 (1987); K. Oshima et al., ibid., 2029 (1987); K. Tamao, E. Nakajo and Y. Ito, J. Org. Chem., 52, 957 (1987); G. Bartoli et al., ibid., 52, 4381 (1987).

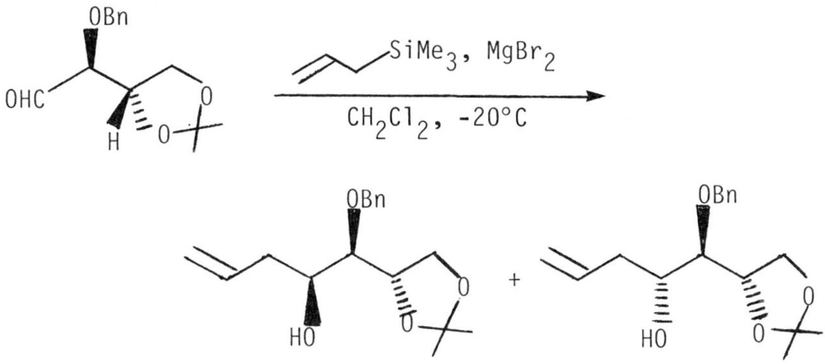

95%, >98 : <2

other Lewis acids also used

I.A.7.a.3-33 R.J.P. Corriu et al., J. Organomet. Chem., 328, C17 (1987); A. Hosomi et al., Chem. Commun., 1517 (1987).

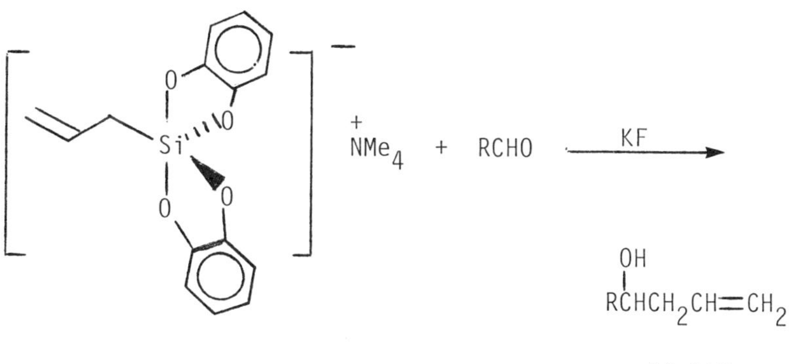

42-71%

I.A.7.a.3-34 K. Hiroi et al., Chem. Pharm. Bull., 35, 1413 (1987); K. Hiroi and H. Sato, Synthesis, 811 (1987); S. Eguchi et al., ibid., 1092 (1987); T. Hayashi, Y. Matsumoto and Y. Ito, Chem. Lett., 2037 (1987).

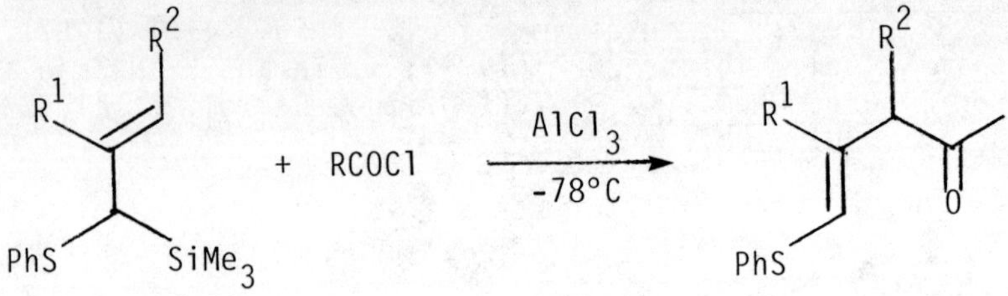

21-92%

I.A.7.a.3-35 A.R. Katritzky and S. Sengupta, Tetrahedron Lett., 28, 1847 (1987).

$$Me_3SiCH_2OH \xrightarrow{1)-5)} RC(=O)CH_2OH$$

42-68%

1) n-BuLi 2) CO_2 3) s-BuLi 4) RCO_2Et 5) H_3O^+

I.A.7.a.3-36 R.M. Williams et al., J. Org. Chem., 52, 2615 (1987); A.B. Holmes et al., J. Chem. Soc., Perkin Trans. 1, 1515 (1987); S. Kano et al., Tetrahedron Lett., 28, 6331 (1987).

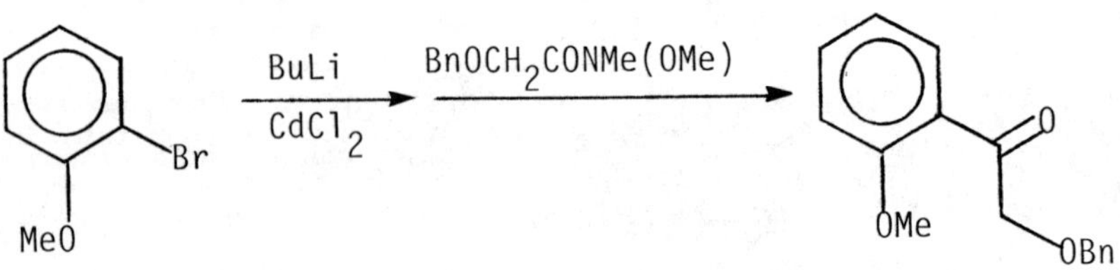

90%

I.A.7.a.3-37 M. Larcheveque and J. Lalande, Bull. Soc. Chim. Fr., 116 (1987); G. Marchese et al., Tetrahedron Lett., 28, 2053 (1987); R.H. Schlessinger, J.R. Tata and J.P. Springer, ibid., 28, 5423 (1987); idem, J. Org. Chem., 52, 708 (1987).

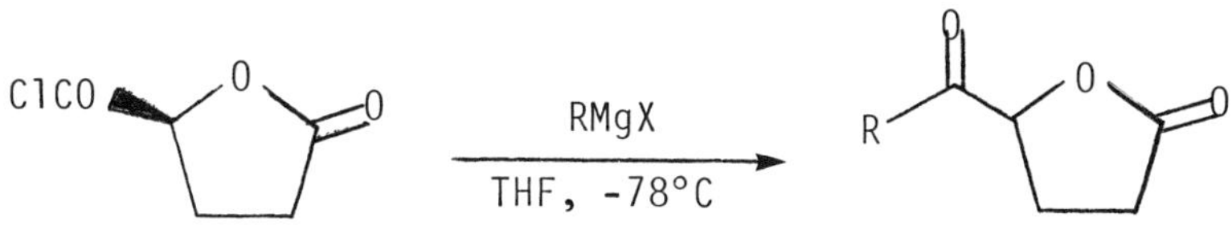

52-76%

similar reaction with $Fe(acac)_3$ catalysis or some lithio species

---

I.A.7.a.3-38 X. Creary, J. Org. Chem., 52, 5026 (1987); H. Uno, H. Suzuki et al., Chem. Lett., 1153 (1987); C. Fehr et al., Helv. Chim. Acta, 70, 1745 (1987).

$$RM + CF_3CO_2Et \xrightarrow[-78°C]{ether} RCOCF_3$$

M = MgBr, Li 63-88%

I.A.7.a.3-39 A. Griesbeck and D. Seebach, Helv. Chim. Acta, 70, 1326 (1987); C.J. Moody and R.J. Taylor, Tetrahedron Lett., 28, 5351 (1987); M.T. Crimmins et al., ibid., 28, 3651 (1987); F.M. Aly et al., Ind. J. Chem., 25B, 1255 (1986).

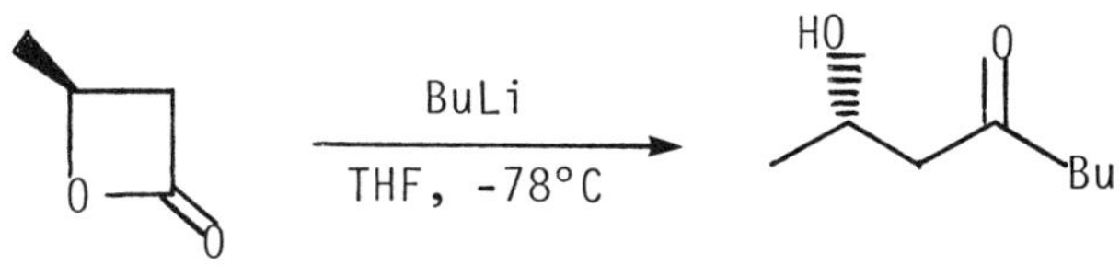

62%

I.A.7.a.3-40 A.I. Meyers and B.A. Lefker, Tetrahedron Lett., 28, 1745 (1987); Y. Honda, A. Ori and G. Tsuchihashi, Chem. Lett., 1259 (1987); S. Nozoe et al., ibid., 2091 (1987); M. Lamant and A. Guignard, Helv. Chim. Acta, 70, 1279 (1987).

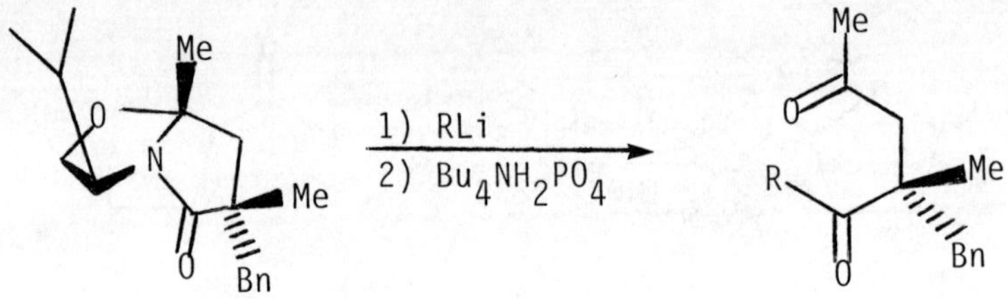

R = Me, Bu 80-91%

I.A.7.a.3-41 F. Clerici et al., Synthesis, 1025 (1987); Y. Tominaga, S. Kohra and A. Hosomi, Tetrahedron Lett., 28, 1529 (1987); S. Collins and Y. Hong, ibid., 28, 4391 (1987).

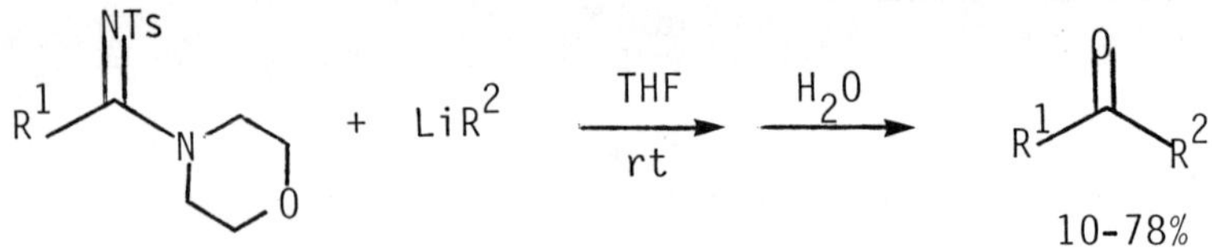

10-78%

similar results with the thioamido analogue and from

$RLa(OTf)_2 \quad + \quad R^1CONR^2_2$

I.A.7.a.4. Other 1,2-Additions

I.A.7.a.4-1 J.A. Prieto, G.L. Larson et al., Synth. Commun., 17, 1047 (1987); J.M. Wyvratt et al., J. Org. Chem., 52, 944 (1987).

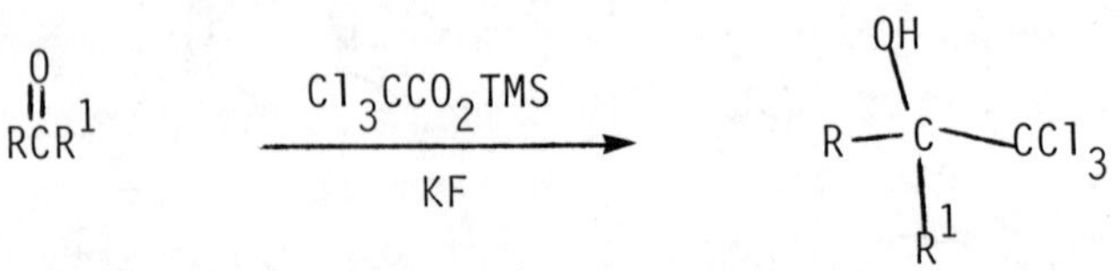

68-75%

I.A.7.a.4-2 A. Solladie-Cavallo et al., Chem. Lett., 1583 (1987); K. Narasaka et al., ibid., 2073 (1987); T. Kurihara et al., Chem. Pharm. Bull., 35, 3850 (1987); K. Sukata, Bull. Chem. Soc. Jpn., 60, 1085 and 3820 (1987); J.H. Babler and C.J. Marcuccilli, Tetrahedron Lett., 28, 4657 (1987).

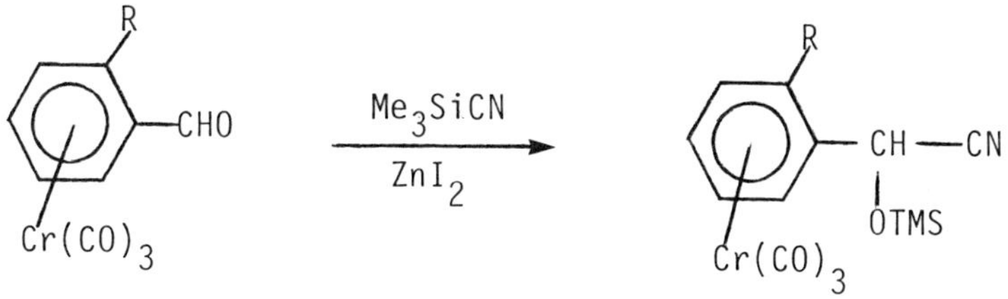

R = Me, CF_3 100%, 90 : 10 to 97 : 3

a chiral Ti Lewis acid and XAD (Amberlite macroporous resin) also used

---

I.A.7.a.4-3 F. Effenberger et al., Angew. Chem., Int. Ed. Engl., 26, 458 (1987).

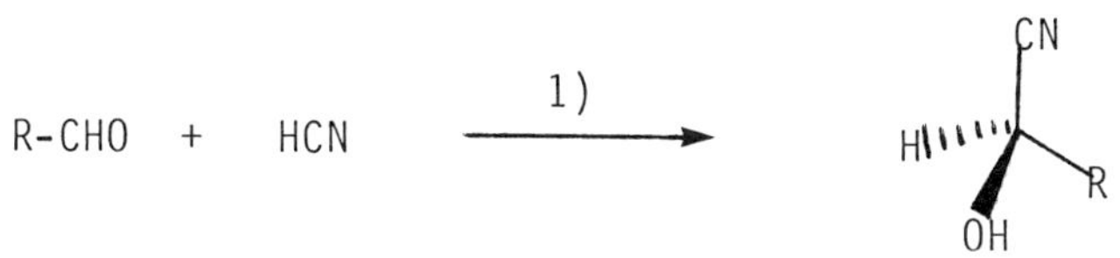

75-99%, 14-99% ee

1) cellulose supported mandelonitrile lyase solution, ethyl acetate

---

I.A.7.a.4-4 T. Hanafusa et al., Chem. Lett., 687 (1987).

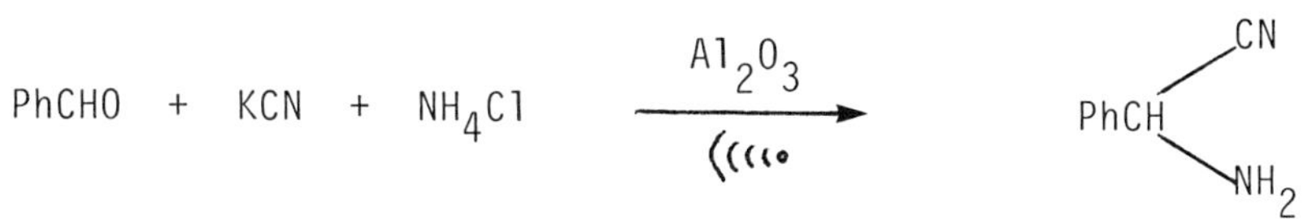

90% (major)

much lower yields and greater complexity without Al_2O_3 or ultrasound

I.A.7.a.4-5 A.B. Smith, III and M. Fukui, J. Am. Chem. Soc., 109, 1269 (1987); G. Quinkert et al., Angew. Chem., Int. Ed. Engl., 26, 61 (1987); O. Achmatowicz et al., Synthesis, 413 (1987); U.R. Nayak et al., Ind. J. Chem., 26B, 8 (1987).

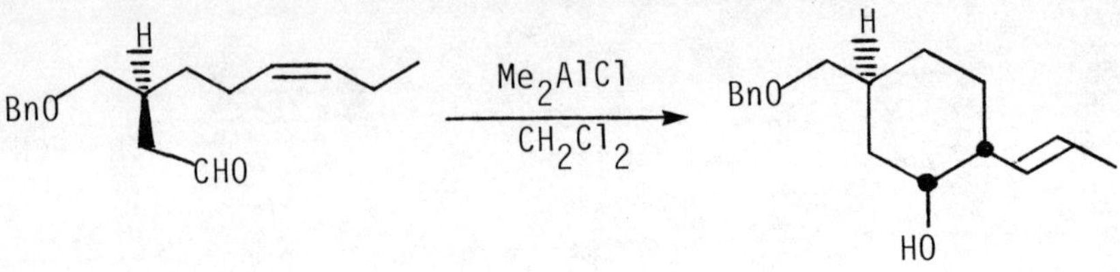

83%

I.A.7.a.4-6 D. Basavaiah and V.V.L. Gowriswari, Synth. Commun., 17, 587 (1987); D. Basavaiah et al., Tetrahedron Lett., 28, 4351 and 5653 (1987); D.S. Watt et al., ibid., 28, 2335 (1987); P.A. Brown and P.R. Jenkins, J. Chem. Soc., Perkin Trans. 1, 1129 (1987); M.A. Tius et al., J. Org. Chem., 52, 2625 (1987).

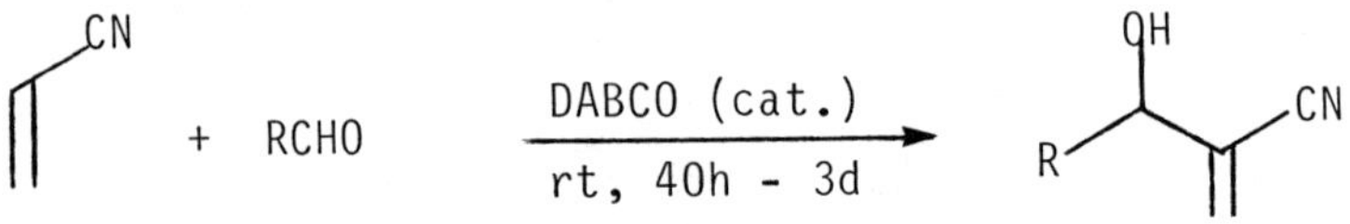

see also I.B.3.

I.A.7.a.4-7 E.J. Roskamp and S.F. Pedersen, J. Am. Chem. Soc., 109, 6551 (1987).

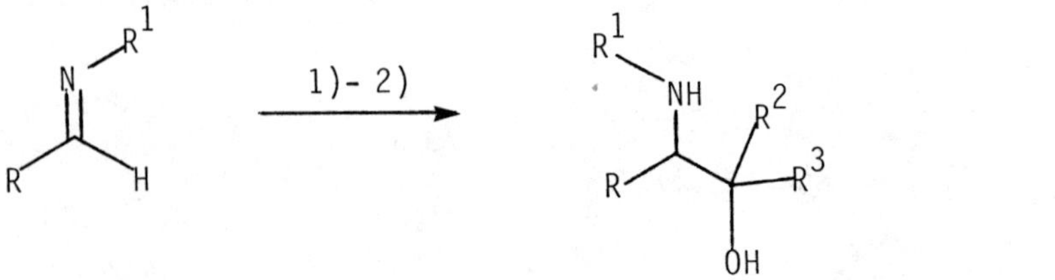

33-97%, 3 : 1 to 83 : 1 ds

1) $NbCl_3$ (DME), THF 2) R^2COR^3

I.A.7.a.4-8 J. Cossy et al., Tetrahedron Lett., 28, 4547 and 4545 (1987).

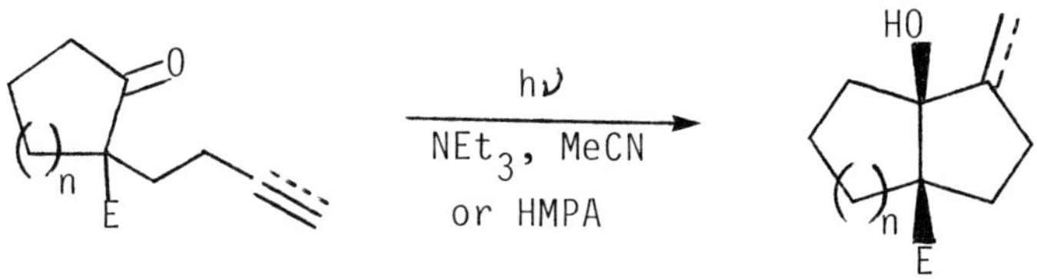

I.A.7.b.1. Conjugate Additions of Enolate-Type Carbanions

I.A.7.b.1-1 M. Hesse et al., Helv. Chim. Acta, 70, 315, 2146 and 2166 (1987); N. Ono et al., Synthesis, 258 (1987); P.G. Baraldi, A. Barco et al., Tetrahedron, 43, 4669 (1987); R.A. Bunce and R.E. Drumright, Org. Prep. Proced. Int., 19, 471 (1987); F. Richter and H.-H. Otto, Tetrahedron Lett., 28, 2945 (1987); J.R. Hwu and N. Wang, Chem. Commun., 427 (1987).

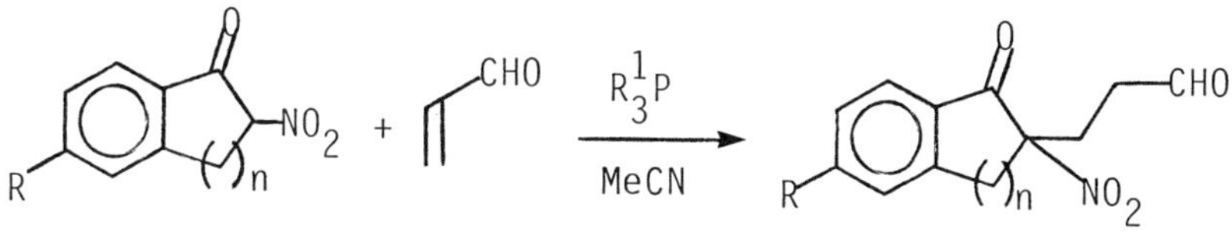

n = 1-3, R = H, MeO 68-94%

TMG, DBU and NaOMe also used

I.A.7.b.1-2 R. Ballini et al., Synth. Commun., 17, 543 (1987); R. Ballini, M. Petrini and G. Rosini, Synthesis, 711 (1987); D.E. Bergbreiter and J.J. Lalonde, J. Org. Chem., 52, 1601 (1987).

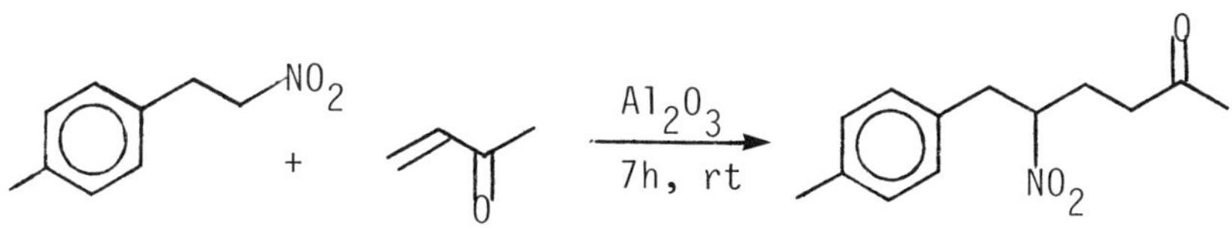

78%

Amberlyst A-21 and KF/Al_2O_3 also used

I.A.7.b.1-3 J. Seyden-Penne et al., Bull. Soc. Chim. Fr., 531 (1987).

Study of Michael Addition in the Presence of Various Agents for Alkaline Cation Complexation in An Aprotic Milieu or PTC.

I.A.7.b.1-4 N. Iwasawa and T. Mukaiyama, Chem. Lett., 463 and 491 (1987); N. Minowa and T. Mukaiyama, ibid., 1719 (1987); T. Mukaiyama et al., ibid., 743 (1987); M. Hayashi and T. Mukaiyama, ibid., 289 (1987); E. Piers et al., Tetrahedron, 43, 5521 (1987); M. Onaka et al., Chem. Commun., 1203 (1987); K. Kanematsu et al., ibid., 102 (1987); T.H. Chan and C.V.C. Prasad, J. Org. Chem., 52, 110 and 120 (1987).

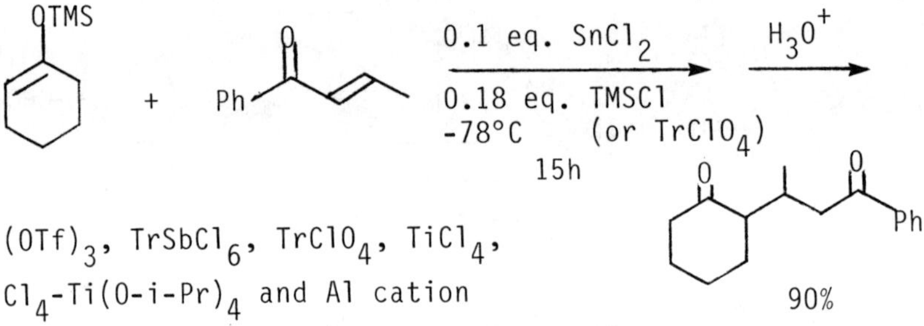

$Al(OTf)_3$, $TrSbCl_6$, $TrClO_4$, $TiCl_4$, $TiCl_4$-$Ti(O\text{-}i\text{-}Pr)_4$ and Al cation exchanged Montmorillonite Clay also used

---

I.A.7.b.1-5 T. Mukaiyama et al., Chem. Lett., 791 (1987); idem, Tetrahedron, 43, 5003 (1987).

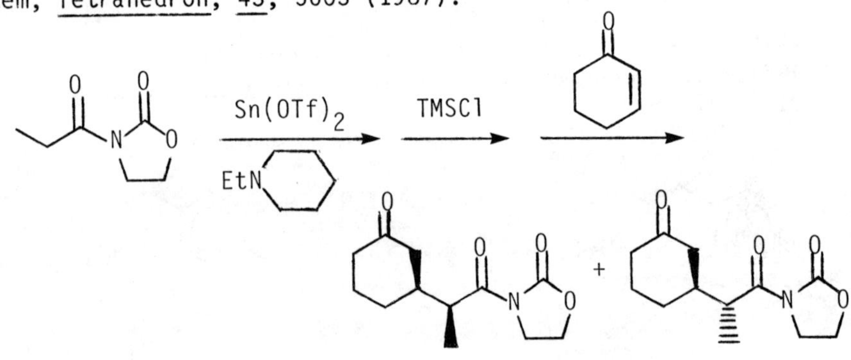

84%, 62 : 38

0% yield with no TMSCl or replacement with $SnCl_4$ or Me_3SnCl

I.A.7.b.1-6 J.-F. Lavallee and P. Deslongschamps, Tetrahedron Lett., 28, 3457 (1987); E.R. Koft et al., ibid., 28, 2799 (1987).

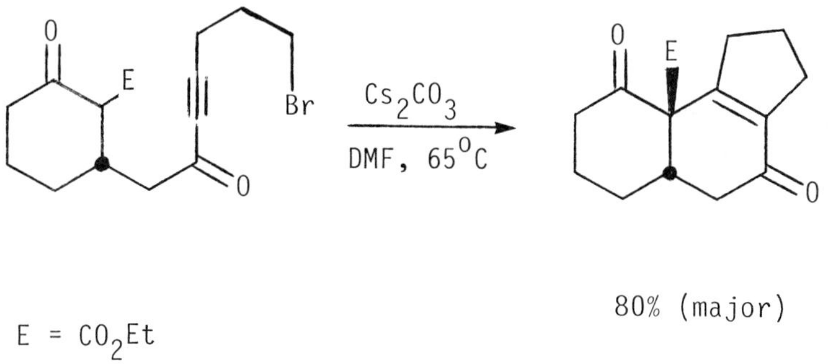

I.A.7.b.1-7 K. Koga et al., Tetrahedron Lett., 28, 6637 (1987); J. d'Angelo et al., ibid., 28, 2367 (1987); A. Gomez-Sanchez et al., J. Chem. Res. (S), 220 and 222 (1987); A. Risaliti et al., Gazz. Chim. Ital., 117, 391 (1987); H. Ahlbrecht and A. von Daacke, Synthesis, 24 (1987); K. Yamada et al., Tetrahedron, 43, 825 (1987); H. Pick-Schliebs and P. Margaretha, Helv. Chim. Acta, 70, 1855 (1987); M.M. Cooper and J.W. Huffman, Chem. Commun., 348 (1987).

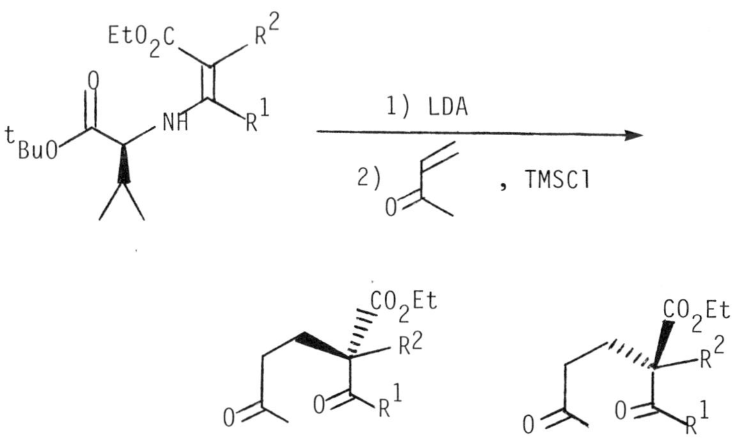

THF, -100°C	66%,	--	87% ee
toluene, HMPA	38%,	50% ee	--

I.A.7.b.1-8 K. Yamamoto et al., Tetrahedron Lett., 28, 6347 (1987); J. Yaozhong et al., Synth. Commun., 17, 1377 (1987); O. Tsuge et al., Bull. Chem. Soc. Jpn., 60, 3347 (1987); M. Westling and T. Livinghouse, Synthesis, 391 (1987).

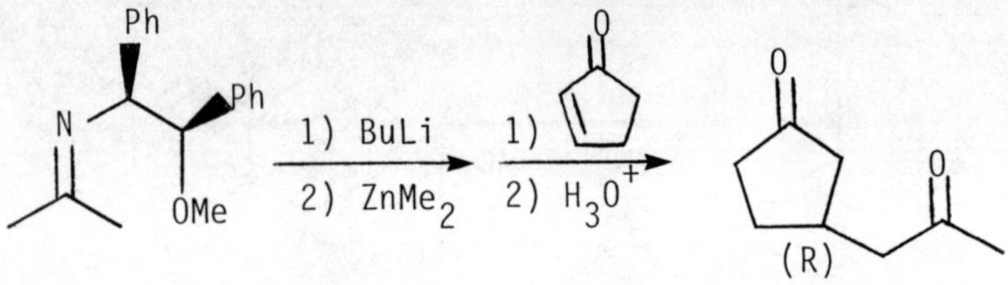

73%, 92% optical yield

I.a.7.b.1-9 K. Shibata et al., Chem. Lett., 519 (1987).

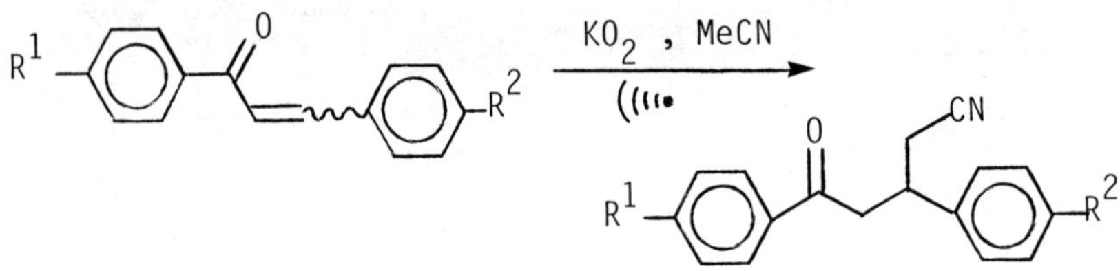

33-67% (major)

I.A.7.b.1-10 Y. Ozaki and S.-W. Kim, Chem. Lett., 1199 (1987); R.K. Haynes et al., Aust. J. Chem., 40, 1211, 1223 and 1249 (1987); M. Mikolajczyk and P. Balczewski, Synthesis, 659 (1987).

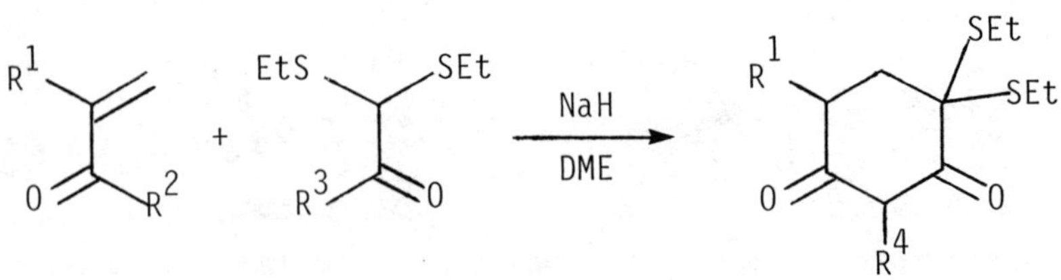

11-91%

R^2 = Me, Et, MeO;

R^3 = MeO, Me; R^4 = H, Me

I.A.7.b.1-11 Y. Yamada et al., Tetrahedron Lett., 28, 2021 and 6641 (1987).

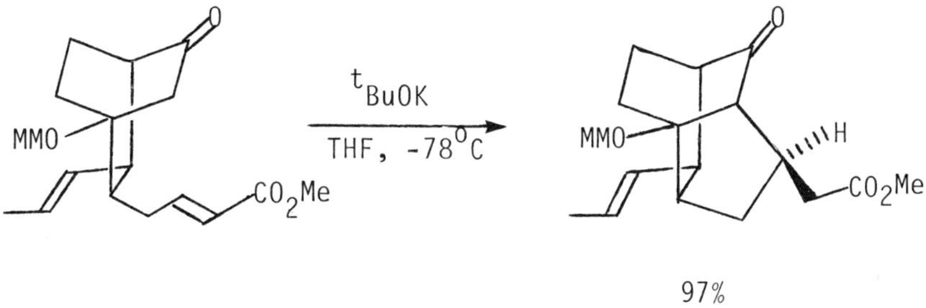

I.A.7.b.1-12 R.A. Bunce et al., J. Org. Chem., 52, 464 (1987); K. Fukumoto et al., Chem. Commun., 721 (1987).

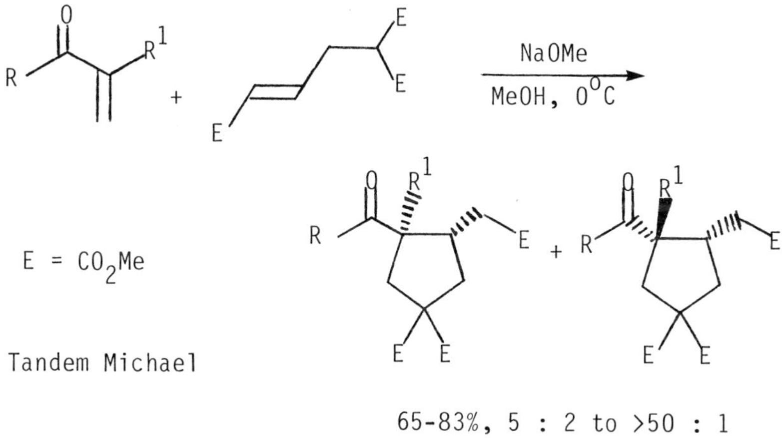

I.A.7.b.1-13 J. Sarrazin et al., Tetrahedron, 43, 5299 (1987).

Michael Additions Induced by Bases Electrogenerated in Contact with a Solid Electrolyte.

I.A.7.b.1-14 K. Koga et al., Chem. Commun., 1345 (1987); J.A. Marshall et al., J. Org. Chem., 52, 2378 (1987); S. Berrada and P. Metzner, Tetrahedron Lett., 28, 409 (1987); E.J. Corey and W. Su, ibid., 28, 5241 (1987); T. Harayama et al., Chem. Pharm. Bull., 35, 1434 (1987); V.A. Gailite et al., J. Org. Chem. (USSR), 22, 1865 (1986).

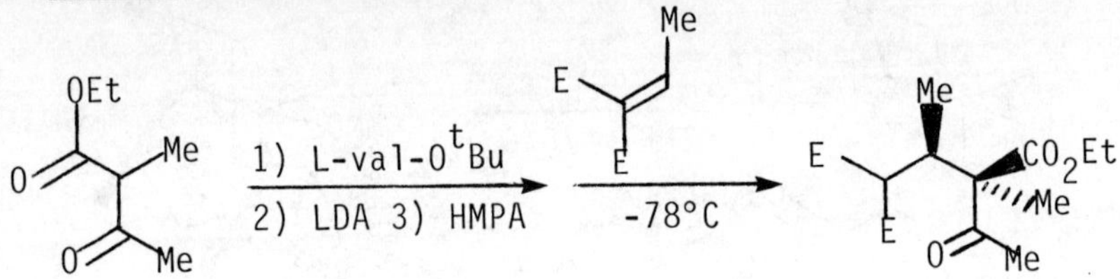

94%, 99% ee

I.A.7.b.1-15 S. Kiyooka and T. Yamashita, Chem. Lett., 1775 (1987); K. Fuji et al., Synthesis, 729 (1987).

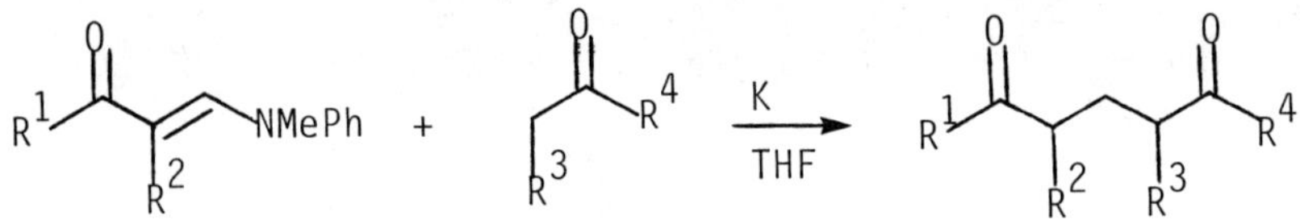

48-78%

I.A.7.b.1-16 M. Yamaguchi et al., Tetrahedron Lett., 28, 1785 (1987).

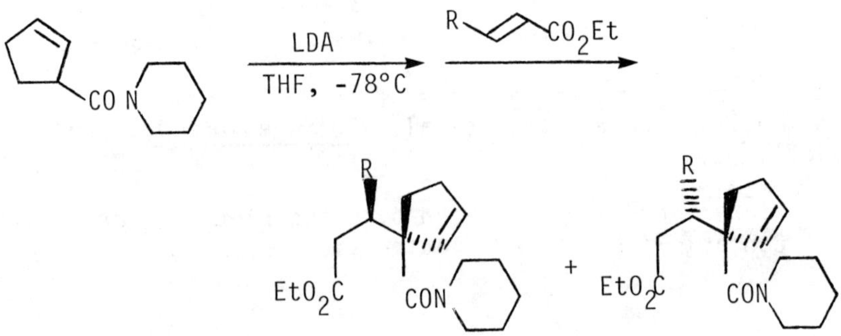

80%, >20 : 1

I.A.7.b.1-17 D.H. Hua et al., J. Org. Chem., 52, 719 (1987); idem, Tetrahedron Lett., 28, 5465 (1987).

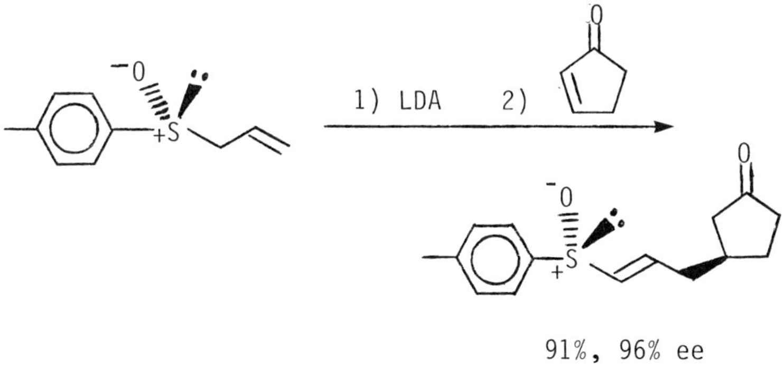

I.A.7.b.1-18 T. Yamazaki, N. Ishikawa et al., Chem. Commun., 1340 (1987).

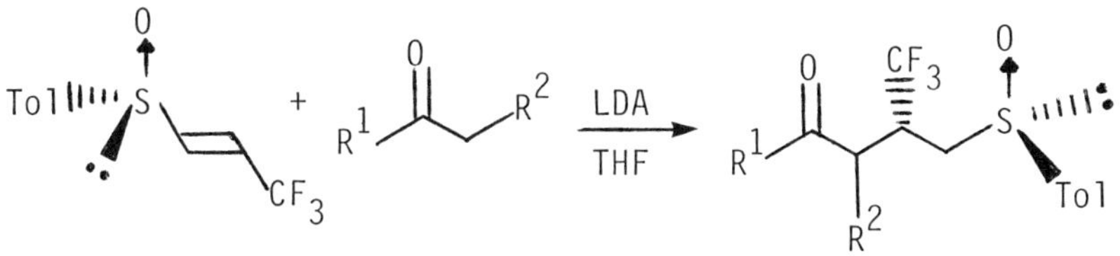

86-99%, 85 to >98% de

I.A.7.b.1-19 D. Enders and B.E.M. Rendenbach, Chem. Ber., 120, 1223 and 1731 (1987).

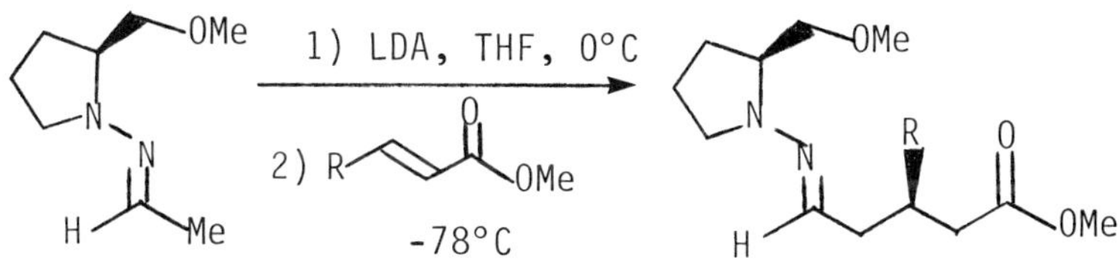

55.5-72%, 90 to ≫96% de

I.A.7.b.1-20 D.H. Hua et al., J. Am. Chem. Soc., 109, 5026 (1987).

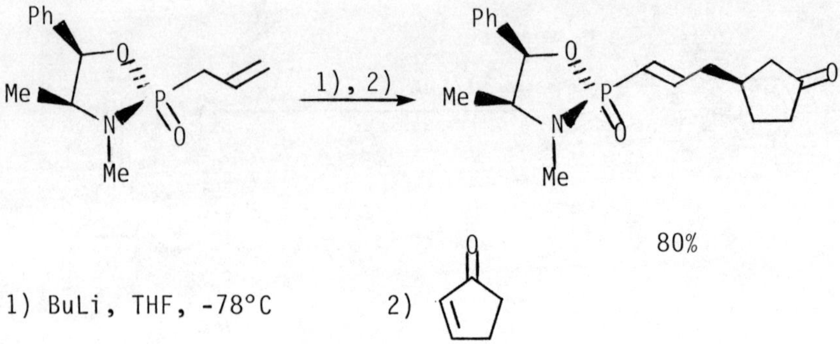

I.A.7.b.2. Conjugate Additions of Organometallic Reagents

I.A.7.b.2-1 B.H. Lipshutz et al., Tetrahedron Lett., 28, 945 (1987); G. Magnusson et al., ibid., 28, 3589 (1987); R.J. Linderman et al., ibid., 28, 3911 (1987); H. Sakata and I. Kuwajima, Tetrahedron Lett., 28, 5719 (1987); M. Nilsson et al., J. Organomet. Chem., 334, 255 (1987); G.A. Tolstikov et al., J. Org. Chem. (USSR), 22, 1464 (1986).

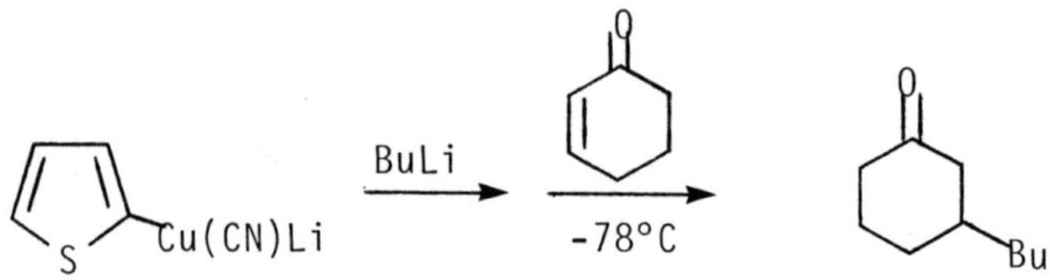

>90%

I.A.7.b.2-2 Y. Yamamoto et al., J. Am. Chem. Soc., 109, 5820 (1987).

Do the Organocopper Conjugate Additions to α,β-Unsaturated Esters Proceed in a 1,4- or a 1,2-Fashion?

I.A.7.b.2-3 J.J.E. Moreau et al., Tetrahedron Lett., 28, 2963 (1987); J.A. Marshall and S.L. Crooks, ibid., 28, 5081 (1987); R.J. Linderman and M.S. Lonikar, ibid., 28, 5271 (1987); C.R. Johnson and T.J. Marren, ibid., 28, 27 (1987); H. Hopf et al., J. Org. Chem., 52, 398 (1987); D.K. Hutchinson and P.L. Fuchs, J. Am. Chem. Soc., 109, 4930 (1987); G. Jommi et al., Gazz. Chim. Ital., 117, 125 (1987); M. Isobe et al., Tetrahedron, 43, 4737, 4749, 4759 and 4767 (1987).

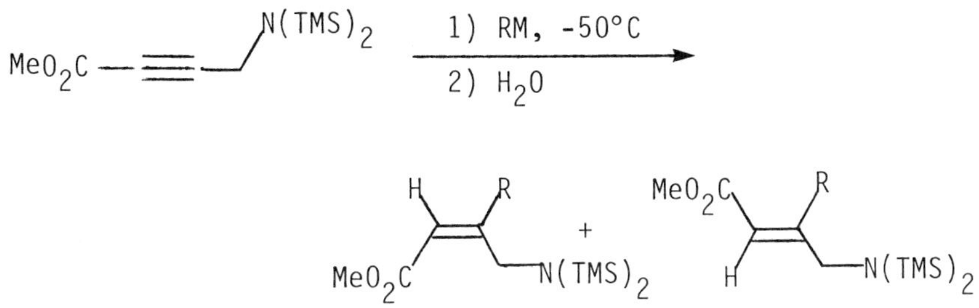

R-M = BuCu 79%, 95 : 5
Bu_2CuLi 60%, 13 : 87
Me(hexynyl)CuLi 65%, 4 : 96

I.A.7.b.2-4 K. Mori and H. Mori, Tetrahedron, 43, 4097 (1987); G.M.P. Giblin and N.S. Simpkins, Chem. Commun., 207 (1987); S. Ikegami et al., ibid., 24 (1987); T.G. Back et al., J. Org. Chem., 52, 4258 (1987); X. Huang and B. Chen, Synthesis, 480 (1987); A. Guingant and M.M. Barreto, Tetrahedron Lett., 28, 3107 (1987).

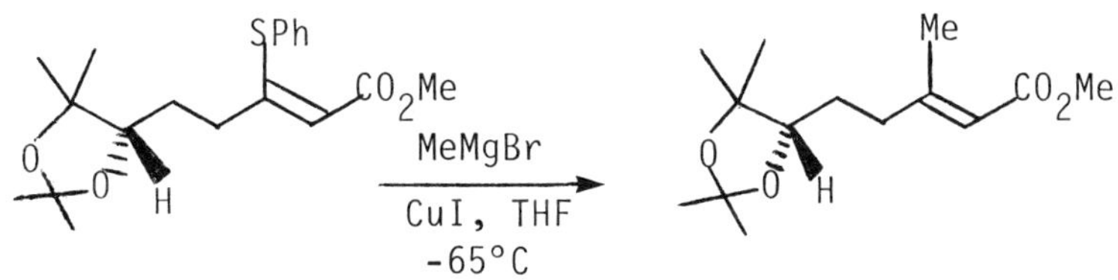

71.3%

towards Juvenile Hormone III

other leaving groups and organometallics also used

I.A.7.b.2-5 L.A. Paquette et al., J. Org. Chem., 52, 2960 (1987); S.W. Baldwin and J.M. McIver, ibid., 52, 320 (1987); F.A.J. Kerdesky et al., J. Med. Chem., 30, 1177 (1987); P.W. Collins et al., ibid., 30, 1952 (1987); A.R. Matlin and W.C. Agosta, J. Chem. Soc., Perkin Trans. 1, 365 (1987); I.W.J. Still and Y. Shi, Tetrahedron Lett., 28, 2489 (1987); N.L. Dovgan' et al., J. Org. Chem. (USSR), 23, 470 (1987); E. Nakamura, I. Kuwajima et al., J. Am. Chem. Soc., 109, 8056 (1987).

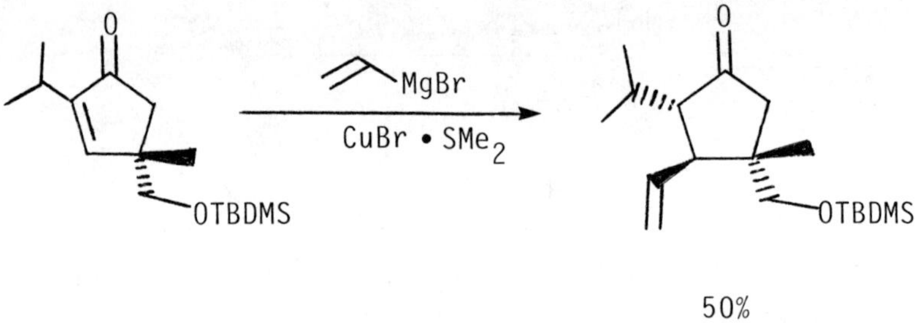

I.A.7.b.2-6 Y. Yamamoto et al., Chem. Commun., 464 and 1572 (1987); C.J. Sih et al., Tetrahedron Lett., 28, 1973 (1987); J. Leonard and G. Ryan, ibid., 28, 2525 (1987); H. Kosugi, H. Uda et al., J. Org. Chem., 52, 1078 (1987); R.D. Rieke et al., ibid., 52, 5056 and 5057 (1987); P.A. Grieco et al., ibid., 52, 3346 (1987); R.L. Funk and G.L. Bolton, ibid., 52, 3173 (1987).

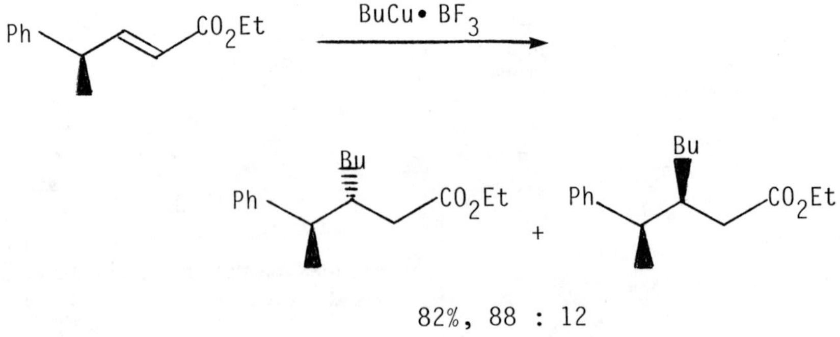

lithio cuprates also used

I.A.7.b.2-7 R.K. Dieter and M. Tokles, J. Am. Chem. Soc., 109, 2040 (1987); S. Kurozumi, R. Noyori et al., Tetrahedron, 43, 813 (1987).

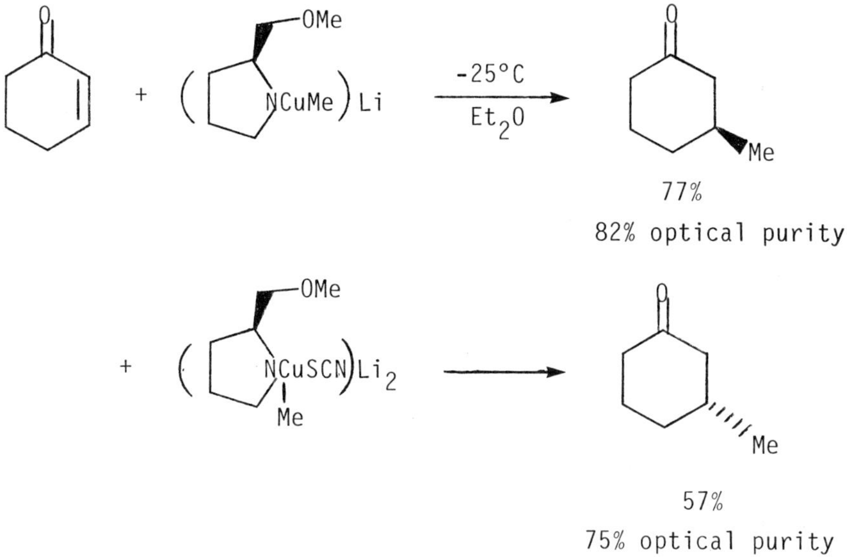

I.A.7.b.2-8 J. Mann et al., J. Chem. Res. (S), 389 (1987); D.K. Hutchinson and P.L. Fuchs, J. Am. Chem. Soc., 109, 4755 (1987).

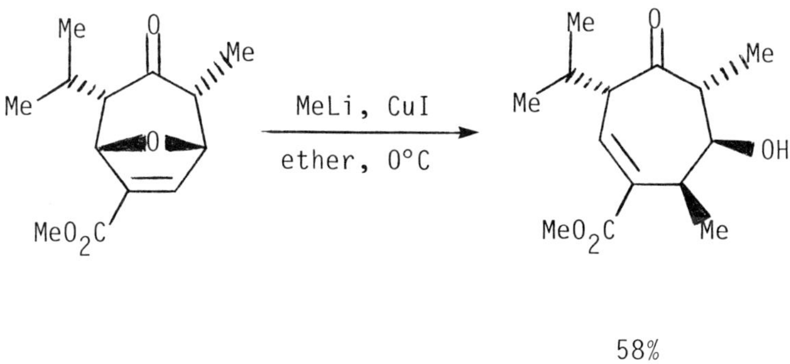

I.A.7.b.2-9 S. Hanessian and P.J. Murray, Tetrahedron, 43, 5055 (1987); idem, J. Org. Chem., 52, 1170 (1987); P. Beak and K.D. Wilson, ibid., 52, 218 (1987); M.R. Binns and R.K. Haynes, Aust. J. Chem., 40, 937 (1987); J. Otera et al., Tetrahedron Lett., 28, 2147 (1987).

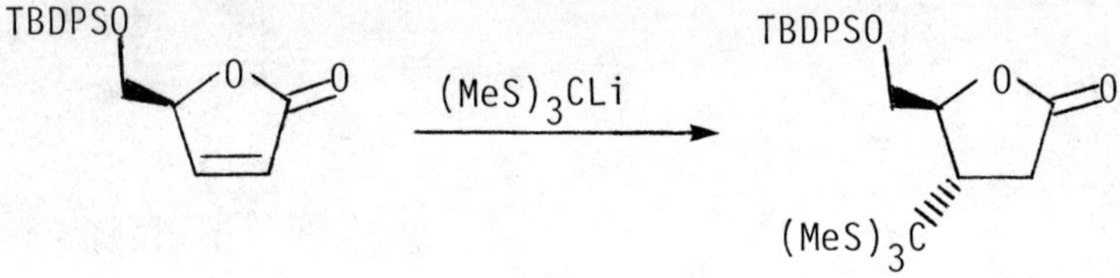

86%

I.A.7.b.2-10 G.H. Posner et al., Synth. Commun., 17, 611 (1987); L.A. Paquette et al., ibid., 17, 1141 (1987); G.H. Posner, Acc. Chem. Res., 20, 72 (1987); S.A. Hardinger and P.L. Fuchs, J. Org. Chem., 52, 2739 (1987); P.L. Fuchs et al., ibid., 52, 586 (1987); M.P. Cooke, Jr. and R.K. Widener, ibid., 52, 1381 (1987); H. Ahlbrecht et al., Synthesis, 251 and 929 (1987).

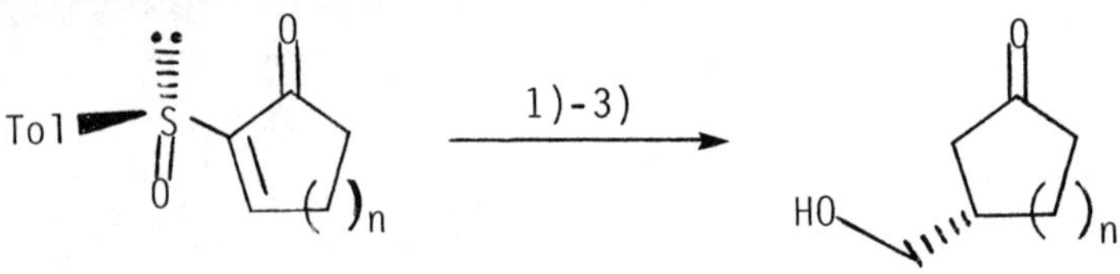

n = 1,2

76-84%, 78-93% ee

1) $PhCH_2OCH_2OCH_2Li$ 2) Ra-Ni 3) H_2, Pd-C

other α, β- unsaturated species also used

I.A.7.b.2-11 H. Yamamoto et al., Tetrahedron Lett., 28, 5723 (1987).

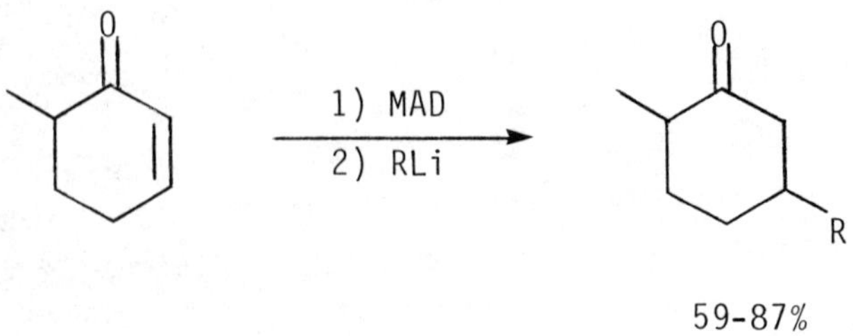

59-87%

MAD = methyl aluminum bis (2,6-di-t-butyl-4-alkylphenoxide)

I.A.7.b.2-12 A.I. Meyers et al., Tetrahedron Lett., 28, 5279 and 5283 (1987); G. Bartoli et al., Tetrahedron, 43, 4221 (1987).

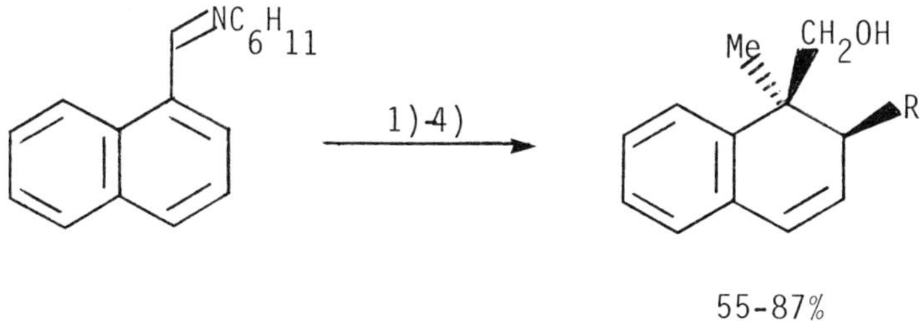

1)RLi 2) MeI 3) aq. HCl 4)$NaBH_4$

I.A.7.b.2-13 M.P. Cooke, Jr., J. Org. Chem., 52, 5729 (1987).

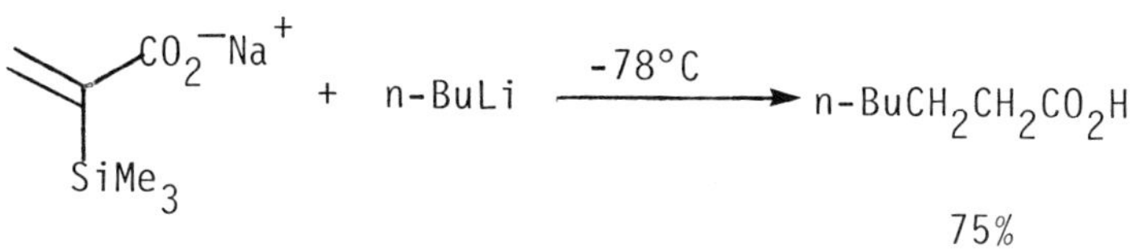

order of addition important

I.A.7.b.2-14 M. Sworin and W.L. Neumann, Tetrahedron Lett., 28, 3217 (1987); T. Cohen et al., J. Am. Chem. Soc., 109, 7923 (1987).

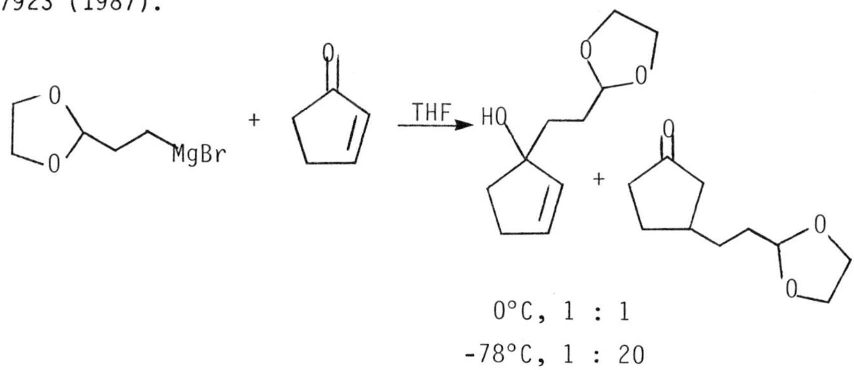

I.A.7.b.2-15 W. Oppolzer et al., Helv. Chim. Acta, 70, 2201 (1987); K. Soai et al., J. Chem. Soc., Perkin Trans. 1, 1909 (1987).

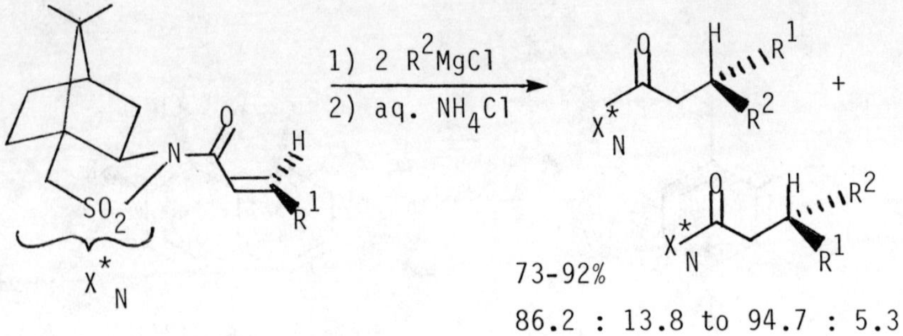

I.A.7.b.2-16 D. Mobilio and B. De Lange, Tetrahedron Lett., 28, 1483 (1987); J.S. Panek and M.A. Sparks, ibid., 28, 4649 (1987); T. Tokoroyama et al., ibid., 28, 6645 (1987); A. Takuwa et al., J. Org. Chem., 52, 1261 (1987); H. Uno, H. Suzuki et al., Synthesis, 471 (1987); G. Majetich and K. Hull, Tetrahedron, 43, 5621 (1987); J.W. Herndon, J. Am. Chem. Soc., 109, 3165 (1987).

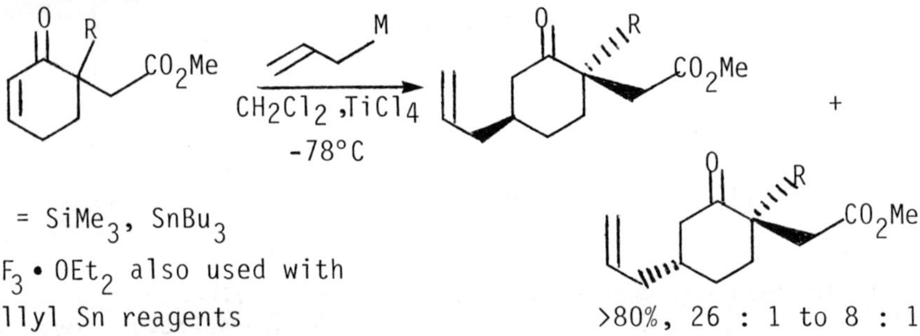

I.A.7.b.2-17 M.M. Joullie et al., Tetrahedron Lett., 28, 6539 (1987); C.H. Heathcock et al., Organometallics, 6, 2069 (1987); A.E. Greene et al., J. Am. Chem. Soc., 109, 4752 (1987).

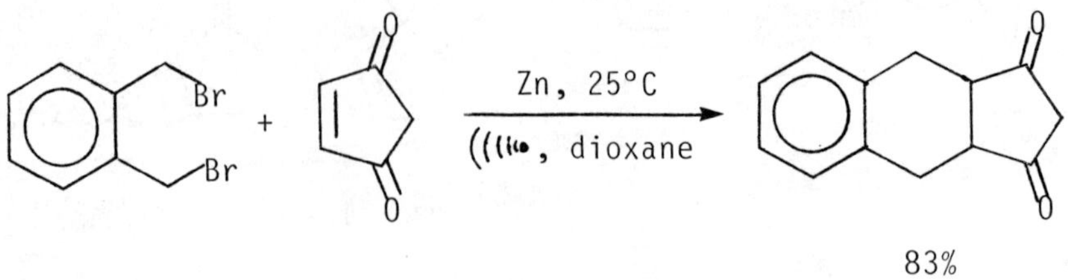

I.A.7.b.2-18 A.V. Kuchin et al., J. Org. Chem. (USSR), 23, 403 (1987).

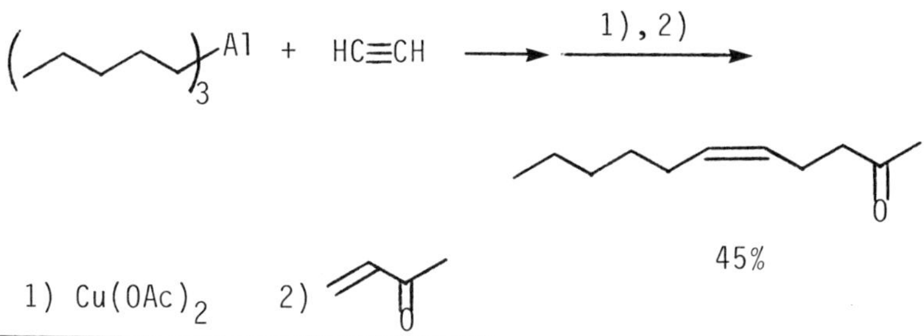

I.A.7.b.2-19 L. Dasaradhi and U.T. Bhalerao, Synth. Commun., 17, 1845 (1987).

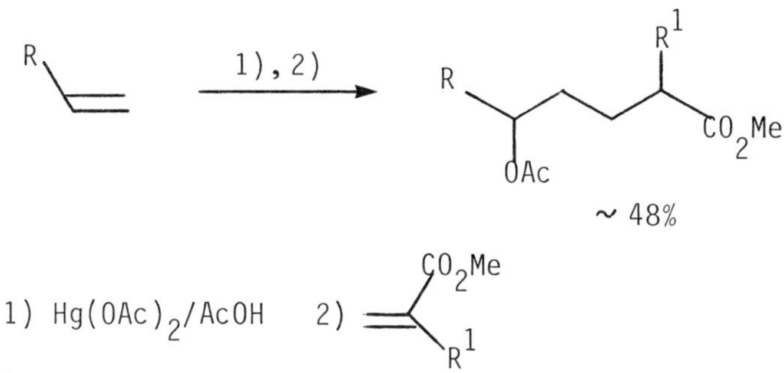

I.A.7.b.2-20 L.H.P. Gommans, L. Main and B.K. Nicholson, Chem. Commun., 761 (1987); S.A. Lebedev et al., J. Org. Chem. (USSR), 22, 1238 (1986); A. Suzuki et al., Bull. Chem. Soc. Jpn., 60, 3471 (1987).

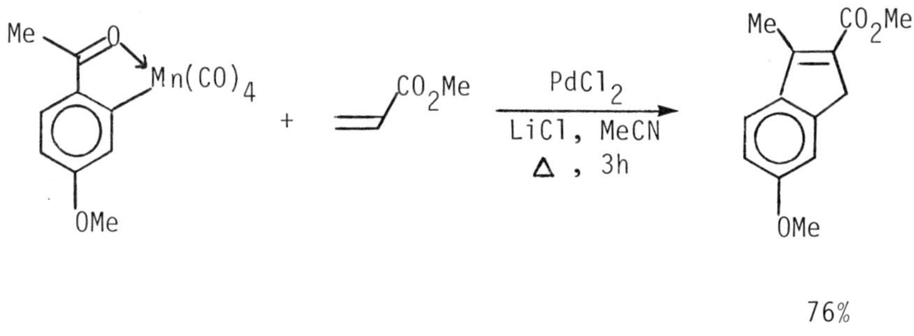

other catalysts and nucleophiles used

I.A.7.b.3. Other Conjugate Additions

I.A.7.b.3-1 H. Stetter et al., Org. Synth., 65, 26 (1987); H. Stetter and H. Skobel, Chem. Ber., 120, 643 (1987).

$Me(CH_2)_5CHO + CH_2{=}CHCOMe \xrightarrow{1)} Me(CH_2)_5C(=O)CH_2CH_2COMe$

71-75%

1) 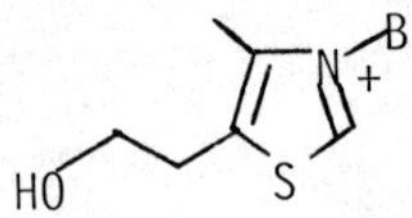

Cl^-, NEt_3, EtOH, 80°C

I.A.7.b.3-2 F. Kuo and P.L. Fuchs, J. Am. Chem. Soc., 109, 1122 (1987).

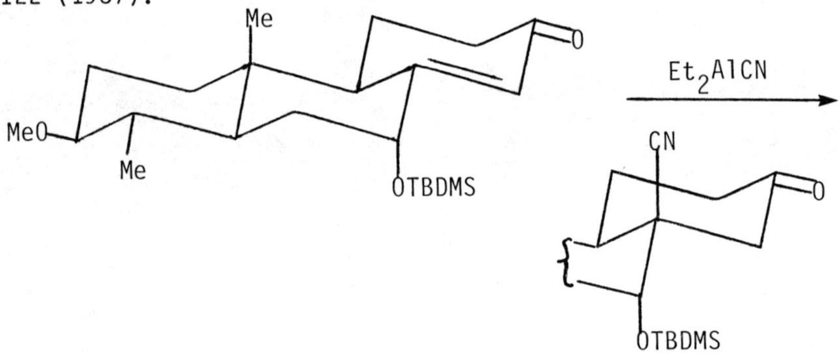

60%

I.A.7.b.3-3 O. Tsuge et al., Bull. Chem. Soc. Jpn., 60, 325 (1987).

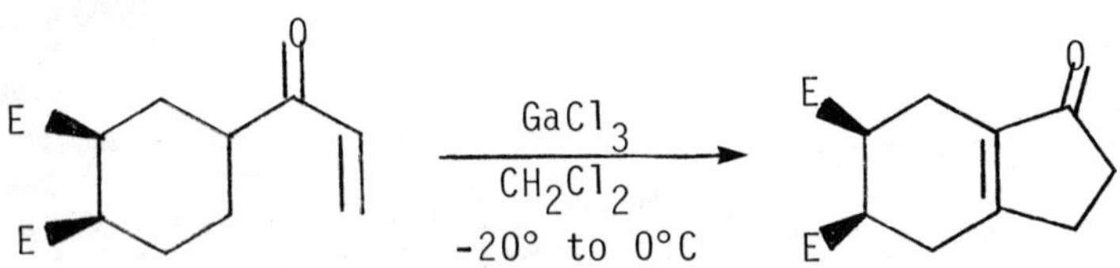

E = CO_2Me 73%

different results with other Lewis acids and dienones

I.A.7.b.3-4 A.R. Katritzky and W. Kuzmierkiewicz, <u>J. Chem. Soc., Perkin Trans. 1</u>, 819 (1987).

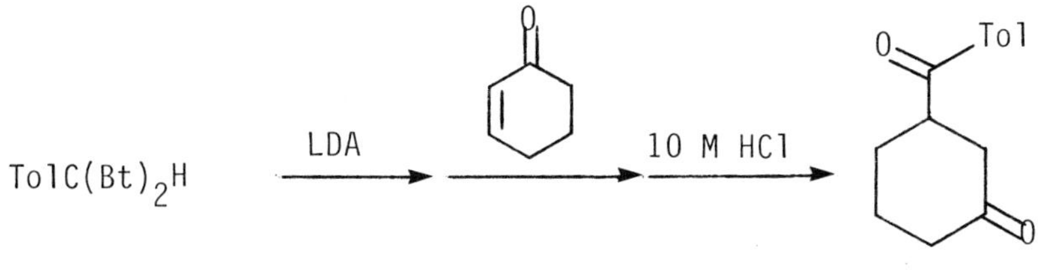

72%

Bt = benzotriazol-1-yl

I.A.7.b.3-5 A. Pecunioso and R. Menicagli, <u>Tetrahedron</u>, <u>43</u>, 5411 (1987).

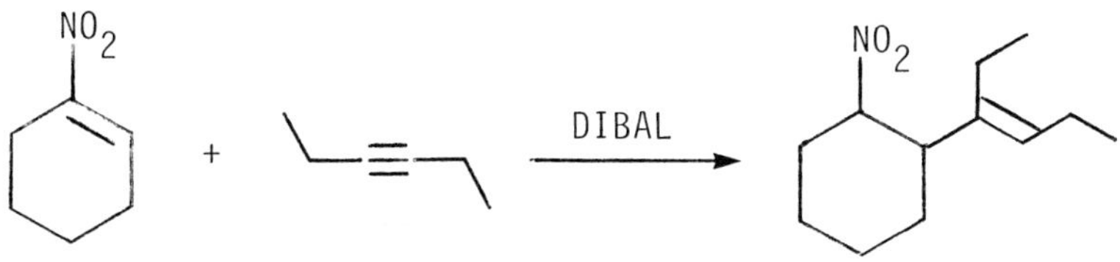

92%

I.A.7.b.3-6 P.S. Mariano et al., <u>J. Am. Chem. Soc.</u>, <u>109</u>, 4421 (1987).

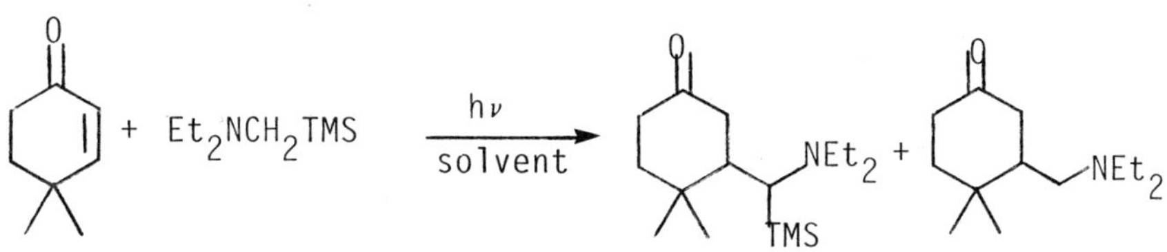

solvent =	MeCN	70%	5%
	MeOH	30%	60%

I.A.7.b.3-7 D.P. Curran and S.-C. Kuo, *Tetrahedron*, 43, 5653 (1987); D.P. Curran and M.-H. Chen, *J. Am. Chem. Soc.*, 109, 6558 (1987); F.L. Harris and L. Weiler, *Tetrahedron Lett.*, 28, 2941 (1987); B. Giese and R. Rupaner, *Liebigs Ann. Chem.*, 231 and 427 (1987).

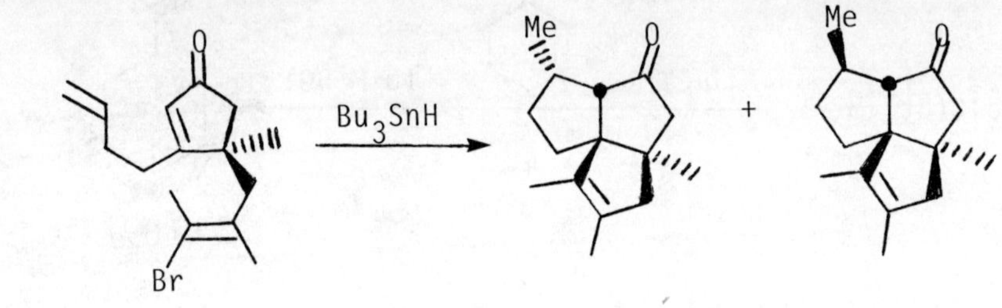

66%, 3 : 1

I.A.8. Other Carbon-Carbon Single Bond Forming Reactions

I.A.8-1 Y. Handa and J. Inanaga, *Tetrahedron Lett.*, 28, 5717 (1987); J. Nakayama et al., *ibid.*, 28, 1799 (1987); L.A. Paquette et al., *ibid.*, 28, 4965 (1987); S.K. Pradhan, K.R. Thakker and A.T. McPhail, *ibid.*, 28, 1813 (1987); Y. Fujiwara et al., *Chem. Lett.*, 2061 (1987); S. Torii et al., *Bull. Chem. Soc. Jpn.*, 60, 3427 (1987); A. Clerici and A. Porta, *J. Org. Chem.*, 52, 5099 (1987); D. Tamarkin and M. Rabinovitz, *ibid.*, 52, 3472 (1987); D.S. Crumrine et al., *ibid.*, 52, 3699 (1987); M. Iyoda, M. Oda et al., *Chem. Commun.*, 1607 (1987); C. Betschart and D. Seebach, *Helv. Chim. Acta*, 70, 2215 (1987).

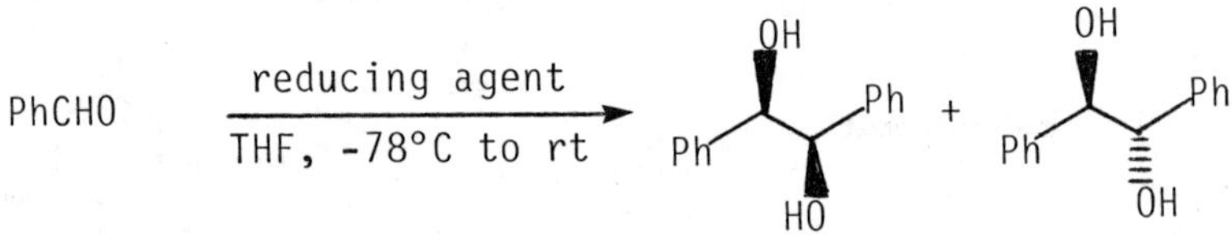

reducing agent =	Cp_2TiCl_2/SmI_2	11 : 1
	$Cp_2TiCl_2/Zn/MgBr_2$	40 : 1
	$Cp_2TiCl_2/^iPrMgI$	80 : 1

$TiCl_4$/Zn, Mg or LAH, Li/NH_3, Yb, Sn/electrolysis used similarly

I.A.8-2 D. Crich, Aldrichimica Acta, 20, 35 (1987).

Review: "O-Acylthiohydroxamates : New and Versatile Sources of Alkyl Radicals for Use in Organic Synthesis."

I.A.8-3 M. Ramaiah, Tetrahedron, 43, 3541 (1987).

Review: "Radical Reactions in Organic Synthesis."

I.A.8-4 W.P. Neumann, Synthesis, 665 (1987).

Review: "Tri-n-butyltin Hydride as a Reagent in Organic Synthesis."

I.A.8-5 D.L.J. Clive et al., J. Org. Chem., 52, 1339 and 4943 (1987); G.E. Keck and D.A. Burnett, ibid., 52, 2958 (1987); B. Delmond et al., Tetrahedron Lett., 28, 2127 (1987); D.P. Curran and C.-T. Chang, ibid., 28, 2477 (1987); D. Crich and S.M. Fortt, ibid., 28, 2895 (1987); D.L.J. Clive et al., Chem. Commun., 353 and 1520 (1987); F. MacCorquodale and J.C. Walton, ibid., 1456 (1987); G. Stork and R. Mook, Jr., J. Am. Chem. Soc., 109, 2829 (1987); K. Oshima et al., ibid., 109, 2547 (1987).

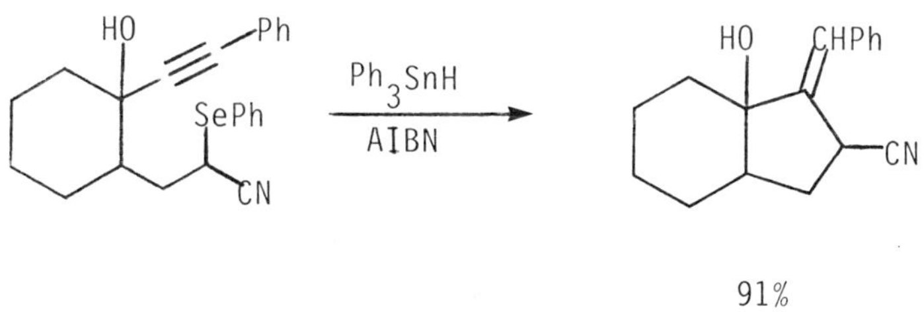

I, Br or a triple bond used instead of SePh

I.A.8-6 R. Tsang, B. Fraser-Reid et al., J. Am. Chem. Soc., 109, 3485 (1987).

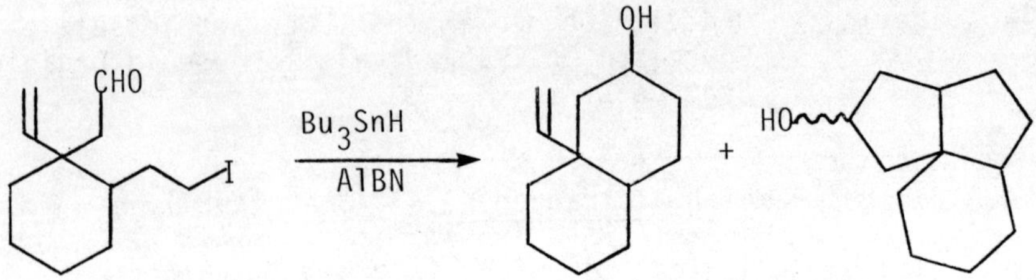

(5 : 1 mixture)

72%, 4 : 1

I.A.8-7 F.E. Ziegler and Z. Zheng, Tetrahedron Lett., 28, 5973 (1987).

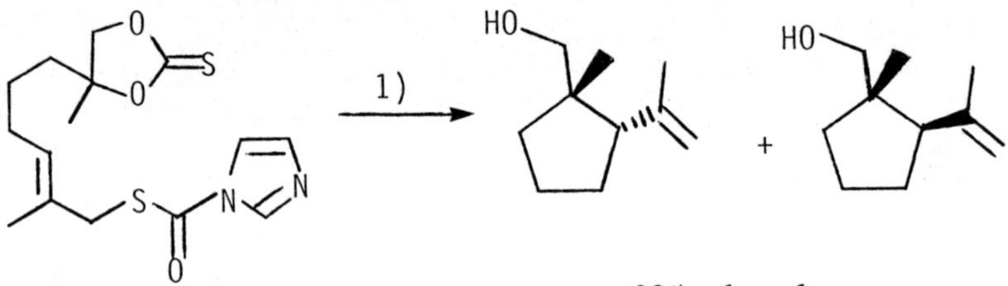

98%,~1 : 1

1) Bu_3SnH, AIBN, toluene, reflux

I.A.8-8 P. Dowd and S.-C. Choi, J. Am. Chem. Soc., 109, 3493 and 6548 (1987).

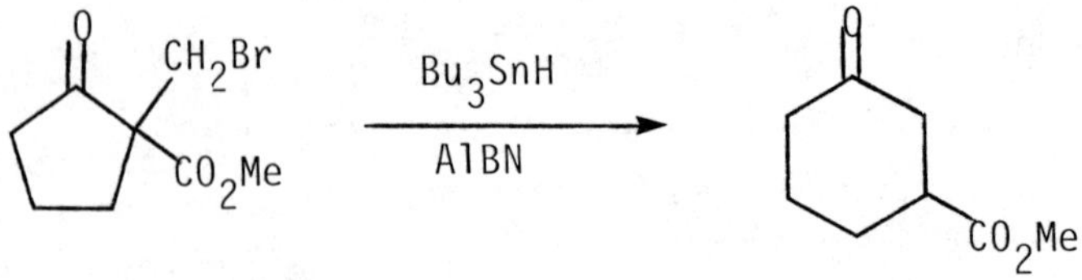

75%

I.A.8-9 G. Maas, Top. Curr. Chem., 137, 75 (1987).

Review: "Transition-Metal Catalyzed Decomposition of Alphatic Diazo Compounds - New Results and Applications in Organic Synthesis".

I.A.8-10 D.F. Taber et al., J. Org. Chem., 52, 28 (1987); S. Ikegami et al., Tetrahedron Lett., 28, 637 (1987); B. Corbel et al., ibid., 28, 6605 (1987); H.J. Monteiro, ibid., 28, 3459 (1987); A. Padwa and P.D. Stull, ibid., 28, 5407 (1987); E. Wenkert et al., Helv. Chim. Acta, 70, 1429 (1987); D.F. Taber and J.L. Schuchardt, Tetrahedron, 43, 5677 (1987); C. Iwata et al., Chem. Ind., 294 (1987); idem, Chem. Pharm. Bull., 35, 544 (1987); R. Bartnik and K. Wawrynko, J. Chem. Soc., Perkin Trans. 1, 715 (1987); G.W. Griffin and K.A. Horn, J. Am. Chem. Soc., 109, 4919 (1987).

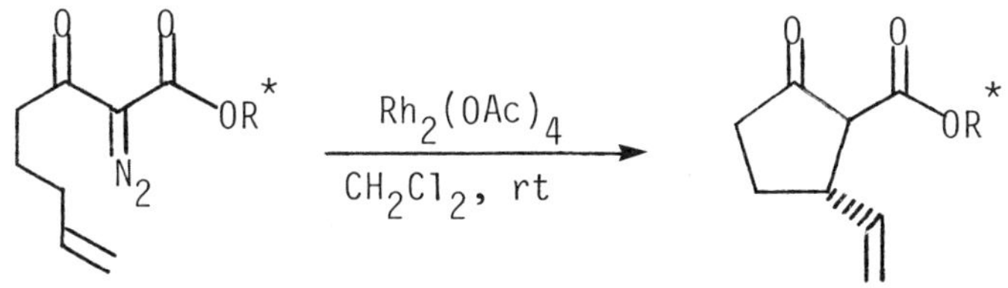

R* = chiral auxiliary 62%, 92 : 8 diastereoselectivity

Cu salts also employed in different systems

I.A.8-11 G.A. Molander et al., J. Am. Chem. Soc., 109, 453 (1987); J.D. White and T.C. Somers, ibid., 109, 4424 (1987); G.A. Molander and J.B. Etter, ibid., 109, 6556 (1987); idem, Synth. Commun., 17, 901 (1987); G.A. Molander and C. Kenny, Tetrahedron Lett., 28, 4367 (1987); H. Suginome and S. Yamada, ibid., 28, 3963 (1987); J. Inanaga et al., ibid, 28, 5877 (1987); H.B. Kagan et al., J. Organomet. Chem., 328, 81 (1987); Y. Butsugan et al., ibid., 333, 329 (1987).

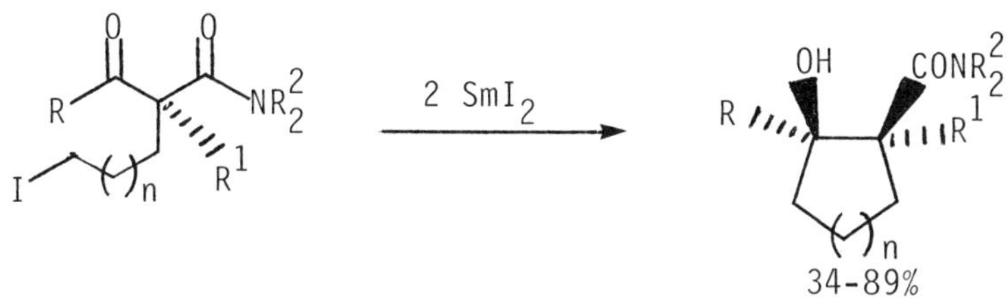

intermolecular reactions and the use of $SmCp_2$ also explored

I.A.8-12 B.B. Snider et al., J. Org. Chem., 52, 307, 1568 and 5413 (1987); A.S. Dreiding et al., Helv. Chim. Acta, 70, 321 and 1302 (1987).

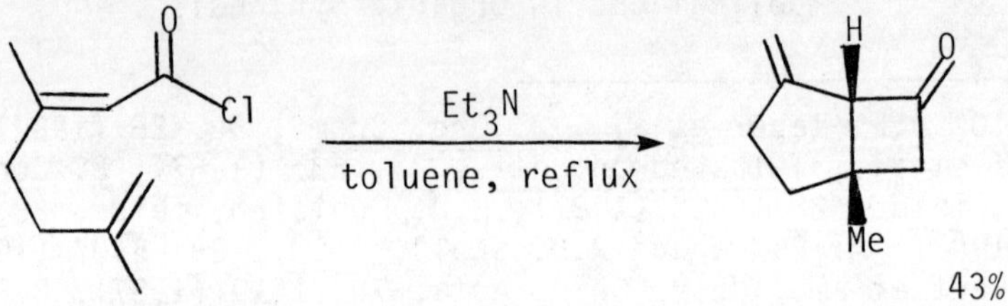

I.A.8-13 K. Weinges and W. Sipos, Angew. Chem., Int. Ed. Engl., 26, 1152 (1987); E.D. Edstrom and T. Livinghouse, J. Org. Chem., 52, 949 (1987); M. Nishizawa et al., ibid., 52, 4878 (1987); E.J. Corey and R.M. Burk, Tetrahedron Lett., 28, 6413 (1987); W.S. Johnson et al., J. Am. Chem. Soc., 109, 2517 and 5852 (1987); E.J. Corey et al., ibid., 109, 918 (1987).

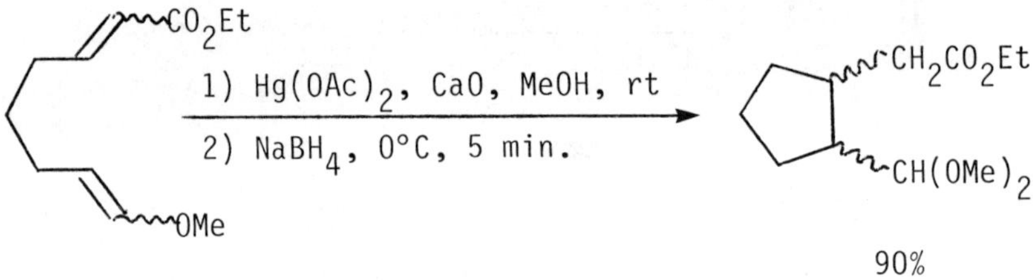

similar cyclizations with N-PhSe succinimide, Hg triflate/ N,N-dimethylaniline or $SnCl_4$

I.A.8-14 B.M. Trost and J.M. Tour, J. Am. Chem. Soc., 109, 5268 (1987); B.M. Trost and F. Rise, ibid., 109, 3161 (1987); B.M. Trost and G.J. Tanoury, ibid., 109, 4753 (1987); J.M. Takacs and L.G. Anderson, ibid., 109, 2200 (1987).

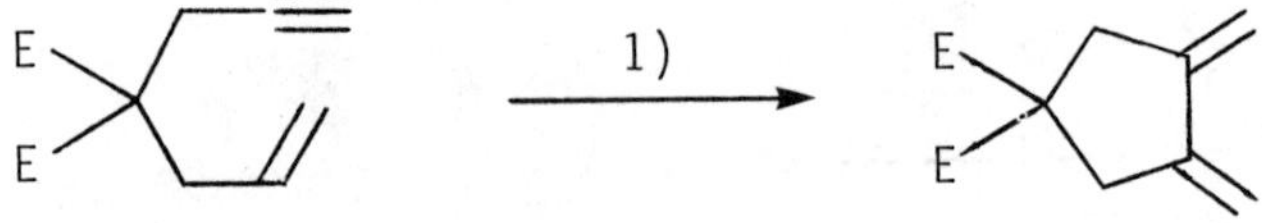

E = CO_2Me

1) Phosphinylated 2% cross-linked polystyrene, $(Ph_3P)_2NiCl_2$
other catalysts used for similar transformations

I.A.8-15 M. Kameyama and N. Kamigata, Bull. Chem. Soc. Jpn., 60, 3687 (1987); R. Grigg et al., J. Chem. Soc., Perkin Trans. 1, 1515 (1987); W.A. Kleschick et al., J. Org. Chem., 52, 3168 (1987).

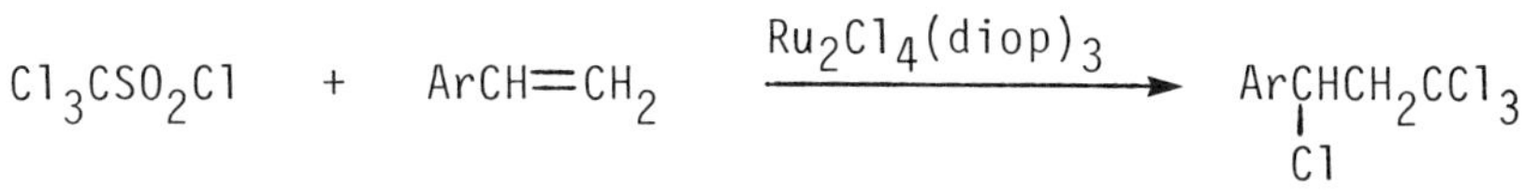

79-100%, 0-13% ee

CCl_4 and other catalysts also used

I.A.8-16 B.M. Trost and P.R. Seoane, J. Am. Chem. Soc., 109, 615 (1987); B.M. Trost and D.T. MacPherson, ibid., 109, 3483 (1987); G.W. Klumpp et al., Tetrahedron Lett., 28, 2889 (1987); P. Breuilles and D. Uguen, ibid., 28, 6053 (1987); D.G. Cleary and L.A. Paquette, Synth. Commun., 17, 497 (1987); W. Oppolzer and J.-M. Gaudin, Helv. Chim. Acta, 70, 1477 (1987); P. Binger et al., Chem. Ber., 120, 1933 (1987).

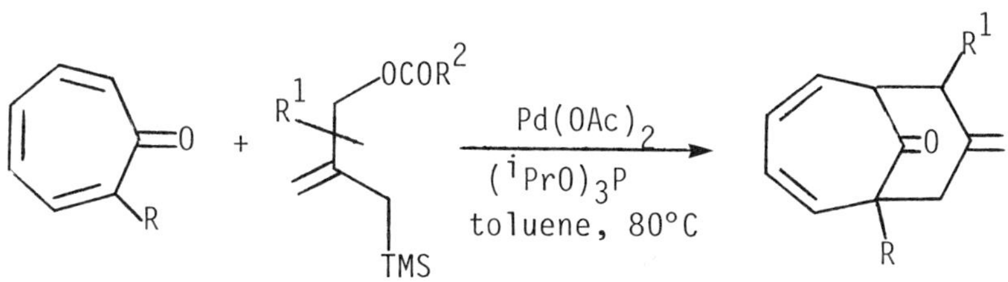

68%

first [6 + 3] cycloaddition

similar reactions with alkenes, dienes and cyclopropanes also explored

I.A.8-17 G.A. Molander and D.C. Shubert, J. Am. Chem. Soc., 109, 576 (1987).

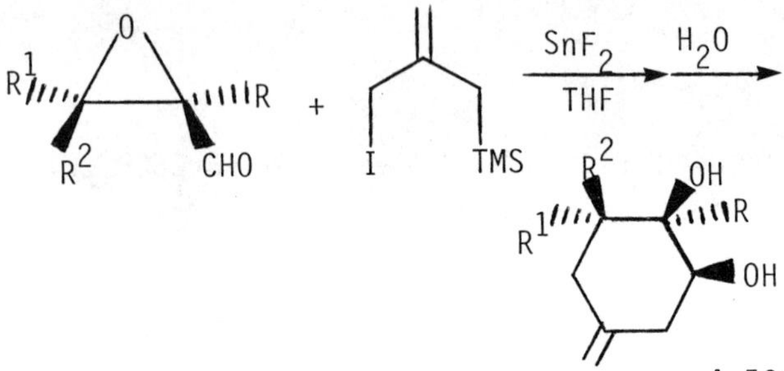

+ diastereomer

0-67%, 4-50 : 1

I.A.8-18 M. Kosugi, T. Migita et al., Chem. Lett., 193 (1987); T. Fuchikami et al., ibid., 521 (1987); R.C. Larock et al., J. Org. Chem., 52, 1364 (1987).

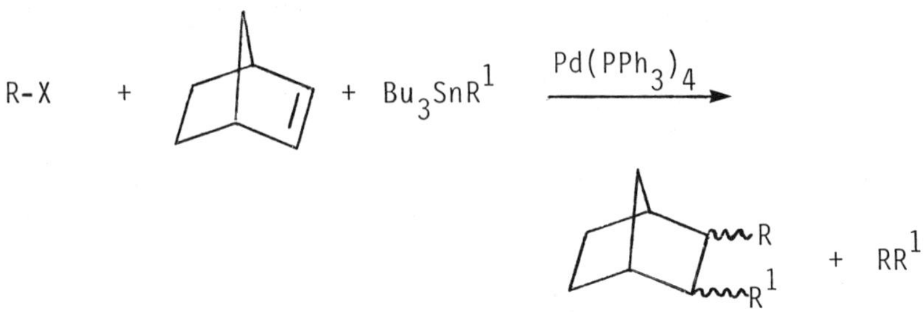

28-96% 0-44%

similar reaction with $AcOHgCH_2CO_2Et$, Li_2PdCl_4 and $CH_2{=}CHCOC_5H_{11}$

I.A.8-19 E. Negishi et al., J. Organomet. Chem., 334, 181 (1987); M.J.M. Abadie et al., Synthesis, 202 (1987).

$$RLi \quad + \quad (Ph_3P)_2PdCl_2 \quad \xrightarrow[22°C]{THF} \quad RR$$

>95%

$CuCl_2$ used similarly

I.A.8-20 H. Meier et al., Chem. Ber., 120, 1607, 1611 and 1617 (1987); G.A. Kraus et al., Chem. Commun., 1849 (1987).

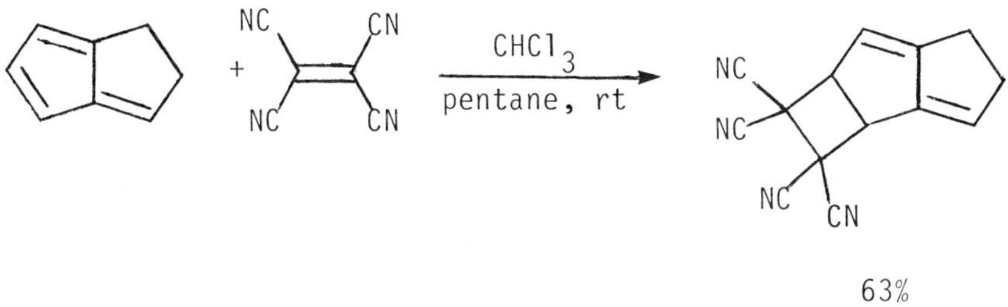

I.A.8-21 M. El Idrissi and M. Santelli, Tetrahedron Lett., 28, 2583 (1987); H. Ishibashi et al., Chem. Pharm. Bull., 35, 2750 (1987).

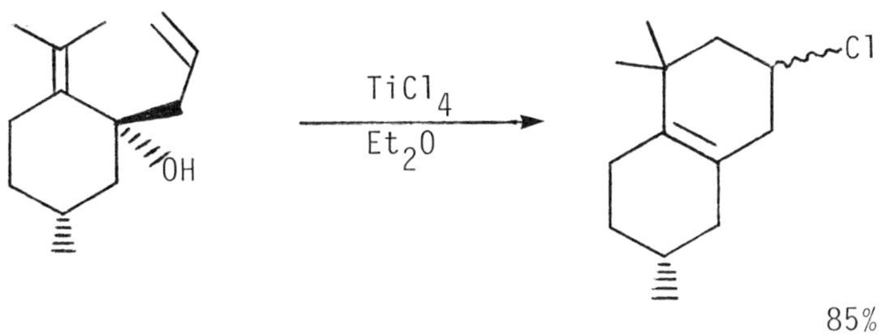

I.A.8-22 D.H.R. Barton, S.Z. Zard et al., Tetrahedron, 43, 2733 (1987); N. Soundararajan and M.S. Platz, Tetrahedron Lett., 28, 2813 (1987).

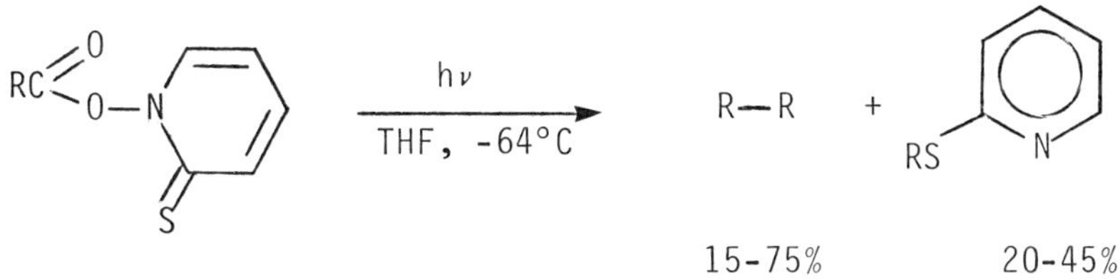

similar reaction with benzyl tributylstannanes

I.A.8-23 G.W. Morrow and J.S. Swenton, Tetrahedron Lett., 28, 5445 (1987); S. Yamamura et al., ibid., 28, 6661 (1987).

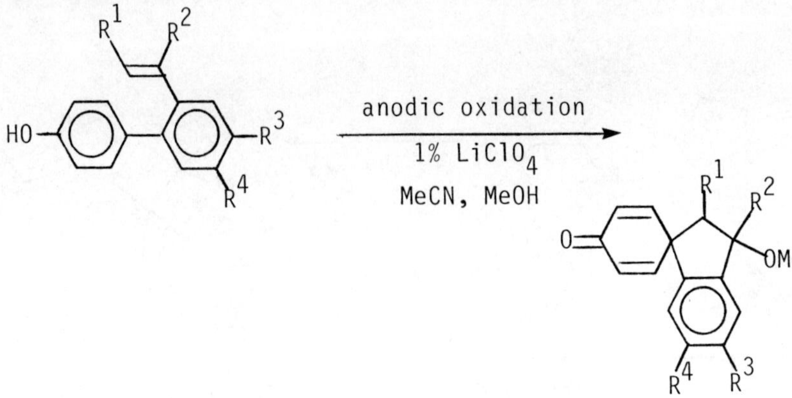

17-92%

I.A.8-24 J.H.P. Utley et al., Tetrahedron Lett., 28, 1349 (1987); J. Berthelot et al., ibid., 28, 1881 (1987).

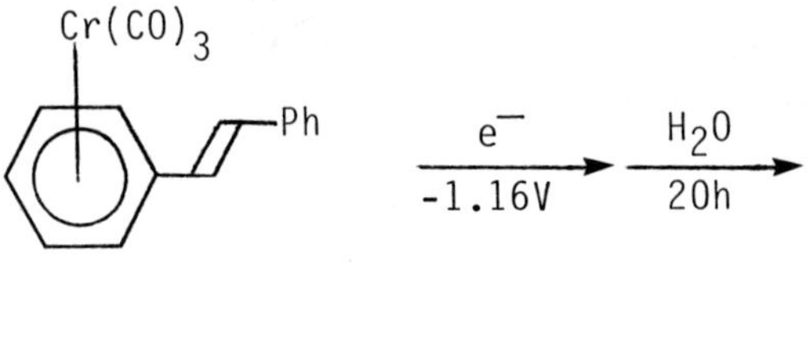

63%

I.A.8-25 T. Kubota et al., Chem. Lett., 1435 (1987).

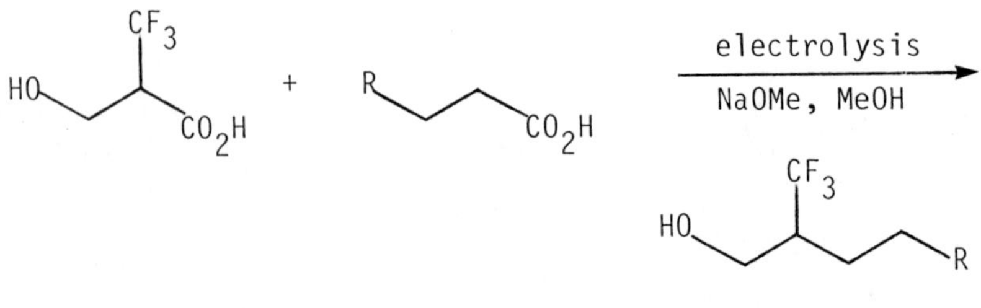

14-70%

I.A.8-26 G. Piancatelli et al., Chem. Commun., 907 (1987); K. Tajima, Chem. Lett., 1319 (1987).

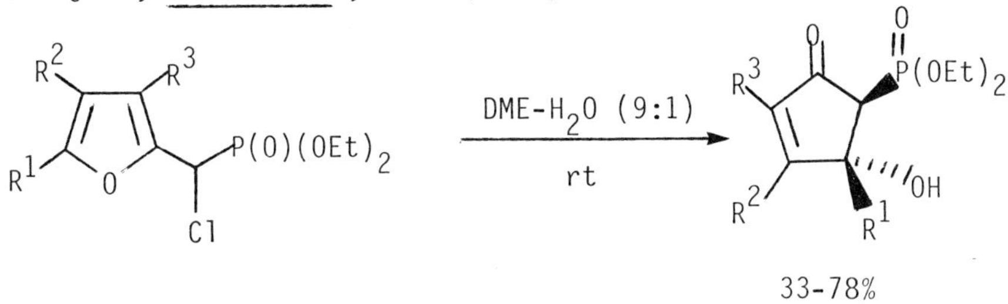

I.A.8-27 K.S. Kim et al., Chem. Lett., 2251 (1987).

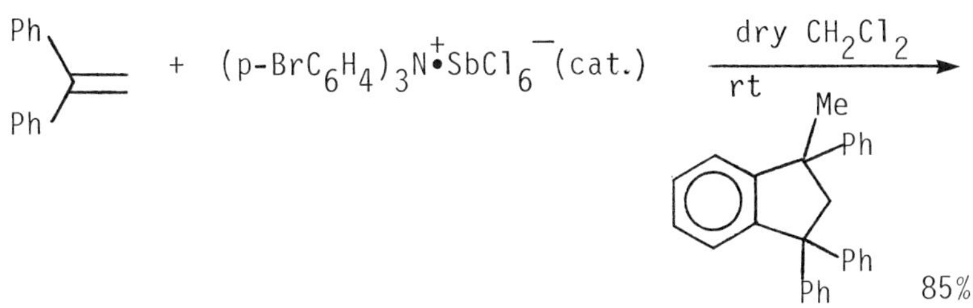

I.A.8-28 M. Julia et al., Tetrahedron Lett., 28, 6273 (1987).

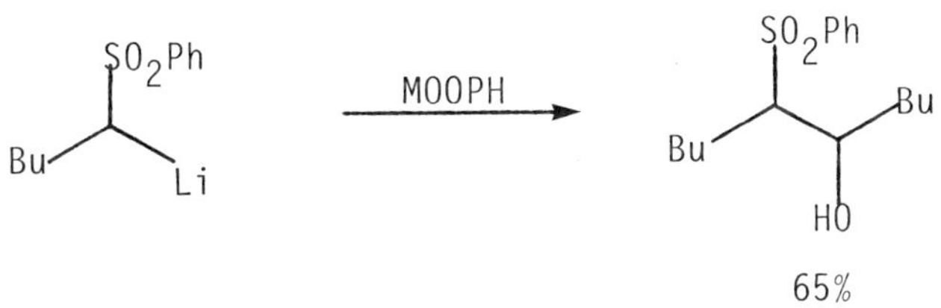

I.A.8-29 Y. Ito et al., Tetrahedron Lett., 28, 2247 (1987).

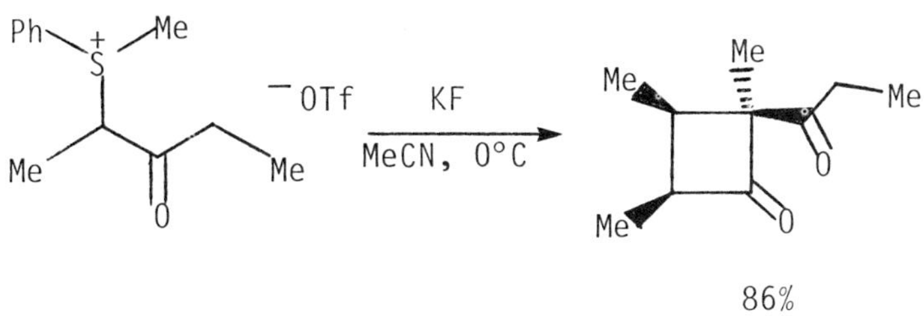

I.B. Carbon-Carbon Double Bonds

(see also: I.E.1, III.G)

I.B.1. Wittig-Type Olefination Reactions

I.B.1-1 Y.Z. Huang et al., *J. Org. Chem.*, *52*, 3558 (1987); idem, *Tetrahedron Lett.*, *28*, 2155 and 2159 (1987); idem, *Synthesis*, 370 (1987); Y. Shen and W. Qiu, *ibid.*, 65 (1987); G. Ourisson et al., *ibid.*, 696 and 1002 (1987); D. Lloyd et al., *Chem. Soc. Rev.*, *16*, 45 (1987).

$$[Ph_3\overset{+}{As}CH_2CHO]Br^- + RCHO \xrightarrow{1)} RCH{=}CHCHO$$

R = alkyl 75-90%, E isomer

1) Et_2O-THF (7:3), trace H_2O/K_2CO_3, 20°C

I.B.1-2 N.S. Isaacs and G.N. El-Din, *Tetrahedron Lett.*, *28*, 2191 (1987).

The application of high pressure to some difficult Wittig reactions

---

I.B.1-3 E. Vedejs et al., *J. Org. Chem.*, *52*, 4637 (1987).

Evidence against reversible Wittig reaction of a stabilized ylide. High (E)-olefin selectivity under kinetic control

---

I.B.1-4 A. Foucaud et al., *Tetrahedron*, *43*, 537 (1987).

Knoevenagel, Wittig and Wittig-Horner Reactions in the Presence of Magnesium Oxide or Zinc Oxide

I.B.1-5 C. Alvarez-Ibarra et al., Chem. Commun., 1509 (1987); J.V. Sinisterra et al., J. Org. Chem., 52, 3875 (1987); idem, Tetrahedron Lett., 28, 2951 (1987); J.C. Carretero and L. Ghosez, ibid., 28, 1101 (1987); J.M. Cassady et al., ibid., 28, 3075 (1987); P. Coutrot et al., Synthesis, 169 (1987); H. Paulsen and W. von Deyn, Liebigs Ann. Chem., 125 and 133 (1987); G. Beck et al., ibid., 321 (1987); G.M. Blackburn and M.J. Parratt, J. Chem. Soc., Perkin Trans. 1, 1417 and 1425 (1987); T. Kawashima, T. Ishii and N. Inamoto, Bull. Chem. Soc. Jpn., 60, 1831 (1987); M.P. Wentland, D.M. Bailey et al., J. Med. Chem., 30, 1482 (1987).

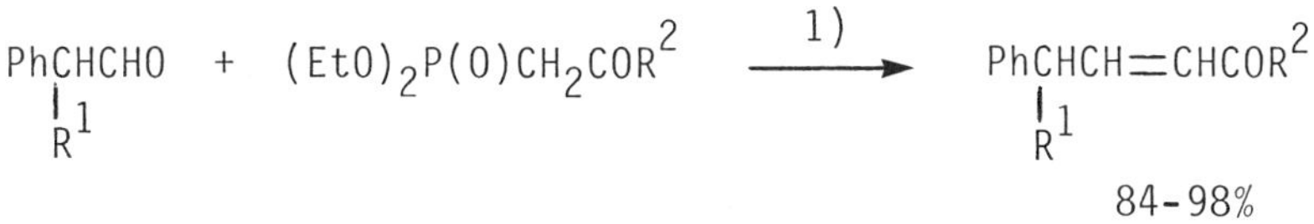

1) C-200, 1,4-dioxane, 70°C, 10-15 min

C-200 = [$Ba(OH)_2$• 0.8 H_2O]

various other bases and substrates used

I.B.1-6 M. Cinquini, F. Cozzi et al., Tetrahedron, 43, 2369 (1987); L. Ghosez et al., ibid., 43, 5125 (1987); M.D. Wittman and J. Kallmerten, J. Org. Chem., 52, 4303 (1987); A. Stambouli et al., Tetrahedron Lett., 28, 5301 (1987); J.E. Forbes and G. Pattenden, ibid., 28, 2771 (1987); S.V. Ley and P.R. Woodward, ibid., 28, 345 (1987); S. Warren et al., J. Chem. Soc., Perkin Trans. 1, 967 and 1883 (1987); H. Fillion et al., Synth. Commun., 17, 929 (1987); T. Koizumi et al., Chem. Pharm. Bull., 35, 3959 (1987); H. Irie et al., ibid., 35, 2598 (1987); P. Coutrot, C. Grison and R. Sauvetre, J. Organomet. Chem., 332, 1 (1987); K. Fukumoto et al., Chem. Commun., 1467 (1987).

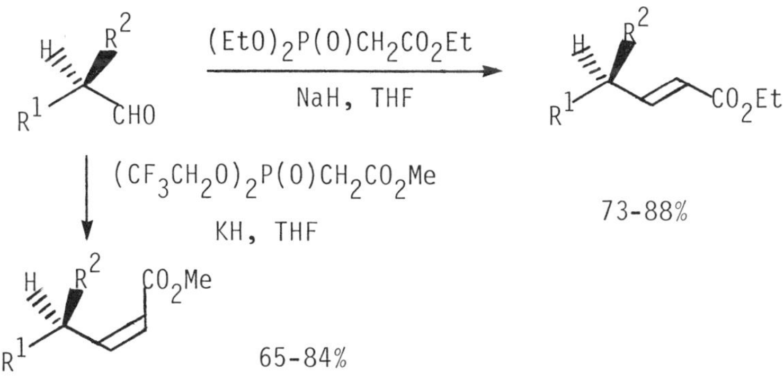

I.B.1-7 B. Costisella and H. Gross, Z. Chem., 27, 143 (1987).

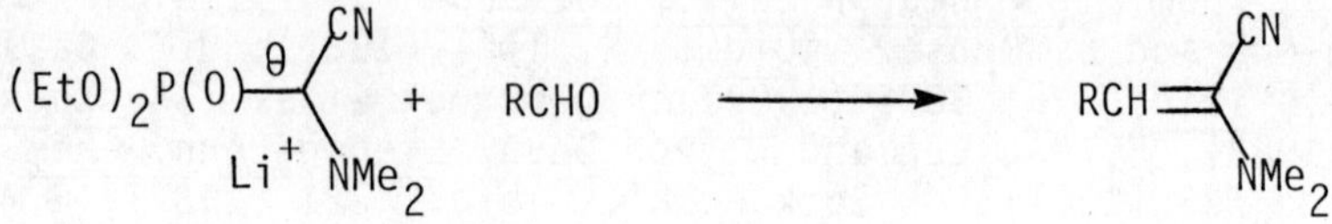

An "Instant-Horner Reagent" 44-71%

I.B.1-8 P. Venturello et al., J. Chem. Soc., Perkin Trans. 1, 713 (1987).

$$RCHO + (EtO)_2P(O)CH_2CO_2Et \xrightarrow[130°C,\ 5\,\tau]{K_2CO_3\ +\ \text{Carbowax 6000}} RCH{=}CHCO_2Et$$

gaseous reactants

50-56%

I.B.1-9 R. Bloch and M. Seck, Tetrahedron Lett., 28, 5819 (1987); A.B. Reitz et al., J. Org. Chem., 52, 4800 (1987).

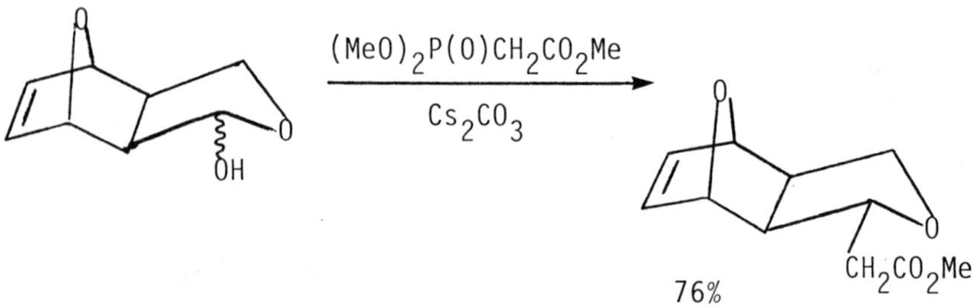

I.B.1-10 R.T. Borchardt et al., J. Org. Chem., 52, 5457 (1987); C.-H. Lin, P.A. Aristoff et al., ibid., 52, 5594 (1987).

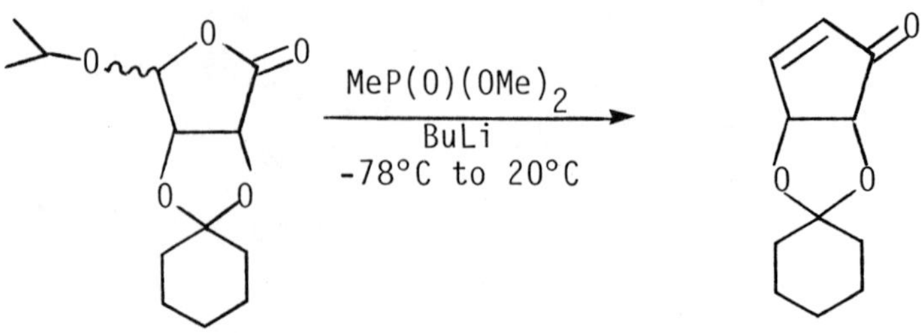

80%

I.B.1-11 I. Yamamoto et al., J. Chem. Soc., Perkin Trans. 1, 1537 (1987); S. Warren et al., ibid., 2569 (1987).

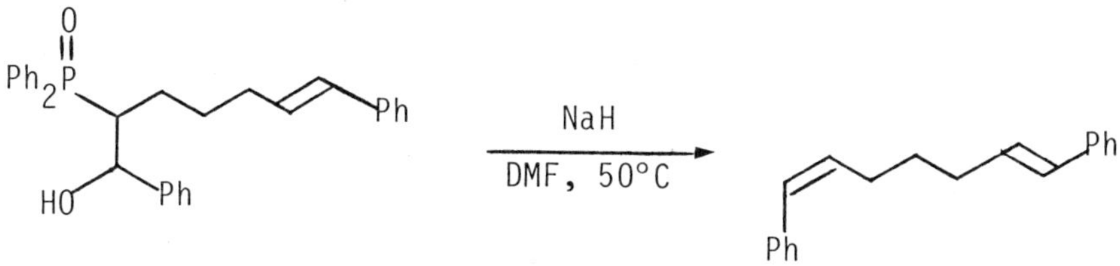

87%

I.B.1-12 H. Hirota et al., Chem. Commun., 1196 (1987); I. Degani et al., Synthesis, 79 (1987); G.J.S. Doad, J. Chem. Res. (S), 370 (1987); M.K. Tay et al., Tetrahedron Lett., 28, 1263 (1987).

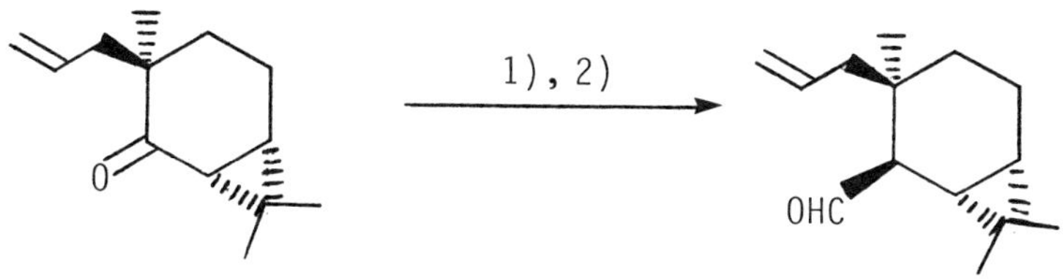

83%

1) $Ph_2P(O)CH_2OMe$, LDA, THF

2) dil. HCl, acetone

I.B.1-13 M.T. Reetz et al., Chem. Ber., 120, 121 (1987); O. Yonemitsu et al., Chem. Pharm. Bull., 35, 2194, 2196, 2203, 2209, 2219 and 2228 (1987).

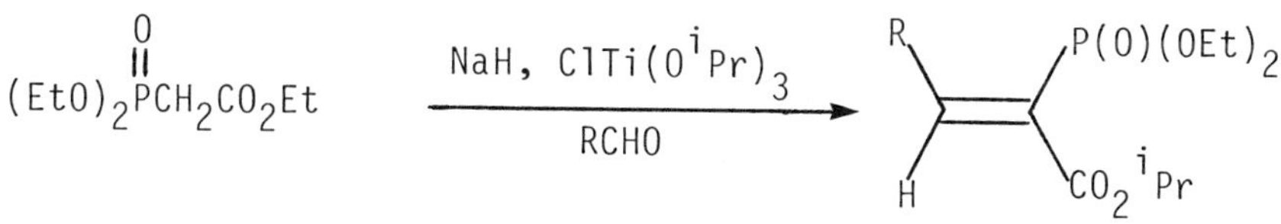

<5% more stable (E) isomer

57-74%

I.B.1-14 Y. Leblanc et al., Tetrahedron Lett., 28, 3449 (1987); G.A. Tolstikov et al., J. Org. Chem. (USSR), 22, 1457, 1597, 1598 and 1875 (1986); V. Guindon and D. Delorme, Can. J. Chem., 65, 1438 (1987); T. Nakamura et al., Chem. Pharm. Bull., 35, 2635 (1987).

Wittig Routes to Leukotrienes

I.B.1-15 A. Focella, D.R. Parrish et al., Synth. Commun., 17, 419 (1987); D.D. Muccio et al., Org. Prep. Proced. Int., 19, 187 (1987); M. Ito et al., J. Chem. Soc., Perkin Trans. 1, 905 (1987); M.C. Whiting et al., ibid., 2447, 2455 and 2465 (1987); R.S.H. Liu et al., Tetrahedron Lett., 28, 259 (1987); J.R. Pougny and P. Rollin, ibid., 28, 2977 (1987); G. Pattenden and D.C. Robson, ibid., 28, 5751 (1987); S. Hanessian and M. Botta, ibid., 28, 1151 (1987); E.J. Corey and R. Nagata, ibid., 28, 5391 (1987); G.B. Reddy and R.B. Mitra, Synth. Commun., 17, 893 (1987); D.R. Buckle et al., J. Chem. Res. (S), 394 (1987); W. Boland et al., Helv. Chim. Acta, 70, 1025 (1987); L. Daradics et al., J. Prakt. Chem., 329, 457 (1987).

Wittig Routes to Retinoic Acid Derivatives and Similar Long Chain Compounds

---

I.B.1-16 J. Einhorn and J.L. Luche, J. Org. Chem., 52, 4124 (1987); B. Miller and J. Baghdadchi, ibid., 52, 2391 (1987); W.G. Dauben and J.J. Takasugi, Tetrahedron Lett., 28, 4377 (1987); C.M. Moorhoff and D.F. Schneider, ibid., 28, 4721 (1987); J.W. Barton and M.K. Shepherd, J. Chem. Soc., Perkin Trans. 1, 1561 (1987); J. Ojima et al., ibid., 831 (1987); H.M. Walborsky et al., J. Am. Chem. Soc., 109, 6719 and 6726 (1987); H. Redlich et al., Tetrahedron, 43, 2029 (1987); M. Stock and H.-R. Schutte, Z. Chem., 27, 65 (1987); B. Dobner et al., ibid., 27, 63 (1987).

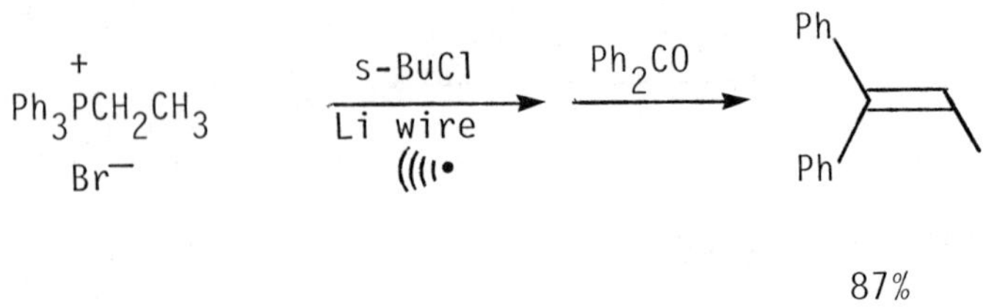

I.B.1-17 A.Kh. Khusid and B.G. Kovalev, J. Org. Chem. (USSR), 23, 62 (1987); J.R. Falck et al., Chem. Commun., 389, (1987); M.P. Schneider et al., ibid., 1434 (1987); F. Cozzi, C. Gennari et al., J. Org. Chem., 52, 4674 (1987); R.K. Boeckman, Jr. et al., ibid., 52, 469 (1987); S. Valverde et al., Can. J. Chem., 65, 332 and 339 (1987); H.-J. Cristau et al., Synthesis, 648 (1987); S. Valverde et al., Tetrahedron, 43, 1895 (1987); F. Scheinmann et al., J. Chem. Res. (S), 344 (1987); H.J. Bestmann, H.J. Schafer et al., Liebigs Ann. Chem., 417 (1987); Y. Kobayashi et al., Tetrahedron Lett., 28, 65 (1987).

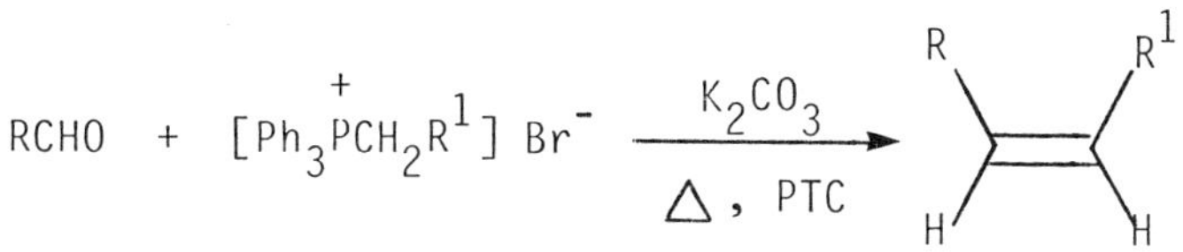

20-83%

cis - stereoselective with saturated aldehydes
cis - stereospecific with unsaturated aldehydes

I.B.1-18 E. Vedejs and C. Marth, Tetrahedron Lett., 28, 3445 (1987); K. Beautement and J.M. Clough, ibid., 28, 475 (1987); R.E. Ireland and R.B. Wardle, J. Org. Chem., 52, 1780 (1987); C.K. Lai and M. Gut, ibid., 52, 685 (1987); P. Garner and S. Ramakanth, ibid., 52, 2629 (1987); E.E. Schweizer et al., ibid., 52, 1324 (1987); C. Monneret et al., Chem. Commun., 615 (1987); P.F. Schuda et al., Synthesis, 1055 (1987); M.A. Rizzacasa and M.V. Sargent, Aust. J. Chem., 40, 1737 (1987); U.P. Dhotke and A.S. Rao, Synth. Commun., 17, 355 (1987).

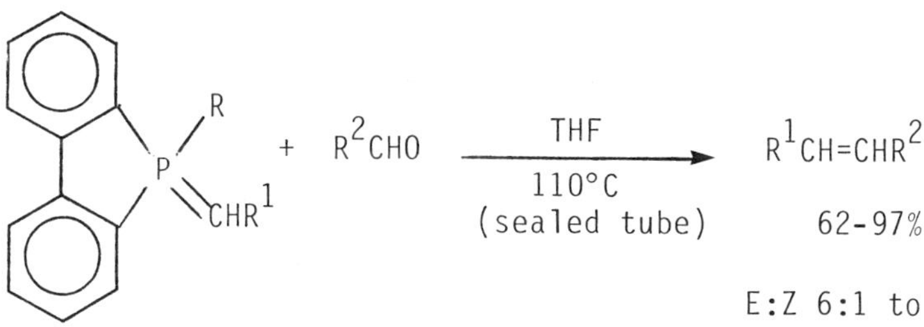

62-97%

E:Z 6:1 to 124:1

I.B.1-19 P. Magnus et al., J. Am. Chem. Soc., 109, 2469 (1987); T.V. RajanBabu, ibid., 109, 609 (1987).

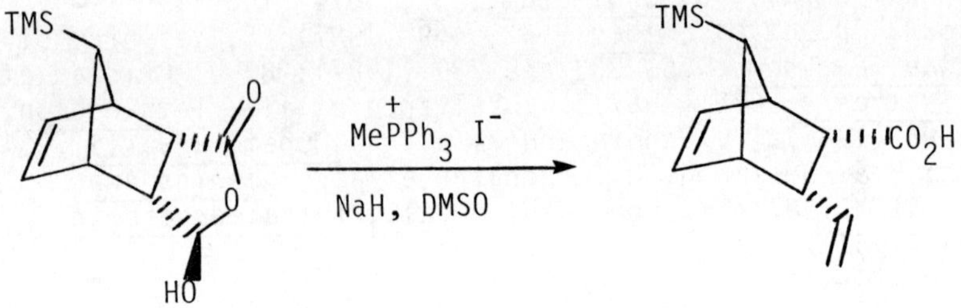

94%

I.B.1-20 H. Daniel and M. Le Corre, Tetrahedron Lett., 28, 1165 (1987).

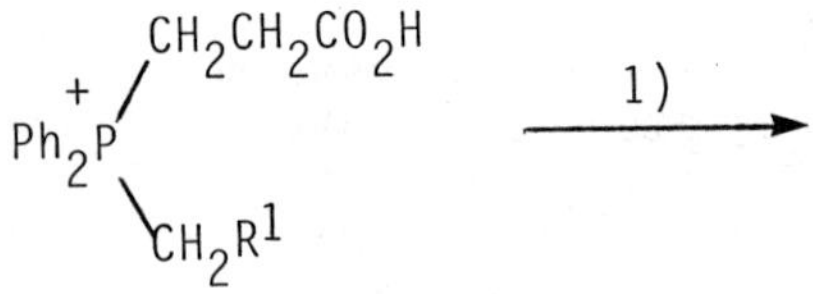

$R\text{-}CH{=}CHR^1$

60-98%

E:Z 82:18 to 97:3

1) 2 t-AmONa or NaOH, $H_2O\text{-}CH_2Cl_2$, RCHO

phosphine oxide by-product water soluble

I.B.1-21 F. Volatron and O. Eisenstein, J. Am. Chem. Soc., 109, 1 (1987).

Wittig vs. Corey-Chaykovsky Reaction. A Theoretical Study of the Reactivity of Phosphonium Methylide and Sulfonium Methylide with Formaldehyde.

I.B.1-22 J.H.P. Utley et al., Tetrahedron Lett., 28, 6081 (1987).

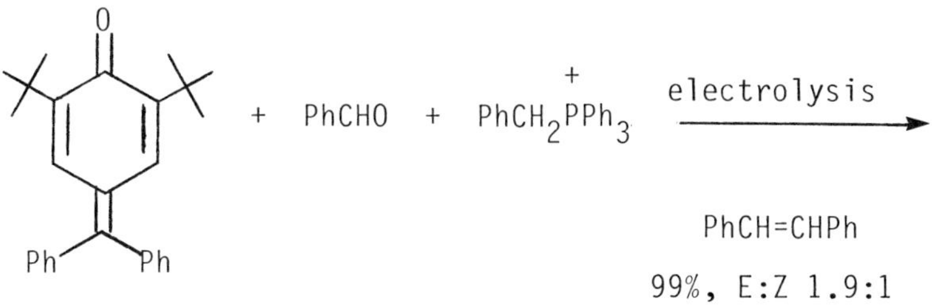

electrogenerated base

I.B.1-23 J.P. Marino et al., J. Org. Chem., 52, 4139 (1987); J.P. Marino and E. Laborde, ibid., 52, 1 (1987); B. O'Connor and G. Just, ibid., 52, 1801 (1987).

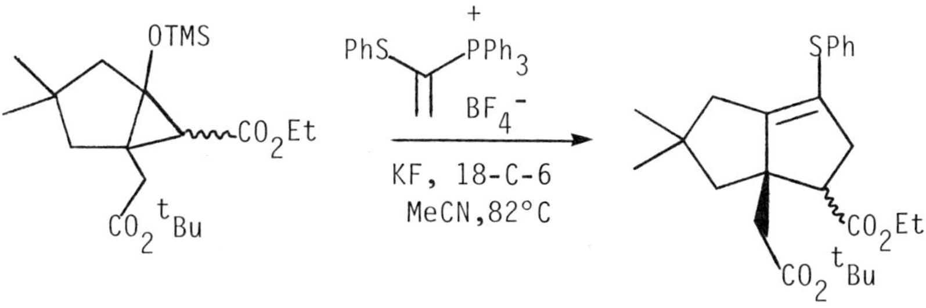

95%

I.B.1-24 R. Tamura et al., J. Org. Chem., 52, 4121 (1987).

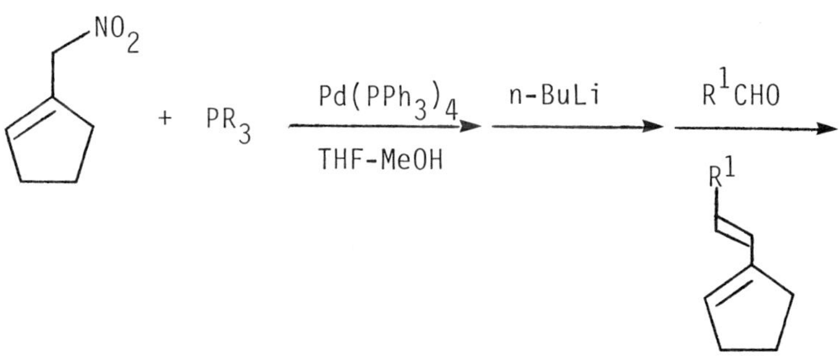

40-83%, E>95%

I.B.1-25 Y. Shen and W. Qiu, *Tetrahedron Lett.*, 28, 449 and 4283 (1987); idem, *Chem. Commun.*, 703 (1987); Y. Shen et al., *Synthesis*, 626 (1987).

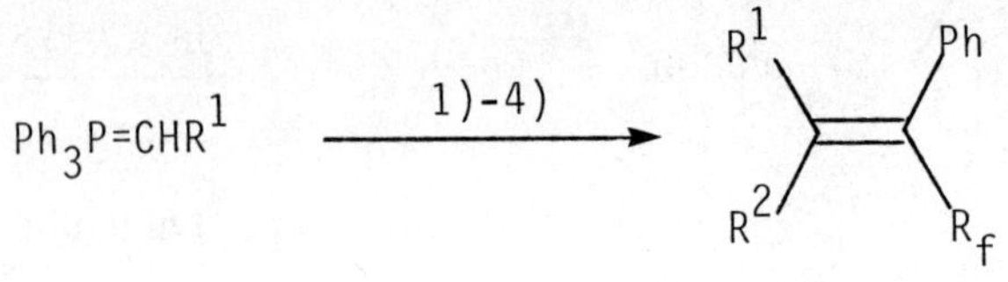

42-71%

1) R^2Li 2) PhI 3) $(R_fCO)_2O$ 4) PhLi

I.B.1-26 M. Hanack and C. Korhummel, *Synthesis*, 944 (1987).

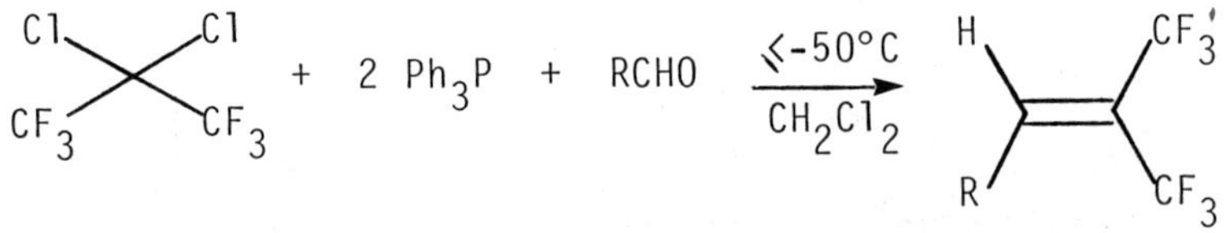

4-88%

I.B.1-27 K. Okuma et al., *Tetrahedron Lett.*, 28, 6649 (1987); L. Fitjer and U. Quabeck, *Synthesis*, 299 (1987).

$$2\ Ph_3P{=}C(R)R^1 + 2\ Se \longrightarrow R^1(R)C{=}C(R)R^1 + 2\ Ph_3P{=}Se$$

45-98%

similar result with O_2 rather than Se

I.B.1-28 H. Yamamoto et al., Tetrahedron, 43, 723, 731 and 743 (1987).

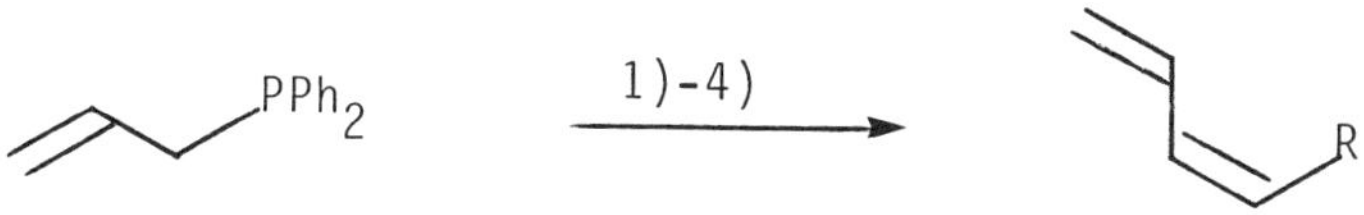

29-89%

E:Z 4:96 to 7:93

1) t-BuLi 2) $Ti(O^iPr)_4$ 3) RCHO 4) MeI

I.B.1-29 H. Iio, T. Tokoroyama et al., Tetrahedron Lett., 28, 2379 and 4561 (1987).

$PhCH_2CH_2CHO + Ar_3P{=}CHCH(CH_3){-}SiR^1R^2R^3 \longrightarrow PhCH_2CH_2CH(OSiR^1R^2R^3){-}CH{=}CHCH_3$

38-77% (major)

E:Z 20:1 to 40:1

I.B.1-30 L. Lombardo, Org. Synth., 65, 81 (1987); F. Leyendecker and M.-T. Comte, Tetrahedron, 43, 85 (1987); T. Moriwake et al., Chem. Lett., 2085 (1987).

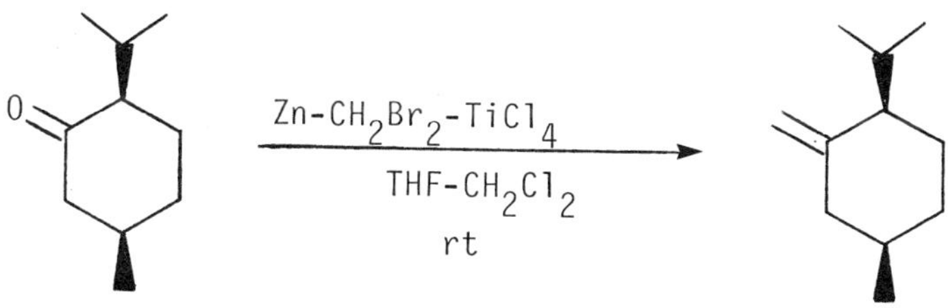

89%

I.B.1-31 X. Huang et al., Tetrahedron Lett., 28, 801 (1987).

$$R\text{-}C_6H_4CHO \;+\; Bu_2\overset{+}{Te}CH_2Y\;X^- \xrightarrow[\text{reflux}]{\text{THF}} R\text{-}C_6H_4CH{=}CHY$$

$Y = CO_2Me$, COPh, CN — 88-97%

I.B.1-32 G.M. Arvantis and J. Schwartz, Organometallics, 6, 421 (1987).

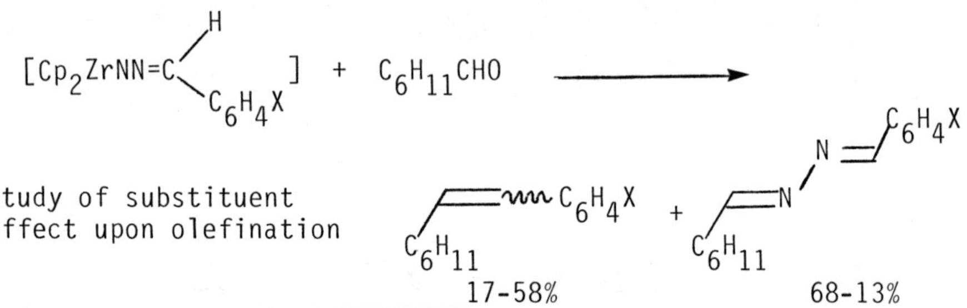

study of substituent effect upon olefination

I.B.1-33 F. Bickelhaupt et al., Tetrahedron Lett., 28, 6493 (1987).

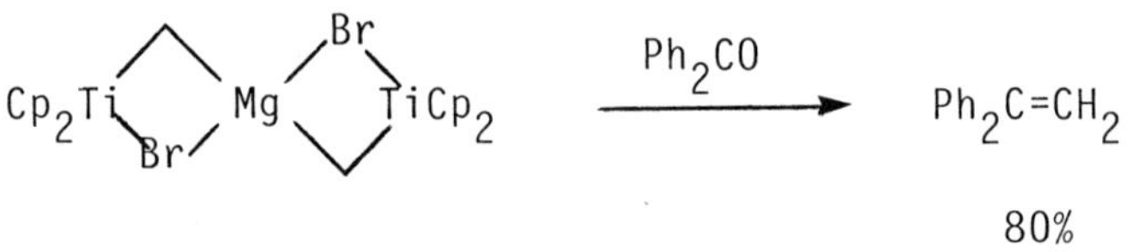

I.B.1-34 A. Dormond et al., J. Org. Chem., 52, 688 (1987).

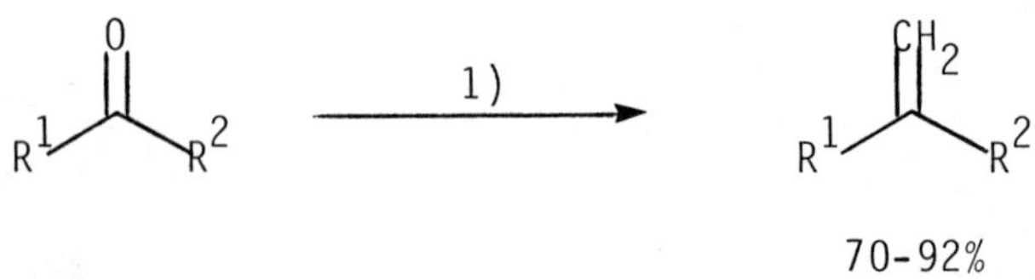

1) $[(TMS)_2N]_2\overline{UCH_2SiMe_2N}TMS$, rt, 1-2min

I.B.2.a. Elimination of Alcohols and Derivatives to Form Double Bonds

I.B.2.a-1 K. Laali et al., Helv. Chim. Acta, 70, 607 (1987); T. Nishiguchi et al., Tetrahedron Lett., 28, 4565 (1987); K.-J. Hwang and E.W. Logusch, ibid., 28, 4149 (1987); R.R. Schmidt and A. Kohn, Angew. Chem., Int. Ed. Engl., 26, 482 (1987); T. Toda et al., ibid., 26, 335 (1987); V.K. Aggarwal and S. Warren, J. Chem. Soc., Perkin Trans. 1, 2579 (1987); A. Kohn and R.R. Schmidt, Liebigs Ann. Chem., 1045 (1987).

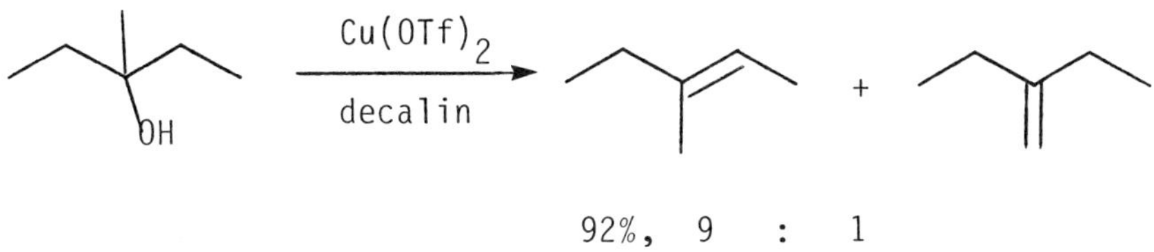

$CuSO_4/SiO_2$, $MeC(OMe)_3$/PTSA, $SOCl_2$/pyr and Ph_3P/DEAD (with or without rearrangement) also effective

---

I.B.2.a-2 E.J. Corey et al., J. Am. Chem. Soc., 109, 6187 (1987); H. Yoshioka et al., Tetrahedron Lett., 28, 663 (1987); R. Sauvetre et al., ibid., 28, 3335 (1987); W. Bretsch and H.-U. Reissig, Liebigs Ann. Chem., 175 and 387 (1987); K. Kojima et al., Chem. Pharm. Bull., 35, 948 and 2286 (1987).

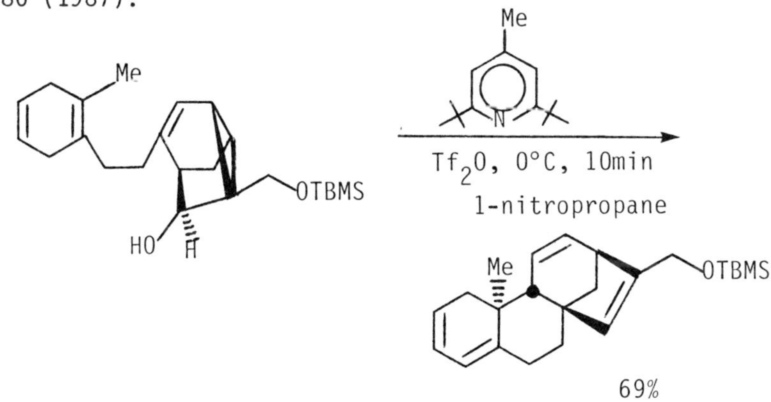

similar results with pyridinium poly(hydrogen fluoride), 2N HCl, $NaBH_4$ (for an allyl alcohol) and Me_3SiI

I.B.2.a-3 N.C. Yang et al., Tetrahedron Lett., 28, 3919 (1987); D. Villemin, Synthesis, 154 (1987); J.C. Sarma and R.P. Sharma, Chem. Ind., 96 (1987); M.K. Gurjar and S.M. Paurar, Ind. J. Chem., 26B, 55 (1987).

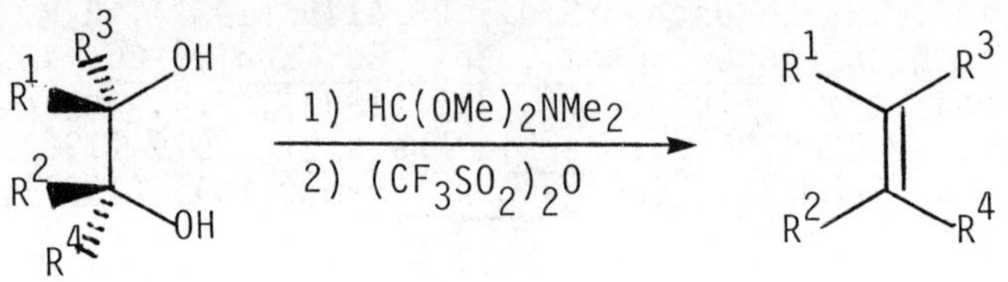

24-94%

PTSA/NaI and NaI/Zn (for a dimesylate) also used

I.B.2.a-4 T.K. Sarkar and S.K. Ghosh, Tetrahedron Lett., 28, 2061 (1987); R.V. Bonnert and P.R. Jenkins, Chem. Commun., 1540 (1987); G. Solladie and J. Hutt, J. Org. Chem., 52, 3560 (1987).

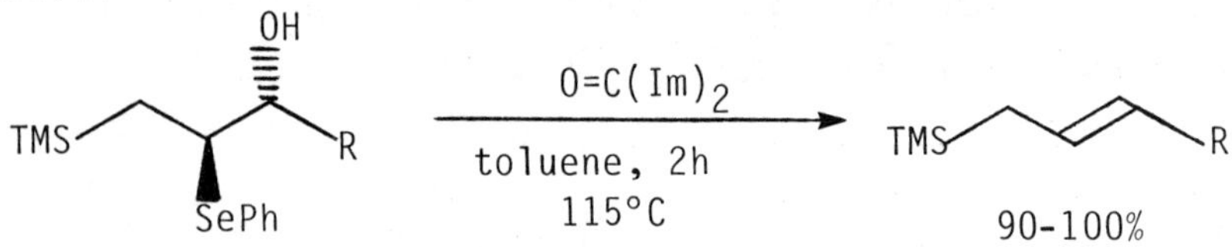

90-100%

no competing Si controlled rearrangement

$SOCl_2/NEt_3$ and $TiCl_3/LAH$ (for methoxy alcohol)

I.B.2.a-5 K. Sakai et al., Tetrahedron Lett., 28, 6489 (1987); M. Shimizu, H. Yoshioka et al., ibid., 28, 1677 (1987).

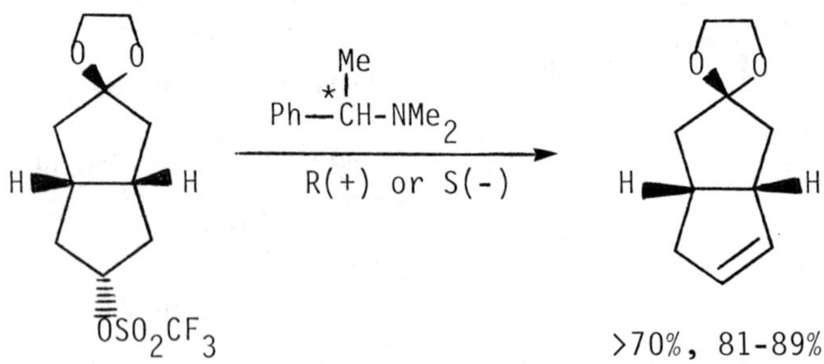

>70%, 81-89% ee

I.B.2.a-6 M. Asami and H. Kirihara, Chem. Lett., 389 (1987); R. Gleiter and B. Kissler, Tetrahedron Lett., 28, 6151 (1987); M. Okahara et al., Bull. Chem. Soc. Jpn., 60, 667 (1987); C. Nussbaumer and G. Frater, J. Org. Chem., 52, 2096 (1987); H. Mayr et al., ibid., 52, 1342 (1987); R. Tamura et al., ibid., 52, 1640 (1987).

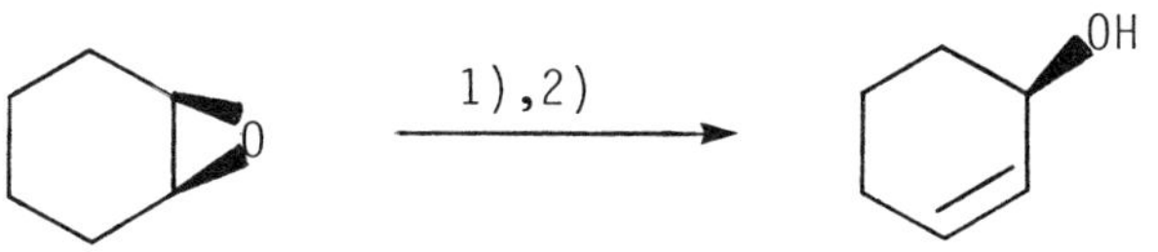

63-85%, 39-62% ee

1) 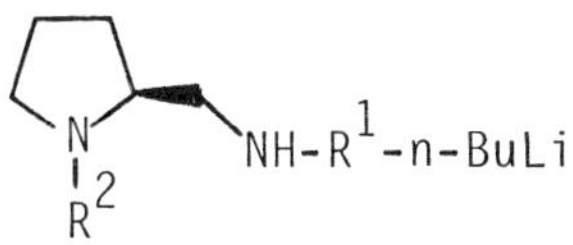

, THF, HMPA, rt, overnight 2) H^+

I.B.2.a-7 J.M. Aurrecoechea and W.H. Okamura, Tetrahedron Lett., 28, 4947 (1987).

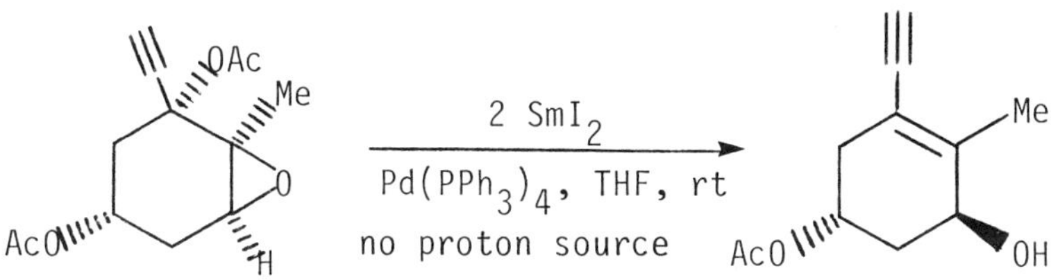

83%

I.B.2.a-8 A.V.R. Rao et al., Tetrahedron Lett., 28, 6497 (1987); K. Mori et al., Tetrahedron, 43, 2249 (1987); A. Molnar and M. Bartok, ibid., 43, 131 (1987).

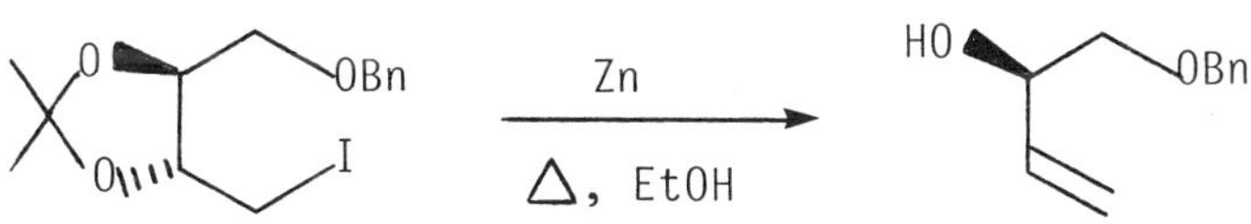

quantitative

I.B.2.a-9 J.G. Duboudin et al., J. Organomet. Chem., 331, 181 (1987); R. Taylor et al., J. Chem. Soc., Perkin Trans. 2, 1683 (1987).

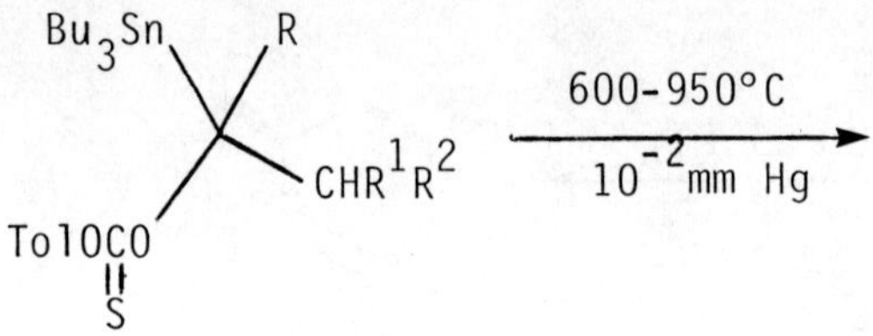

17-75%

I.B.2.a-10 D. Hoppe et al., Tetrahedron Lett., 28, 785 and 789 (1987).

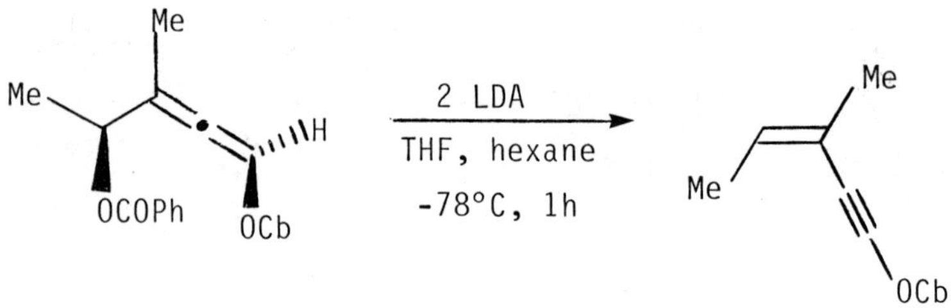

Cb = CON^iPr_2 88%

I.B.2.a-11 P.G. McDougal and J.G. Rico, J. Org. Chem., 52, 4817 (1987).

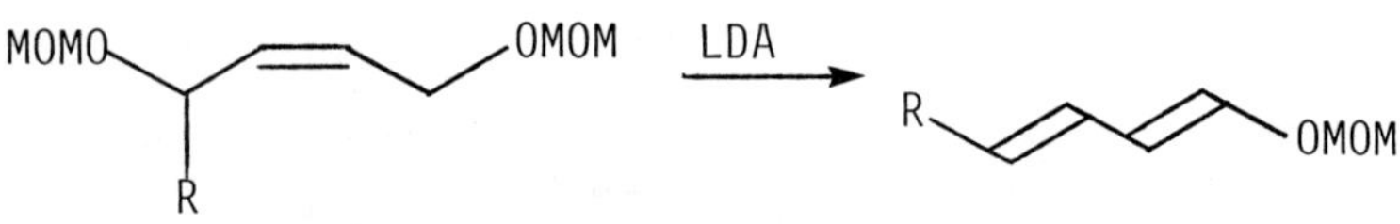

52-80%

I.B.2.a-12 M.A. Pericas et al., Tetrahedron, 43, 2311 (1987); J.L. LaMattina and D.E. Muse, J. Org. Chem., 52, 3479 (1987).

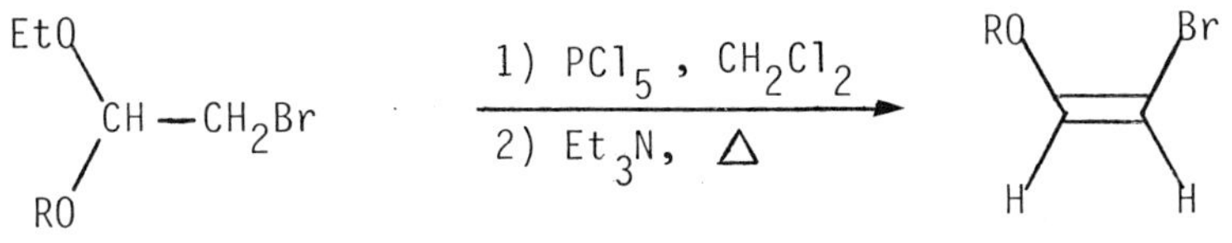

72-76%

R = t-Bu, adamantyl

I.B.2.b. Elimination of Halides to Form Double Bonds

I.B.2.b-1 C. Wakselman et al., J. Org. Chem., 52, 290 (1987); A.R. Suarez and M.R. Mazzieri, ibid., 52, 1145 (1987); S. Dev et al., Tetrahedron, 43, 2543 (1987); J.L.G. Ruano et al., ibid., 43, 4417 (1987); M.C. Cabaleiro and R.O. Garay, J. Chem. Soc., Perkin Trans. 2, 1473 (1987); C. Portella and M. Iznaden, Tetrahedron Lett., 28, 1655 (1987).

$$CF_2BrCH_2CO_2Et \xrightarrow[CH_2Cl_2,\ 0°C]{NEt_3} CF_2{=}CHCO_2Et$$

SiO_2, K_2CO_3, hν, and NH_4OH used similarly 74%

F eliminated in a similar case with heat

I.B.2.b-2 L. Duhamel et al., Tetrahedron Lett., 28, 5517 (1987).

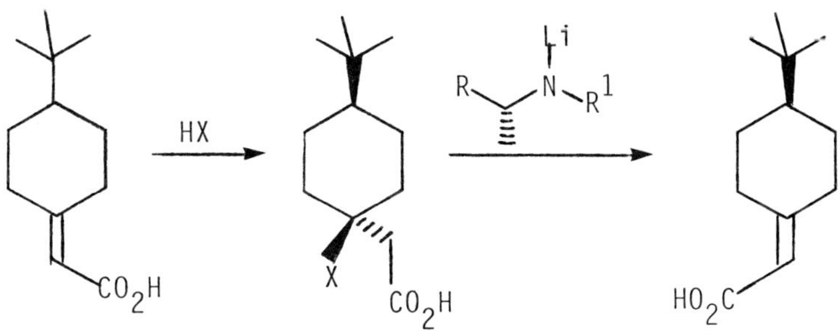

racemic

60-95%, 8-82% ee

X = Br, Cl

I.B.2.b-3 E. Block and M. Aslam, Org. Synth., 65, 90 (1987).

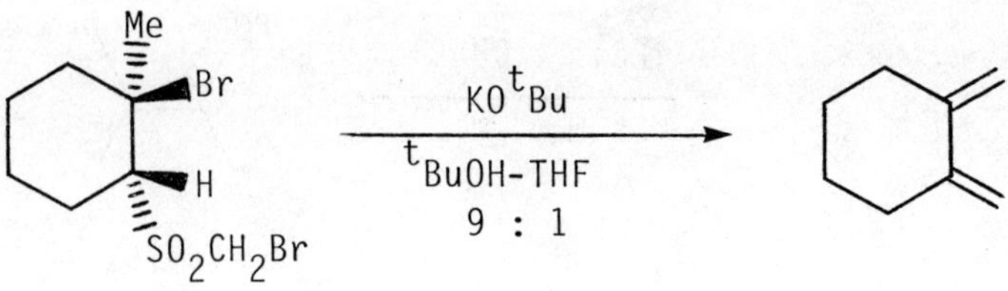

65%

I.B.2.b-4 John E. Baldwin and K.D. Belfield, J. Org. Chem., 52, 4772 (1987).

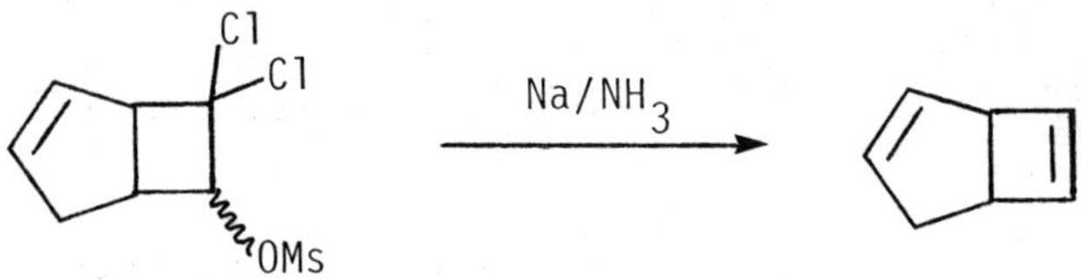

67%

I.B.2.b-5 M.T. Crimmins and S.W. Mascarella, Tetrahedron Lett., 28, 5063 (1987).

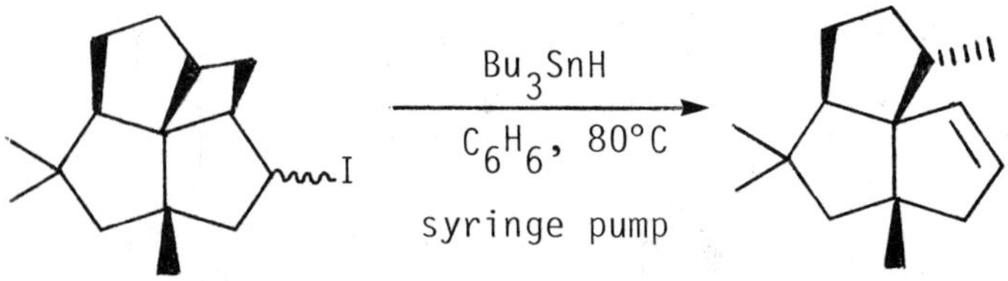

95%

I.B.2.b-6 K. Takaki et al., J. Org. Chem., 52, 1256 (1987).

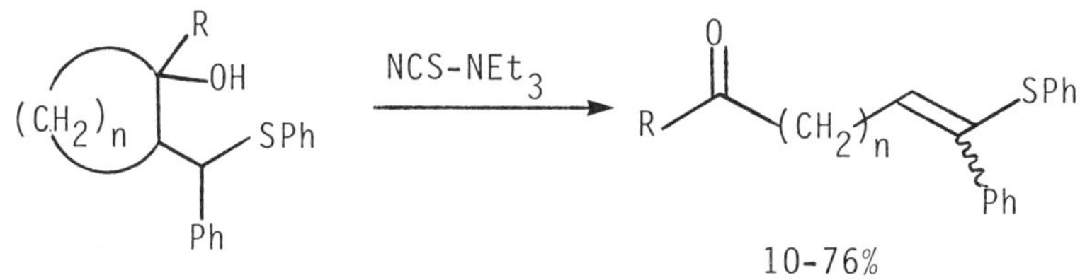

n = 3-5, R = H, Me

10-76%

I.B.2.c. Other Eliminations to Form Double Bonds

I.B.2.c-1 F. Huet et al., Tetrahedron, 43, 317 (1987); K. Yamakawa et al., Bull. Chem. Soc. Jpn., 60, 301 and 1839 (1987); S.T. Purrington and J.H. Pittman, Tetrahedron Lett., 28, 3901 (1987); D. Crich and L.B.L. Lim, J. Chem. Res. (S), 353 (1987); E.J. Thomas et al., Chem. Commun., 880 (1987); M.I. Al-Hassan, Gazz. Chim. Ital., 117, 187 (1987); S. Maemoto and K. Mori, Chem. Lett., 109 (1987).

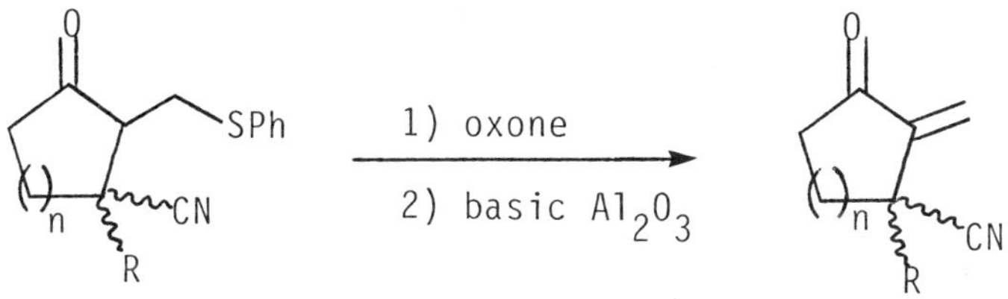

much lower yield by usual methods
MCPBA and selenides also used

20-58%

I.B.2.c-2 T. Chou et al., J. Org. Chem., 52, 2224 and 5082 (1987); S.P. Chou et al., ibid., 52, 4468 (1987); P.L. Fuchs et al., Synth. Commun., 17, 39 (1987); H. Takayama et al., Chem. Commun., 332 (1987).

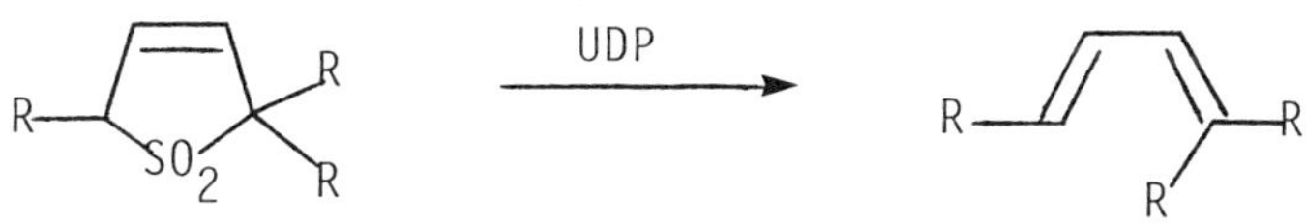

UDP = ultrasonically dispersed K

90-91%

similar results with heat or different bases

I.B.2.c-3 H. Matsuyama et al., J. Org. Chem., 52, 1703 (1987); K. Weinges and S. Haremsa, Liebigs Ann. Chem., 679 (1987).

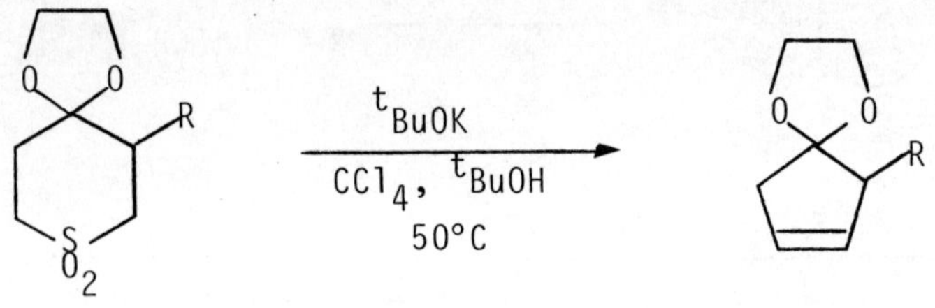

50-84%

I.B.2.c-4 C.R. Johnson and T. Imamoto, J. Org. Chem., 52, 2170 (1987).

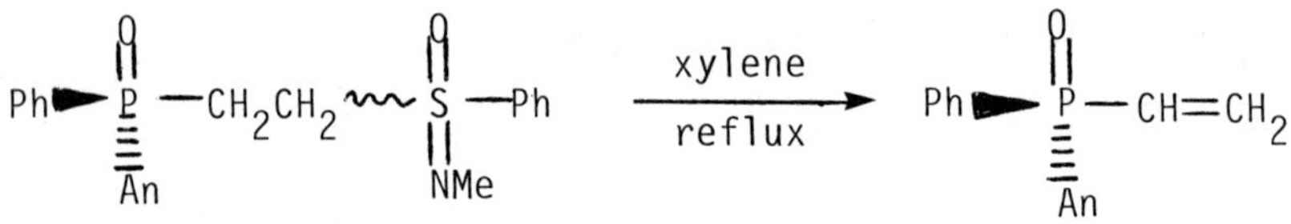

An = o-anisyl 82%

I.B.2.c-5 K. Inomata et al., Chem. Lett., 707 (1987).

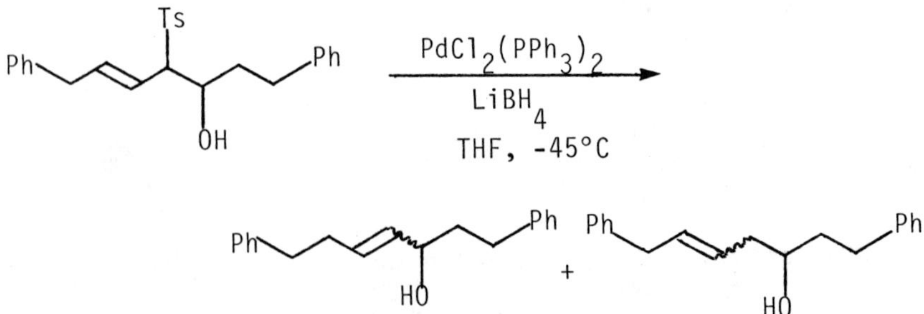

quantitative, 1 : 99

E/Z = 89 : 11

I.B.2.c-6 N. Ono et al., J. Org. Chem., 52, 5111 (1987).

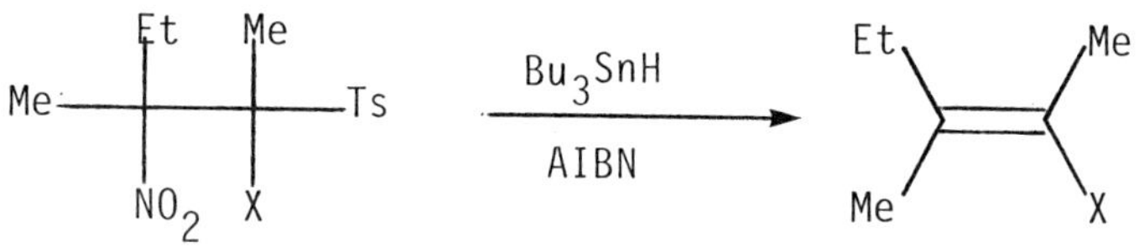

X = CN, CO_2Et 76-85%

I.B.2.c-7 A. Eschenmoser et al., Chem. Lett., 219 (1987).

On the Stereochemistry of E' and E'' - Reactions

I.B.2.c-8 R. Hunter et al., Tetrahedron Lett., 28, 2985 (1987); K.M. Nsunda and L. Hevesi, Chem. Commun., 1518 (1987).

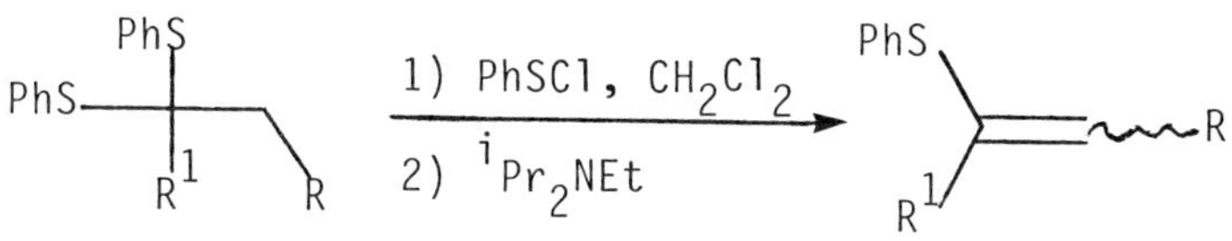

63-97%

I.B.2.c-9 L. Engman, Tetrahedron Lett., 28, 1463 (1987); idem, J. Org. Chem., 52, 4086 (1987).

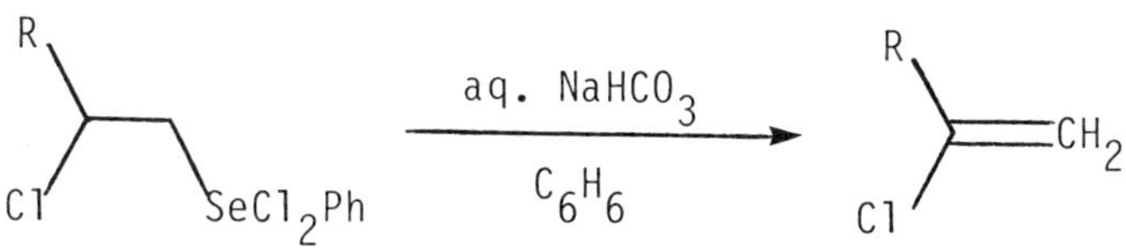

64-96%

I.B.2.c-10 J. Blagg and S.G. Davies, Tetrahedron, 43, 4463 (1987); A. Hosomi et al., Chem. Pharm. Bull., 35, 1663 (1987).

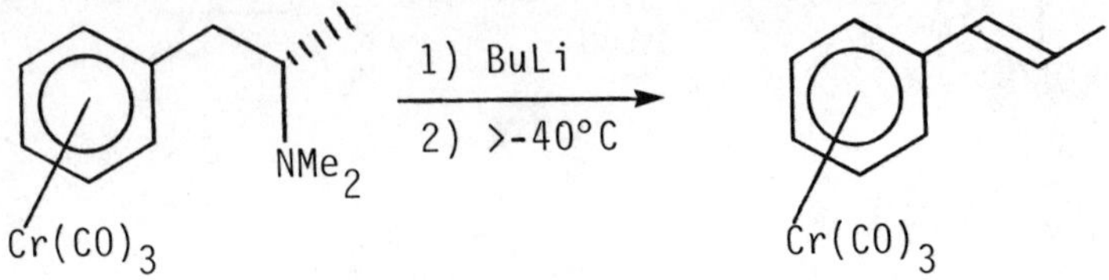

52%

I.B.2.c-11 N.E. Alexandrou et al., Liebigs Ann. Chem., 537 (1987).

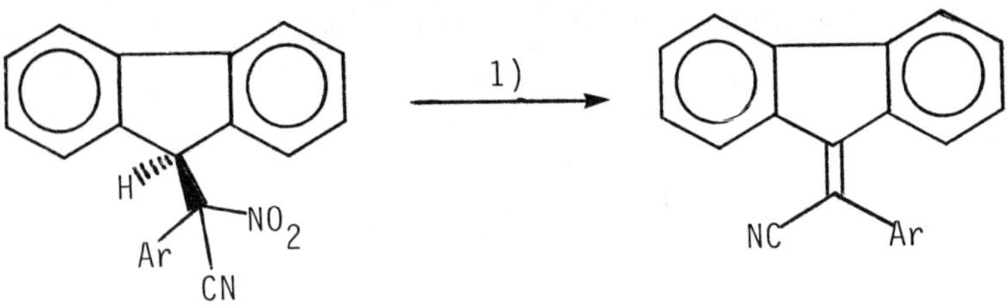

1) NaOH, EtOH, 55°C, 7h

~80%

I.B.2.c-12 O. De Lucchi et al., Gazz. Chim. Ital., 117, 401 (1987).

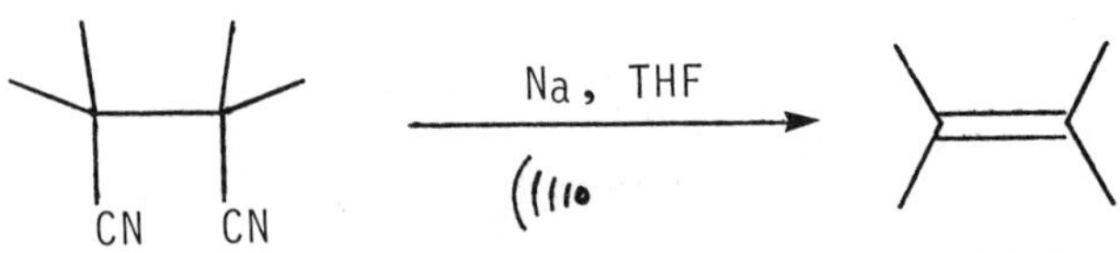

>90%

I.B.2.c-13 J.R. Hwu and B.A. Gilbert, J. Organomet. Chem., 332, 53 (1987).

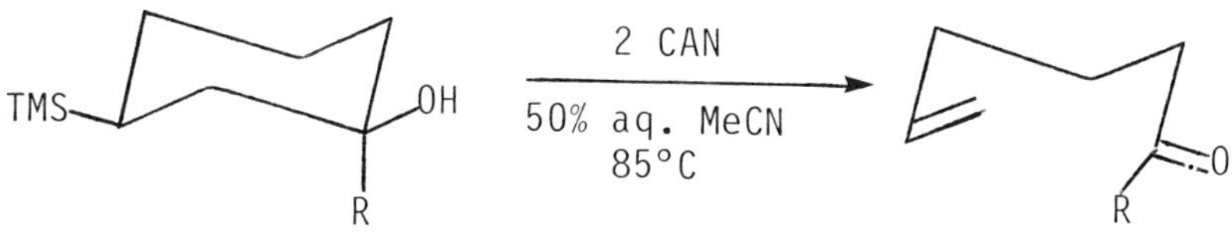

R = Me, Ph

75-88%

I.B.2.c-14 H.N.C. Wong et al., Heterocycles., 26, 1345 (1987).

Review: "Stereospecific Deoxygenation of Epoxides to Olefins."

I.B.2.c-15 J. Inanaga et al., Chem. Lett., 2101 (1987); L. Garlaschelli and G. Vidari, Gazz. Chim. Ital., 117, 251 (1987).

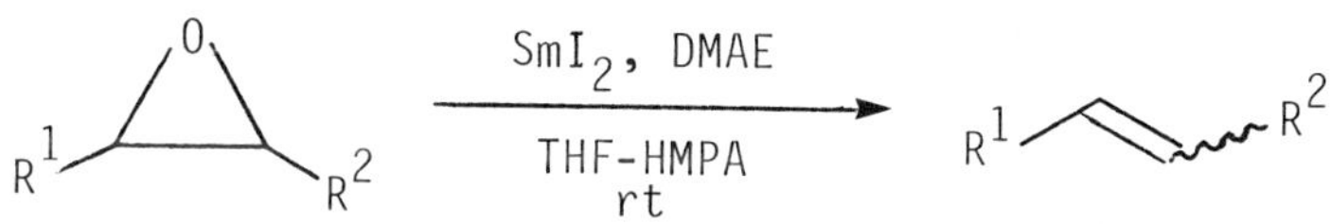

88-95%

DMAE = N,N-dimethylaminoethanol

similar result with Ph_3P/I_2 in moist MeCN

I.B.2.c-16 J. Tsuji et al., Tetrahedron, 43, 4107 (1987); idem, Chem. Lett., 23 (1987).

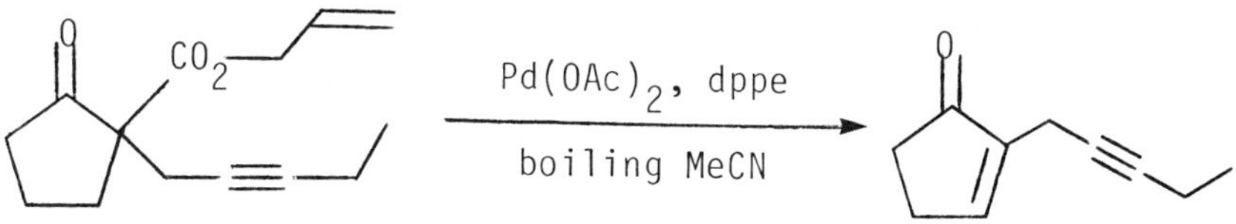

85% (major)

I.B.2.c-17 D.E. Johnson et al., J. Org. Chem., 52, 3688 (1987).

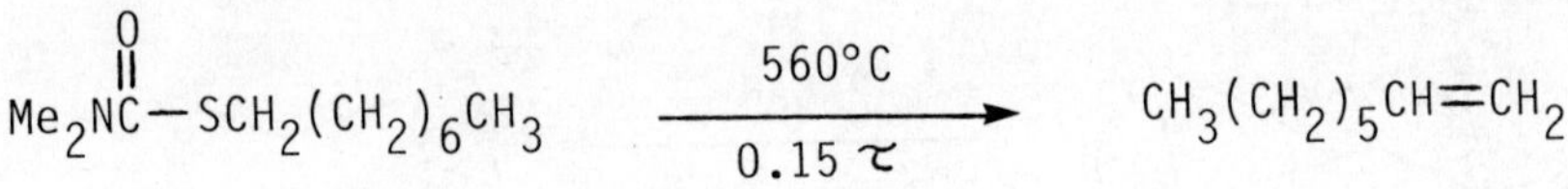

84%

I.B.3. Other Carbon-Carbon Double Bond Forming Reactions

I.B.3-1 J.A. Miller et al., Tetrahedron Lett., 28, 235 and 689 (1987); B.D. Gray and J.A. Miller, Chem. Commun., 1136 (1987); idem, J. Chem. Res. (S), 206 (1987); H. Mayr et al., J. Org. Chem., 52, 1989 (1987); A.B. Smith, III et al., ibid., 52, 5280 (1987); G. Melloni et al., Gazz. Chim. Ital., 117, 219 (1987).

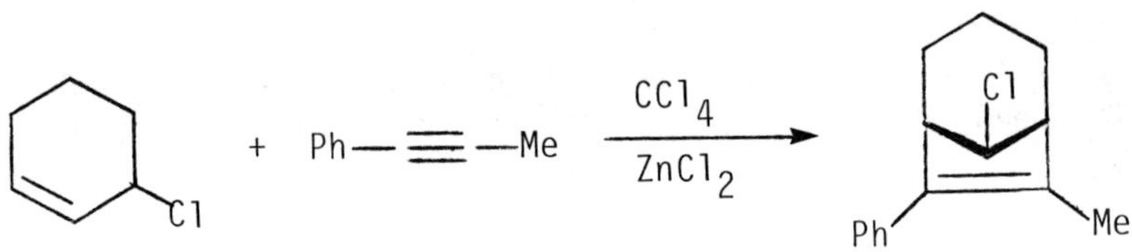

76%

I.B.3-2 E. Negishi et al., Tetrahedron Lett., 28, 917 (1987); B.M. Trost and D.J. Jebaratnam, ibid., 28, 1611 (1987); C.A. Broka and D.E.C. Reichert, ibid., 28, 1503 (1987).

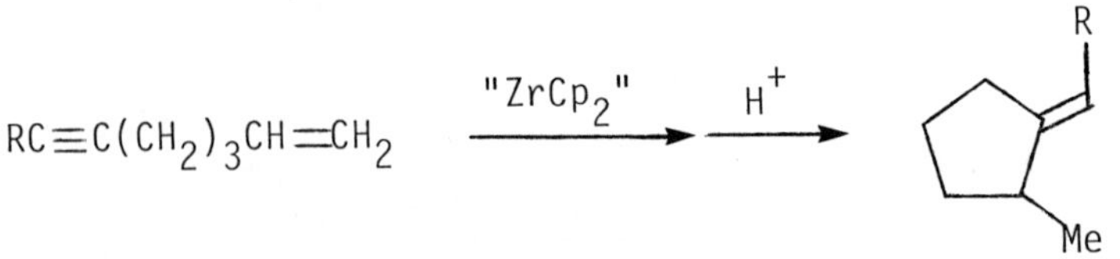

55-70%

$Pd(OAc)_2$, N,N -bis(phenylmethylene)-1,2-ethylenediamine or PhSH, AIBN also used similarly

I.B.3-3 K. Suzuki, T. Ohkuma and G. Tsuchihashi, J. Org. Chem., 52, 2929 (1987).

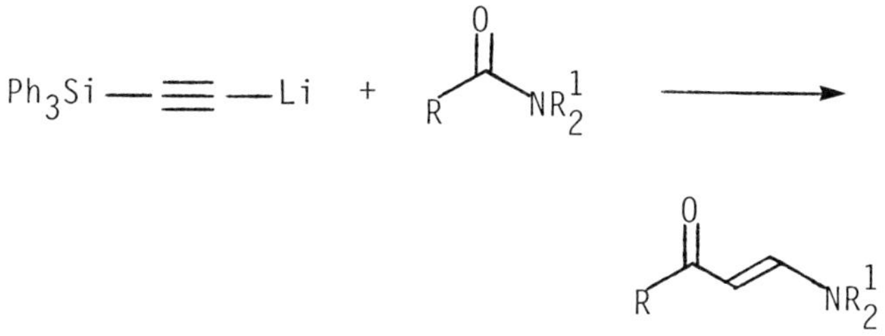

28-82%

I.B.3-4 P.J. Kocienski et al., J. Chem. Soc., Perkin Trans. 1, 2171, 2183 and 2189 (1987).

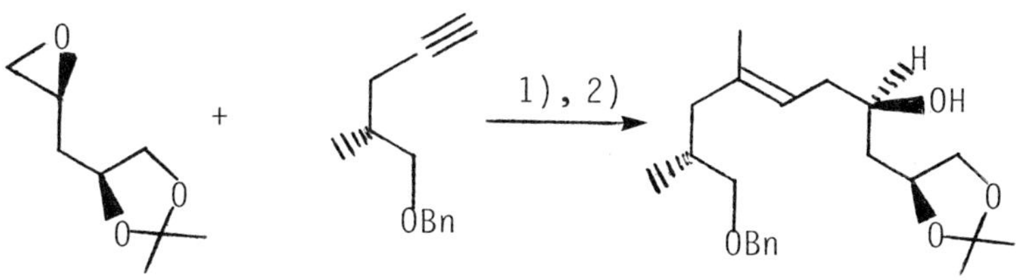

79-91%

towards (+)—Milbemycin β_3

1) Me_3Al, $(C_5H_5)_2ZrCl_2$ 2) BuLi

I.B.3-5 S.A. Rao and M. Periasamy, Chem. Commun., 495 (1987).

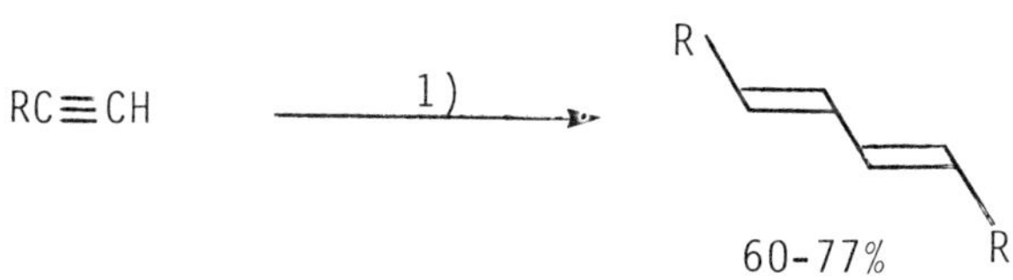

60-77%

1) $NaH-MgBr_2-CuCl$, THF, -20°C to rt

I.B.3-6 N.H.T. Huy et al., J. Organomet. Chem., 327, 211 (1987); H. Fischer et al., Chem. Commun., 559 (1987).

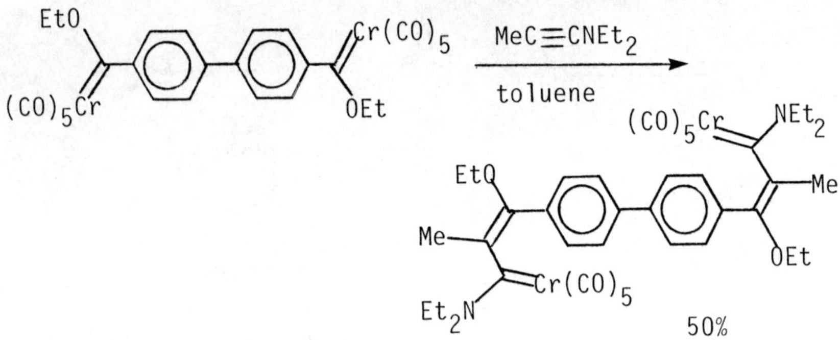

I.B.3-7 R. Aumann and H. Heinen, Chem. Ber., 120, 537 (1987).

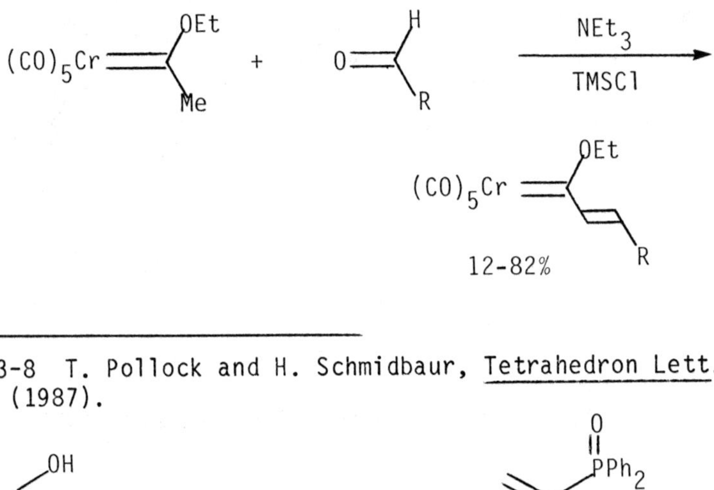

I.B.3-8 T. Pollock and H. Schmidbaur, Tetrahedron Lett., 28, 1085 (1987).

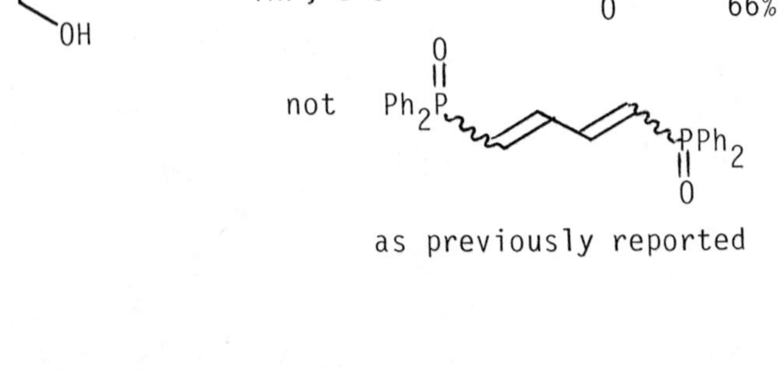

I.B.3-9 J.V. Sinisterra et al., Tetrahedron Lett., 28, 1515 and 4541 (1987); W.B. Kover et al., Tetrahedron, 43, 3199 (1987); E. Suzuki et al., Chem. Lett., 1843 (1987); M.T. Ayoub et al., Synthesis, 653 (1987); B.P. Murphy et al., J. Heterocycl. Chem., 24, 941 (1987); D. Dauzonne and R. Royer, Synthesis, 1020 (1987); G. Seitz et al., Chem. Ber., 120, 1691 (1987); I.P. Beletskaya et al., J. Org. Chem. (USSR), 23, 657 (1987); T. Hiyama et al., Bull. Chem. Soc. Jpn., 60, 2127 and 2139 (1987); A. Guingant et al., J. Org. Chem., 52, 4788 (1987).

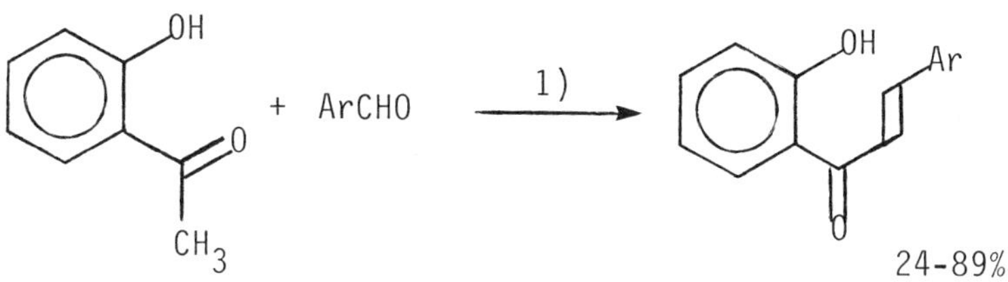

1) C-200 (catalyst), interfacial solid-liquid, EtOH

C-200 = a new barium hydroxide catalyst

various other activated methylene compounds, bases and electrophiles used

---

I.B.3-10 W. Preston Reeves and B.L. Williamson, Synth. Commun., 17, 1721 (1987); S. Eguchi et al., J. Org. Chem., 52, 496 (1987).

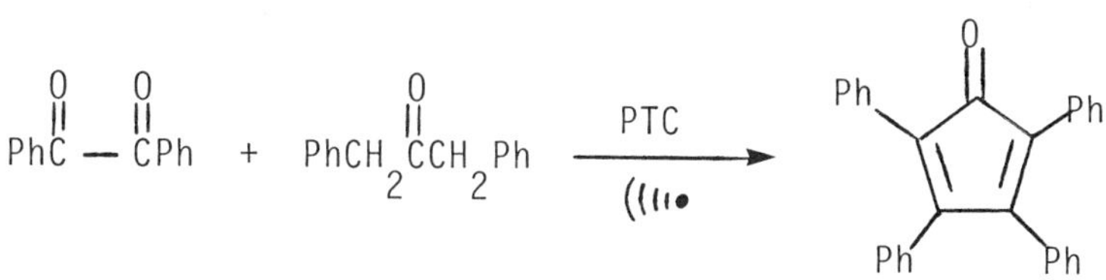

87.5%

better yields and lower temperatures than in the conventional process

I.B.3-11 N. Ragoussis, Tetrahedron Lett., 28, 93 (1987); J.-M. Lehn et al., Chem. Commun., 1424 (1987); E. Koepp and F. Vogtle, Synthesis, 177 (1987).

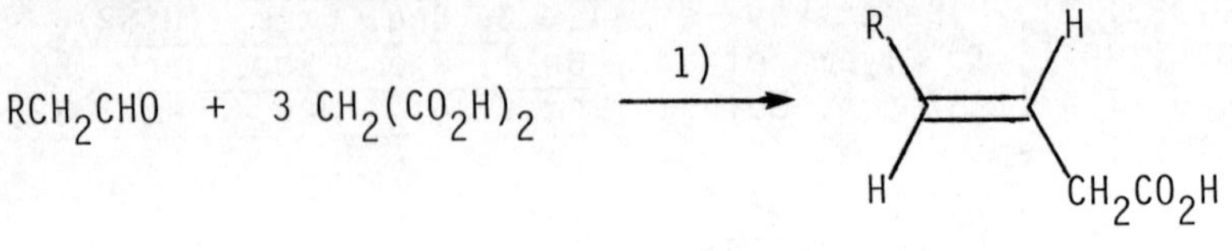

25-85%, >95% E

1) piperidinium acetate (cat.), heat, xylene

I.B.3-12 M. Makosza and A. Tyrala, Synthesis, 1142 (1987).

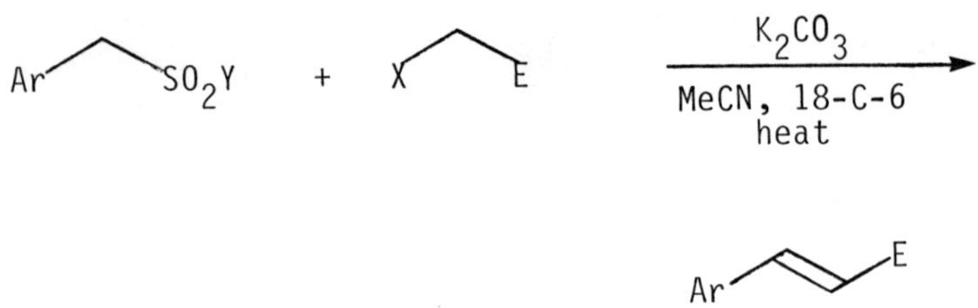

E = electron withdrawing

74-90%

I.B.3-13 M.T. Reetz and M. von Itzstein, J. Organomet. Chem., 334, 85 (1987); M.T. Reetz et al., Chem. Ber., 120, 119 (1987).

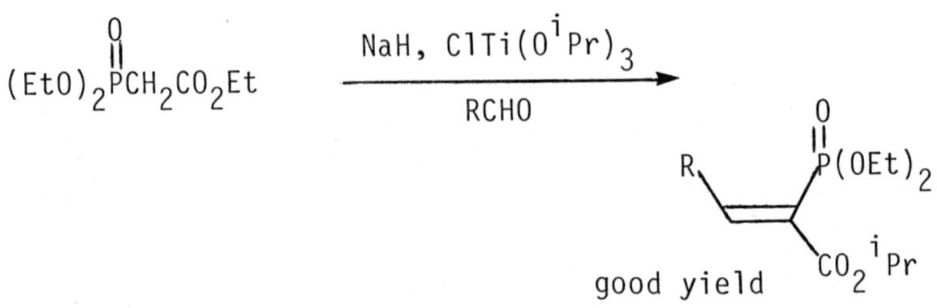

good yield

>95% stereoselectivity

I.B.3-14 J. Nokami et al., Tetrahedron Lett., 28, 649 (1987).

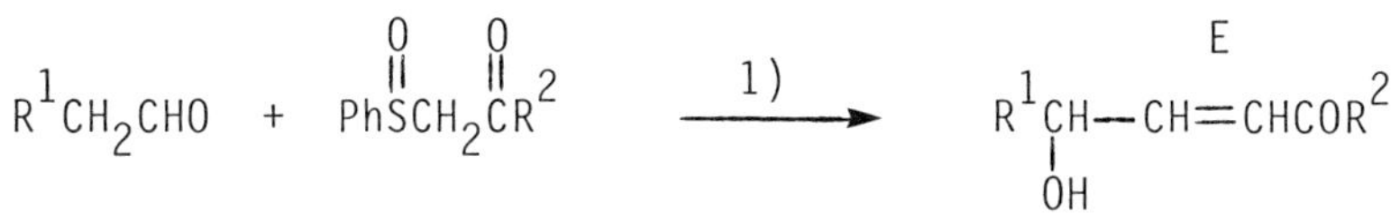

57-82%

1) $Et_2NH(AcOH)$, MeCN, 60°C

I.B.3-15 M. Julia et al., Tetrahedron, 43, 859 (1987).

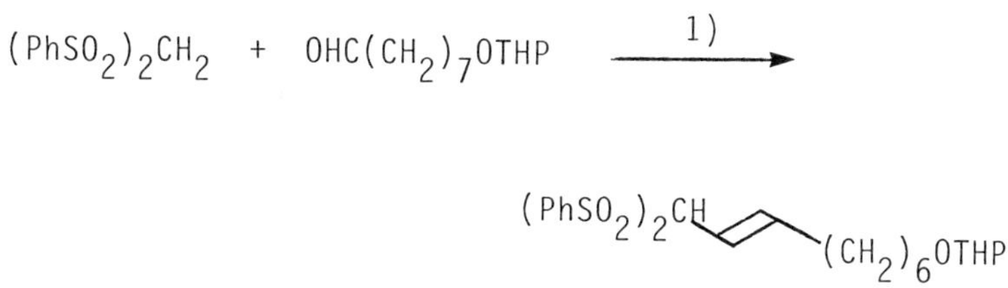

85%

1) piperidinium acetate

I.B.3-16 M. Ogata et al., Synth. Commun., 17, 809 (1987).

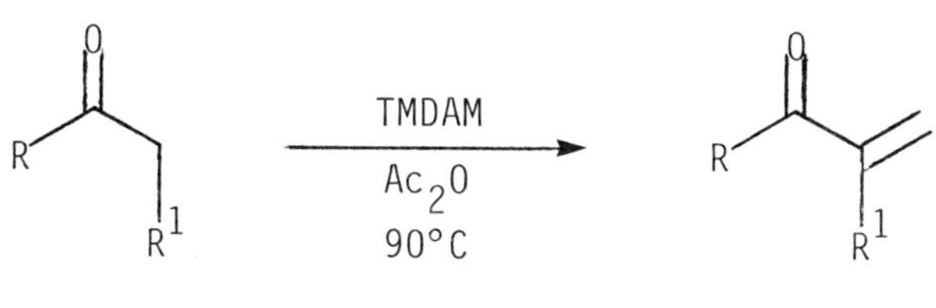

64-92%

TMDAM = N,N,N',N'-tetramethyldiaminomethane

I.B.3-17 S. Apparao and R.R. Schmidt, Synthesis, 896 (1987); M. Mittelbach, Monatsh. Chem., 118, 617 (1987); R. Menicagli et al., Tetrahedron, 43, 171 (1987).

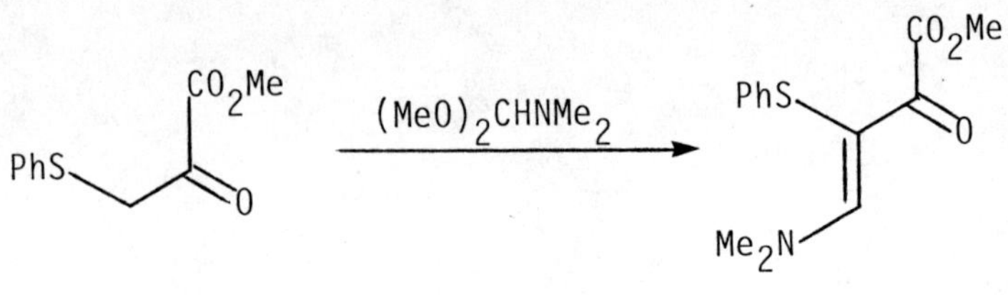

100%

I.B.3-18 I. Kuwajima et al., Tetrahedron Lett., 28, 1299 (1987).

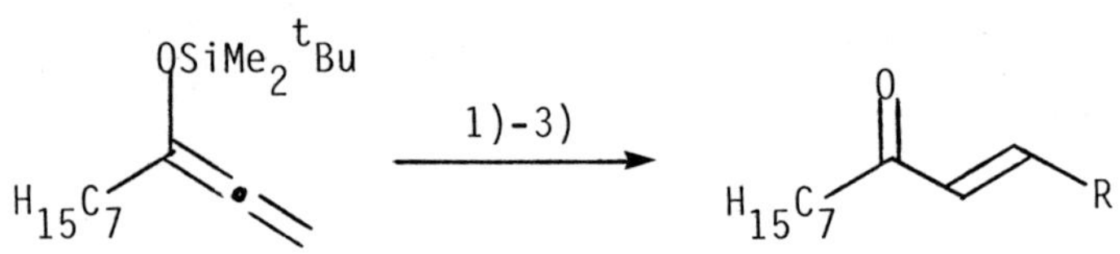

71-87%

1) BuLi 2) RX 3) H_3O^+

I.B.3-19 H. Mayr et al., Tetrahedron, 43, 4119 (1987).

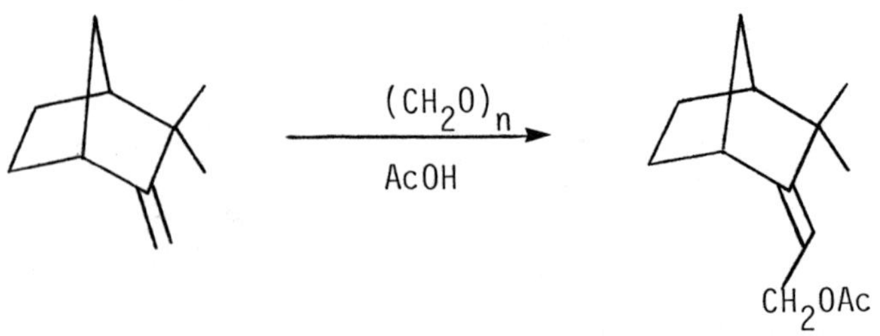

94%

I.B.3-20 N. Hussain and J. Leonard, Tetrahedron Lett., 28, 4871 (1987); B. Fohlisch et al., Liebigs Ann. Chem., 1 (1987).

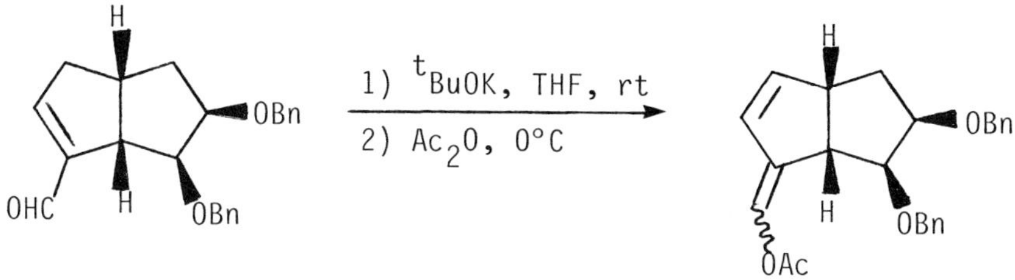

>56%, unstable

I.B.3-21 G. Zweifel and W. Leong, J. Am. Chem. Soc., 109, 6409 (1987); E. Piers and R.T. Skerlj, J. Org. Chem., 52, 4421 (1987).

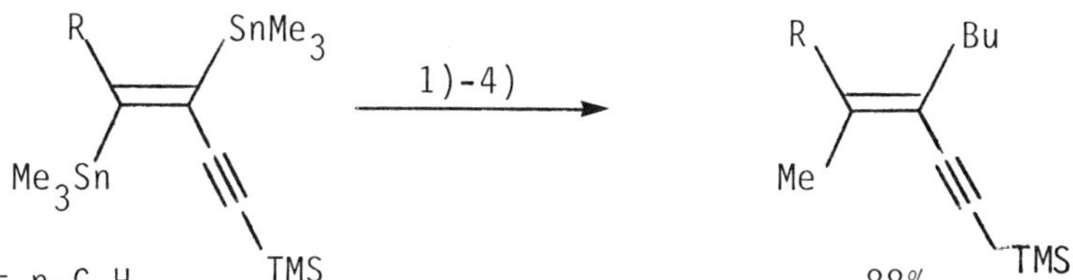

$R = n\text{-}C_6H_{13}$ 88%

1) MeLi 2) BuI 3) MeLi 4) MeI

I.B.3-22 T. Uyehara, N. Asao and Y. Yamamoto, Chem. Commun., 1410 (1987); S.W. McCombie, A. Padwa et al., Tetrahedron Lett., 28, 4127 (1987); C. Najera and M. Yus, ibid., 28, 6709 (1987); M. Yamamoto et al., Bull. Chem. Soc. Jpn., 60, 1523 (1987); T. Sugimura and L.A. Paquette, J. Am. Chem. Soc., 109, 3017 (1987); H.D. Verkruijsse, L. Brandsma and P. von R. Schleyer, J. Organomet. Chem., 332, 99 (1987); L. Duhamel et al., Tetrahedron, 43, 4339 and 4349 (1987); W.R. Dolbier, Jr. et al., Synthesis, 956 (1987).

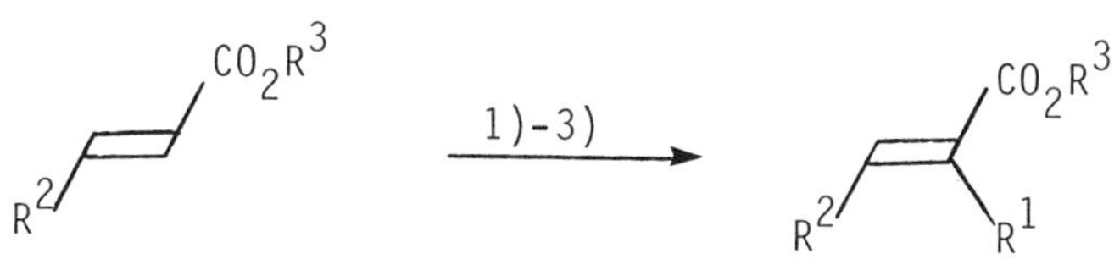

1) R_2NLi 2) R^1X 3) SiO_2

33-74%

E:Z 85:15 to 95:5

I.B.3-23 G.A. Cross and R.M. Kellogg, Chem. Commun., 1746 (1987); T.M. Bargar et al., J. Org. Chem., 52, 678 (1987); L.I. Zakharin and E.V. Anikina, J. Org. Chem. (USSR), 22, 2217 (1986); F. Sato et al., Tetrahedron Lett., 28, 6351 (1987); V. Ratovelomanana and G. Linstrumelle, Bull. Soc. Chim. Fr., 174 (1987).

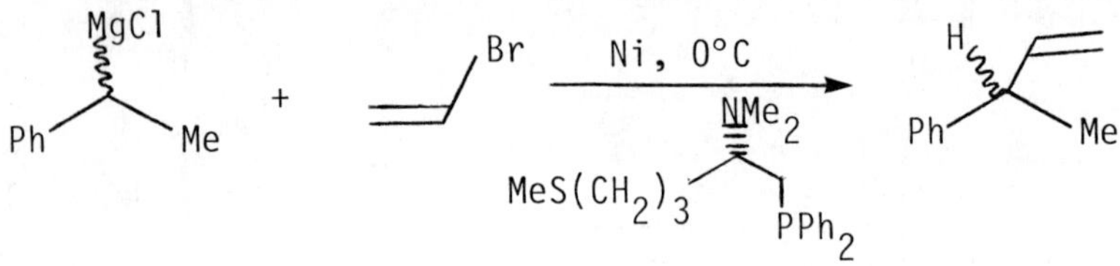

without additive 95%, optical yield 61% (S)

with additive 71%, optical yield 67% (R)

other Ni and Pd catalysts used

I.B.3-24 P. Kocienski et al., Chem. Commun., 241 and 429 (1987).

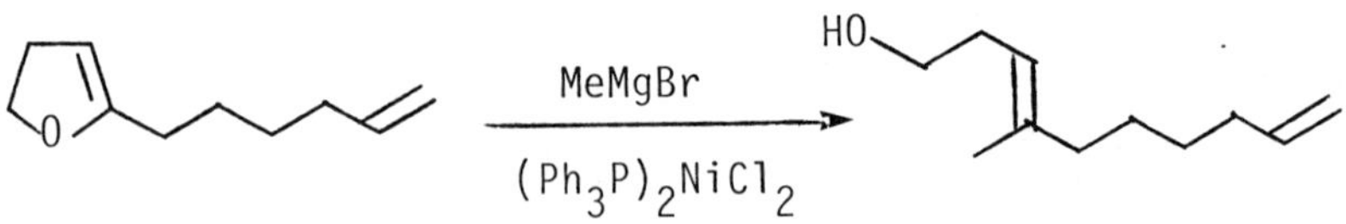

96%

I.B.3-25 Z.-J. Ni and T.-Y. Luh, Chem. Commun., 1515 (1987).

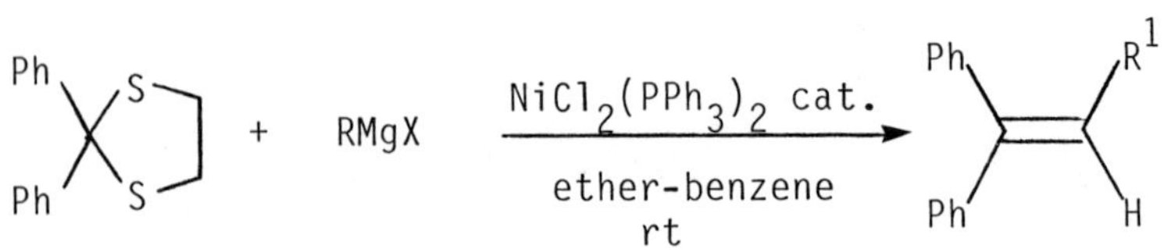

70-84%

R = Me, Et, Bu R^1 = H, Me, Pr

I.B.3-26 T. Norin et al., J. Org. Chem., 52, 292 (1987); A. Suzuki et al., Chem. Lett., 25 (1987).

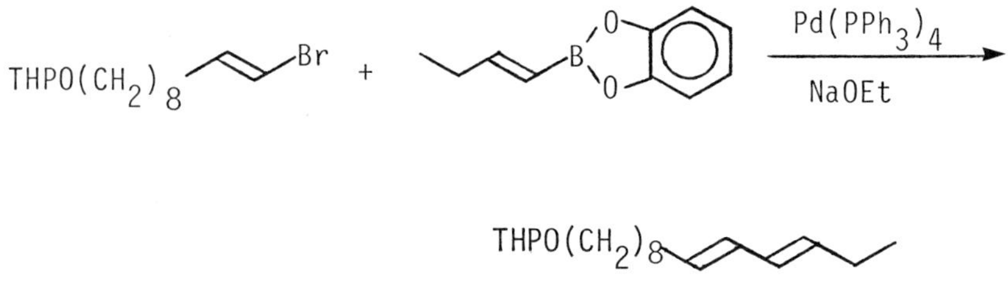

87%

similarly for the cis vinyl bromide

I.B.3-27 A. Suzuki et al., Chem. Lett., 1757 (1987).

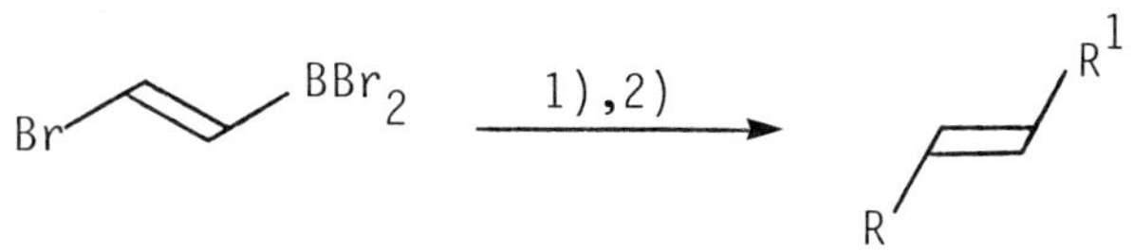

61-88%

93-99% isomeric purity

1) RZnCl, $PdCl_2(PPh_3)_2$ 2) R^1X, LiOH

I.B.3-28 E. Negishi et al., J. Am. Chem. Soc., 109, 2393 (1987).

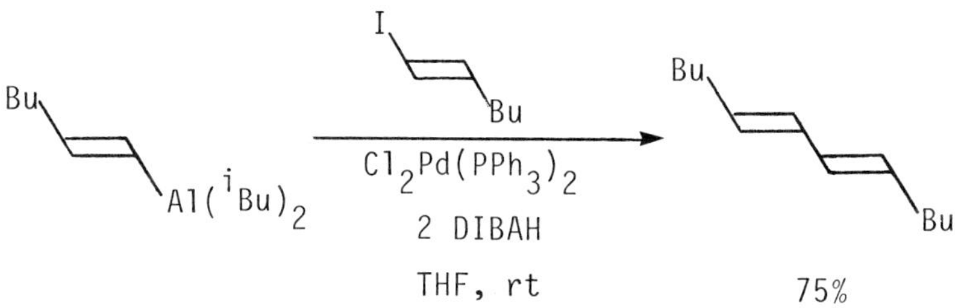

75%

≥98% E,E

I.B.3-29 J.K. Stille and B.L. Groh, J. Am. Chem. Soc., 109, 813 (1987); M.S. Miftakhov et al., J. Org. Chem. (USSR), 22, 2007 (1986); T.N. Mitchell et al., J. Org. Chem., 52, 4868 (1987).

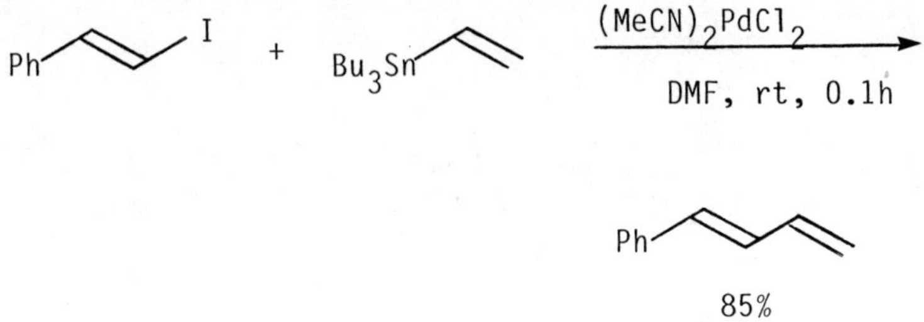

I.B.3-30 K. Oshima et al., Chem. Lett., 5 (1987).

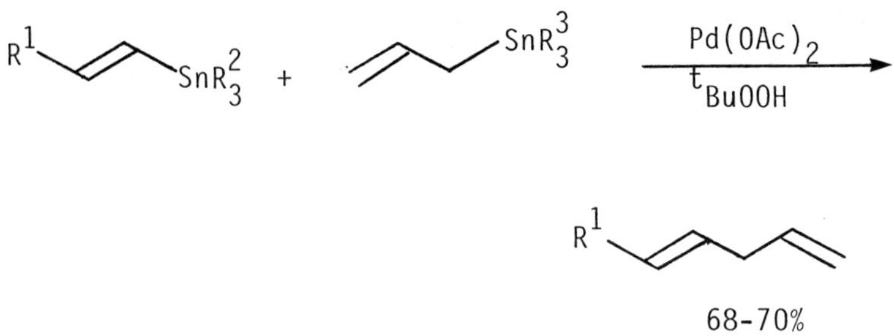

I.B.3-31 G.A. Russell and P. Ngoviwatchai, Tetrahedron Lett., 28, 6113 (1987); R.C. Larock et al., ibid., 28, 4977 (1987).

$$\mathrm{PhCH{=}CHI} + {}^{t}\mathrm{BuHgCl} \xrightarrow[\mathrm{Me_2SO},\ 35\text{-}45^{\circ}\mathrm{C}]{h\nu} \mathrm{PhCH{=}CH}^{t}\mathrm{Bu}$$

100%

96:4, E:Z

I.B.3-32 L. Crombie et al., Tetrahedron Lett., 28, 4875 and 4879 (1987).

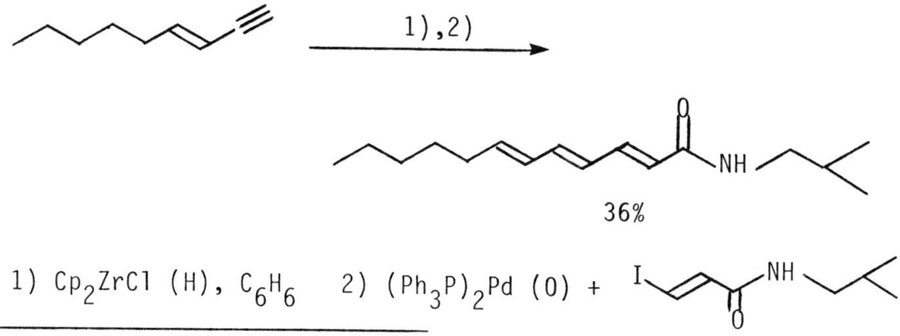

I.B.3-33 Y.-Z. Huang and Q.-L. Zhou, J. Org. Chem., 52, 3552 (1987).

$$Cl(CF_2)_5CF_2I + NEt_3 \xrightarrow[\text{rt, 0.5h}]{Ni(PPh_3)_4} Cl(CF_2)_5CF_2CH{=}CHNEt_2 + Cl(CF_2)_5CF_2H$$

50% 50%

I.B.3-34 C. Schmidt et al., Synth. Commun., 17, 85 (1987); T. Jeffery, Synthesis, 70 (1987); A. de Meijere et al., Chem. Ber., 120, 1667 (1987).

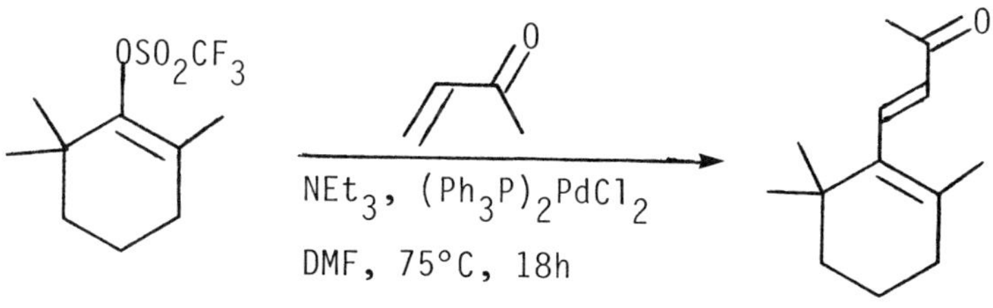

88%

various catalysts and leaving groups used

I.B.3-35 I. Erden and M.A. Amputch, Tetrahedron Lett., 28, 3779 (1987).

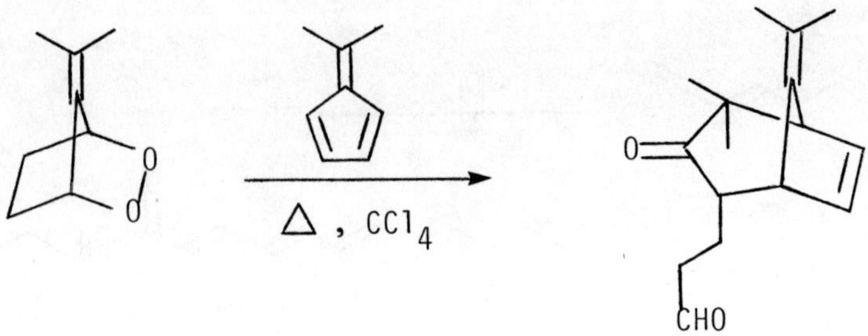

68%

I.B.3-36 K. Sakai et al., Tetrahedron Lett., 28, 3373 (1987).

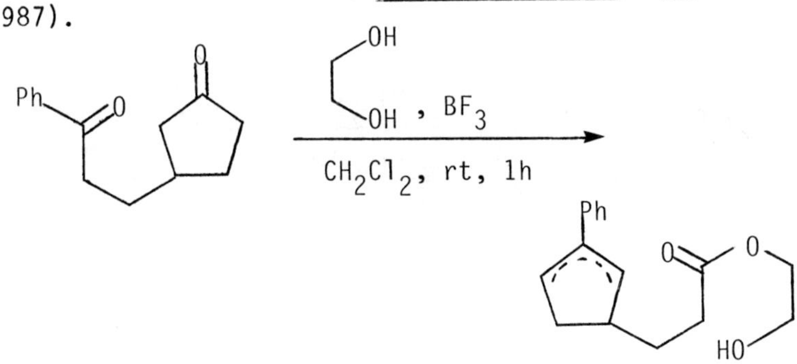

87%

I.B.3-37 D.H.R. Barton et al., Tetrahedron, 43, 5031 (1987).

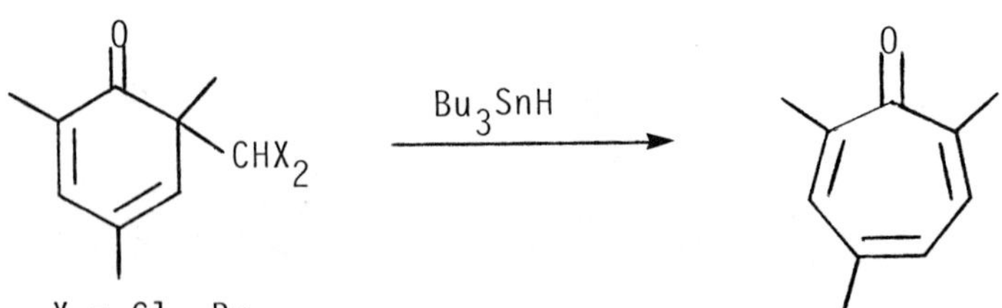

X = Cl, Br

near quantitative

I.B.3-38 K. Takai et al., Tetrahedron Lett., 28, 1443 (1987).

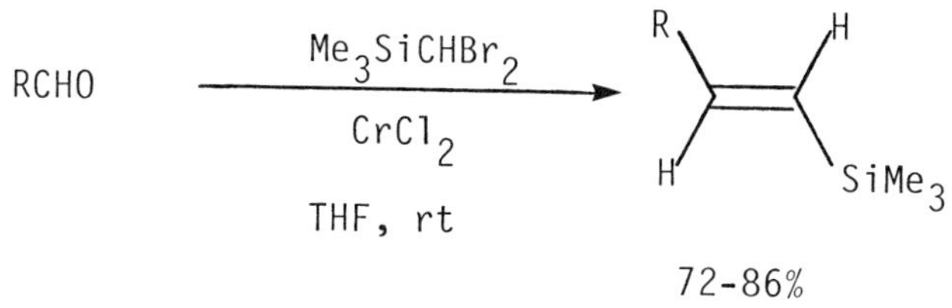

I.B.3-39 J. Salaun and B. Karkour, Tetrahedron Lett., 28, 4669 (1987).

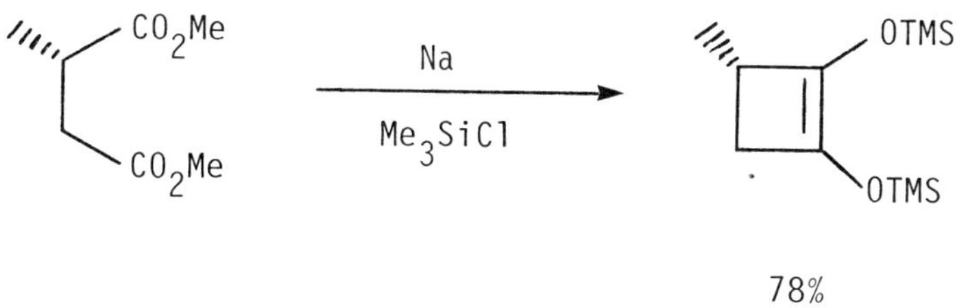

I.B.3-40 G.H. Whitham et al., J. Chem. Soc., Perkin Trans. 1, 2707 (1987).

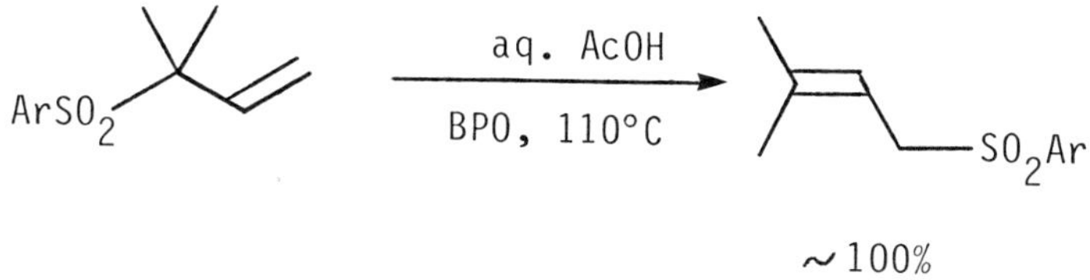

BPO = dibenzoyl peroxide

I.B.3-41 G. Pattenden et al., Tetrahedron Lett., 28, 5949 (1987); N. Takaishi et al., ibid., 28, 5883 (1987); C.-M. Andersson and A. Hallberg, ibid., 28, 4215 (1987).

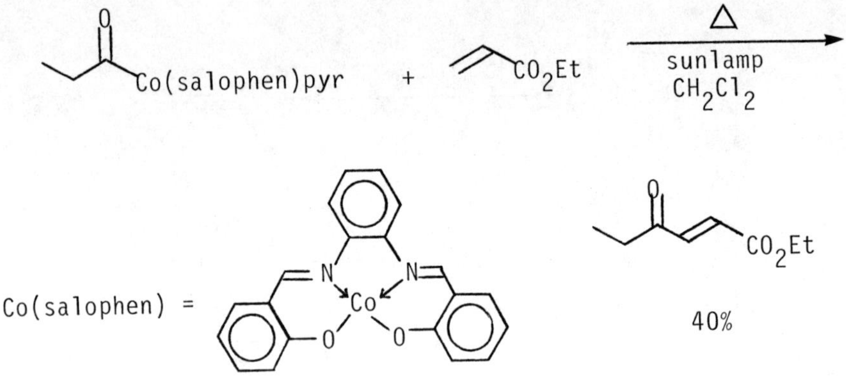

similar results with acid chlorides and $Pd(OAc)_2$

I.B.3-42 A.F. Renaldo and H. Ito, Synth. Commun., 17, 1823 (1987).

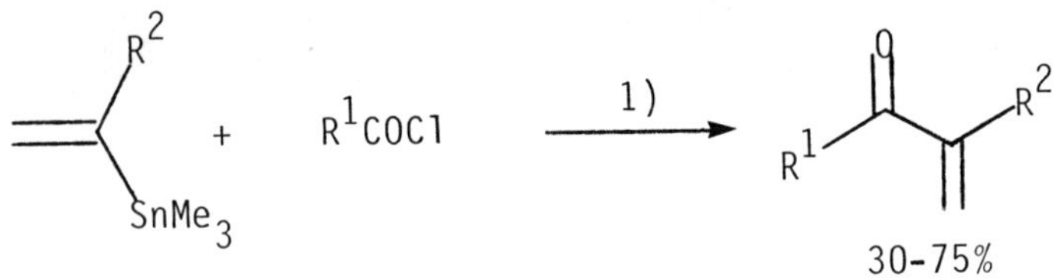

1) $Pd(PPh_3)_4$, toluene or 1,2,3,4-$Me_4C_6H_2$, 1.5-5.5h

I.B.3-43 T. Hudlicky et al., Tetrahedron Lett., 28, 167 (1987).

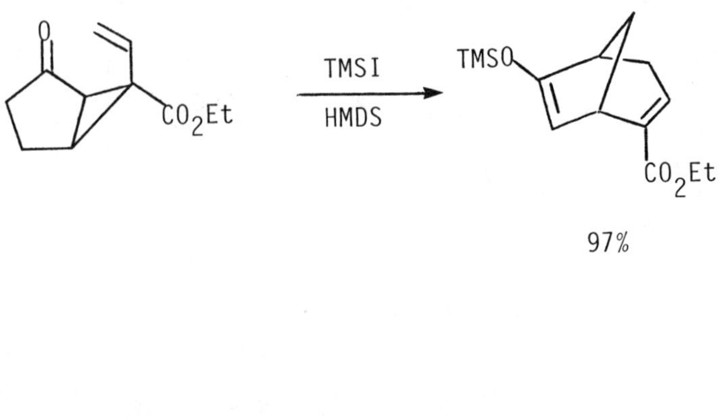

I.B.3-44 A. Padwa et al., Tetrahedron Lett., 28, 913 (1987).

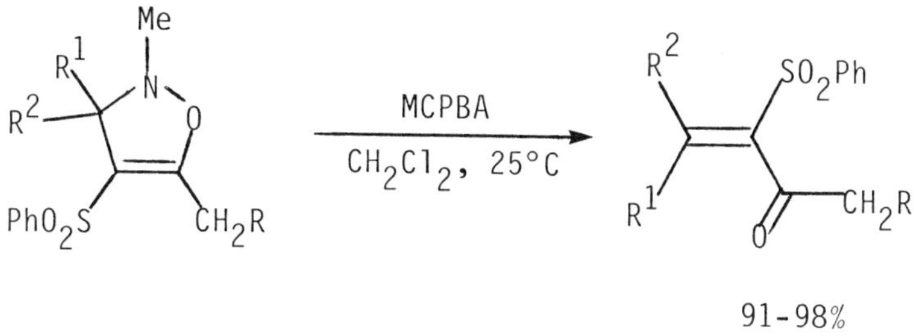

I.B.3-45 H.M.L. Davies et al., Tetrahedron, 43, 4265 (1987).

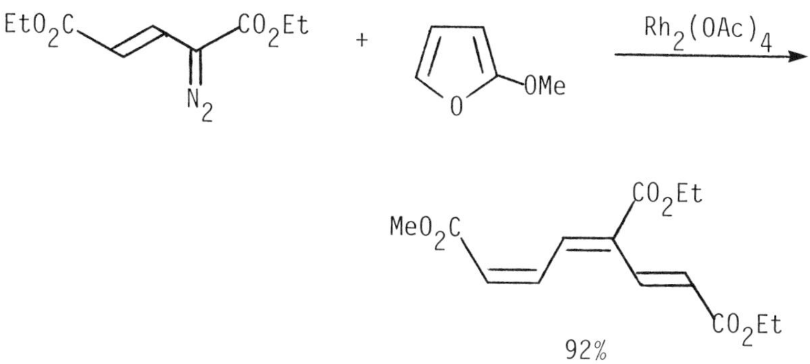

I.B.3-46 K. Ishiguro, Y. Sawaki and H. Iwamura, Chem. Lett., 1853 (1987); M.P. Doyle et al., J. Org. Chem., 52, 1619 (1987).

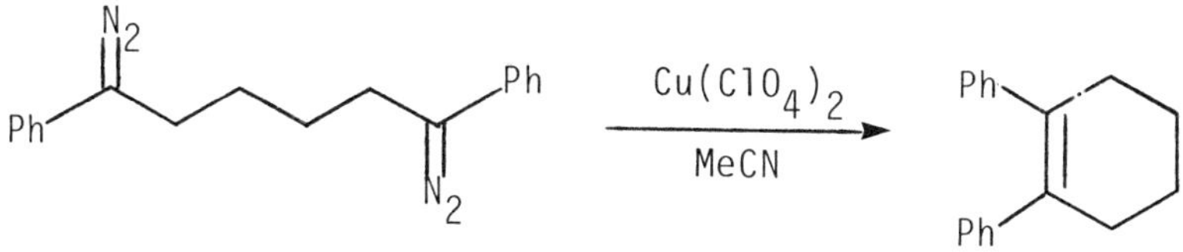

84%

similar result on heating a diazirine and a diazo compound

I.B.3-47 W. Freund and S. Hunig, Helv. Chim. Acta, 70, 929 (1987).

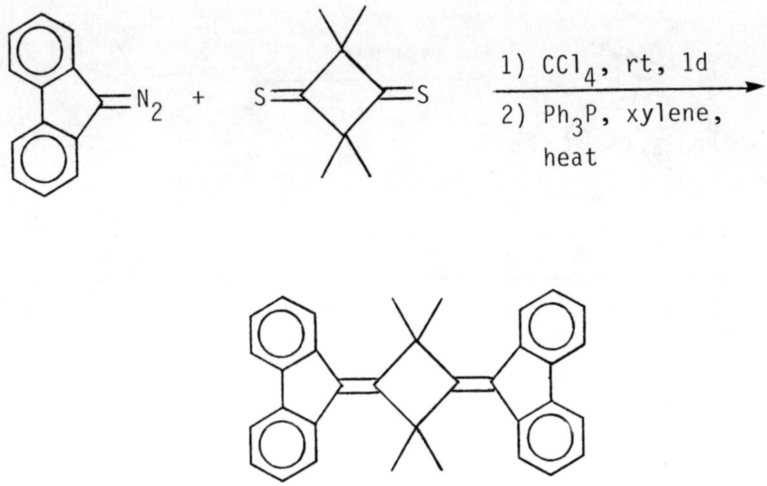

76%

I.B.3-48 J.E. McMurry et al., Tetrahedron, 43, 5489 (1987); J.E. McMurry and G.K. Bosch, J. Org. Chem., 52, 4885 (1987); J.E. McMurry and R. Swenson, Tetrahedron Lett., 28, 3209 (1987); A. Furstner and H. Weidmann, Synthesis, 1071 (1987).

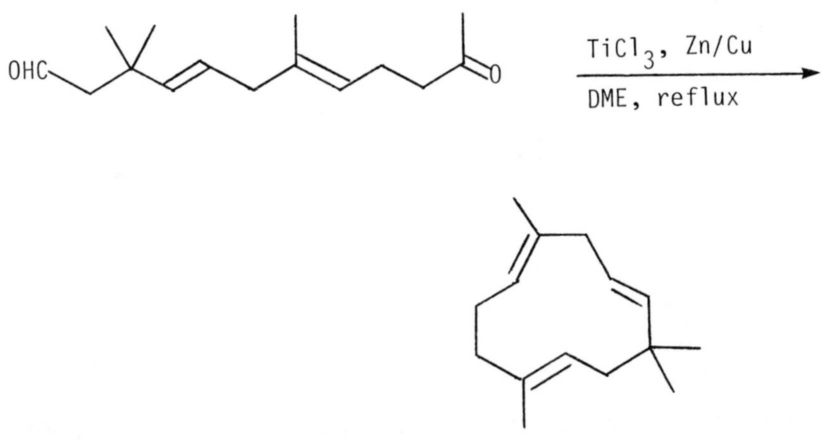

60%

slow addition over 18h

I.B.3-49 S.-I. Murahashi et al., Bull. Chem. Soc. Jpn., 60, 3285 (1987); A. de Groot et al., Synth. Commun., 17, 1607 (1987); S.B. Mahato et al., Tetrahedron Lett., 28, 5315 (1987).

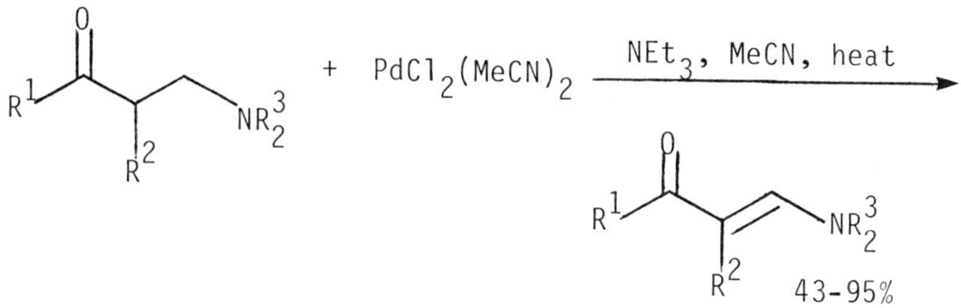

similar results with chloreal and Arthrobacter simplex (11CB-321)

I.B.3-50 A.E. Greene et al., J. Org. Chem., 52, 1169 (1987).

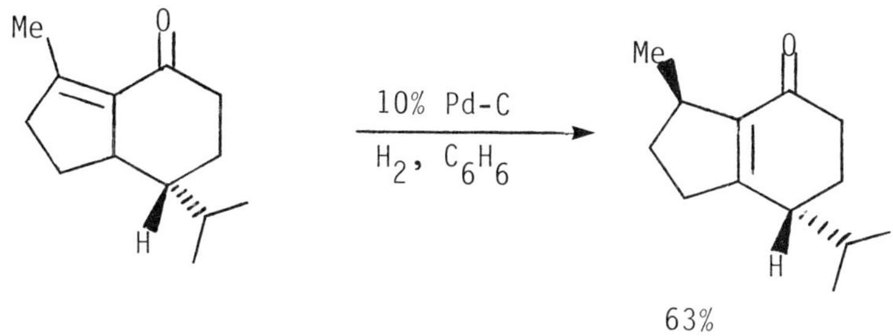

I.B.3-51 J.-P. Pete et al., Tetrahedron Lett., 28, 4825 (1987); K. Inomata et al., Bull. Chem. Soc. Jpn., 60, 1767 (1987); T. Chou et al., J. Org. Chem., 52, 3394 (1987).

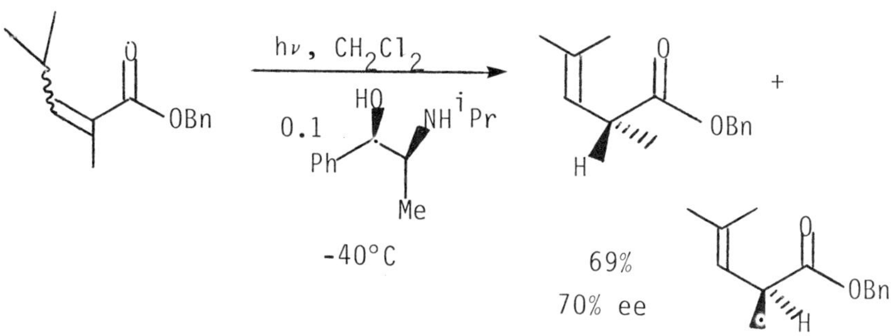

deconjugation also observed with vinyl sulfones and DBN or DBU

I.B.3-52 P.H. Buist et al., Tetrahedron Lett., 28, 857 (1987).

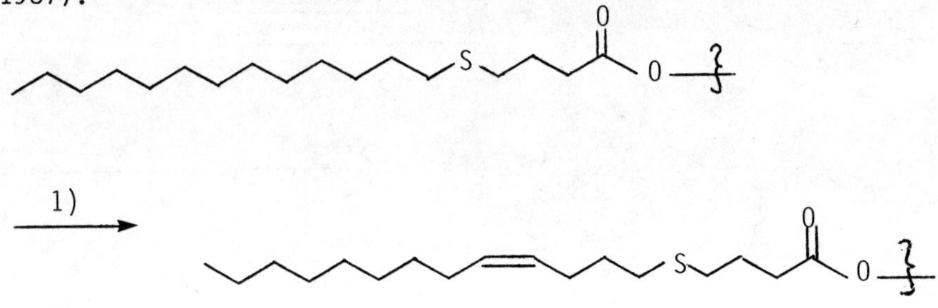

~ 20%

1) O_2, S•cerevisiae

I.B.3-53 S. Warwel et al., Angew. Chem., Int. Ed. Engl., 26, 702 (1987); S. Warwel and H. Katker, Synthesis, 935 (1987).

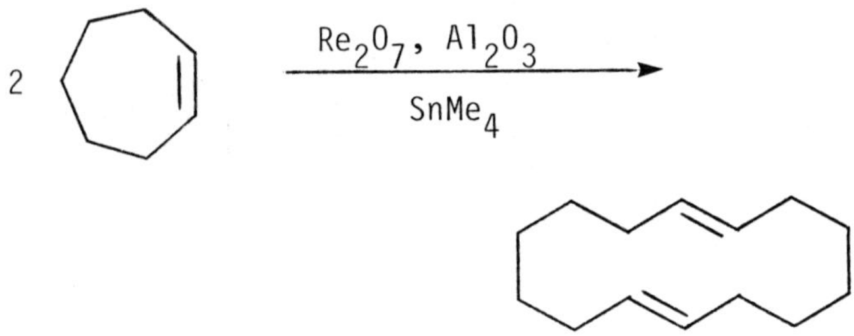

68%

I.B.3-54 A. Takeda et al., Bull. Chem. Soc. Jpn., 60, 1103 (1987); K. Mori and S. Maemoto, Liebigs Ann. Chem., 863 (1987).

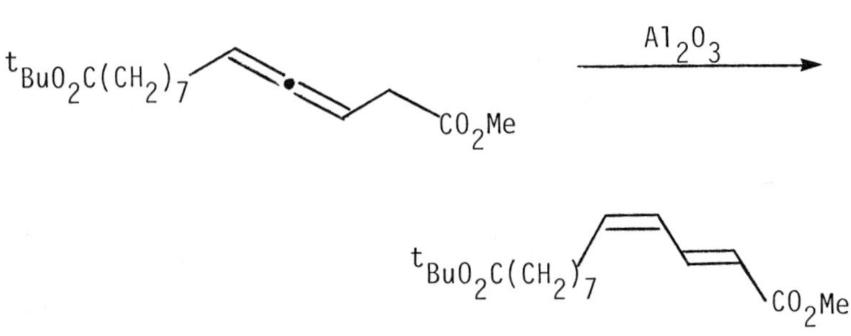

91%

I.B.3-55 J.-E. Backval, Bull. Soc. Chim. Fr., 665 (1987).

Colloquium seminar: "The use of metal reagents in stereo- and regio- selective functionalizations of conjugated dienes."

I.B.3-56 W.R. Jackson et al., Synthesis, 1032 (1987); idem, Aust. J. Chem., 40, 1053 (1987).

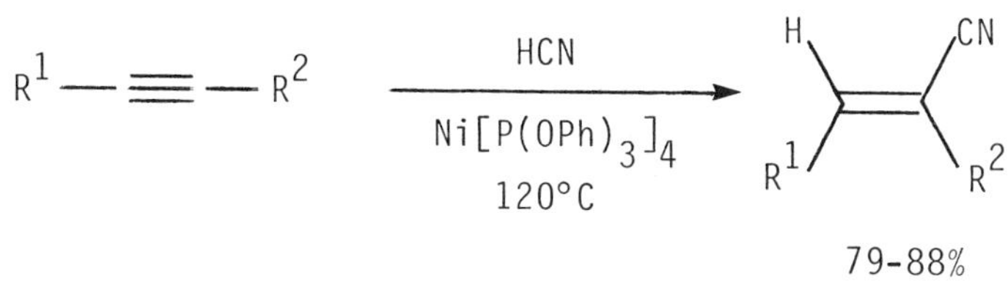

I.B.3-57 G. Pattenden et al., Tetrahedron Lett., 28, 4749 (1987).

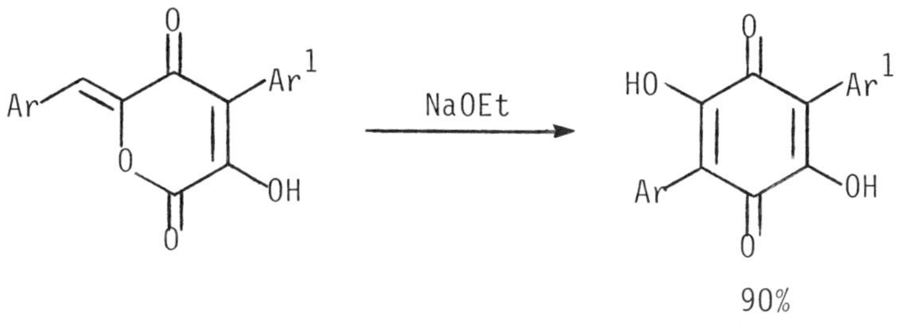

I.B.3-58 Y. Lin et al., Tetrahedron Lett., 28, 3249 (1987).

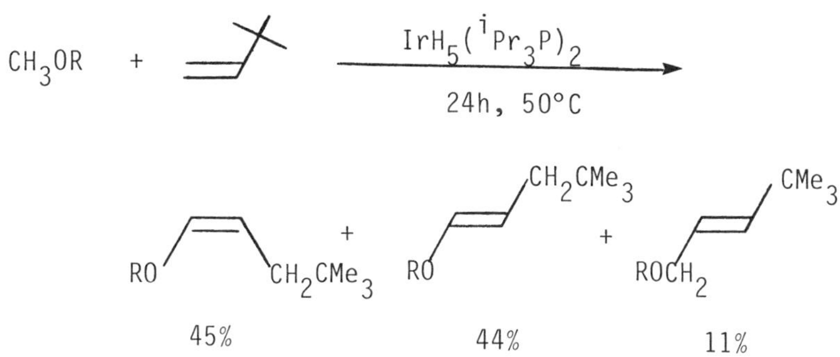

I.B.3-59 T.-Y. Luh et al., Chem. Commun., 981 (1987).

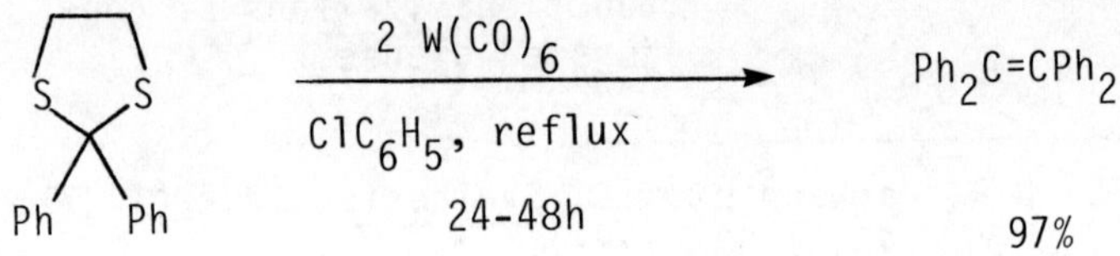

I.B.3-60 R.S. Threlkel, R.G. Bergman et al., Org. Synth., 65, 42 (1987).

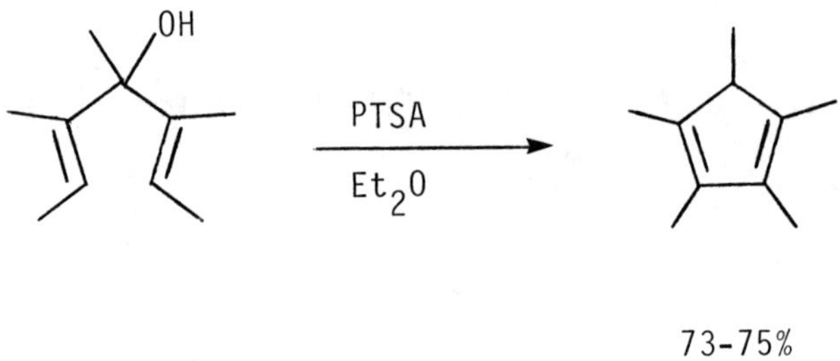

I.B.4. Allene Forming Reactions

I.B.4-1 N.G. Khusainova and A.N. Pudovik, Russ. Chem. Rev., 56, 564 (1987).

Review: "Phosphorylated Allenes. Methods of Synthesis and Properties."

I.B.4-2 C.J. Elsevier and H.H. Mooiweer, J. Org. Chem., 52, 1536 (1987); J.R. McCarthy et al., Tetrahedron Lett., 28, 2207 (1987); R.A. Gibbs and W.H. Okamura, ibid., 28, 6021 (1987); I. Kuwajima et al., ibid., 28, 2281 (1987); I. Fleming et al., J. Chem. Soc., Perkin Trans. 1, 2269 (1987); A. Doutheau et al., Tetrahedron, 43, 2721 (1987); N.O. Nilsen et al., Acta Chem. Scand., B41, 459 (1987).

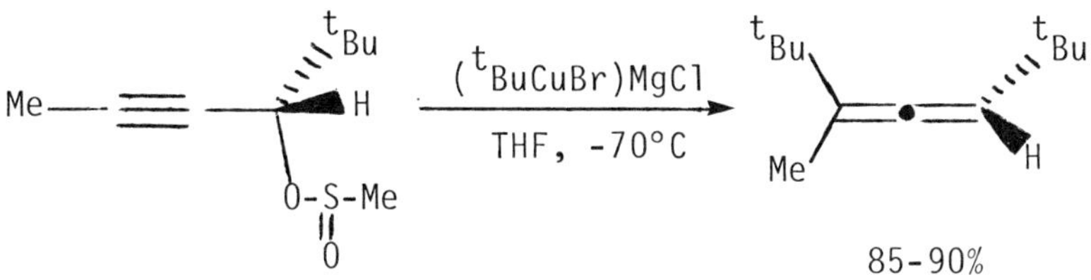

85-90%

other nucleophiles and leaving groups employed

I.B.4-3 A.M. Caporusso et al., Tetrahedron Lett., 28, 6073 (1987); idem, J. Org. Chem., 52, 3920 (1987).

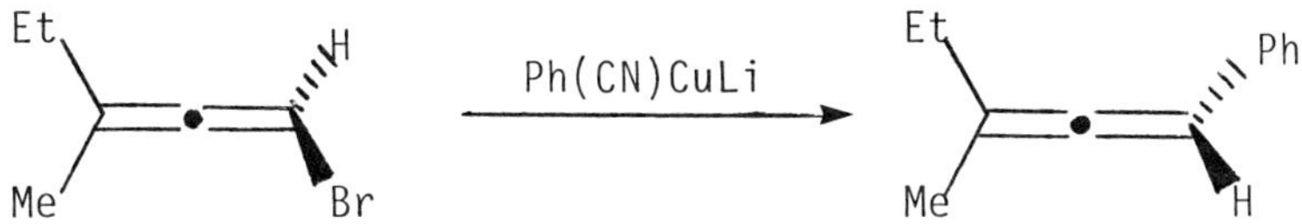

70-80%, 100% ee

retention of configuration with i-Bu(CN)CuLi

I.B.4-4 J. Tsuji et al., Synthesis, 603 (1987); J. Gore et al., Bull. Soc. Chim. Fr., 165 (1987); A.L. Castelhano and A. Krantz, J. Am. Chem. Soc., 109, 3491 (1987).

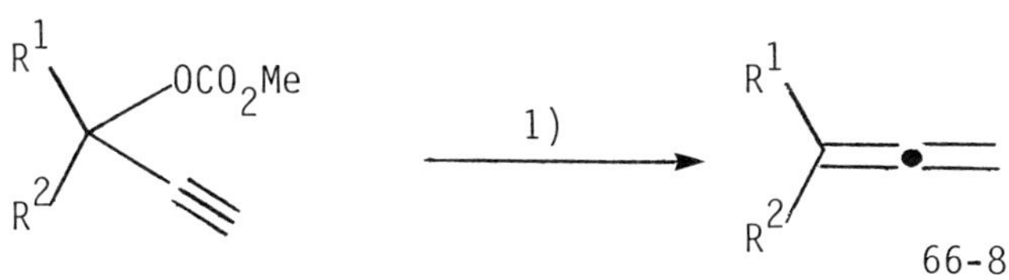

66-87%

1) HCO_2NH_4, DMF, $Pd_2(dba)_3$•$CHCl_3$, Bu_3P, 70°C

different leaving groups and reductants also used

I.B.4-5 D. Hoppe, E. Egert et al., Tetrahedron, 43, 2457 (1987).

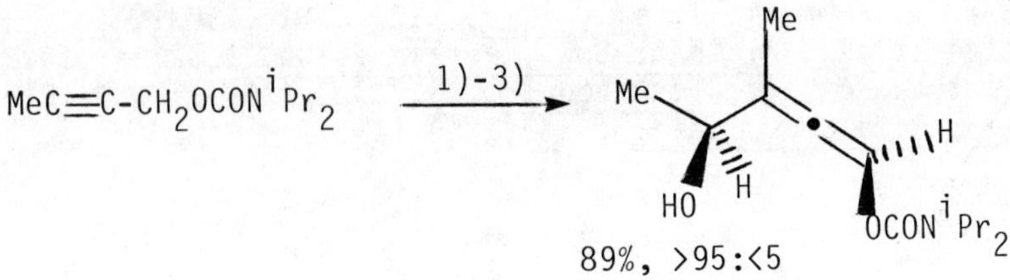

1) n-BuLi 2) $Ti(O^iPr)_4$ or $Ti(NEt_2)_3Cl$ 3) MeCHO

I.B.4-6 W.R. Roth et al., Angew. Chem., Int. Ed. Engl., 26, 256 (1987).

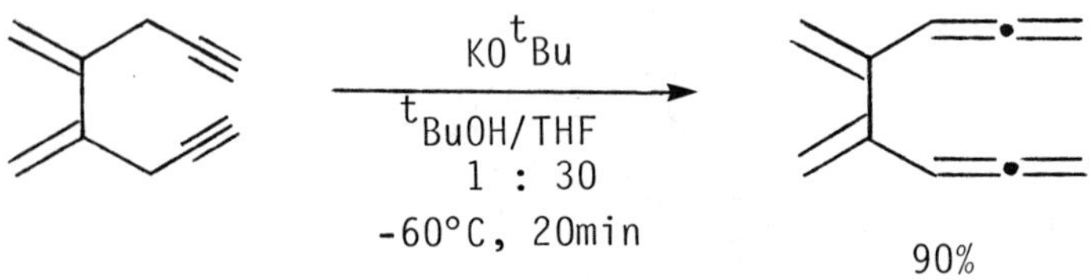

I.B.4-7 M. Taddei et al, Tetrahedron Lett., 28, 2751 (1987).

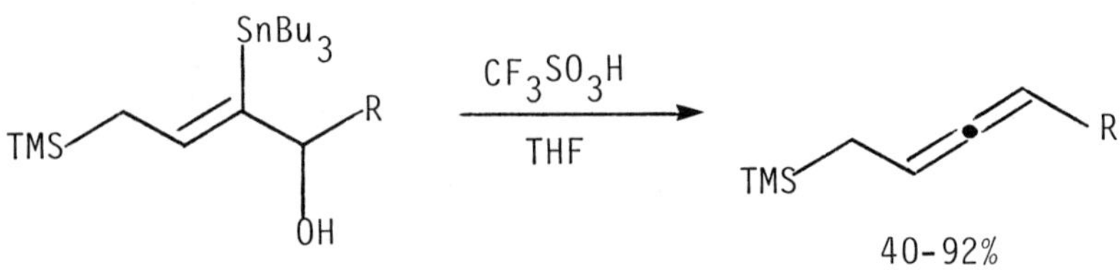

I.B.4-8 J. Inanaga et al., Chem. Lett., 2275 (1987).

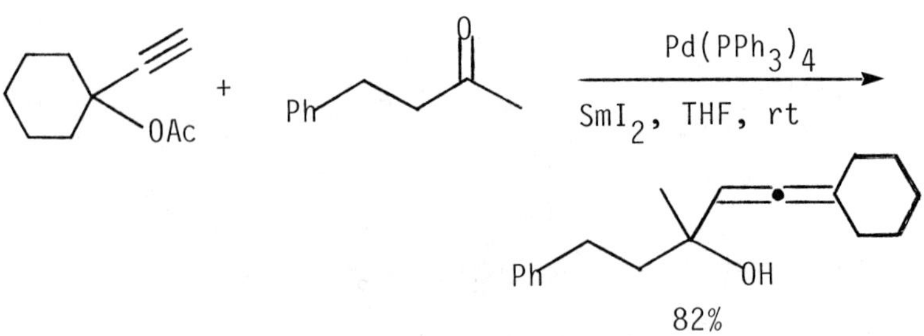

I.B.4-9 T.G. Back et al., <u>Tetrahedron Lett.</u>, <u>28</u>, 1737 (1987).

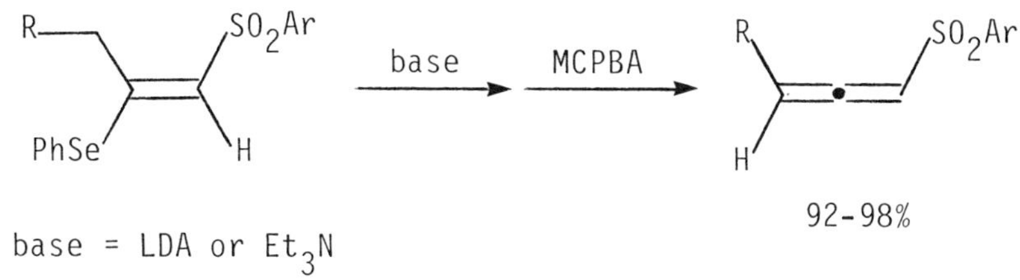

I.B.4-10 N. Slougui and G. Rousseau, <u>Tetrahedron Lett.</u>, <u>28</u>, 1651 (1987).

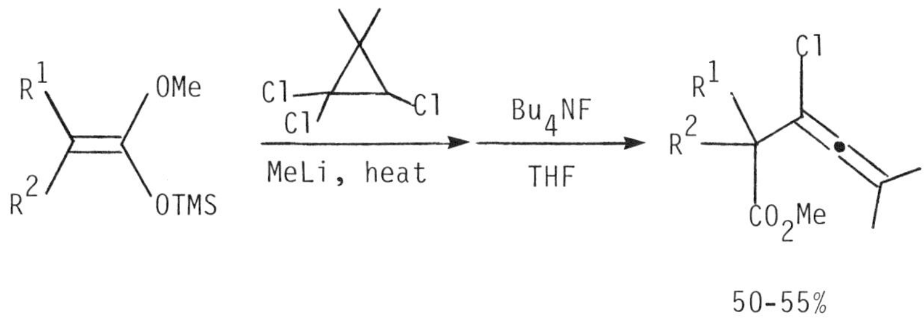

I.B.4-11 H. Fillion et al., <u>Synthesis</u>, 708 (1987).

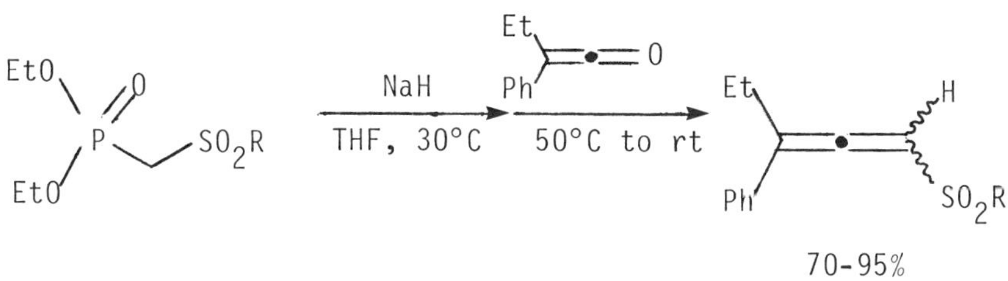

I.C. Carbon-Carbon Triple Bonds

I.C-1 B.A. Trofimov, <u>J. Org. Chem. (USSR)</u>, <u>22</u>, 1788 (1986).

Mini-Review: "Superbasic Media in the Chemistry of Acetylene."

I.C-2 S. Takano et al., Chem. Lett., 1905 (1987); idem, Chem. Commun., 555 (1987); S. Takano, Synthesis, 139 (1987); P. Margaretha et al., Helv. Chim. Acta, 70, 339 (1987); L.A. Remizova et al., J. Org. Chem. (USSR), 22, 2209 (1986).

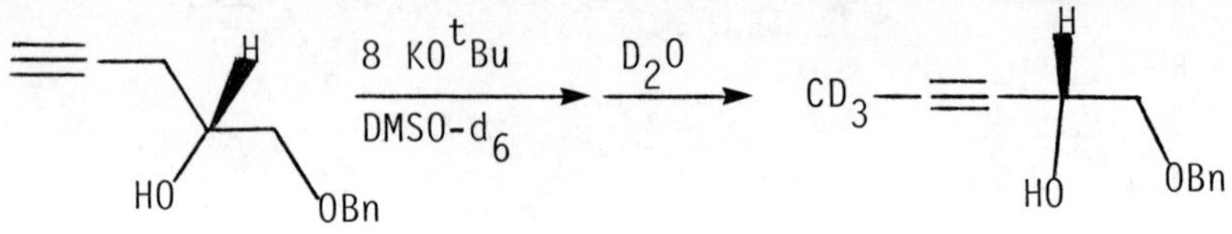

89%

93% D incorporation

I.C-3 J. Fried et al., J. Am. Chem. Soc., 109, 3684 (1987); K.C. Nicolaou et al., ibid., 109, 2205 and 2208 (1987); R.M. Soll and S.P. Seitz, Tetrahedron Lett., 28, 5457 (1987); R.J.K. Taylor et al., J. Chem. Soc., Perkin Trans. 1, 1573 (1987); A.V.R. Rao et al., Tetrahedron, 43, 4385 (1987); L.Yu. Ukhin et al., J. Org. Chem. (USSR), 23, 1197 and 1198 (1987).

28-52%

toward difluoroarachidonic acids

alkylation with epoxides, aldehydes and imines also examined

I.C-4 T. Mukaiyama et al., Chem. Lett., 1975 (1987); T. Katsuki et al., Tetrahedron Lett., 28, 6205 (1987).

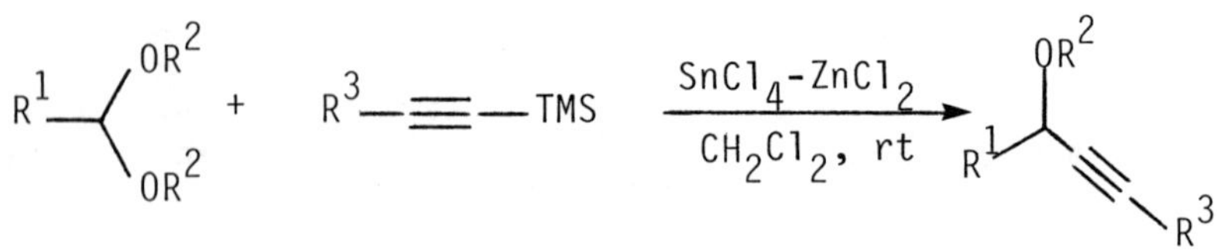

44-85%

I.C-5 G. Himbert and L. Henn, Liebigs Ann. Chem., 771 (1987).

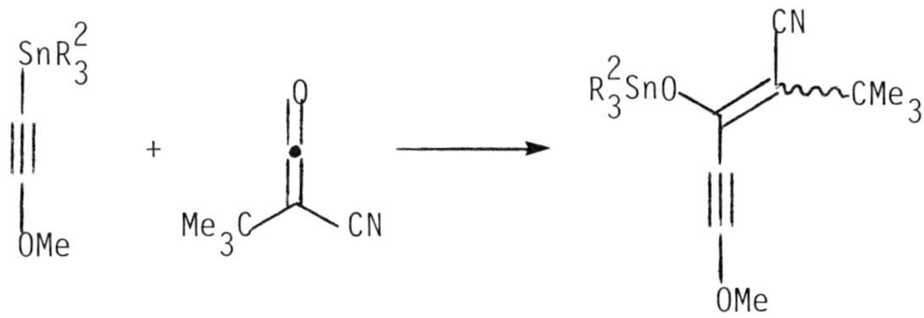

~48%

I.C-6 R. Rossi et al., Tetrahedron, 43, 4591 (1987); P.W. Collins et al., J. Med. Chem., 30, 193 (1987); J.K. Stille and J.H. Simpson, J. Am. Chem. Soc., 109, 2138 (1987).

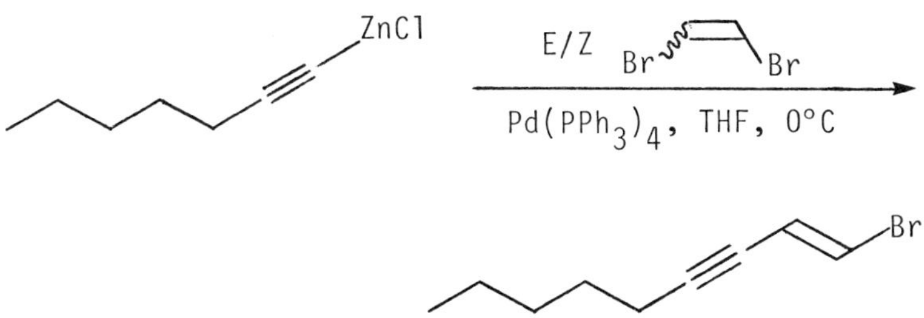

58% (major)

other catalysts and Me_3Sn species also used

I.C-7 G. Linstrumelle et al., Tetrahedron Lett., 28, 1649 (1987); F. Sato et al., ibid., 28, 3959 (1987); R. Rossi et al., Gazz. Chim. Ital., 117, 481 and 503 (1987).

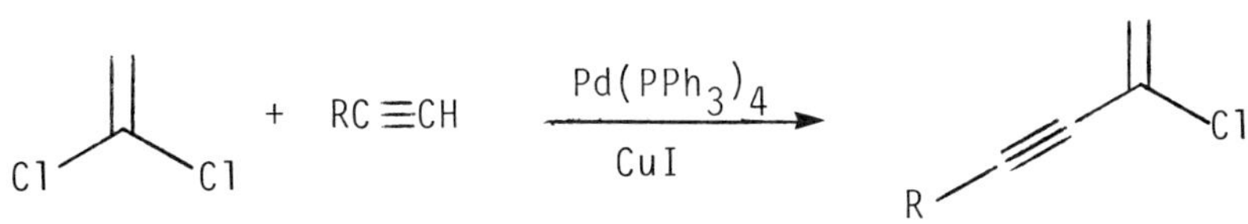

72-90%

I.C-8 P.J. Stang and T. Kitamura, J. Am. Chem. Soc., 109, 7561 (1987).

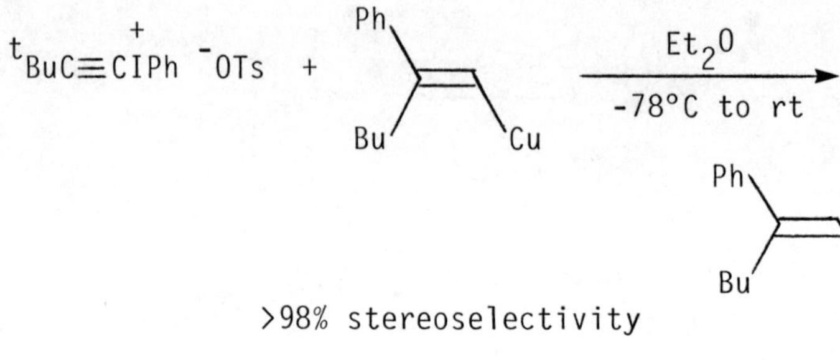

94%

>98% stereoselectivity

I.C-9 R.C. Larock and S. Babu, Tetrahedron, 43, 2013 (1987); B.M. Trost et al., J. Am. Chem. Soc., 109, 3486 (1987).

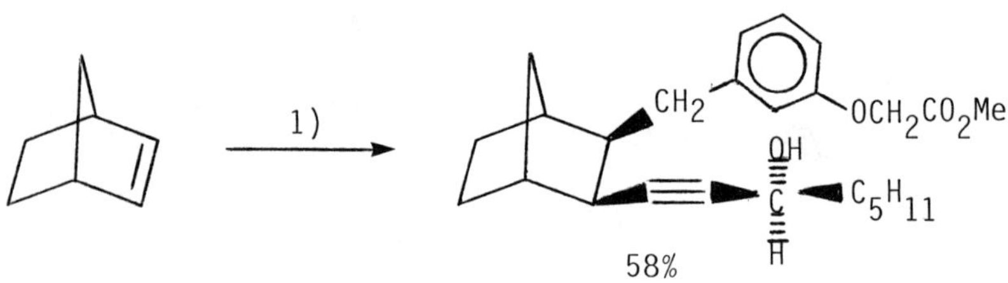

58%

inseparable mixture with regioisomer

1) 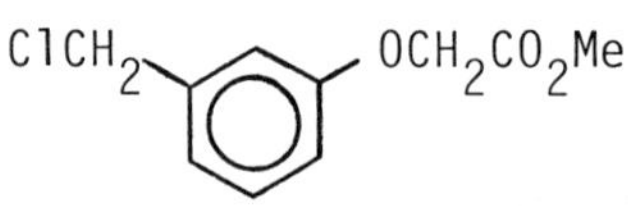

, (S)-1-octyn-3-ol, 8% $Pd(PPh_3)_4$, 70°C, 1 day

I.C-10 H. Hart and K. Shahlai, Tetrahedron Lett., 28, 5437 (1987).

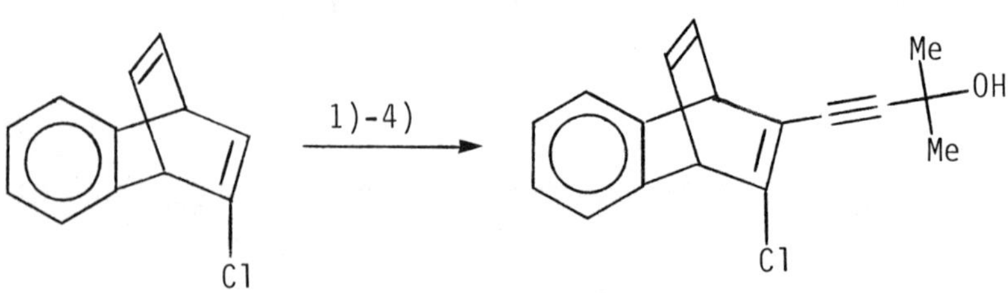

88% (+ naphthalene)

1) BuLi, THF, -42°C 2) reflux
3) acetone 4) aqueous acid

I.C-11 K. Musigmann, H. Mayr and A. de Meijere, Tetrahedron Lett., 28, 4517 (1987); M.R. Detty et al., J. Org. Chem., 52, 3662 (1987).

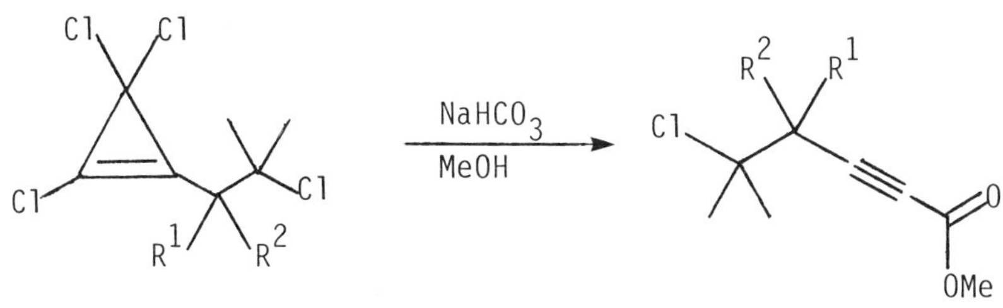

55-86%

I.C-12 C.B. Ziegler, Jr. et al., J. Org. Chem., 52, 443 (1987); J. Chenault and J.-F.E. Dupin, Synthesis, 498 (1987); Y. Shen and W. Qiu, ibid., 42 (1987); S. Hanessian et al., Tetrahedron Lett., 28, 1143 (1987); H. Rehman and J.M. Rao, Synth. Commun., 17, 1119 (1987).

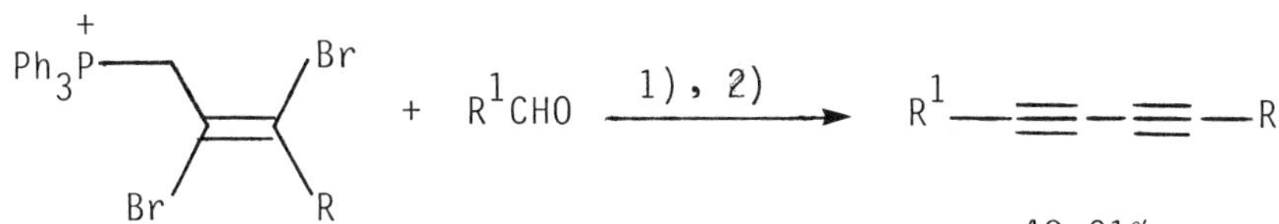

48-91%

1) base 2) -60 to 20°C

I.C-13 G. Himbert et al., Synthesis, 73 (1987); A. Greene et al., J. Org. Chem., 52, 3461 (1987).

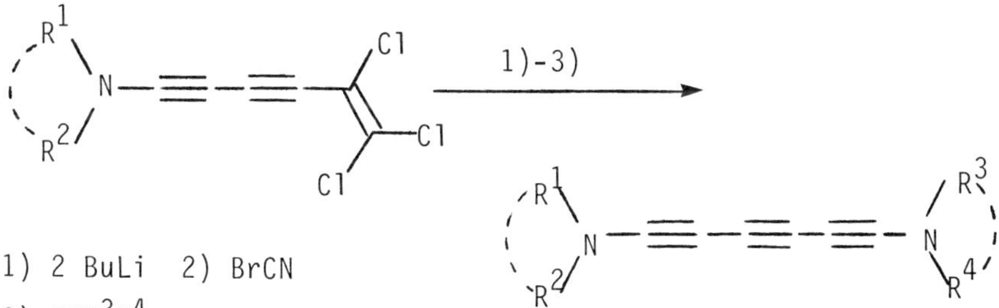

1) 2 BuLi 2) BrCN

3) HNR^3R^4

15-64%

a carbanion also used as a nucleophile

I.C-14 E.J. Corey et al., Tetrahedron Lett., 28, 4921 (1987).

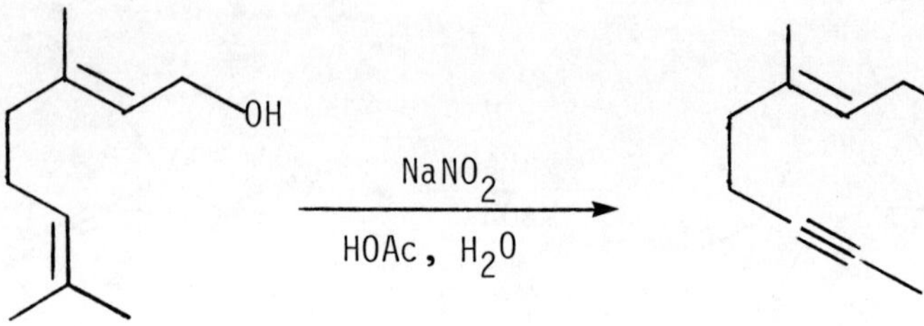

a mechanistic study

I.C-15 R.M. Moriarty et al., Chem. Commun., 711 (1987).

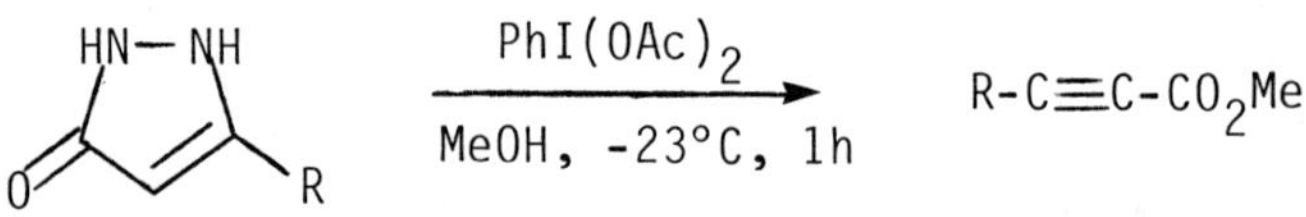

59-63%

I.C-16 G. Bertrand et al., J. Am. Chem. Soc., 109, 4711 (1987).

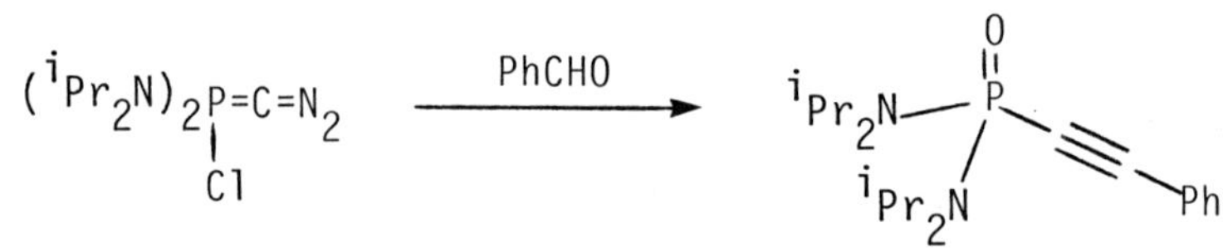

quantitative

I.C-17 A.B. Holmes et al., Org. Synth., 65, 52 (1987).

$$\text{TMS-C}\equiv\text{CH} \xrightarrow[(\text{Me}_2\text{NCH}_2)_2]{\text{O}_2,\ \text{CuCl}} \text{TMS-C}\equiv\text{C-C}\equiv\text{C-TMS}$$

68-76%

I.C-18 R.W. Saalfrank et al., Z. Naturforsch., 42b, 97 (1987).

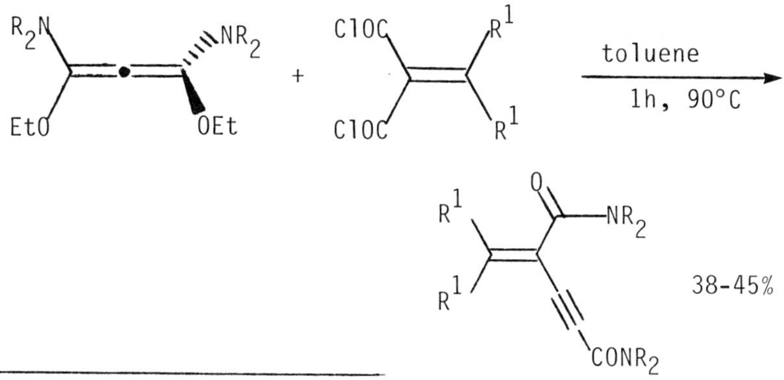

I.D. Cyclopropanations

I.D.1. Carbene or Carbenoid Additions to a Multiple Bond

I.D.1-1 M. Brookhart and W.B. Studabaker, Chem. Rev., 87, 411 (1987).

Review: "Cyclopropanes from Reactions of Transition-Metal-Carbene Complexes with Olefins."

I.D.1-2 M.P. Doyle et al., Tetrahedron Lett., 28, 833 (1987); H.M.L. Davies et al., ibid., 28, 1853 (1987).

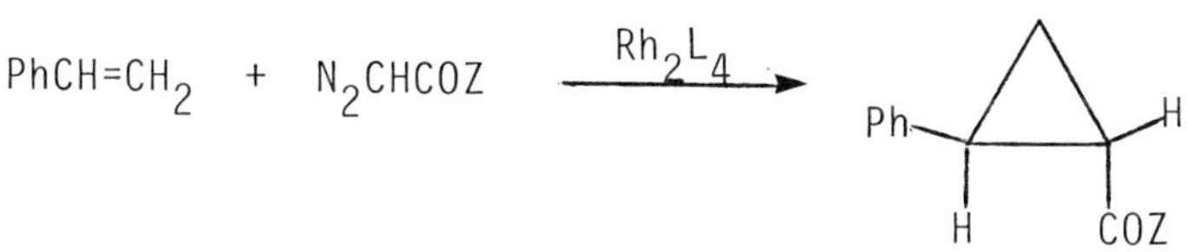

generally high yield

Z = OR, NR^1_2 , L = OCOPr, OCOMe, NHCOMe

I.D.1-3 N.S. Zefirov et al., J. Org. Chem. (USSR), 22, 1400 (1986); X. Creary et al., J. Org. Chem., 52, 3254 (1987); T. Hudlicky et al., Tetrahedron, 43, 5685 (1987).

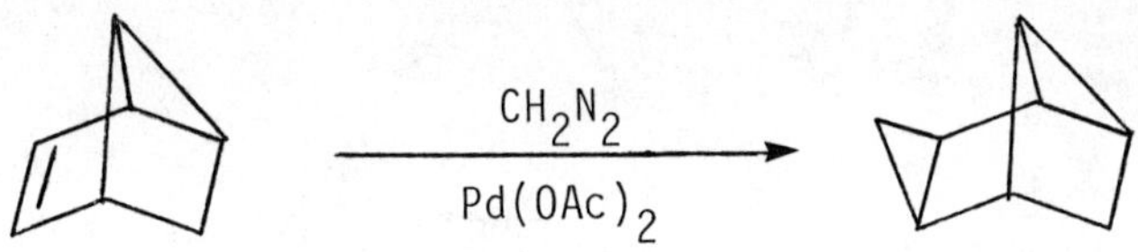

78%

$Cu(OTf)_2$ with an allene and $Cu(acac)_2/CuSO_4$ for intramolecular diazoketone cyclization also used

I.D.1-4 I.E. Dolgii and G.P. Okonnishnikova, J. Org. Chem. (USSR), 22, 1811 (1986).

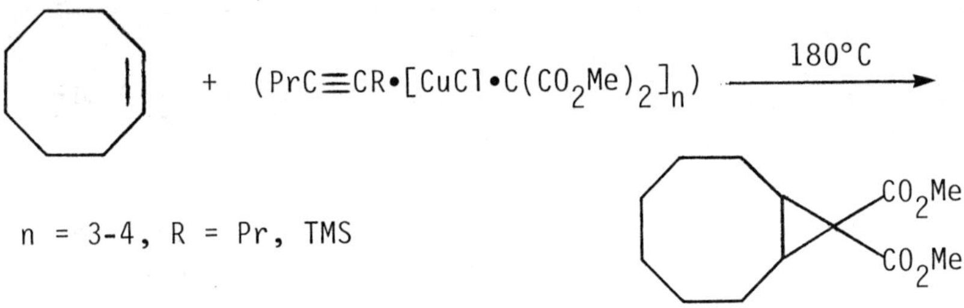

n = 3-4, R = Pr, TMS

30%

I.D.1-5 A. Padwa et al., J. Org. Chem., 52, 4760 (1987); H. Durr, N.J. Turro et al., ibid., 52, 429 and 434 (1987); G.B. Schuster et al., J. Am. Chem. Soc., 109, 5003 (1987); M. Oda et al., Chem. Lett., 93 (1987); P. Ceccherelli, E. Wenkert et al., Synth. Commun., 17, 1261 (1987).

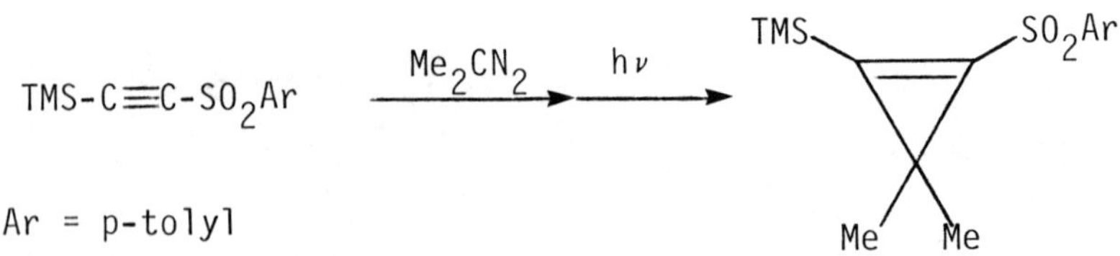

Ar = p-tolyl

99%

similar result with thermolysis or hν /copper oxide

I.D.1-6 E.A. Mash and K.A. Nelson, Tetrahedron, 43, 679 (1987); E.A. Mash et al., Tetrahedron Lett., 28, 1865 (1987); E.A. Mash and J.A. Fryling, J. Org. Chem., 52, 3000 (1987); G. Neef et al., ibid., 52, 4143 (1987); I. Fleming et al., Chem. Commun., 157 (1987).

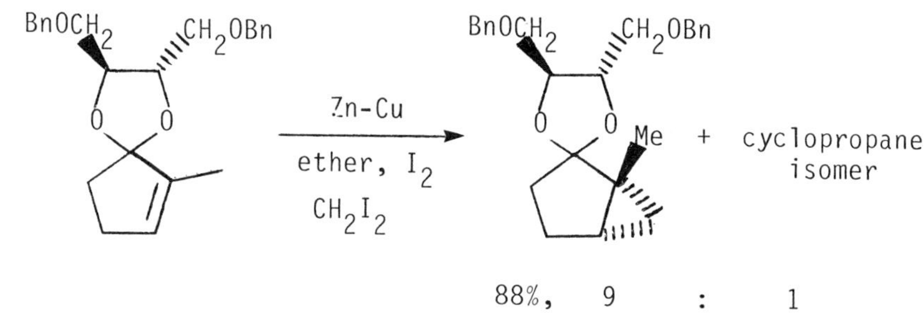

I.D.1-7 G.A. Molander and J.B. Etter, J. Org. Chem., 52, 3942 (1987); T. Imamoto and N. Takiyama, Tetrahedron Lett., 28, 1307 (1987).

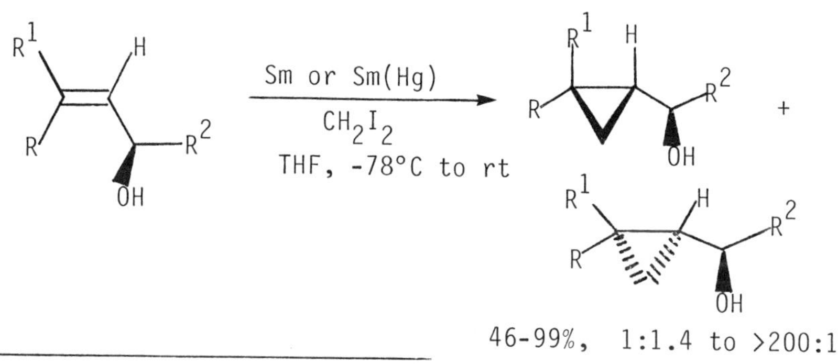

I.D.1-8 K.B. Wiberg and J.V. McClusky, Tetrahedron Lett., 28, 5411 (1987); R.C. Haddon et al., J. Org. Chem., 52, 711 (1987).

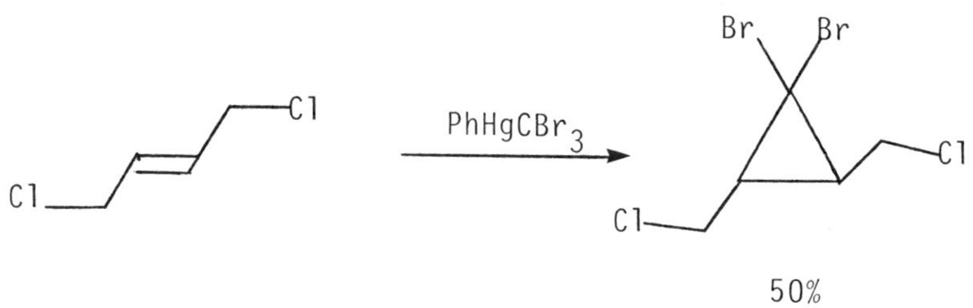

I.D.1-9 R.F. Cunico and K.S. Chu, Synth. Commun., 17, 271 (1987).

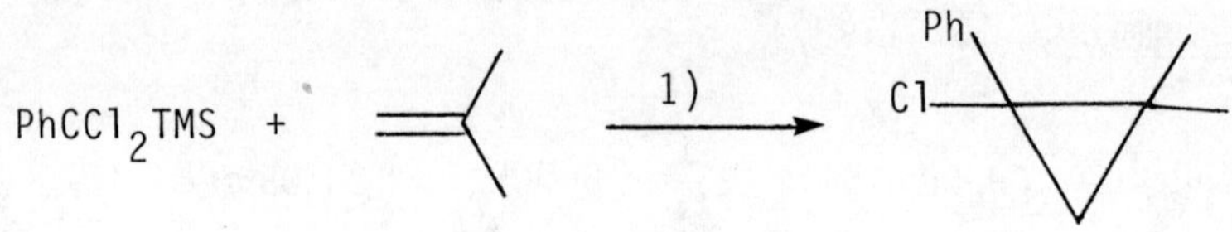

1) KF, 18-C-6, diglyme, 25°C

I.D.1-10 P. Helquist et al., J. Am. Chem. Soc., 109, 3739 (1987).

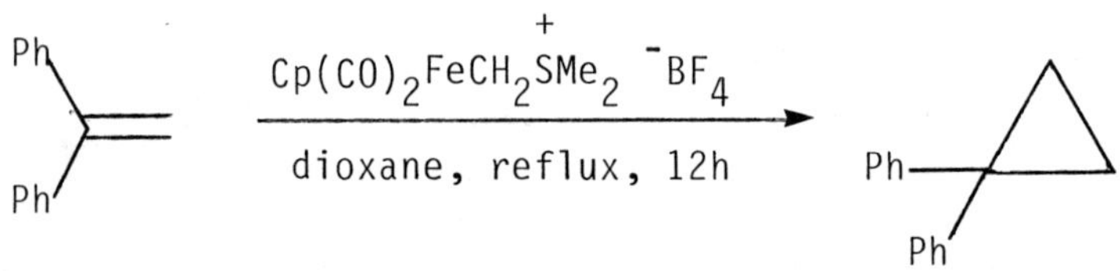

88%

I.D.2. Other Cyclopropanations

I.D.2-1 B. Halton and P.J. Stang, Acc. Chem. Res., 20, 443 (1987).

Review: "Alkylidenecycloproparenes and Related Compounds."

I.D.2-2 A. de Meijere, Chem. Brit., 23, 865 (1987).

Review: "Cyclopropyl Building Blocks for Organic Synthesis."

I.D.2-3 C. Cativiela et al., Synth. Commun., 17, 1549 (1987); M. Franck-Neumann et al., Tetrahedron, 43, 845 and 853 (1987); W.R. Dolbier, Jr. and M.J. Seabury, ibid., 43, 2437 (1987); R.J. Bushby and S. Mann, J. Chem. Soc., Perkin Trans. 1, 2497 (1987).

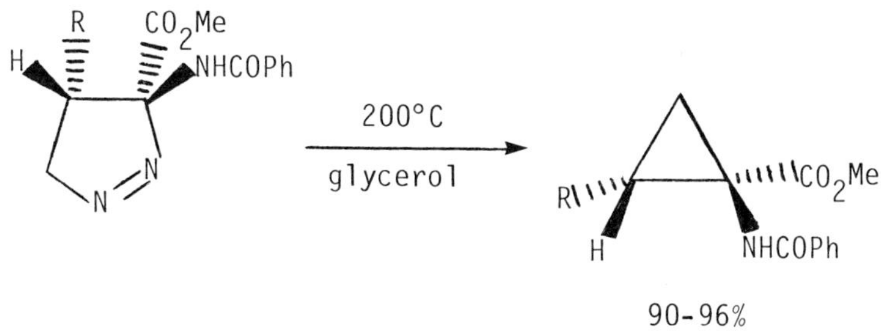

I.D.2-4 C.J.M. Stirling et al., J. Chem. Soc., Perkin Trans. 2, 1253, 1265, 1273 and 1281 (1987).

Eliminative Fission of Cyclopropanes

I.D.2-5 T. Kurihara et al., Synth. Commun., 17, 921 (1987); D.B. Reddy et al., Synthesis, 74 (1987); A. Foucaud et al., ibid., 364 (1987); Y. Takikawa et al., Chem. Lett., 1939 (1987); M. Makosza et al., ibid., 61 (1987); A. de Meijere et al., Chem. Ber., 120, 395 (1987); A.A. Vardapetyan et al., J. Org. Chem. (USSR), 22, 2030 and 2034 (1986); N.N. Magdesieva et al., ibid., 22, 1894 (1986); F. Farina et al., Tetrahedron, 43, 4007 (1987).

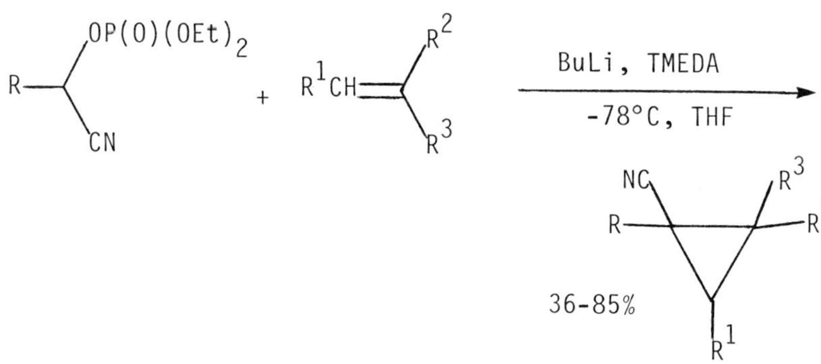

various other bases and leaving groups used

I.D.2-6 V.I. Ognyanov and M. Hesse, *Helv. Chim. Acta*, **70**, 1393 (1987); A.Kh. Khusid, *J. Org. Chem. (USSR)*., **23**, 112 (1987); P.K. Subramanian and R.W. Woodward, *J. Org. Chem.*, **52**, 15 (1987).

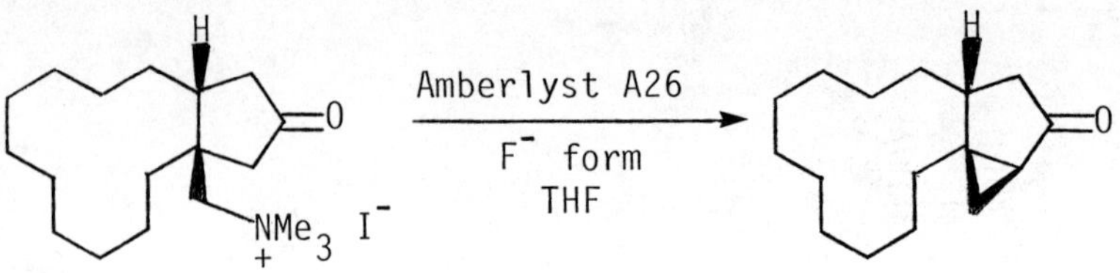

86%

no ring enlarged product

other bases and leaving groups used

I.D.2-7 G. Fournet et al., *Tetrahedron Lett*., **28**, 4533 (1987).

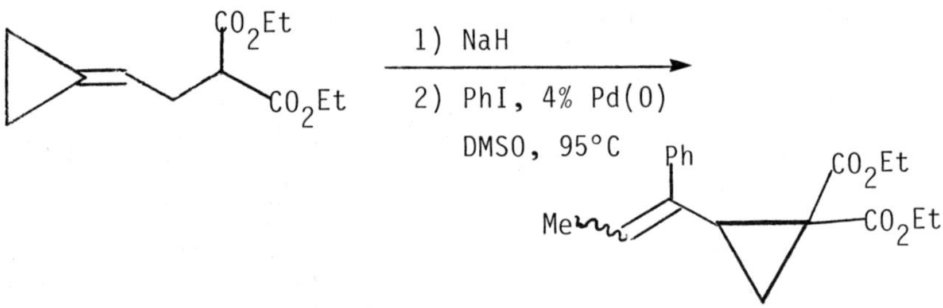

43%

I.D.2-8 P.J. Garratt and J.R. Porter, *J. Chem. Res. (S)*, 108 (1987); L. Toke et al., *Synthesis*, 738 (1987).

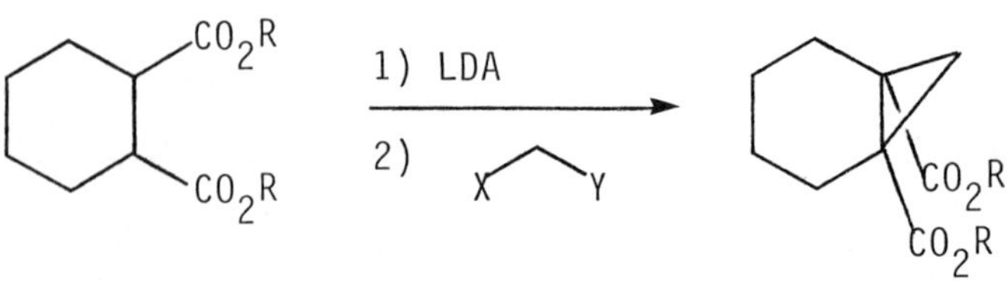

X = Cl, Br, I; Y = Br, I 74%

lower yields with smaller rings

I.D.2-9 J.E. Backvall et al., J. Org. Chem., 52, 5430 (1987); idem, Chem. Commun., 159 (1987); J.P. Genet and J.M. Gaudin, Tetrahedron, 43, 5315 (1987).

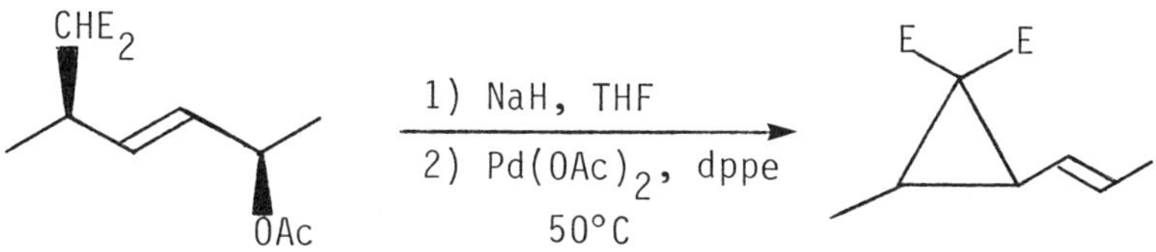

$E = CO_2Me$

>95% (major)

I.D.2-10 J. Barluenga et al., Synthesis, 584 (1987).

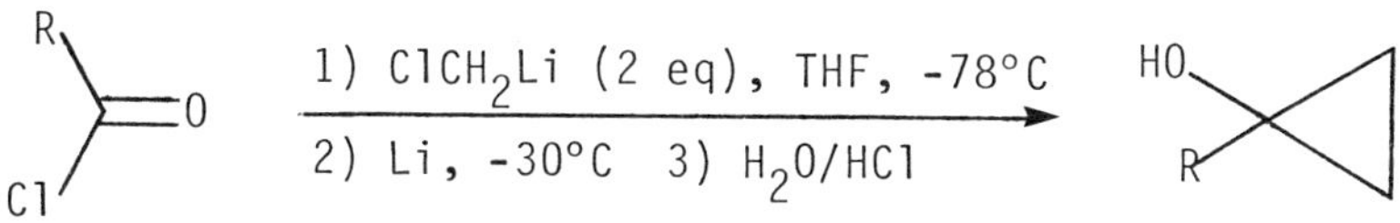

37-80%

I.D.2-11 K. Tanaka et al., Chem. Lett., 809 (1987).

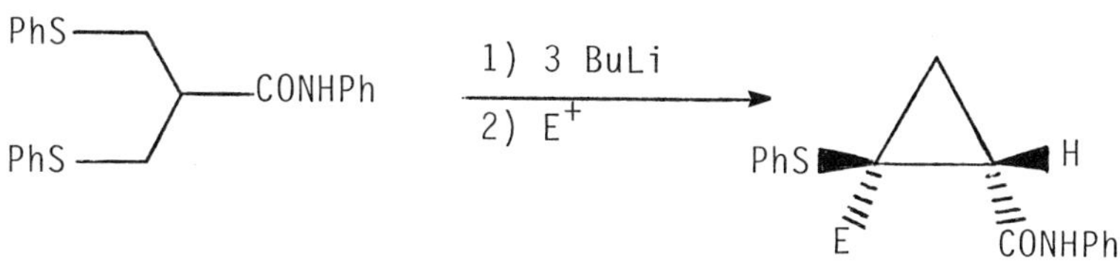

55-86%

I.D.2-12 P.A. Wade et al., J. Am. Chem. Soc., 109, 5452 (1987).

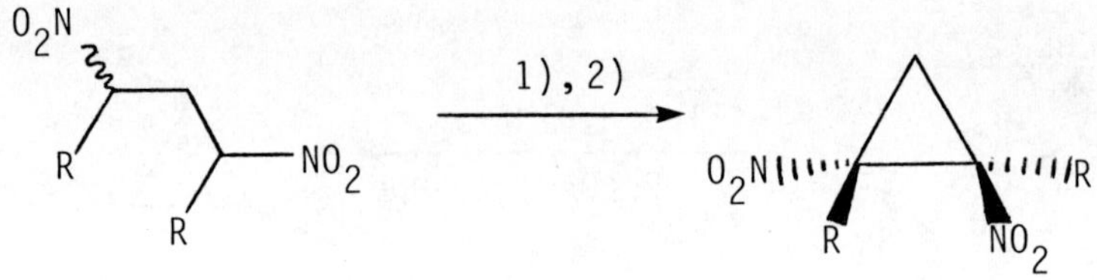

23-36%

1) dimsyl Na, DMSO 2) I_2, DMSO

rate of iodine addition important

I.D.2-13 I. Fleming and M. Rowley, J. Chem. Soc., Perkin Trans. 1, 2259 (1987); T. Sato et al., Synth. Commun., 17, 781 (1987).

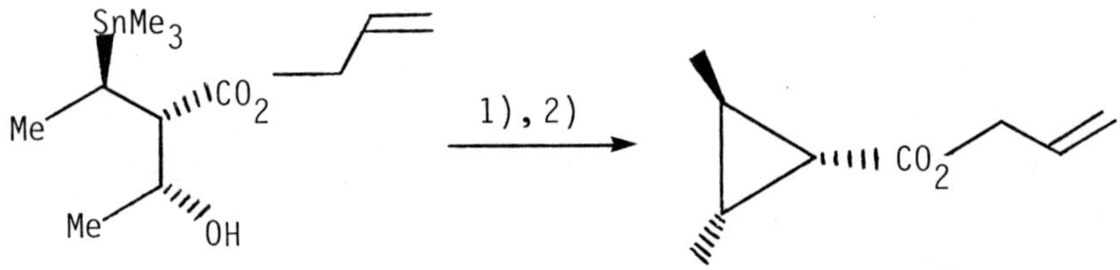

63%

1) MsCl, Et_3N, Et_2O, 0°C 2) 110°C, DMSO

I.D.2-14 E. Vilsmaier et al., Monatsh. Chem., 118, 1147 (1987).

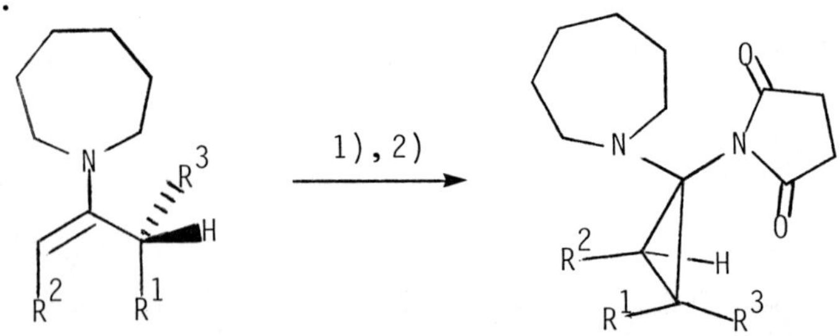

70-74%

1) 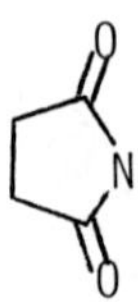 N-$\overset{+}{S}Me_2Cl^-$, 0°C 2) EtN^iPr_2, 20°C

I.D.2-15 M. Brookhart et al., Organometallics, 6, 1141 (1987).

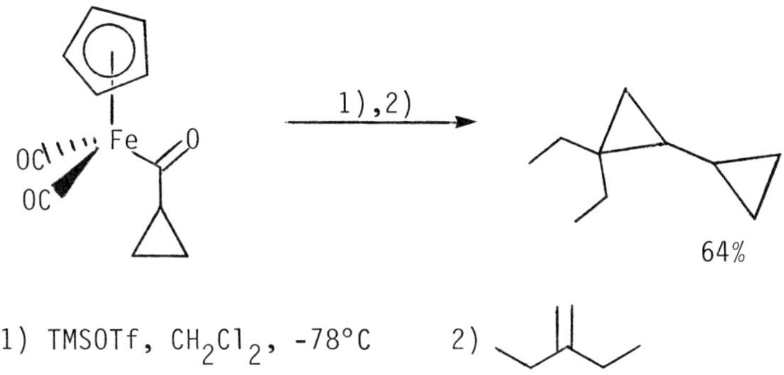

I.D.2-16 T.W. Hanks and P.W. Jennings, J. Am. Chem. Soc., 109, 5023 (1987).

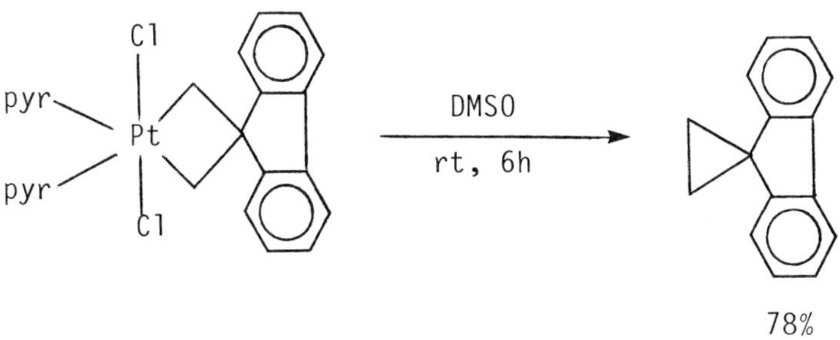

I.D.2-17 P. Binger and B. Biedenbach, Chem. Ber., 120, 601 (1987).

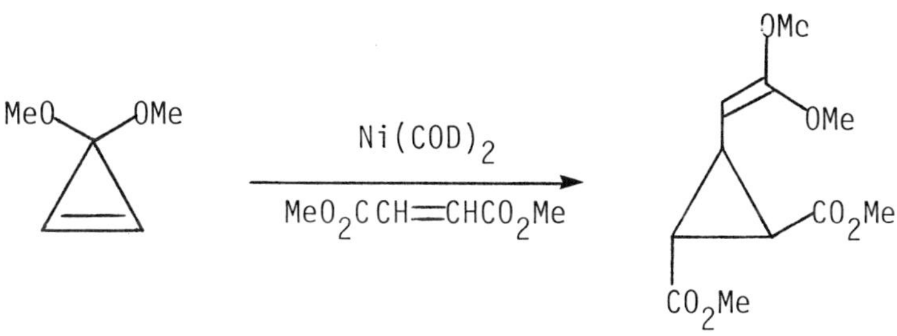

cis-alkene 80%

trans-alkene 59%

I.D.2-18 P.K. Das, M.V. George et al., J. Org. Chem., 52, 5512 (1987); A.G. Schultz et al., J. Am. Chem. Soc., 109, 3991 (1987); D. Armesto et al., Chem. Commun., 1874 (1987).

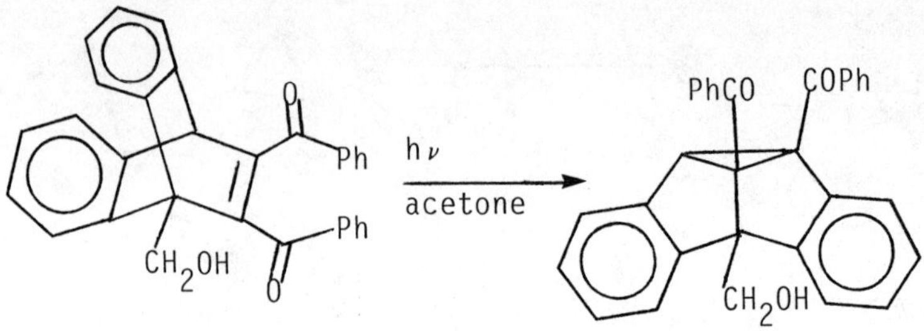

82%

I.D.2-19 E. Piers et al., Tetrahedron Lett., 28, 5075 (1987).

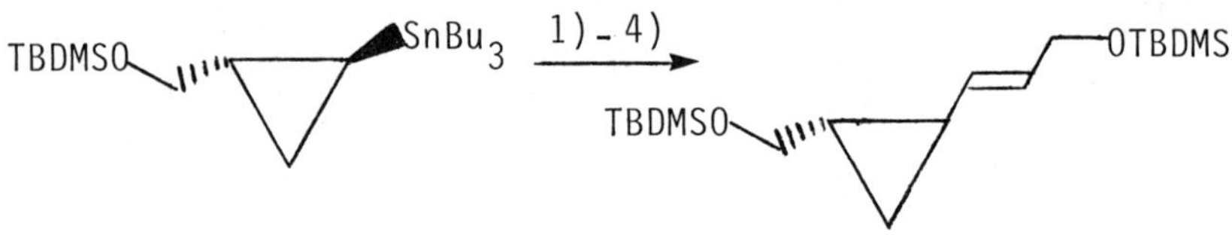

1) BuLi, THF, -78°C 2) $ZnCl_2$, -40°C 3) $Pd(PPh_3)_4$

4) 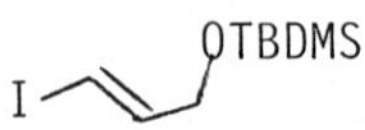

I.E. Thermal and Photochemical Reactions

I.E.1. Cycloadditions

I.E.1-1 M.S. Salakhov and S.A. Ismailov, Russ. Chem. Rev., 55, 1145 (1986).

Review: "Intramolecular [4+2] Cycloaddition."

I.E.1-2 D. Craig, Chem. Soc. Rev., 16, 187 (1987).

Review: "Stereochemical Aspects of the Intramolecular Diels-Alder Reaction."

I.E.1-3 A.J. Fatiadi, Synthesis, 749 (1987).

Review: "Addition and Cycloaddition Reactions of Tetracyanoethylene (TCNE) in Organic Chemistry."

I.E.1-4 J.L. Charlton and M.M. Alauddin, Tetrahedron, 43, 2873 (1987).

Review: "Orthoquinodimethanes."

I.E.1-5 N.L. Bauld et al., Acc. Chem. Res., 20, 371 (1987).

Review: "Cation Radical Pericyclic Reactions."

I.E.1-6 A. Ichihara, Synthesis, 207 (1987).

Review: "Retro-Diels-Alder Strategy in Natural Product Synthesis."

I.E.1-7 S. Misumi, Pure Appl. Chem., 59, 1627 (1987).

Symposium lecture: "Cycloaddition Reactions of Cyclophanes."

I.E.1-8 K.V. Radha Kishan and G.R. Desiraju, J. Org. Chem., 52, 4641 (1987).

"Crystal Engineering a Solid-State Diels-Alder Reaction."

I.E.1-9 R. Sustmann et al., Chem. Ber., 120, 1315 and 1323 (1987).

"Charge Transfer Complexes and Reactivity in Diels-Alder Cycloadditions."

I.E.1-10 P. Nanjappan and A.W. Czarnik, J. Am. Chem. Soc., 109, 1826 (1987).

"Metal Ion Catalyzed Reactions of Acrylonitrile, Acrylamide and Ethyl Acrylate by Way of their Diels-Alder Cycloadducts."

I.E.1-11 H.-J. Schneider and N.K. Sangwan, Angew. Chem., Int. Ed. Engl., 26, 896 (1987).

"Changes of Stereoselectivity in Diels-Alder Reactions by Hydrophobic Solvent Effects and by β-Cyclodextrin."

I.E.1-12 N.L. Bauld et al., J. Am. Chem. Soc., 109, 4960 (1987).

"Mechanistic Diagnosis of Aminium Salt Initiated Diels-Alder Cycloadditions in the Diene/Diene Format."

I.E.1-13 T. Koizumi et al., Tetrahedron Lett., 28, 3689 (1897).

"A Disproof of Kahn-Hehre's Proposal on the Ground State Conformations and the Steric Course of the Diels-Alder Reaction of Vinyl Sulfoxides."

I.E.1-14 J.J. Gajewski et al., J. Am. Chem. Soc., 109, 5545 (1987).

"Transition State Structure Variation in the Diels-Alder Reaction from Secondary Deuterium Kinetic Isotope Effects: The Reaction of a Nearly Symmetrical Diene and Dienophile is Nearly Synchronous."

I.E.1-15 J. Mlcoch and E. Steckhan, Tetrahedron Lett., 28, 1081 (1987).

"Electrochemically Induced [4+2]-Cycloadditions-A Mechanistic Interpretation of the Cation Radical Diels-Alder Reaction Based on Preparative Results."

I.E.1-16 R.M. Ottenbrite et al., J. Org. Chem., 52, 391 (1987).

"An Ab Initio Molecular Orbital Evaluation of Lewis Acid Catalysis on Diels-Alder Reactions of Acrolein."

I.E.1-17 M.A. Fox et al., J. Org. Chem., 52, 1469 (1987).

"Steric Effects vs. Secondary Orbital Overlap in Diels-Alder Reactions. MNDO and AM1 Studies."

I.E.1-18 R.N. Warrener et al., Tetrahedron Lett., 28, 6503 and 6507 (1987); W. Grimme et al., Angew. Chem., Int. Ed. Engl., 26, 1268 (1987); T. Eicher and M. Abel, Synthesis, 542 (1987).

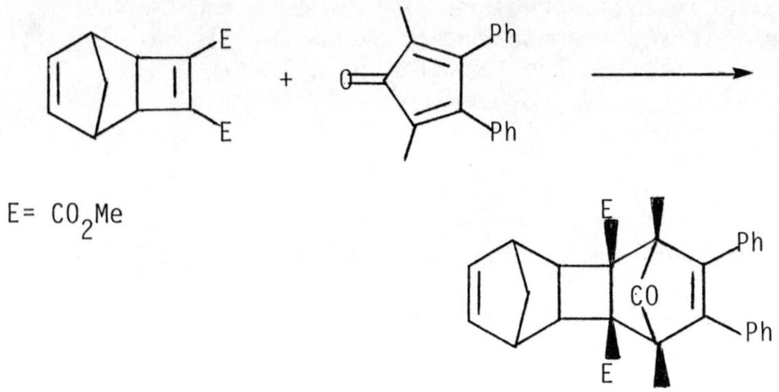

Diels-Alder "Transfer Technology" - stable molecules as synthetic equivalents for synthons or highly reactive/ unstable molecules

---

I.E.1-19 R.V. Williams and C.-L.A. Sung, Chem. Commun., 590 (1987); V.K. Singh et al., Synth. Commun., 17, 115 (1987); S.S. Roy and S. Ghosh, Tetrahedron, 43, 5995 (1987); J.M. Coxon, M.J. O'Connell and P.J. Steel, J. Org. Chem., 52, 4726 (1987); Y. Langlois et al., Tetrahedron Lett., 28, 6183 (1987); N. Ono et al., J. Chem. Soc., Perkin Trans. 1, 1929 (1987).

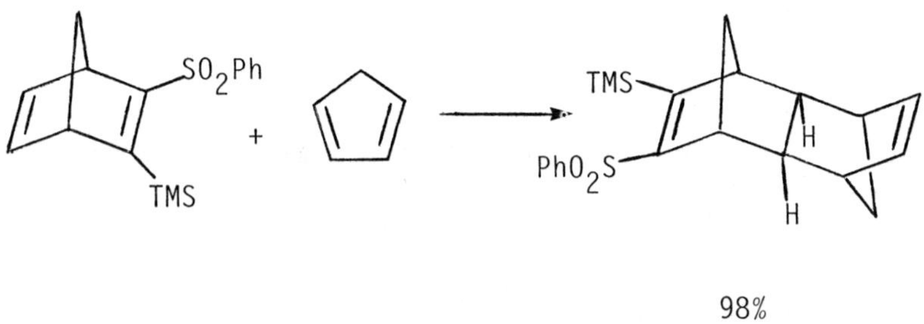

various other dienes and dienophiles also give endo selectivity

I.E.1-20 M. Vaultier, R.W. Hoffmann et al., Tetrahedron Lett., 28, 4169 (1987); N. Katagiri, C. Kaneko et al., Chem. Commun., 1422 (1987).

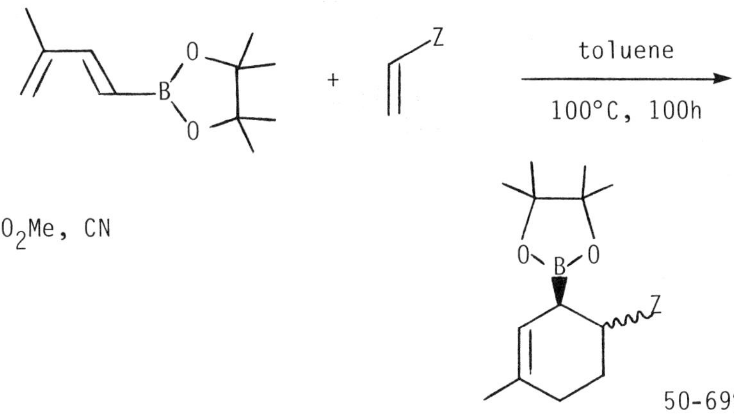

Z = CO_2Me, CN

50-69%

endo : exo 1 : 1.8 to 1 : 1.6

I.E.1-21 B. Potthoff and E. Breitmaier, Chem. Ber., 120, 255 (1987); H. Fillion et al., Tetrahedron Lett., 28, 4529 (1987).

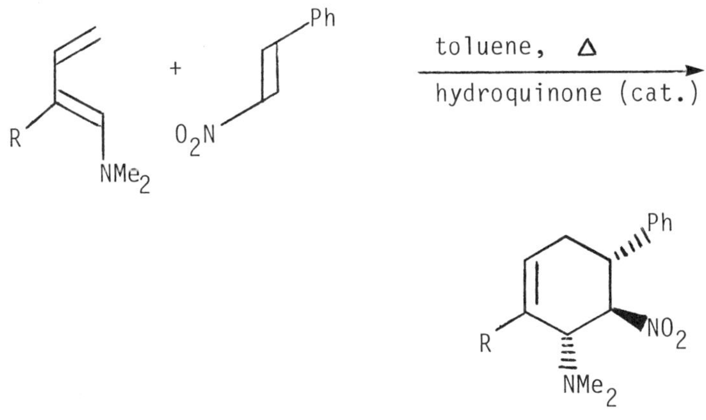

26-71%

I.E.1-22 B. Demarchi and P. Vogel, Tetrahedron Lett., 28, 2239 (1987); J. Tamariz et al., ibid., 28, 865 (1987); R.V. Bonnert and P.R. Jenkins, ibid., 28, 692 (1987); H.-J. Liu and E.N.C. Browne, Can. J. Chem., 65, 1262 (1987); C. Cativiela et al., ibid., 65, 2182 (1987); C.R. Johnson and J.F. Kadow, J. Org. Chem., 52, 1493 (1987); M. Asaoka et al., Bull. Chem. Soc. Jpn., 60, 2259 (1987); H.-J. Liu and W.M. Feng, Synth. Commun., 17, 1777 (1987); H. Hagiwara, H. Uda et al., Chem. Commun., 1333 (1987); J. Jurczak et al., Tetrahedron, 43, 3063 (1987).

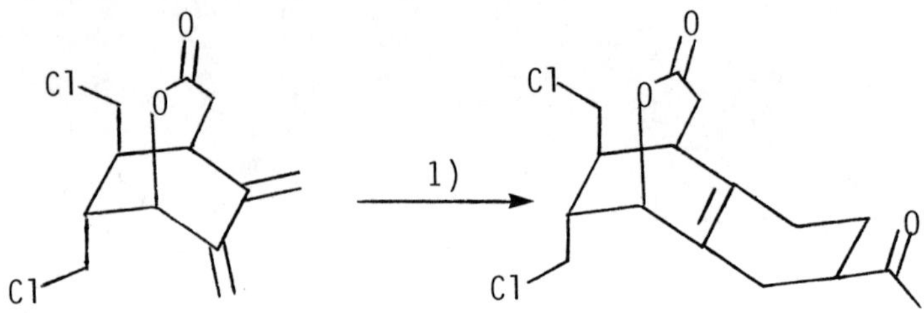

1) 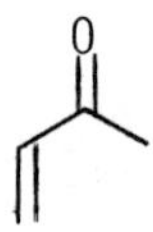 , $BF_3 \cdot OEt_2$, -78°C, CH_2Cl_2, 3h

a wide variety of dienes, dienophiles and Lewis acids used

I.E.1-23 L. Ghosez et al., Tetrahedron Lett., 28, 397 (1987); F. Lehrich and H. Hopf, ibid., 28, 2697 (1987); A. Battaglia et al., J. Org. Chem., 52, 3289 (1987).

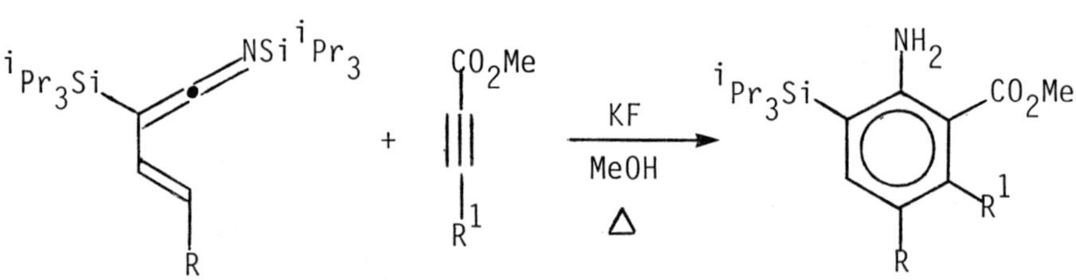

52-89%

similar reactions with allene analogues

I.E.1-24 H. Mayr and U.W. Heigl, Chem. Commun., 1804 (1987).

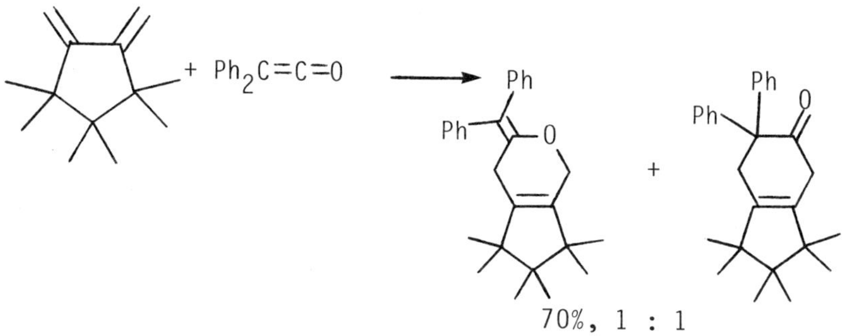

I.E.1-25 B. Fuchs et al., Synthesis, 850 (1987); E. Piers and R.W. Friesen, Can. J. Chem., 65, 1681 (1987); Y. Thebtaranonth et al., Tetrahedron Lett., 28, 6671 (1987).

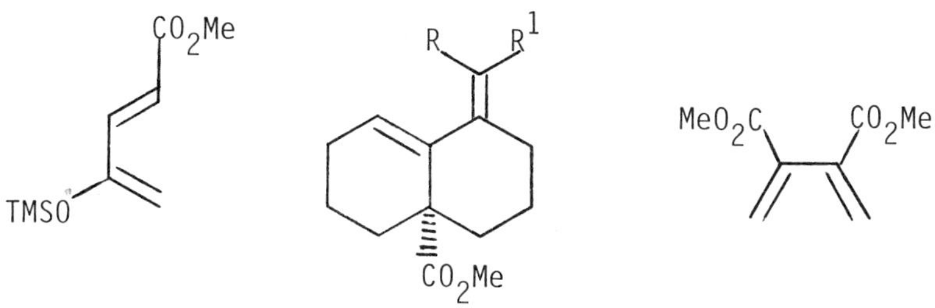

Diels-Alder dienes

I.E.1-26 J. Ipaktschi et al., Tetrahedron Lett., 28, 3467 (1987); D.J. Chadwick and A. Plant, ibid., 28, 6085 (1987); J. Mann et al., Tetrahedron, 43, 4207 (1987).

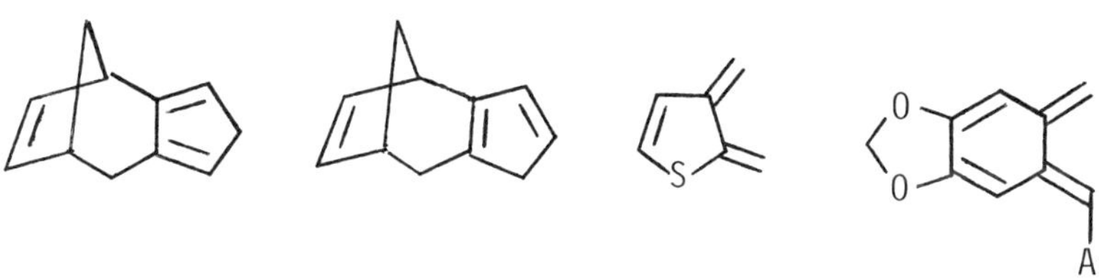

Diels-Alder dienes

I.E.1-27 M. Sakaino and J. Meinwald, Tetrahedron Lett., 28, 3201 (1987); P. Vogel et al., Helv. Chim. Acta, 70, 1231 and 1638 (1987).

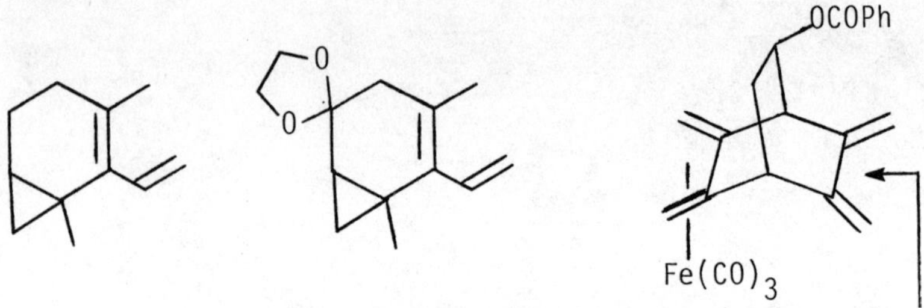

Diels-Alder dienes

reaction here

I.E.1-28 F. Bickelhaupt et al., Tetrahedron, 43, 2955 (1987); A. Lubineau and Y. Queneau, J. Org. Chem., 52, 1001 (1987).

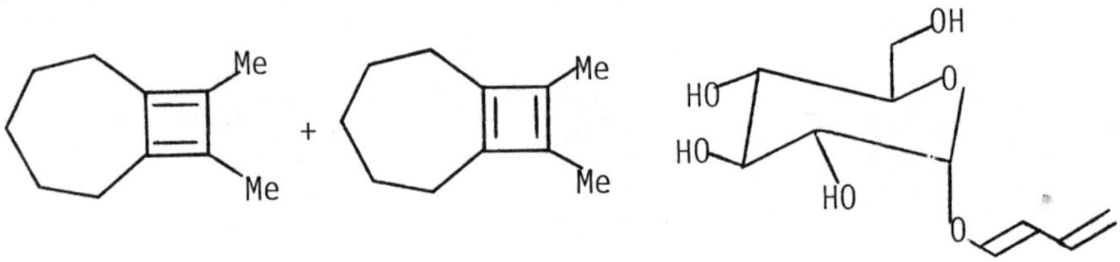

for reaction under aqueous conditions

Diels-Alder dienes

I.E.1-29 Z. Valenta et al., Can. J. Chem., 65, 154 (1987); G.S. Bates, M.D. Fryzuk and C. Stone, ibid., 65, 2612 (1987); G. Cardinale et al., Rec. Trav. Chim., 106, 62 (1987).

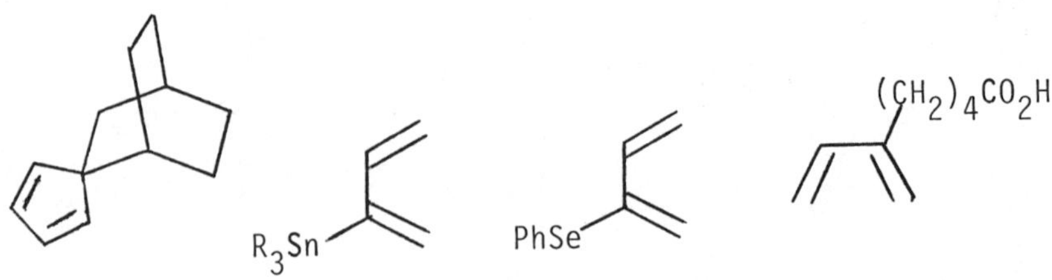

Diels-Alder dienes

I.E.1-30 D.W. Jones and A.M. Thompson, *Chem. Commun.*, 1797 (1987); F. Effenberger et al., *Chem. Ber.*, *120*, 1339 and 1347 (1987); G.R. Tian, H. Takeshita et al., *Chem. Lett.*, 1557 (1987).

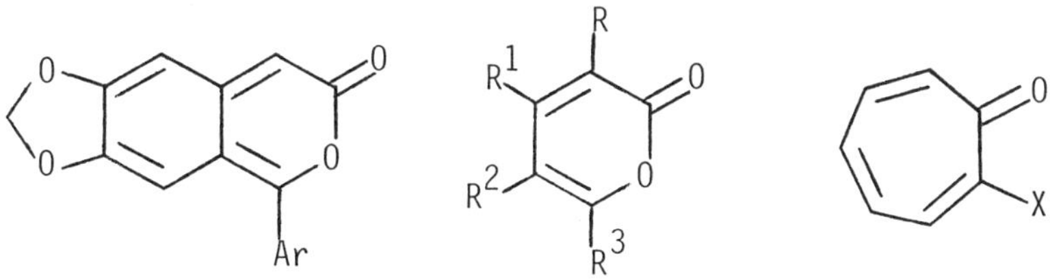

Diels-Alder dienes

I.E.1-31 F. Kienzle and R.E. Minder, *Helv. Chim. Acta*, *70*, 1537 (1987); M.H. Nantz and P.L. Fuchs, *J. Org. Chem.*, *52*, 5298 (1987); S.R. Samanta and A.K. Mukherjee, *Ind. J. Chem.*, *26B*, 26 (1987); V. Singh et al., *ibid*, *26B*, 301 (1987).

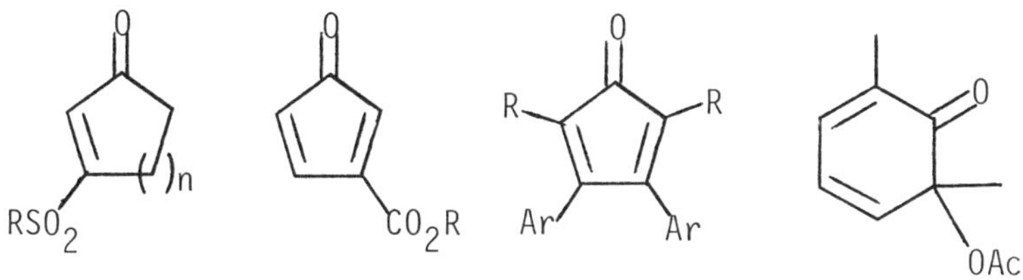

Diels-Alder dienophiles

I.E.1-32 R.W. Read et al., *Aust. J. Chem.*, *40*, 1831 (1987); A.B. Holmes et al., *J. Chem. Soc., Perkin Trans. 1*, 1507 (1987); U. Gruseck and M. Heuschmann, *Tetrahedron Lett.*, *28*, 2681 and 6027 (1987); idem, *Chem. Ber.*, *120*, 2065 (1987).

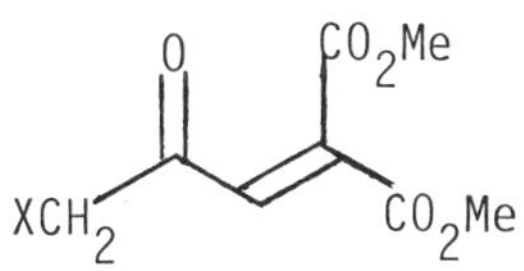

X = H, Br

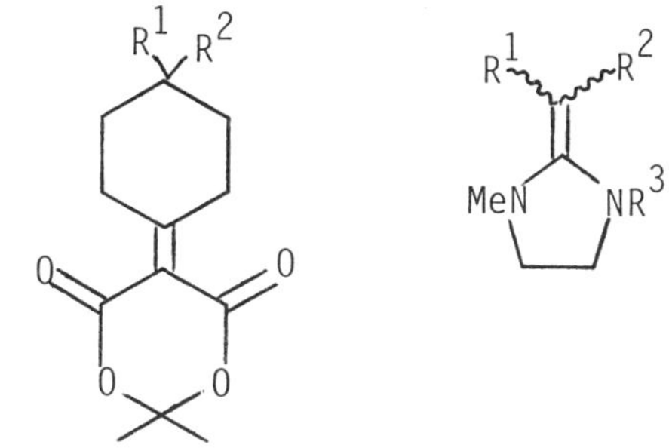

Diels-Alder dienophiles

I.E.1-33 O. De Lucchi et al., Tetrahedron Lett., 28, 107 (1987); T.K.M. Shing and P. Lloyd-Williams, Chem. Commun., 423 (1987).

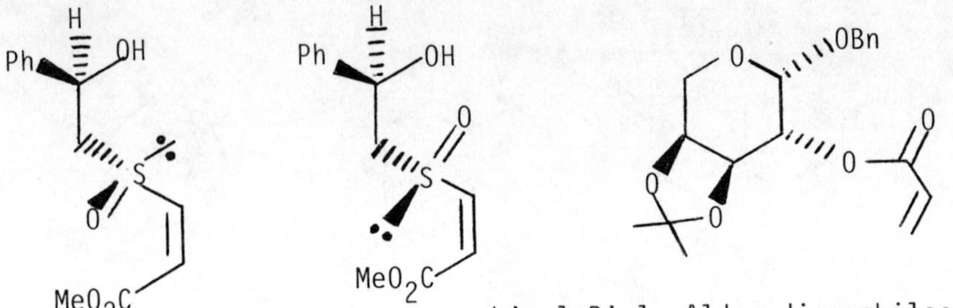

chiral Diels-Alder dienophiles

I.E.1-34 C.M. Lukehart and L. Sacksteder, Organometallics, 6, 150 (1987); M.A. Battiste et al., J. Organomet. Chem., 330, 437 (1987).

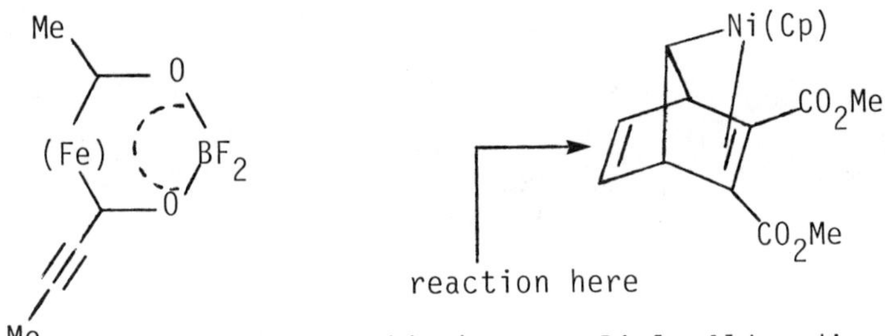

[Fe] = (η-C_5H_5)(OC)Fe Diels-Alder dienophiles

I.E.1-35 T. Kametani et al., J. Chem. Soc., Perkin Trans. 1, 1373 (1987); C. Kaneko et al., Chem. Pharm. Bull., 35, 3647 (1987); M. Oda et al., Bull. Chem. Soc. Jpn., 60, 3801 (1987); J. Luo and H. Hart, J. Org. Chem., 52, 4833 (1987).

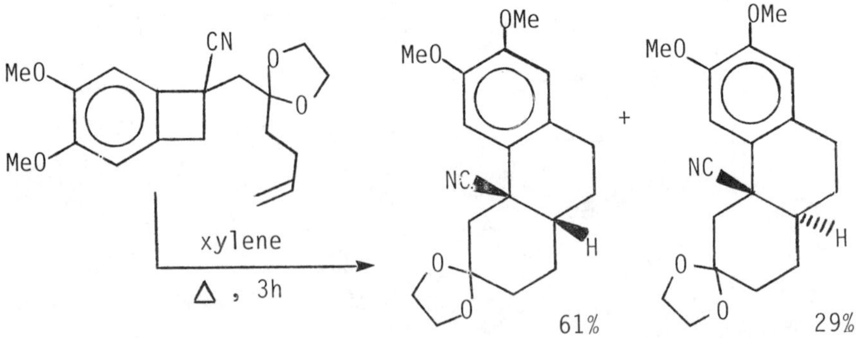

intermolecular examples also studied

I.E.1-36 G.H. Posner et al., J. Org. Chem., 52, 4836 (1987); J.S. Swenton et al., ibid., 52, 713 (1987); A. Ichihara, K. Murakami and S. Sakamura, Tetrahedron, 43, 5245 (1987); W.C. Christopfel and L.L. Miller, ibid., 43, 3681 (1987); D.L. Boger and M.D. Mullican, Org. Synth., 65, 98 (1987); W. Adam et al., J. Am. Chem. Soc., 109, 7074 (1987); Y. Tamura et al., Tetrahedron Lett., 28, 5709 (1987); idem, Chem. Pharm. Bull., 35, 1405 (1987); R.A. Russell, R.N. Warrener et al., Aust. J. Chem., 40, 311 (1987); A.R. Mehendale et al., Ind. J. Chem., 26B, 205 (1987); H. Kotsuki et al., Heterocycles, 26, 2347 (1987).

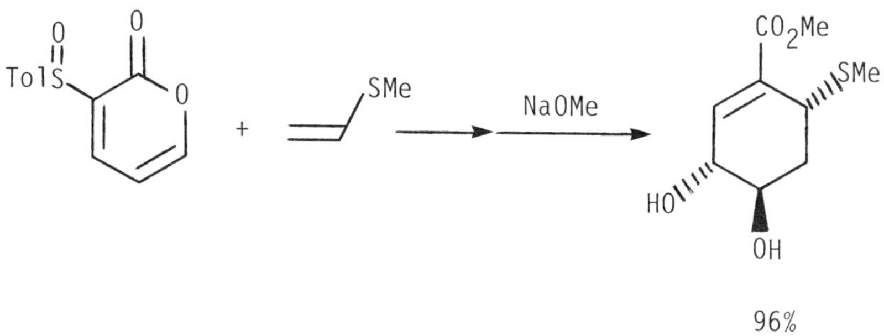

similar reactions with γ-lactones and cyclic anhydrides

I.E.1-37 M.R. Bryce et al., Tetrahedron Lett., 28, 577 (1987); G.P. Stahly, Synth. Commun., 17, 1053 (1987); K. Mackenzie et al., Tetrahedron, 43, 5981 (1987); H. Prinzbach, E. Heilbronner et al., Helv. Chim. Acta, 70, 1816 (1987).

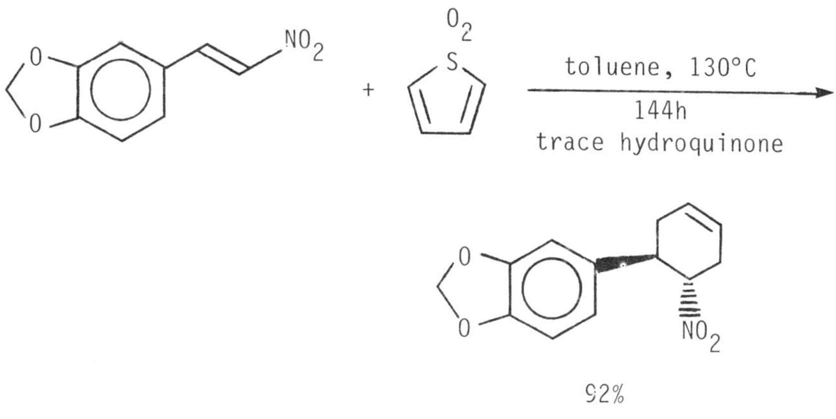

I.E.1-38 V. Guay and P. Brassard, <u>Synthesis</u>, 294 (1987); T.C. McKenzie et al., <u>Tetrahedron Lett.</u>, <u>28</u>, 5435 (1987).

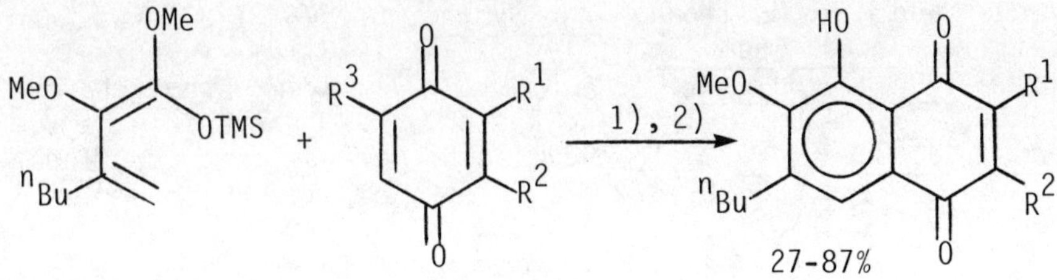

1) C_6H_6 or CH_2Cl_2, rt 2) THF, 6N HCl or HBr, rt

I.E.1-39 V.K. Singh et al., <u>Synth. Commun.</u>, <u>17</u>, 593 and 1103 (1987).

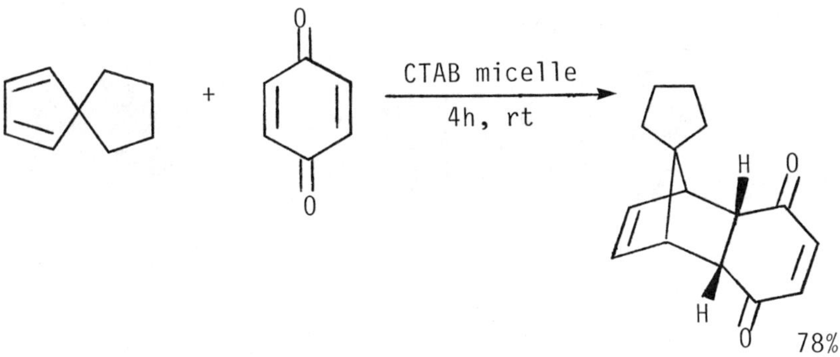

I.E.1-40 R.J. Giguere, G. Majetich et al., <u>Tetrahedron Lett.</u>, <u>28</u>, 6553 (1987); G. Jenner, <u>ibid.</u>, <u>28</u>, 3927 (1987); S. Kirchmeyer and A. de Meijere, <u>Chem. Ber.</u>, <u>120</u>, 2083 (1987).

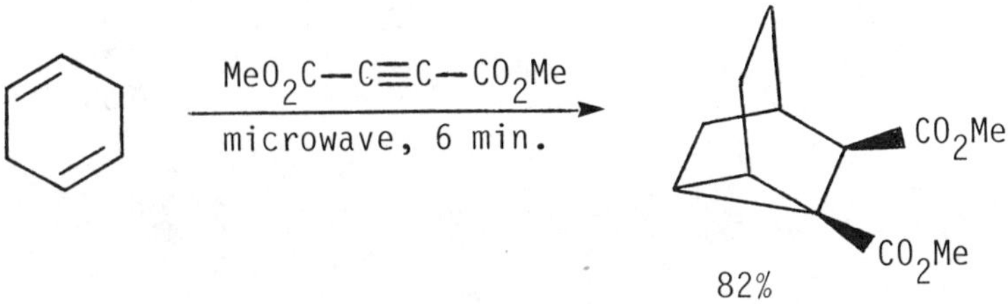

conventional route : 185°C, 20h ⟶ 40%

similar results with pressure or TCNE, Δ

I.E.1-41 S. Toma et al., Coll. Czech. Chem. Commun., 52, 391 (1987).

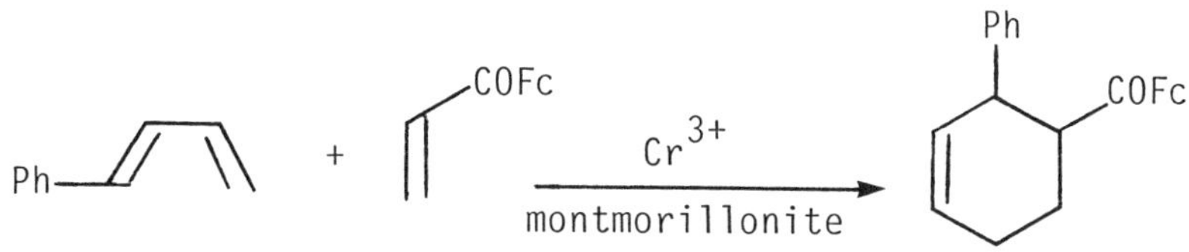

95%

Fc = ferrocenyl

66% with $AlCl_3$

I.E.1-42 L.A. Paquette and U.S. Racherla, J. Org. Chem., 52, 3250 (1987); T.A. Engler et al., Tetrahedron Lett., 28, 5267 (1987).

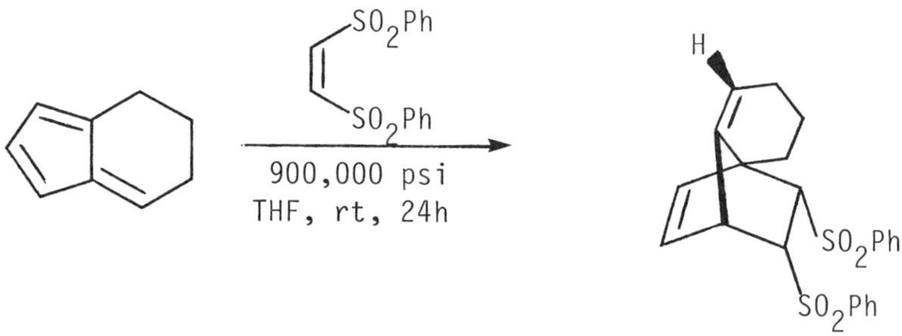

60%

I.E.1-43 I. Matsuda et al., Tetrahedron Lett., 28, 3361 (1987).

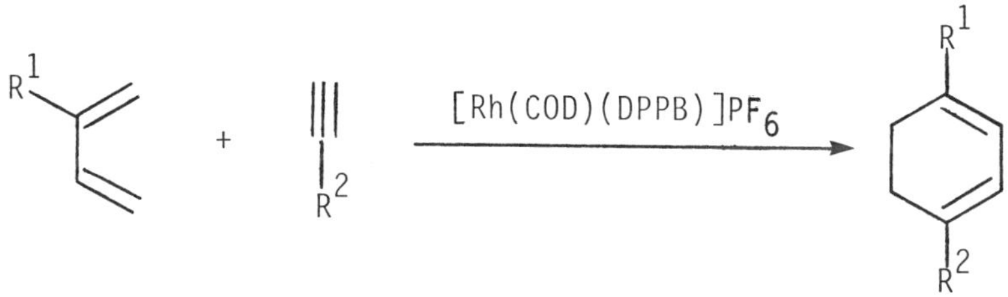

44-85%

I.E.1-44 W.H. Hersh et al., J. Org. Chem., 52, 5293 (1987).

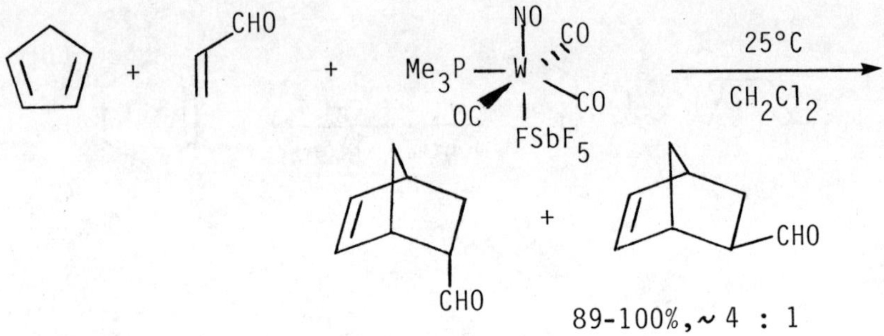

89-100%, ~ 4 : 1

I.E.1-45 A.P. Kozikowski et al., J. Am. Chem. Soc., 109, 5167 (1987).

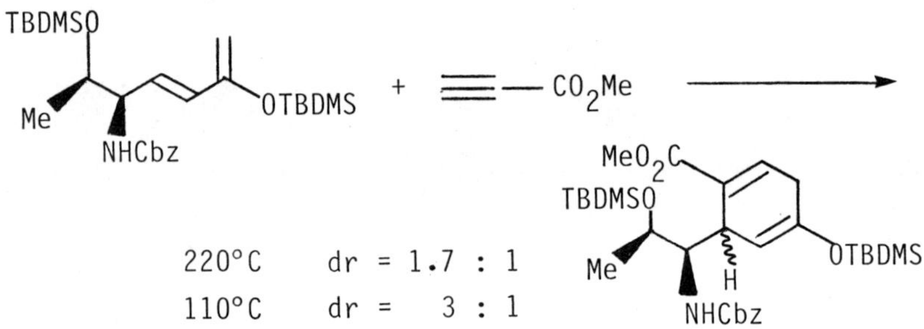

220°C	dr = 1.7 : 1
110°C	dr = 3 : 1
rt, 6 Kbar	dr = 10 : 1

good yield

I.E.1-46 K. Koga et al., Tetrahedron Lett., 28, 5687 (1987); G. Bir and D. Kaufmann, ibid., 28, 777 (1987); C.J. Northcott and Z. Valenta, Can. J. Chem., 65, 1917 (1987); C. Chapius and J. Jurczak, Helv. Chim. Acta, 70, 436 (1987).

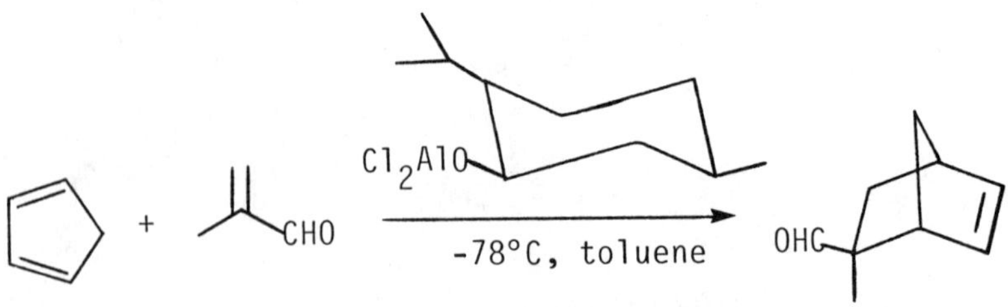

80%, 61% ee

other chiral Lewis acids also used

I.E.1-47 G. Helmchen et al., Angew. Chem., Int. Ed. Engl., 26, 1143 (1987); D.A. Evans et al., ibid., 26, 1184 (1987); H. Kunz et al., ibid., 26, 267 (1987); G. Helmchen et al., Tetrahedron Lett., 28, 183 (1987); H. Yamamoto et al., ibid., 28, 5841 (1987); T. Koizumi et al., Chem. Lett., 185 (1987); H.-D. Scharf et al., Chem. Ber., 120, 1737 (1987); E. Jakel and M.P. Schneider, Chem. Commun., 733 (1987); D. Dopp and M. Pies, ibid., 1734 (1987).

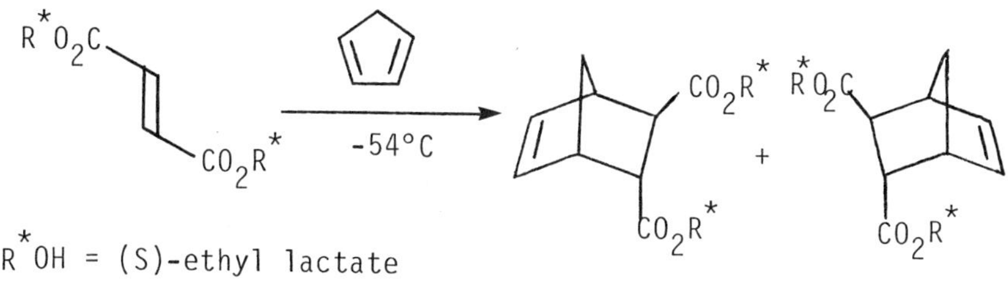

R^*OH = (S)-ethyl lactate

99%, 98 : 2

with $TiCl_4$ (CH_2Cl_2) 99%, 5 : 95

various other chiral dienophiles employed

I.E.1-48 W. Grimme and G. Wiechers, Tetrahedron Lett., 28, 6035 (1987); A. Ichihara, S. Sakamura et al., ibid., 28, 1175 (1987); J.K. Sutherland et al., Chem. Commun., 1449 and 1728 (1987); Z.-Y. Liu et al., ibid., 1868 (1987); K.A. Parker and T. Iqbal, J. Org. Chem., 52, 4369 (1987); D. Caine and B. Stanhope, Tetrahedron, 43, 5545 (1987); M. Yoshida and K. Kanematsu, Heterocycles, 26, 3093 (1987); L.S. Trifonov and A.S. Orahovats, Helv. Chim. Acta, 70, 1732 (1987).

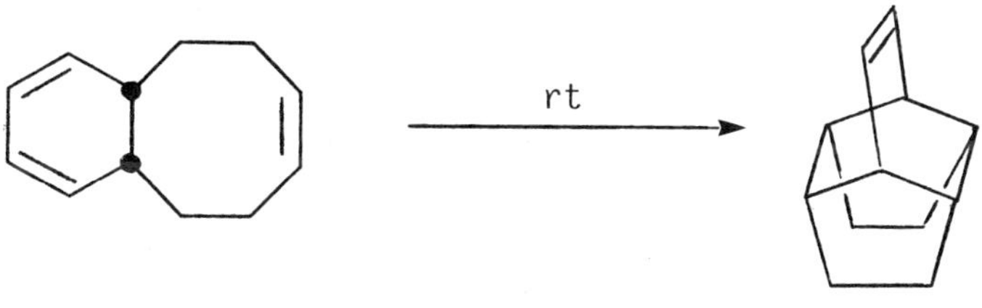

no yield given

faster than intermolecular reaction
with maleic anhydride
various other intramolecular Diels-Alder reactions examined

I.E.1-49 W.R. Roush and J.W. Coe, Tetrahedron Lett., 28, 931 (1987); P. Deslongchamps et al., ibid., 28, 5249, 5253 and 5255 (1987); E.J. Corey and P.A. Magriotis, J. Am. Chem. Soc., 109, 287 (1987); A.G. Fallis et al., Can. J. Chem., 65, 114 (1987).

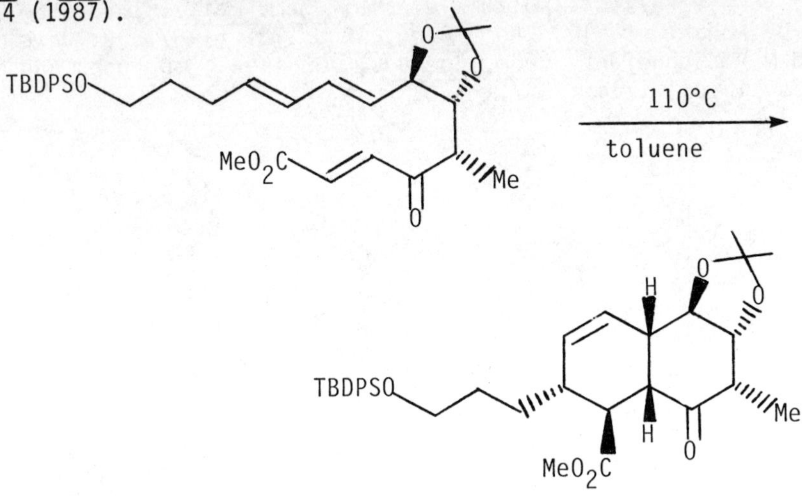

>78%

I.E.1-50 W.R. Roush et al., Tetrahedron Lett., 28, 2447 (1987); G. Stork and N.A. Saccomano, ibid., 28, 2087 (1987); J.A. Marshall et al., J. Org. Chem., 52, 1236 (1987); idem, J. Am. Chem. Soc., 109, 1186 (1987); E. Winterfeldt et al., Chem. Ber., 120, 2007 (1987).

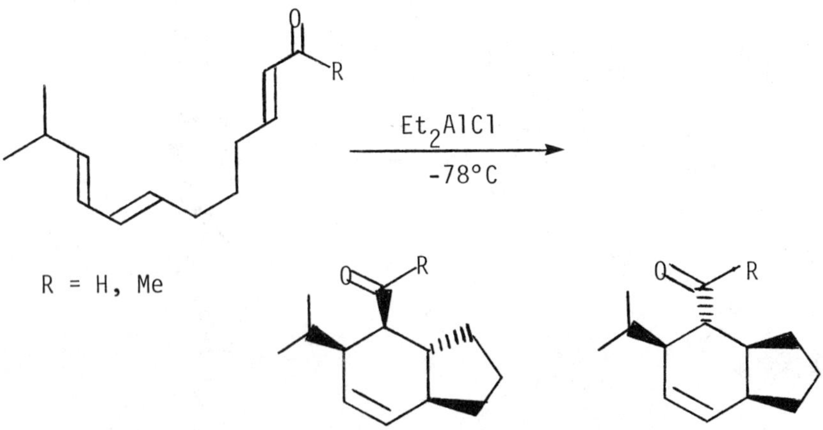

53%, >99 : 1

Me_2AlCl, $EtAlCl_2$ and TFA also used similarly

I.E.1-51 B. Harirchian and N.L. Bauld, <u>Tetrahedron Lett.</u>, <u>28</u>, 927 (1987).

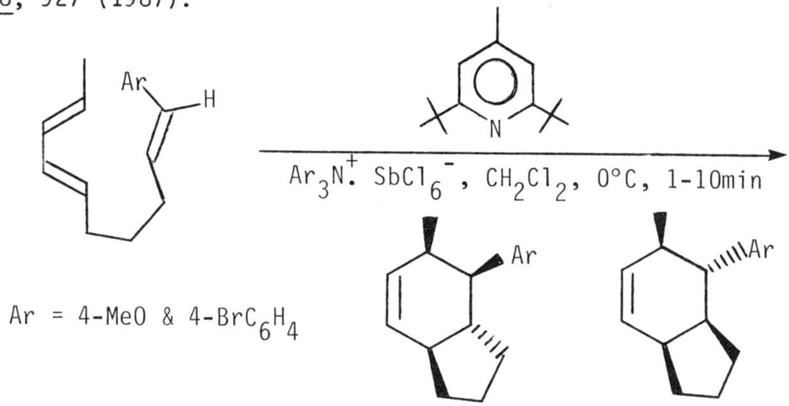

83%, 98 : 2

first intramolecular cation radical Diels-Alder

I.E.1-52 H.C. van der Plas et al., <u>Rec. Trav. Chim.</u>, <u>106</u>, 547 (1987).

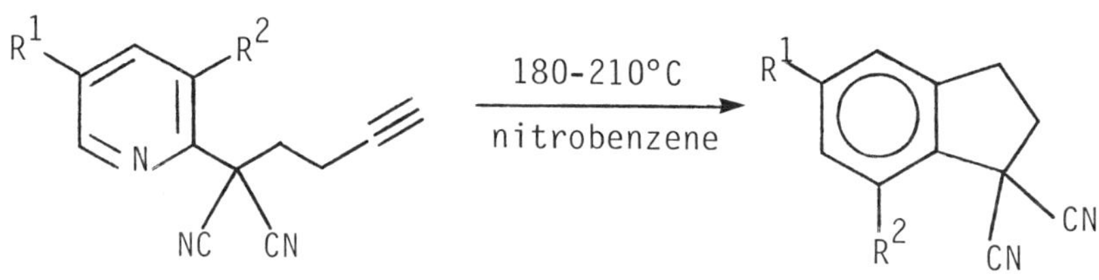

72-82%

I.E.1-53 H.-J. Wu and K. Pan, <u>Chem. Commun.</u>, 899 (1987).

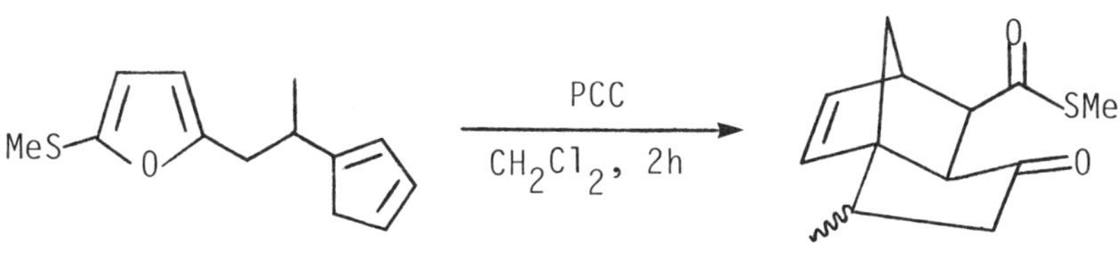

40%

I.E.1-54 G.A. Kraus and S.H. Woo, J. Org. Chem., 52, 4841 (1987); K.J. Shea et al., J. Am. Chem. Soc., 109, 447 (1987).

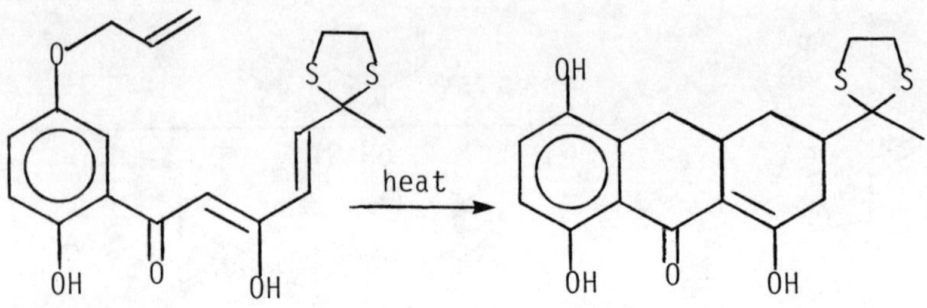

85%

tandem Claisen-Diels-Alder

I.E.1-55 P.G. Gassman et al., J. Am. Chem. Soc., 109, 2182 (1987); P.G. Gassman and D.A. Singleton, Tetrahedron Lett., 28, 5969 (1987).

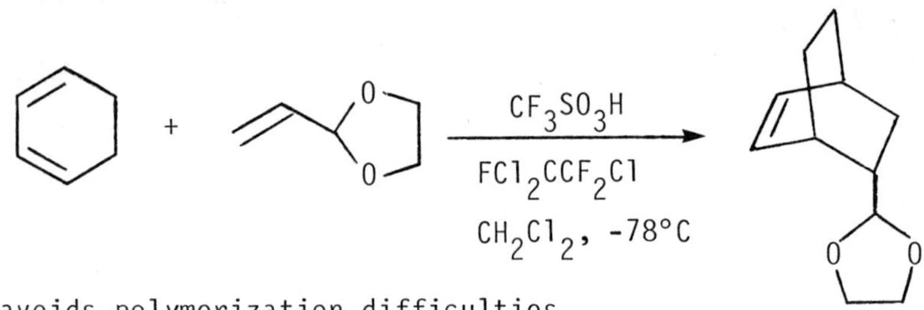

avoids polymerization difficulties with acrolein itself

94%

I.E.1-56 M.G. Banwell and J.H. Knight, Chem. Commun., 1082 (1987); Y. Apeloig et al., Tetrahedron Lett., 28, 5917 (1987).

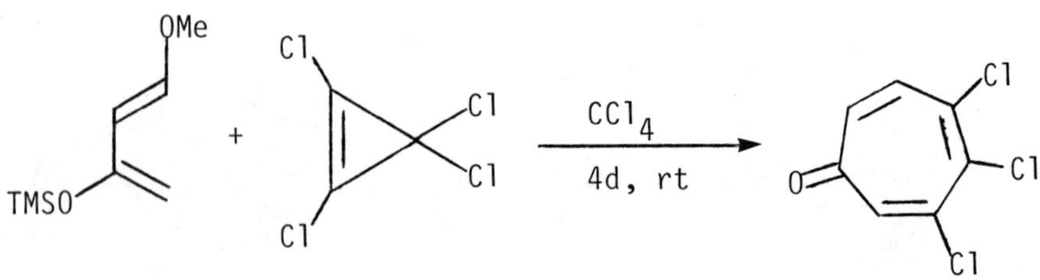

56%

I.E.1-57 D. Spitzner, A. de Meijere, A. Simon et al., Tetrahedron, 43, 3213 (1987).

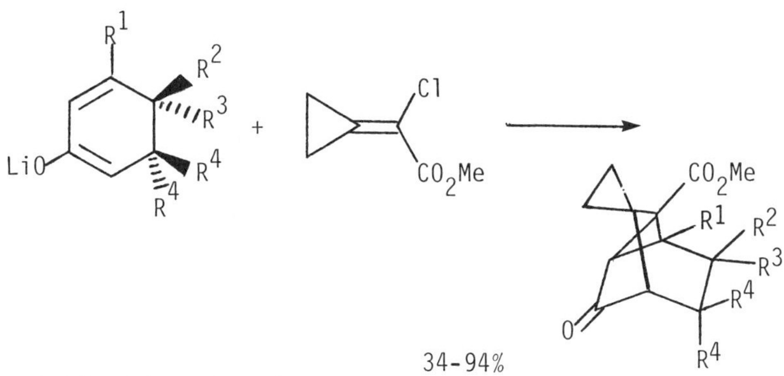

I.E.1-58 H.M.R. Hoffmann et al., Angew. Chem., Int. Ed. Engl., 26, 1015 (1987).

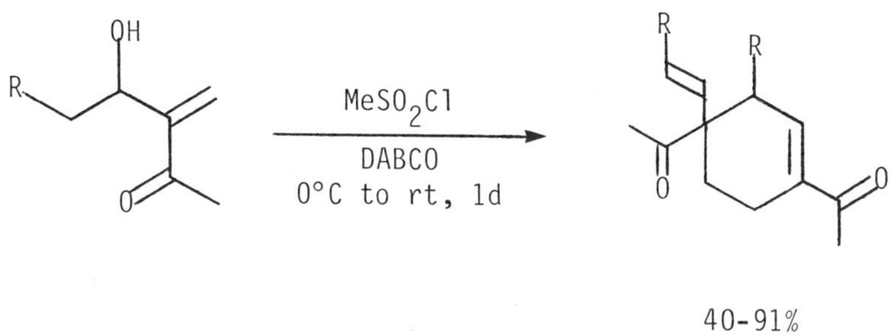

I.E.1-59 K. Banert, Angew. Chem., Int. Ed. Engl., 26, 879 (1987).

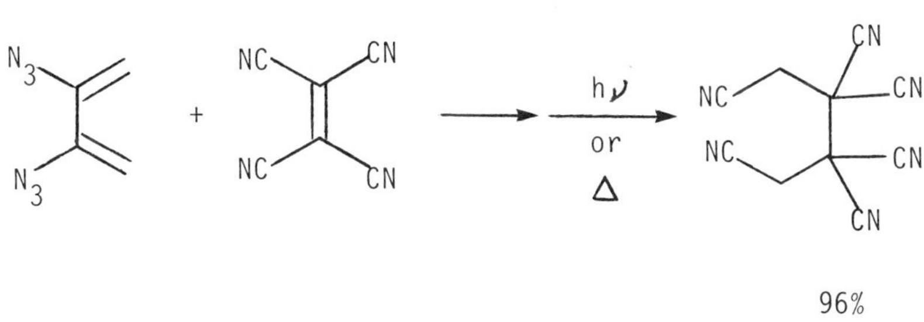

I.E.1-60 H. Hart et al., Tetrahedron, 43, 5203 (1987).

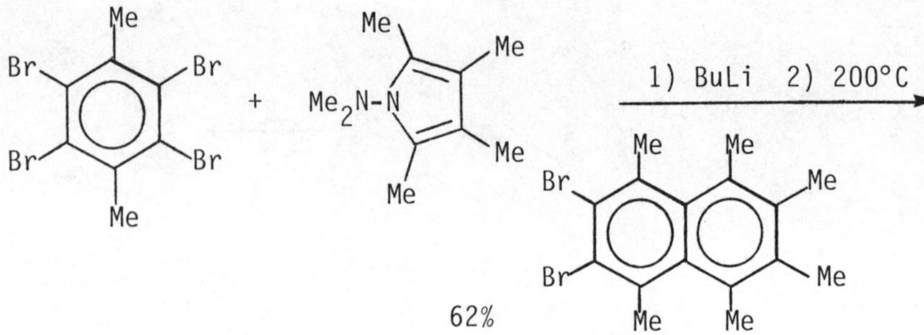

I.E.1-61 S. Nishida et al., Chem. Lett., 863 (1987).

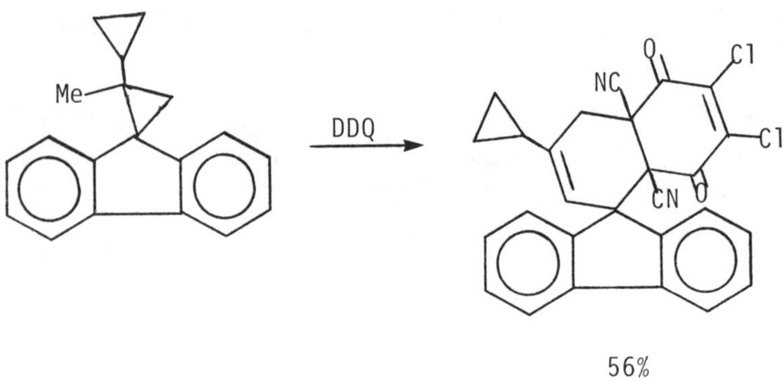

I.E.1-62 K. Mullen et al., Angew. Chem., Int. Ed. Engl., 26, 1290 (1987); E. Winterfeldt et al., Liebigs Ann. Chem., 1095 (1987).

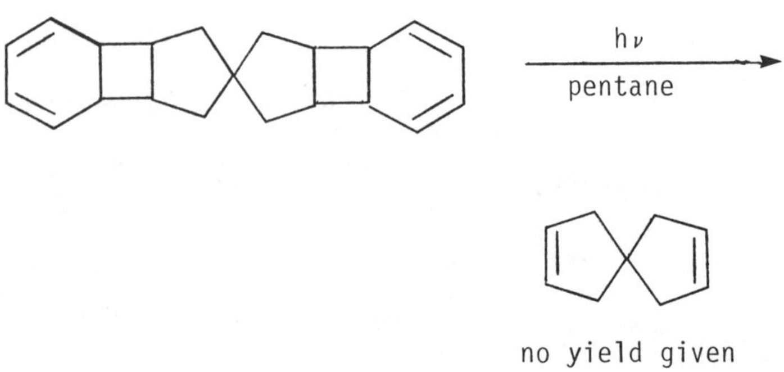

I.E.1-63 D.L. Boger et al., Org. Synth., 65, 32 (1987).

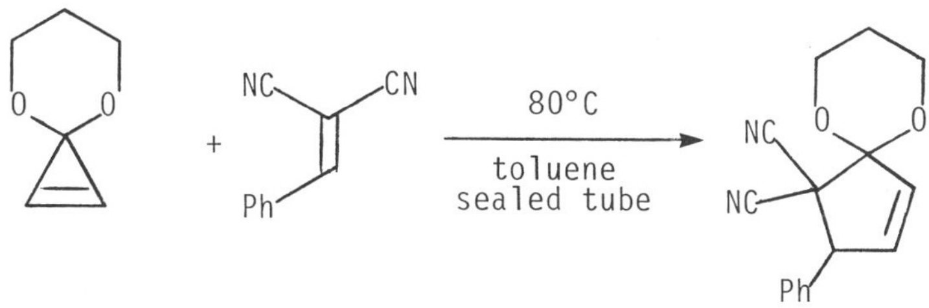

60-64%

I.E.1-64 W.R. Dolbier, Jr. and M.J. Seabury, J. Am. Chem. Soc., 109, 4393 (1987).

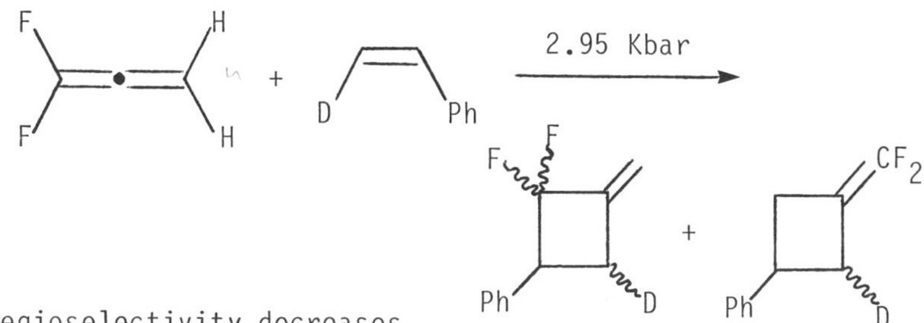

regioselectivity decreases with increased pressure

88%, 7.3 : 1

I.E.1-65 P.A. Wender and M.L. Snapper, Tetrahedron Lett., 28, 2221 and 2451 (1987).

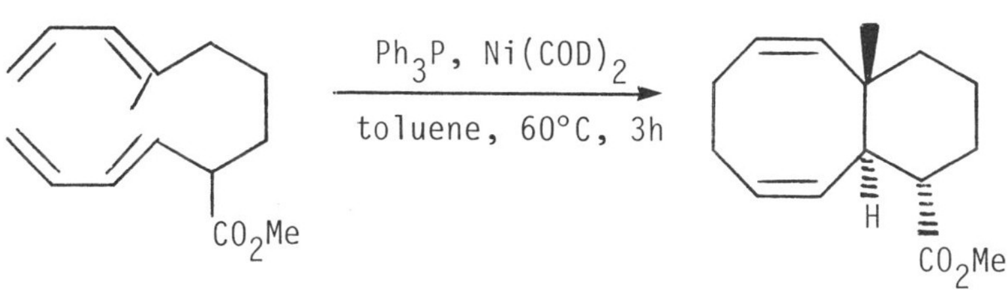

94% conversion, 92%, >97% ds

I.E.1-66 C.-Y. Liu and K.N. Houk, Tetrahedron Lett., 28, 1371 (1987).

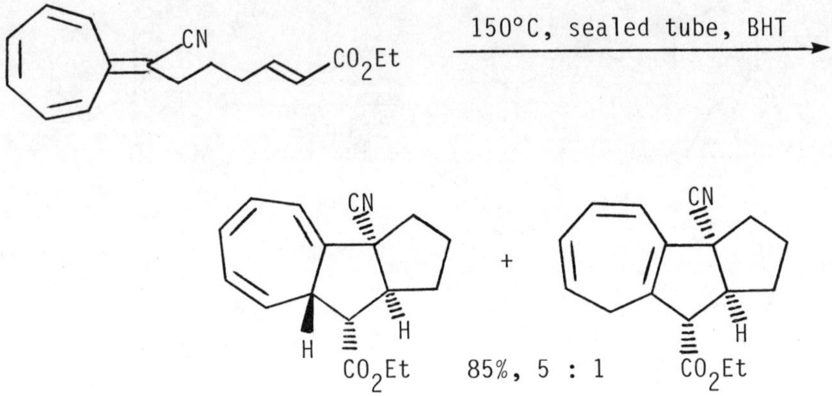

I.E.2. Other Thermal Reactions

I.E.2-1 W.R. Dolbier, Jr., D.J. Burton et al., J. Am. Chem. Soc., 109, 219 (1987).

Kinetic and Thermodynamic Studies of the Thermal Electrocyclic Interconversions of Perfluorinated Dienes and Cyclobutenes

I.E.2-2 R. Sauvetre et al., J. Organomet. Chem., 328, 1 (1987); Y. Tobe et al., Tetrahedron Lett., 28, 2861 (1987).

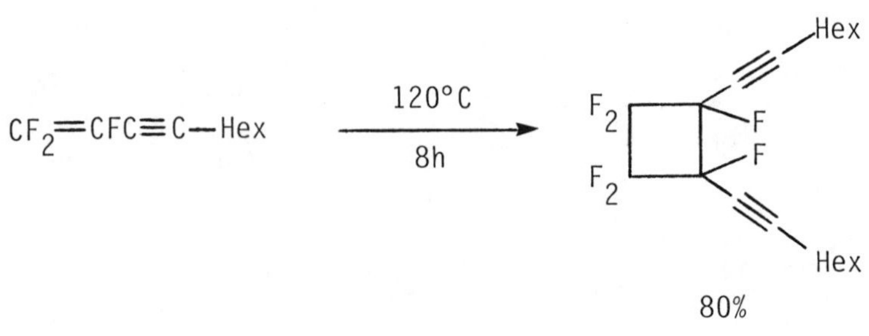

I.E.2-3 G.L. Lange and M. Lee, J. Org. Chem., 52, 325 (1987).

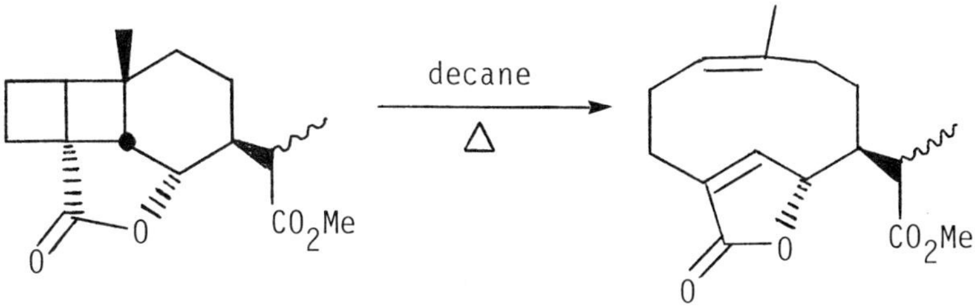

60% (major)

I.E.2-4 J.W. Barton and M.K. Shepherd, J. Chem. Soc., Perkin Trans. 1, 961 (1987).

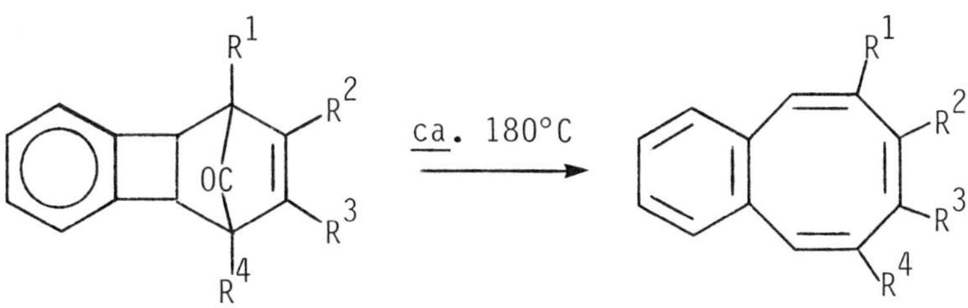

ca. 100%

I.E.2-5 T. Hudlicky et al., J. Org. Chem., 52, 4641 (1987).

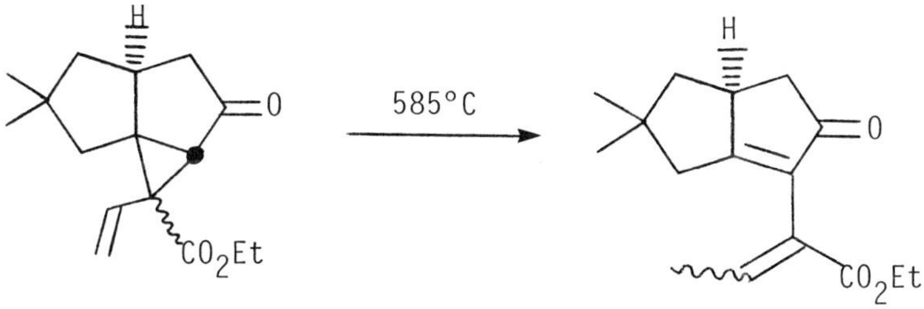

>90%

I.E.2-6 W. Boland and D.G. Muller, Tetrahedron Lett., 28, 307 (1987); E. Piers and G.L. Jung, Can. J. Chem., 65, 1668 (1987).

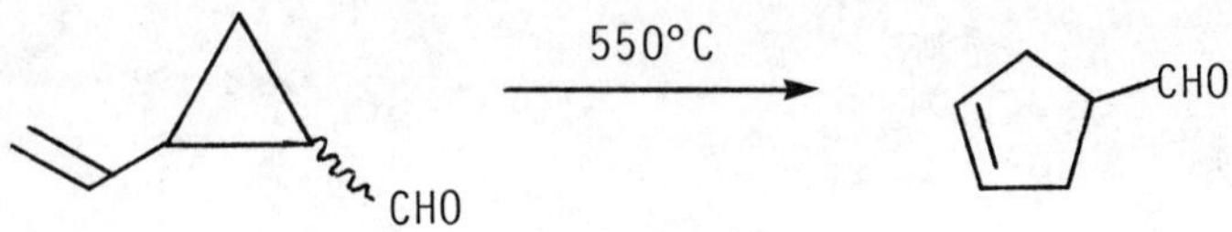

78%

I.E.2-7 K. Fukumoto et al., Chem. Commun., 1360 (1987).

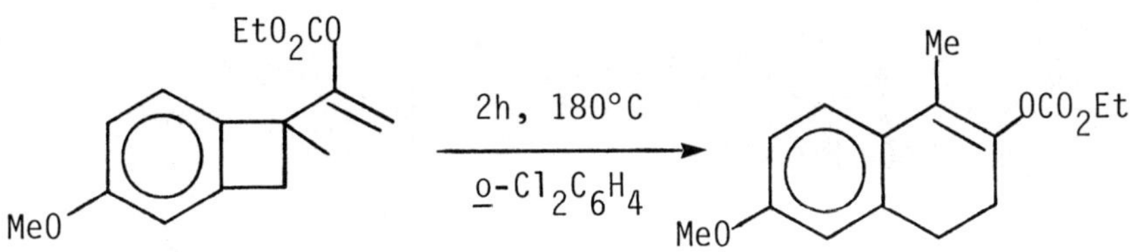

94%

I.E.2-8 O.H.W. Decker and H.W. Moore, J. Org. Chem., 52, 1174 (1987).

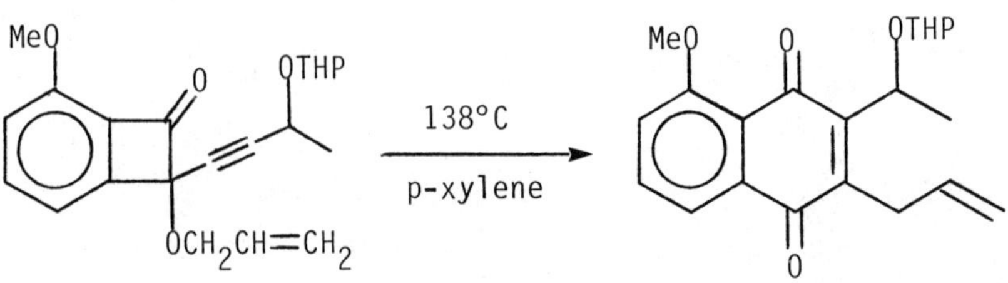

71%

I.E.2-9 R.F.C. Brown et al., Aust. J. Chem., 40, 107 and 1687 (1987).

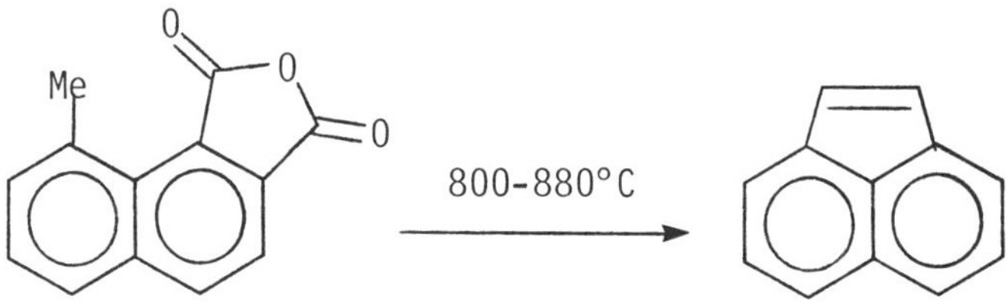

50-56%

I.E.2-10 K.P.C. Vollhardt et al., J. Am. Chem. Soc., 109, 8105 (1987).

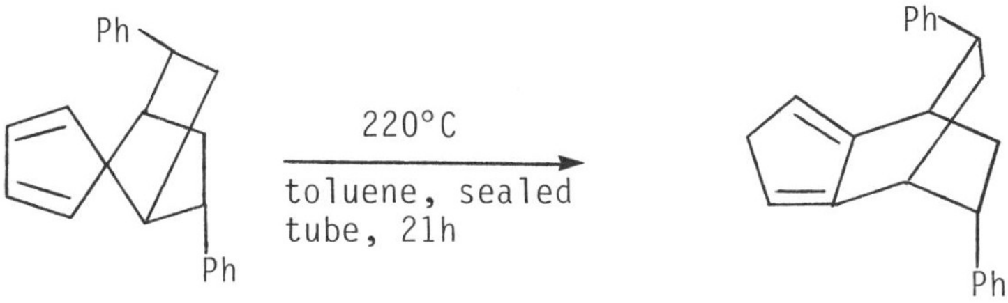

98%

I.E.2-11 D. Liotta and W. Ott, Synth. Commun., 17, 1655 (1987).

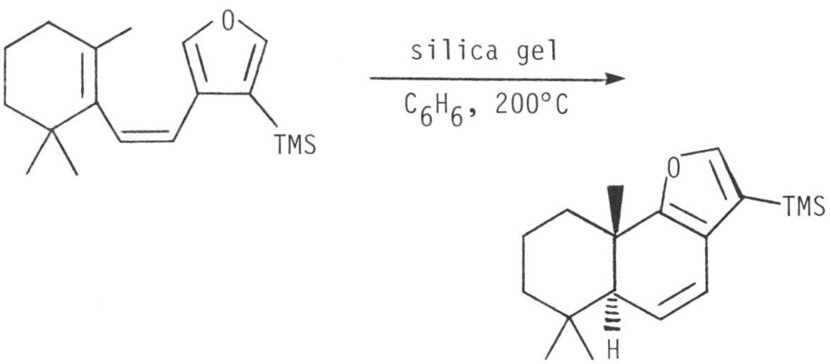

towards Pallescensin A

quantitative

I.E.2-12 A.S. Dreiding et al., Tetrahedron, 43, 4875 (1987).

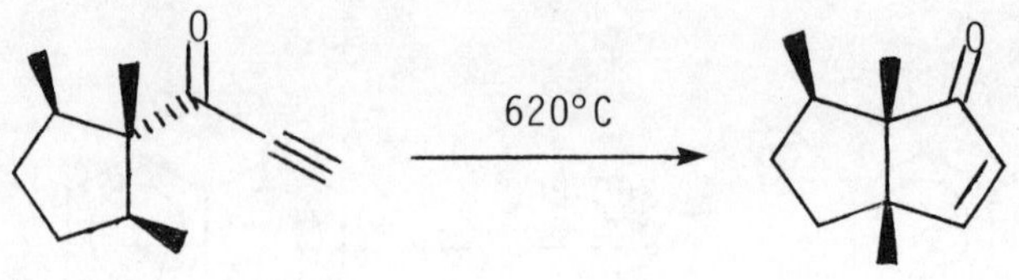

76%

I.E.2-13 G.L. Larson et al., Organometallics, 6, 2250 (1987).

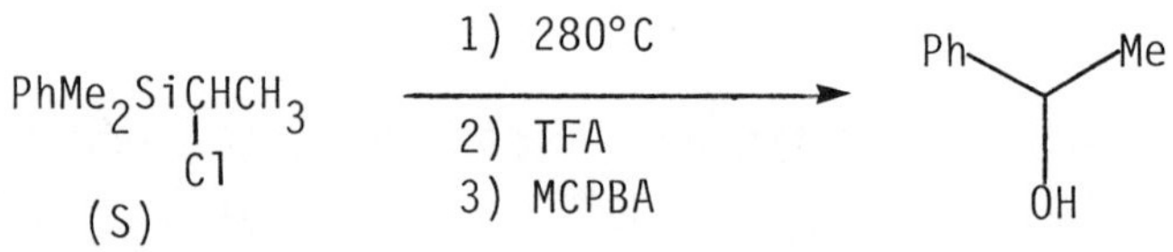

88% ee, no yield given

I.E.3. Photochemical Reactions

I.E.3-1 V. Ramamurthy and K. Venkatesan, Chem. Rev., 87, 433 (1987).

Review: "Photochemical Reactions of Organic Crystals."

I.E.3-2 J.J. McCullough, Chem. Rev., 87, 811 (1987).

Review: "Photoadditions of Aromatic Compounds."

I.E.3-3 T.A. Anglea and A.R. Pinder, Tetrahedron, 43, 5537 (1987); M. Vandewalle et al., ibid., 43, 5593 (1987); L.A. Paquette et al., Tetrahedron Lett., 28, 5017 (1987); G.L. Lange et al., ibid., 28, 2833 (1987); W.R. Dolbier, Jr. and M. Seabury, ibid., 28, 1491 (1987); N.W.A. Geraghty et al., ibid., 28, 4729 (1987); B.F. Plummer and S.F. Singleton, ibid., 28, 4801 (1987); W.F. Berkowitz et al., J. Org. Chem., 52, 1119 (1987); P. Sarti-Fantoni, G. Renzi et al., ibid., 52, 5124 (1987); L.K. Sydnes and H.L. Meling, Acta Chem. Scand., B41, 660 (1987); M.T.M. Clements and T.B.H. McMurry, Can. J. Chem., 65, 1810 (1987); M. Cavazza, M. Zandomeneghi et al., Gazz. Chim. Ital., 117, 17 and 255 (1987).

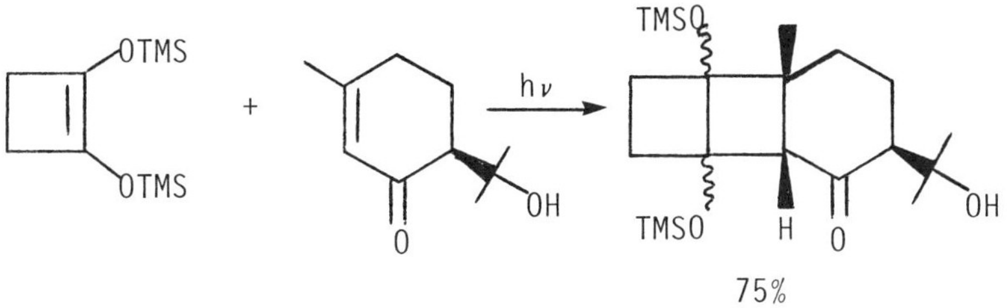

photolysis in the solid state and circularly polarized light also used

---

I.E.3-4 M.T. Crimmins and L.D. Gould, J. Am. Chem. Soc., 109, 6199 (1987); G. Mehta and S. Padma, ibid., 109, 2212 and 7230 (1987); J. Nishimura et al., ibid., 109, 5293 (1987); Y. Yamashita et al., ibid., 109, 6682 (1987); R.G. Salomon et al., J. Org. Chem., 52, 83 (1987); K. Hirao et al., J. Chem. Res. (S), 162 (1987); R. Gleiter and U. Steuerle, Tetrahedron Lett., 28, 6159 (1987).

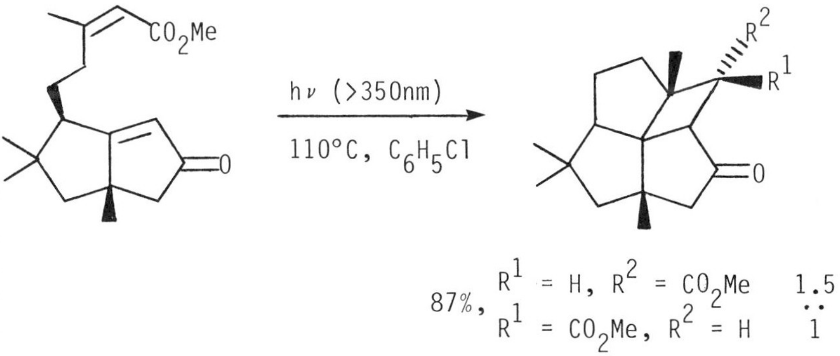

I.E.3-5 D.H.R. Barton et al., Tetrahedron, 43, 4297 (1987); D. Crich and J.W. Davies, Tetrahedron Lett., 28, 4205 (1987).

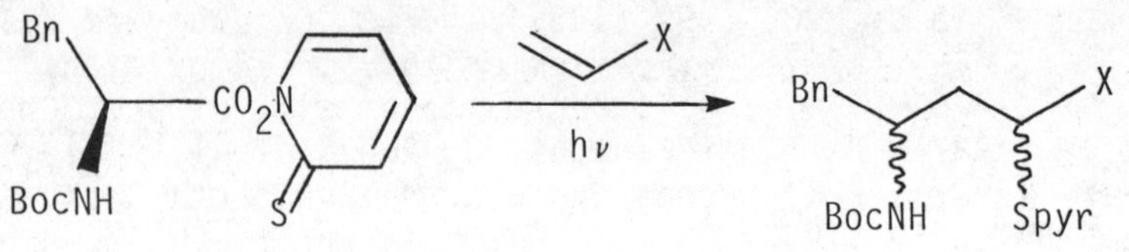

X = CO_2Me, CN, NO_2 63-83%

I.E.3-6 R. Lomolder and H.J. Schafer, Angew. Chem., Int. Ed. Engl., 26, 1253 (1987).

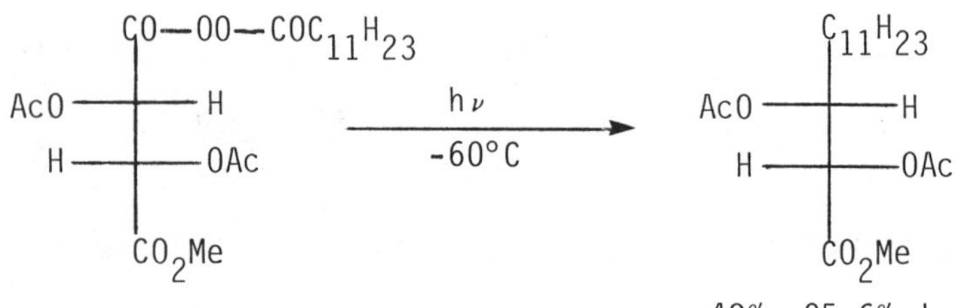

48%, 95.6% de

I.E.3-7 N.R. Ayyangar et al., Ind. J. Chem., 26B, 195 (1987); W.C. Agosta, R.A. Caldwell, J.C. Scaiano et al., J. Am. Chem. Soc., 109, 3050 (1987); M. Kijima et al., J. Org. Chem., 52, 706 (1987).

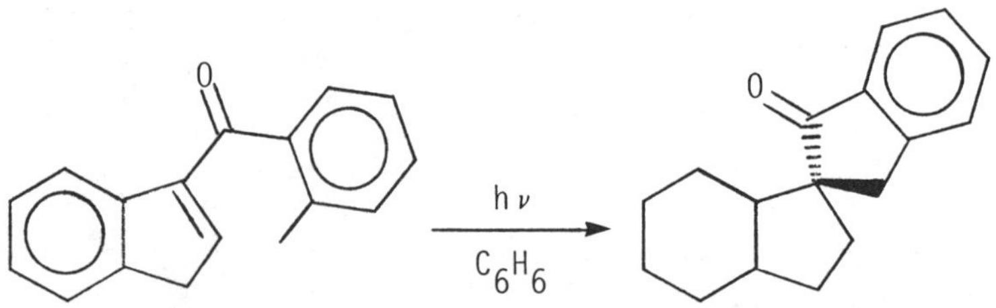

85%

similar reaction with methyl methacrylate/hν in the presence of benzyl bis(dimethylglyoximato)pyridine cobalt (III)

I.E.3-8 M.C. Pirrung, J. Org. Chem., 52, 1635 (1987); A. Gilbert and P. Heath, Tetrahedron Lett., 28, 5909 (1987).

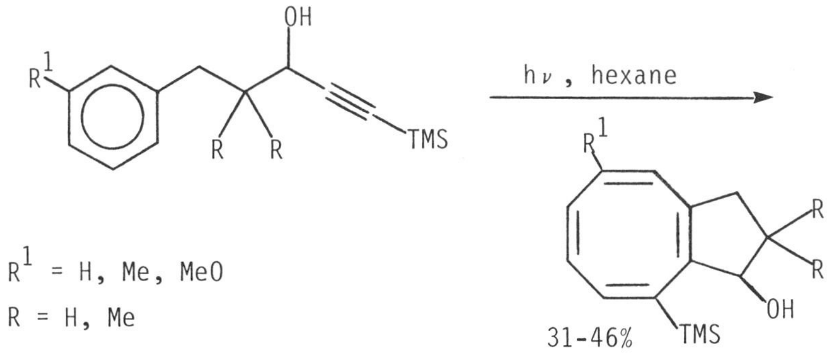

I.E.3-9 K.S. Feldman et al., Tetrahedron Lett., 28, 607 (1987).

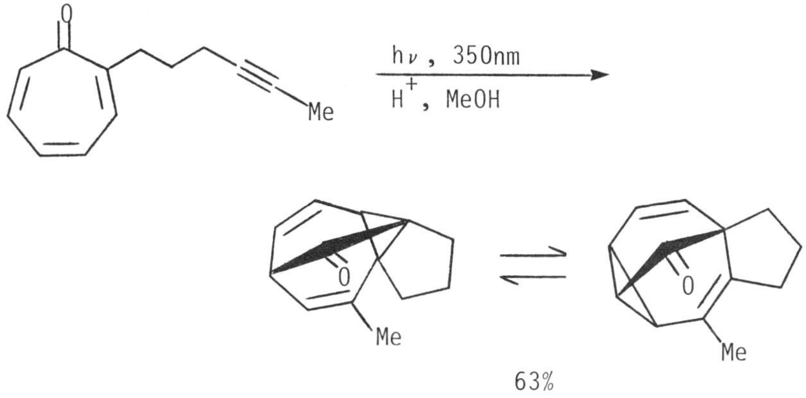

I.E.3-10 J.D. Winkler et al., J. Am. Chem. Soc., 109, 2850 (1987).

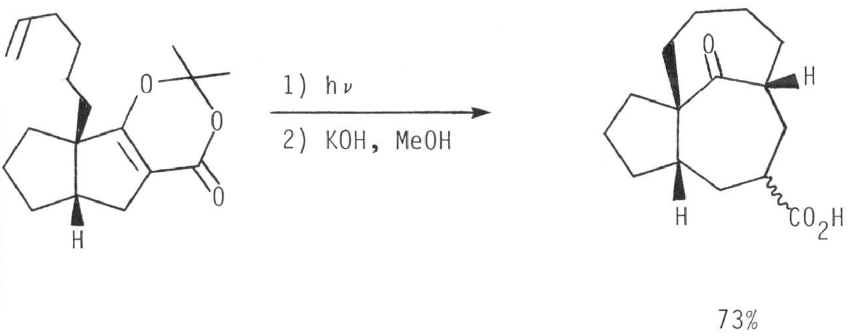

I.E.3-11 R.D. Little et al., J. Org. Chem., 52, 4647 (1987).

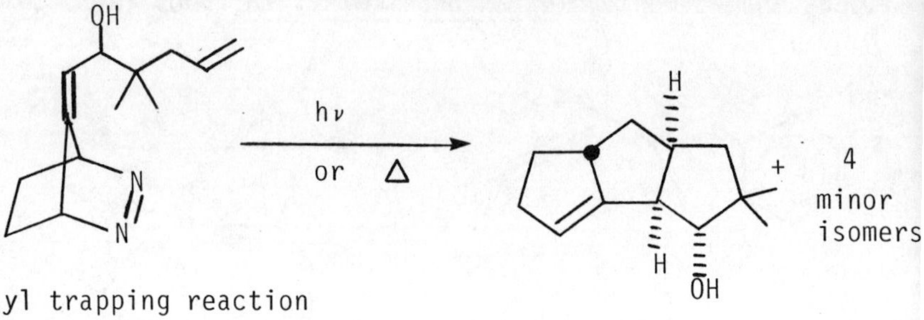

diyl trapping reaction

87-96%, 4 : 1 to 30 : 1

I.E.3-12 R. Noyori et al., J. Org. Chem., 52, 5583 (1987).

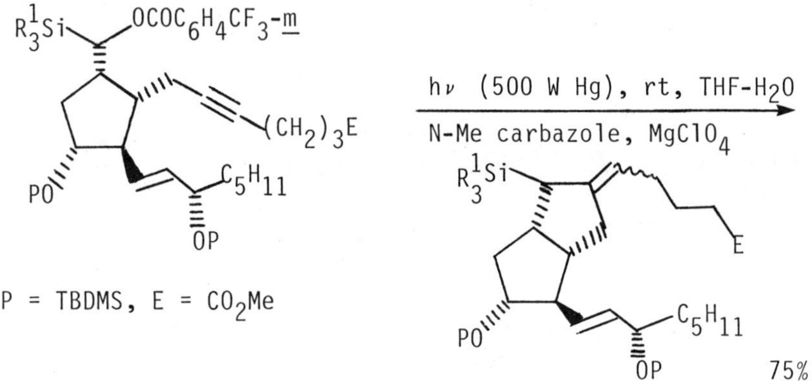

P = TBDMS, E = CO_2Me

75%

I.E.3-13 Z. Goldschmidt and E. Genizi, Tetrahedron Lett., 28, 4867 (1987).

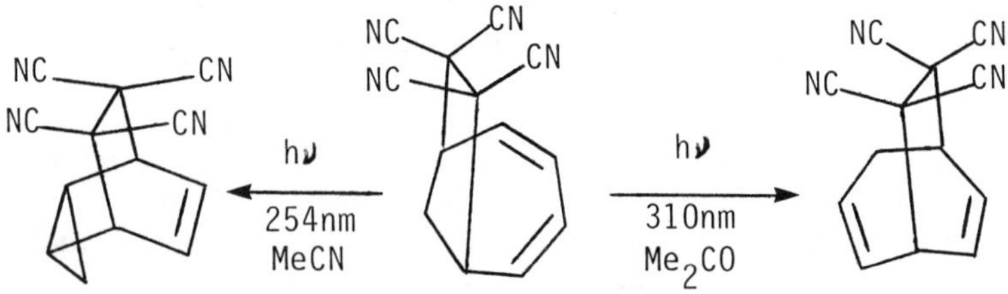

60%

I.E.3-14 P.A. Wender and C.R.D. Correia, J. Am. Chem. Soc., 109, 2523 (1987); J.K. Whitesell and M.A. Minton, ibid., 109, 6403 (1987).

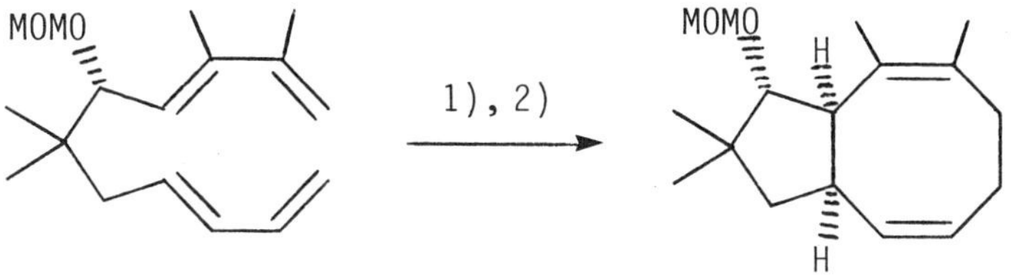

60%

1) hν, PhCOPh, hexane 2) 200°C, 12h, hexane

similar triene cyclization to give a cis junction

I.E.3-15 M. Fetizon et al., Synth. Commun., 17, 913 (1987).

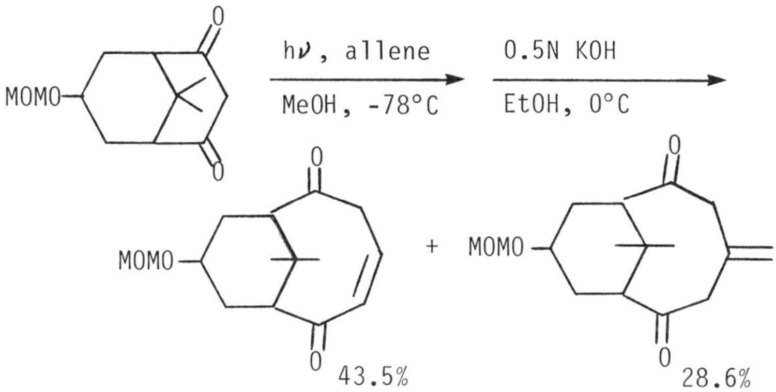

43.5% 28.6%

I.E.3-16 M.C. Pirrung and N.J.G. Webster, J. Org. Chem., 52, 3603 (1987).

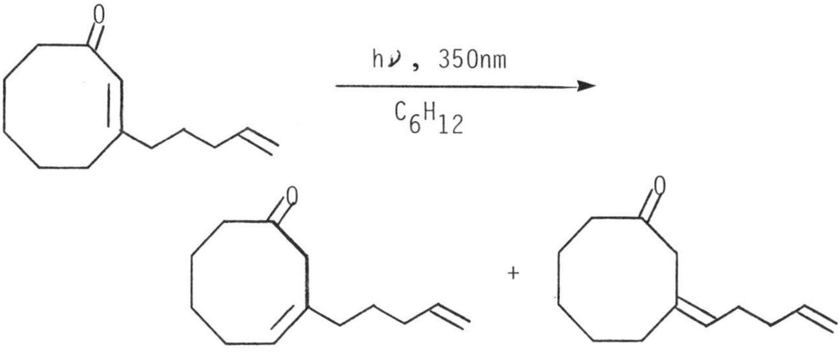

35% 57%

I.E.3-17 S.H. Brown and R.H. Crabtree, Tetrahedron Lett., 28, 5599 (1987).

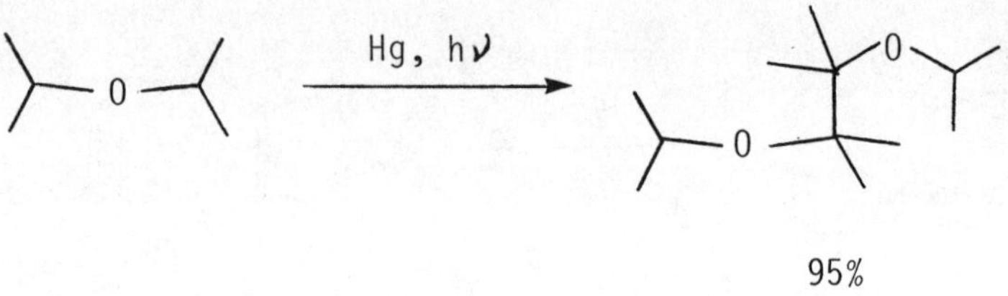

I.F. Aromatic Substitutions Forming A New Carbon - Carbon Bond

I.F.1. Friedel - Crafts Type Aromatic Substitution Reactions

I.F.1-1 P. Laszlo and A. Mathy, Helv. Chim. Acta, 70, 577 (1987); G.A. Olah et al., J. Am. Chem. Soc., 109, 3708 (1987); V.V. Mezheritskii and L.G. Minyaeva, J. Org. Chem. (USSR), 22, 2150 (1986); M.C. Rezende et al., Synth. Commun., 17, 457 (1987).

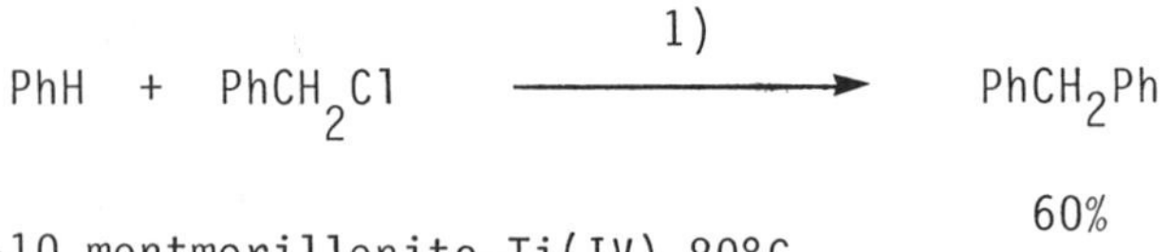

1) K-10 montmorillonite-Ti(IV),80°C

$^tBuF/BF_3$, gem di- and tri- halides/$AlCl_3$ also used

I.F.1-2 M. Tashiro et al., J. Chem. Soc., Perkin Trans. 1, 1 (1987).

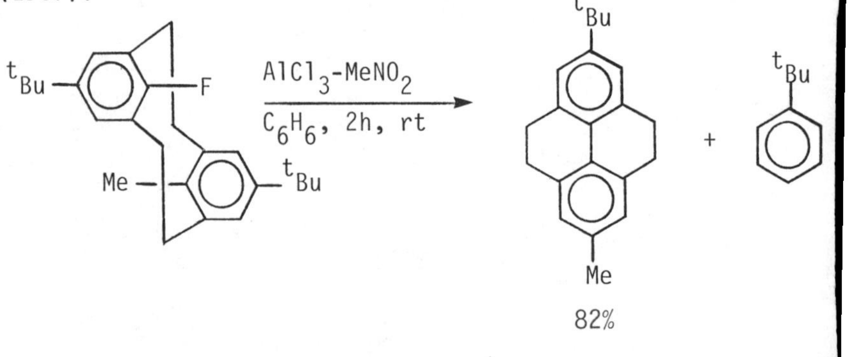

I.F.1-3 O. Piccolo et al., J. Org. Chem., 52, 10 (1987).

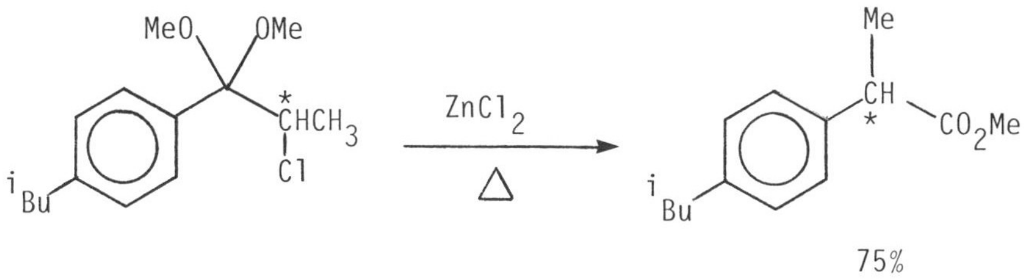

I.F.1-4 M. Nimtz and G. Hafelinger, Liebigs Ann. Chem., 765 (1987).

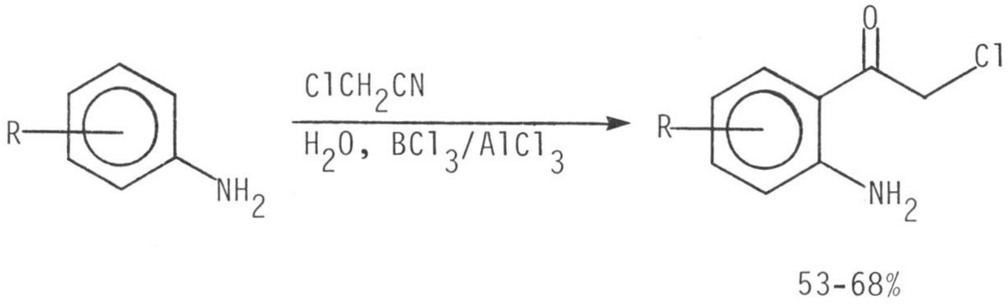

I.F.1-5 H. Ishibashi et al., J. Chem. Soc., Perkin Trans. 1, 589 and 1095 (1987); K. Sato et al., J. Org. Chem., 52, 5495 (1987); S. Veeraraghavan et al., ibid., 52, 1355 (1987); T.H. Kim et al., Synth. Commun., 17, 755 (1987); I. Stahl, Chem. Ber., 120, 135 (1987).

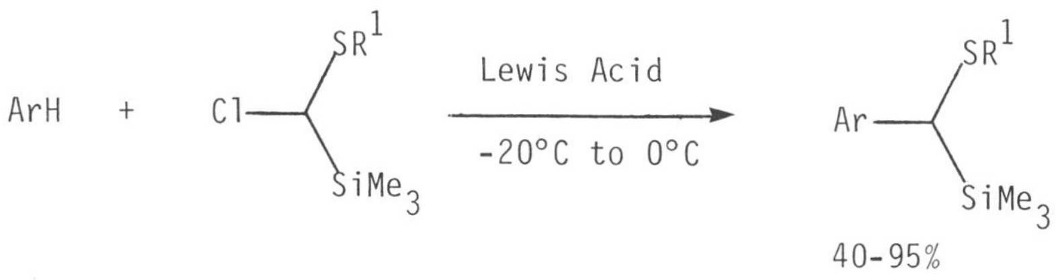

sulfoxides/TFAA/$SnCl_4$ and a sulfide/SO_2Cl_2 also used

I.F.1-6 A.G. Martinez et al., Chem. Ber., 120, 1255 (1987); F. Effenberger and T. Weber, Angew. Chem., Int. Ed. Engl., 26, 142 (1987).

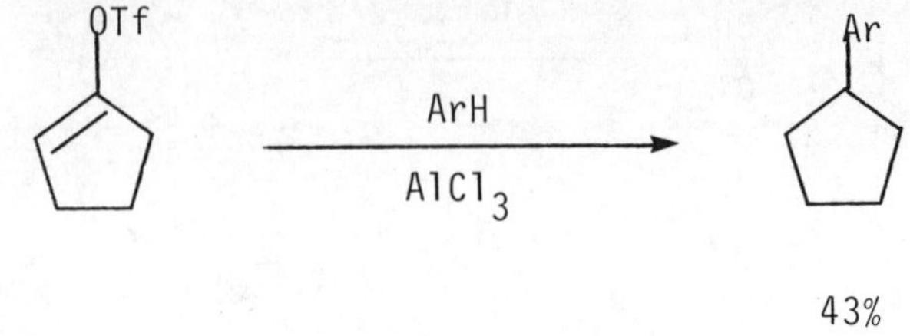

I.F.1-7 N. Ono et al., J. Org. Chem., 52, 4133 (1987).

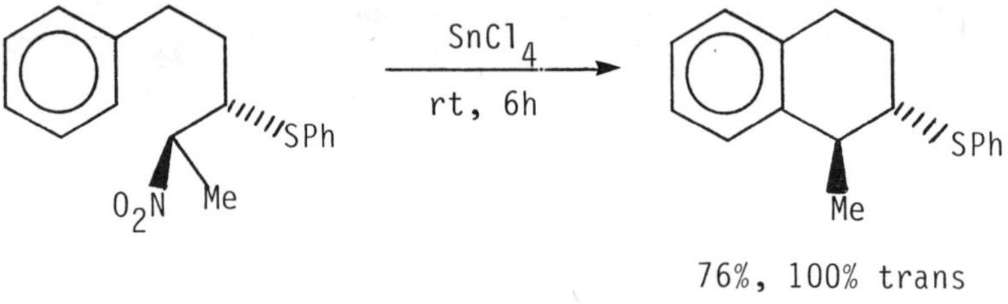

I.F.1-8 W.E. Fristad et al., Tetrahedron Lett., 28, 1493 (1987); Y. Lepage et al., Bull. Soc. Chim. Fr., 309 (1987).

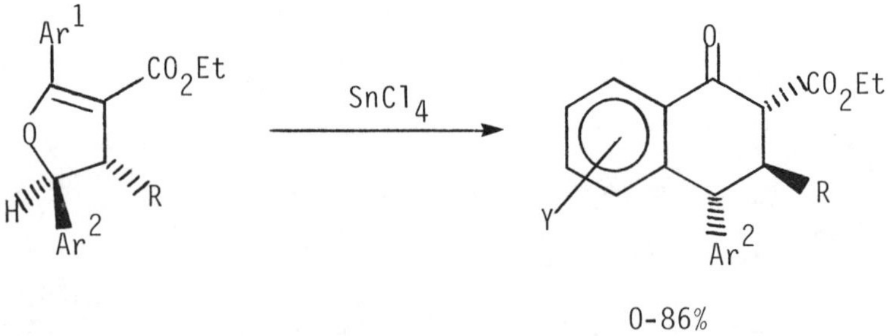

Y = Ar^1 substituents

I.F.1-9 I.D. Rae and M.L. Woolcock, Aust. J. Chem., 40, 1023 (1987).

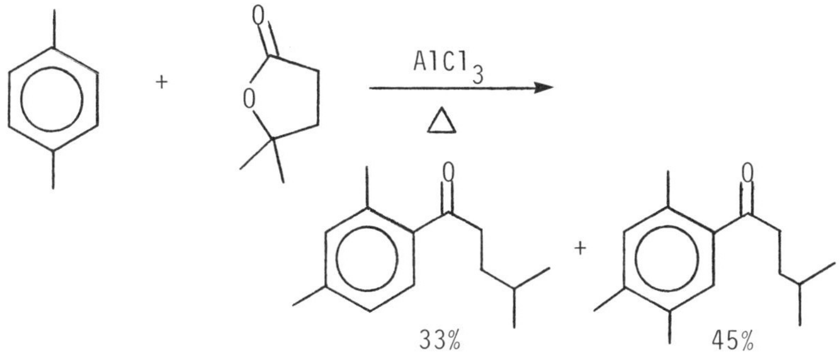

I.F.1-10 K. Shudo et al., J. Am. Chem. Soc., 109, 7036 (1987); T. Ohwada et al., Tetrahedron, 43, 297 (1987).

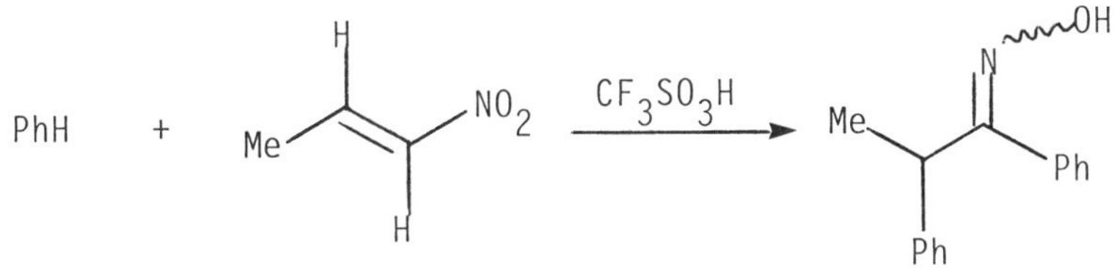

100%, anti:syn 3:1

I.F.1-11 I.V. Bodrikov et al., J. Org. Chem. (USSR), 22, 1699 (1986); R.F.C. Brown et al., Aust. J. Chem., 40, 1675 (1987).

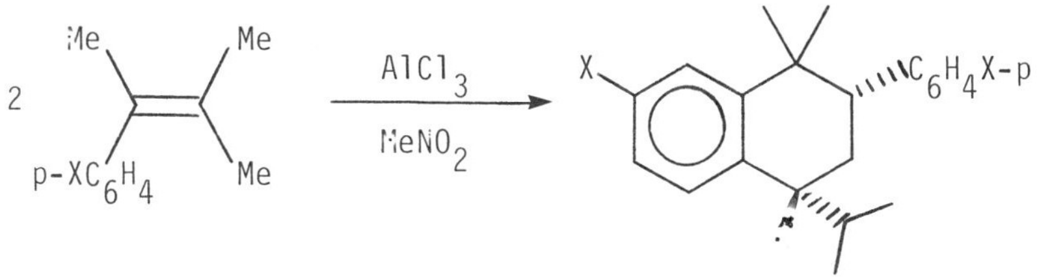

50-78%

I.F.1-12 M. Gruselle et al., J. Organomet. Chem., 328, C12 (1987).

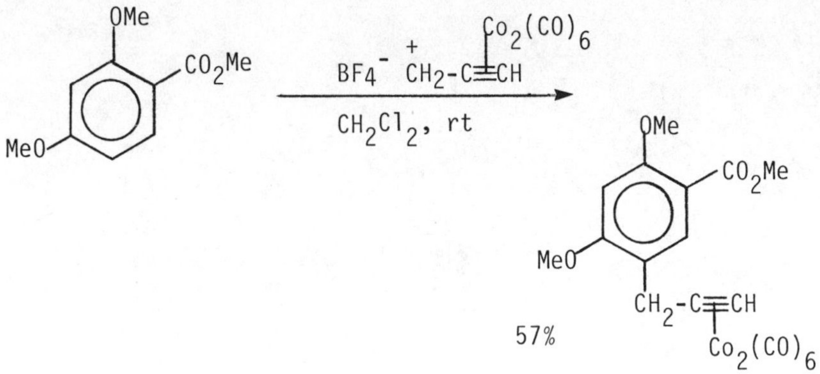

I.F.1-13 T. Kametani et al., J. Org. Chem., 52, 5233 (1987); M. Braun et al., Chem. Ber., 120, 265 (1987).

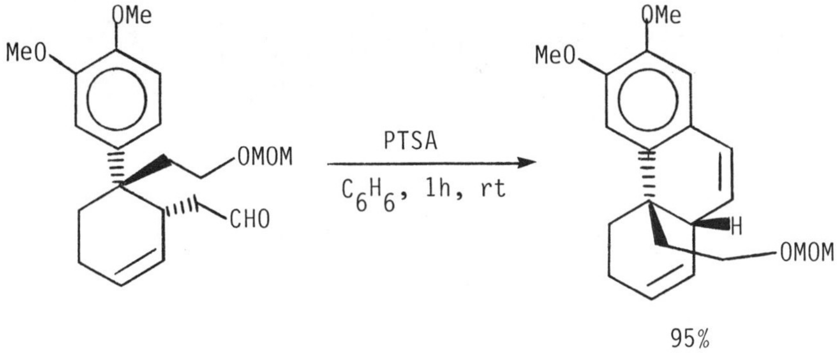

I.F.1-14 A.R.B. Manas and R.A.J. Smith, Tetrahedron, 43, 1847 (1987).

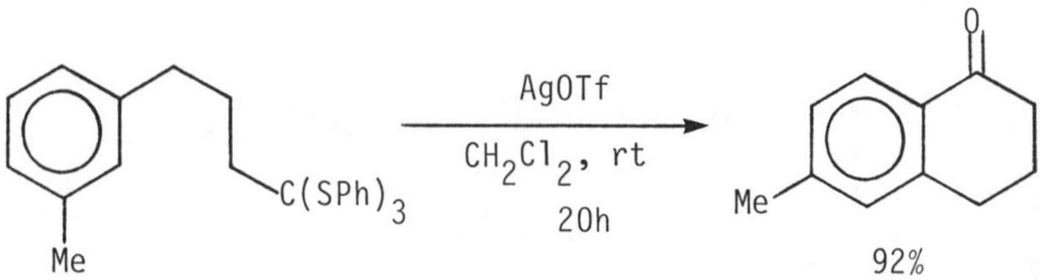

I.F.1-15 A.R. Balaban et al., Synthesis, 577 (1987); S.D. Samant et al., Ind. J. Chem., 26B, 75 (1987); I. Hashimoto and M. Tashiro, Org. Prep. Proced. Int., 19, 447 (1987); R.H. Burnell et al., Can. J. Chem., 65, 775 (1987); M. Hojo et al., Tetrahedron Lett., 28, 6199 (1987).

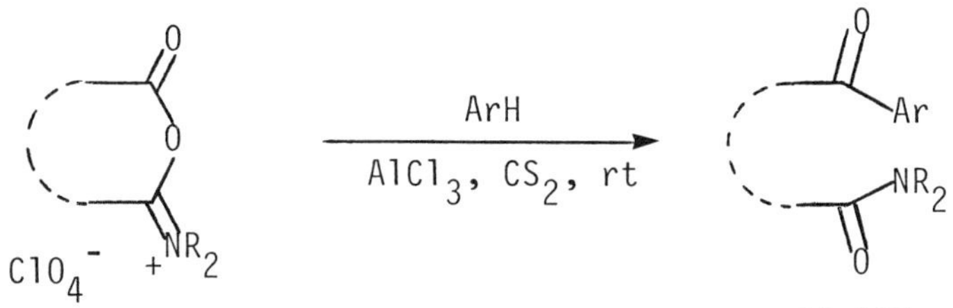

23-53%

similar results with anhydrides

I.F.1-16 J.L. von Rosenberg and A.R. Pinder, J. Chem. Soc., Perkin Trans. 1, 747 (1987).

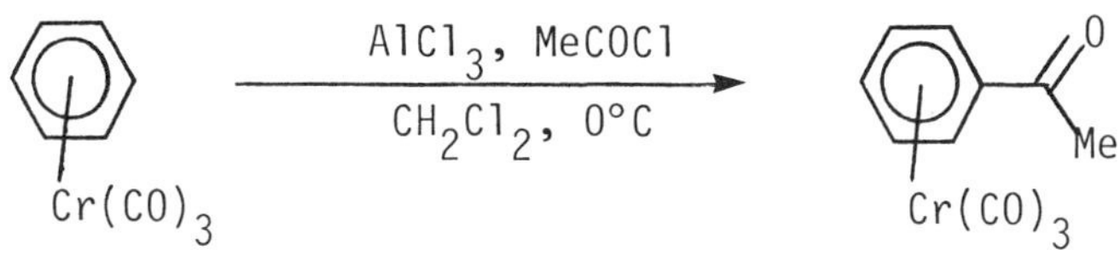

66%

I.F.1-17 S.B. Mahato et al., Ind. J. Chem., 25B, 1263 (1986); G. Sartori et al., Tetrahedron Lett., 28, 1533 (1987); H. Ishii et al., J. Chem. Soc., Perkin Trans. 1, 2415 (1987); M. Psiorz and R. Schmid, Chem. Ber., 120, 1825 (1987); A. Haas and W. Wanzke, ibid., 120, 429 (1987); R. Sangaiah and A. Gold, J. Org. Chem., 52, 3205 (1987); D.G. Melillo, R.D. Larsen, D.J. Mathre et al., ibid., 52, 5143 (1987); X.L. Hou and H.N.C. Wong, J. Am. Chem. Soc., 109, 1868 (1987).

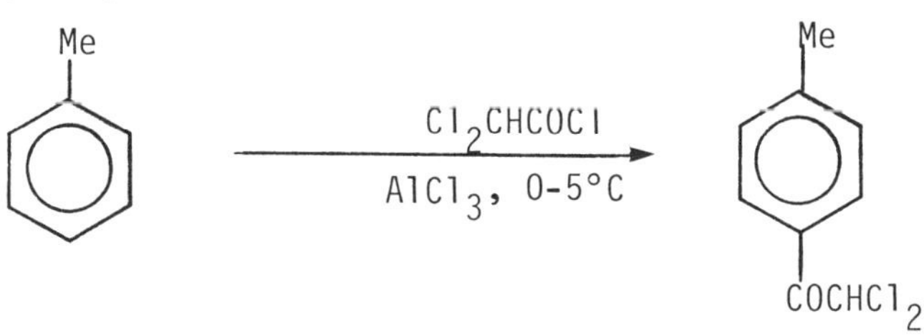

95%

Friedel-Craft's type acylation also with F_3CSCSF/SbF_5, aralkyl $CO_2H/TiCl_4$ and a biaryl ester/PPA at 150°C

I.F.2. Coupling Reactions to Form an Aromatic Carbon-Carbon Bond

I.F.2-1 P. Kovacic and M.B. Jones, Chem. Rev., 87, 357 (1987).

Review: "Dehydro Coupling of Aromatic Nuclei by Catalyst-Oxidant Systems: Poly (p-phenylene)."

I.F.2-2 J.K. Stille et al., J. Org. Chem., 52, 422 (1987); A.M. Echavarren and J.K. Stille, J. Am. Chem. Soc., 109, 5478 (1987); W.J. Scott, Chem. Commun., 1755 (1987); M. Kosugi, T. Migita et al., Bull. Chem. Soc. Jpn., 60, 767 (1987); G. Dumartin et al., Tetrahedron Lett., 28, 3935 (1987).

$$\mathrm{ArBr} + \mathrm{CH_2{=}CHSnBu_3} \xrightarrow[\text{toluene, 110°C}]{\mathrm{Pd(PPh_3)_4}} \mathrm{ArCH{=}CH_2}$$

62-85%

similar results with allyl tin and chloroarene chromium tricarbonyl species

---

I.F.2-3 B.M. Trost and R. Walchli, J. Am. Chem. Soc., 109, 3487 (1987).

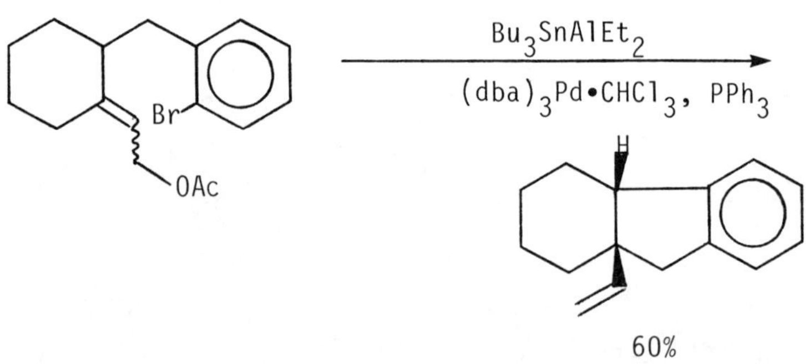

60%

I.F.2-4 C. Mellin and U. Hacksell, Tetrahedron., 43, 5443 (1987); V. Kvita and H.-P. Sauter, Angew. Chem., Int. Ed. Engl., 26, 790 (1987); C. Georgoulis et al., Organometallics, 6, 166 (1987); L.E. Overman et al., J. Org. Chem., 52, 4130 (1987); K. Ikenaga, K. Kikukawa and T. Matsuda, ibid., 52, 1276 (1987); S.W. McCombie and T.L. Nagabhushan, Tetrahedron Lett., 28, 5395 (1987); G. Just and R. Singh, ibid., 28, 5981 (1987); R.J. Blade et al., ibid., 28, 3857 (1987); G. Ortar et al., ibid., 28, 3039 (1987); K. Horn et al., J. Organomet. Chem., 332, 271 (1987); B.M. Choudary, Ind. J. Chem., 25B, 1159 (1987)

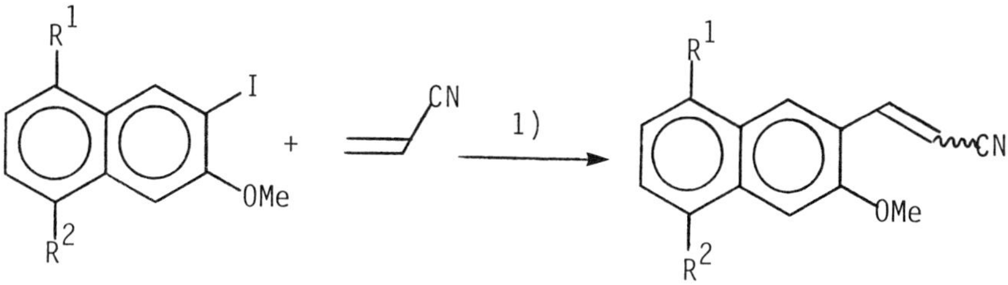

1) NEt_3, $Pd(OAc)_2$, MeCN

90-96%, 4 : 1 E : Z

similar results with different Pd catalysts or halides, with alkynes and with PhN(NO)COMe

---

I.F.2-5 E. Negishi et al., Chem. Commun., 477 (1987); E. Negishi and K. Akiyoshi, Chem. Lett., 1007 (1987); R. Sauvetre et al., J. Organomet. Chem., 331, 281 (1987); Q.-Y. Chen and Y.-B. He, Tetrahedron Lett., 28, 2387 (1987); R. McCague, ibid., 28, 701 (1987); T.L. Gilchrist and R.J. Summersell, ibid., 28, 1469 (1987); M. Iyoda et al., ibid., 28, 625 (1987); M.I. Al-Hassan, Synth. Commun., 17, 1247 and 1787 (1987); T. Frejd and T. Klingstedt, Synthesis, 40 (1987).

$$\mathrm{PhI} + \mathrm{ClZnC{\equiv}C{}^tBu} \xrightarrow[\text{THF, rt, 3h}]{\mathrm{Pd(PPh_3)_4}} \mathrm{PhC{\equiv}C{}^tBu}$$

90%

similar results with alkenyl or aryl zinc halides, with triflate leaving groups and with $Ni(acac)_2$ catalysis

I.F.2-6 M.I. Al-Hassan, Synth. Commun., 17, 1413 (1987); idem, Synthesis, 815 and 816 (1987).

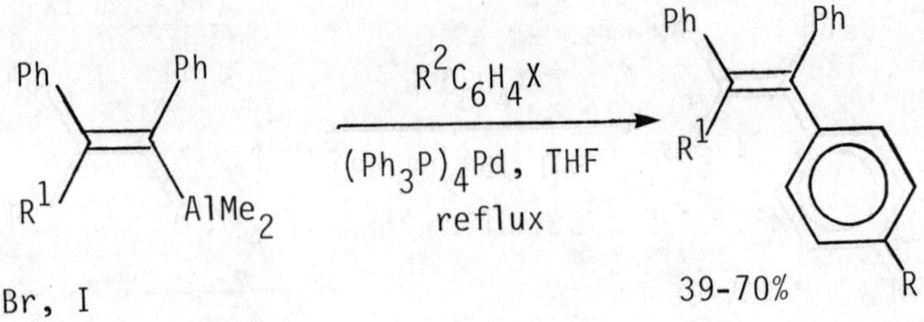

X = Br, I

I.F.2-7 A. Suzuki et al., Synthesis, 373 (1987); V. Snieckus et al., Tetrahedron Lett., 28, 5093 and 5097 (1987); K. Nilsson and A. Hallberg, Acta Chem. Scand., B41, 569 (1987).

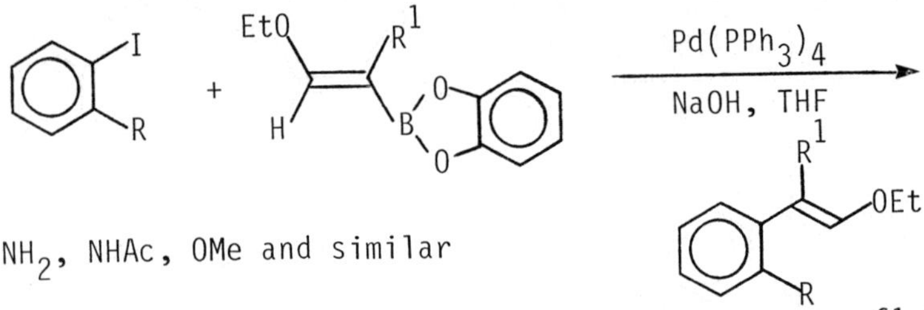

R = NH_2, NHAc, OMe and similar

$(EtO\frown\!\!\frown)_3B$ also used

I.F.2-8 D.J. Cram et al., Chem. Lett., 19 (1987); A.I. Meyers et al., J. Am. Chem. Soc., 109, 5446 (1987); A. Minato, K. Suzuki and K. Tamao, ibid., 109, 1257 (1987); J.E. Rice et al., J. Org. Chem., 52, 849 (1987); A.I. Meyers and K. Higashiyama, ibid., 52, 4592 (1987).

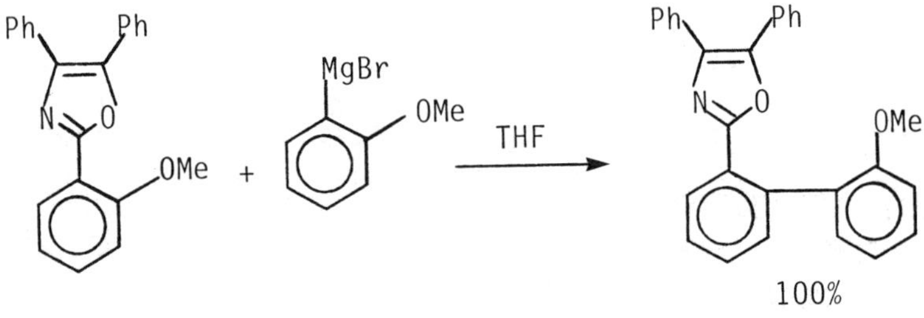

similar results with a vinyl chloride/$PdCl_2$(dppb)
loss of aromaticity when no ortho MeO and vinyl lithium utilized

I.F.2-9 C.F. Du and H. Hart, J. Org. Chem., 52, 4311 (1987); N. Furukawa et al., Tetrahedron Lett., 28, 2727 (1987).

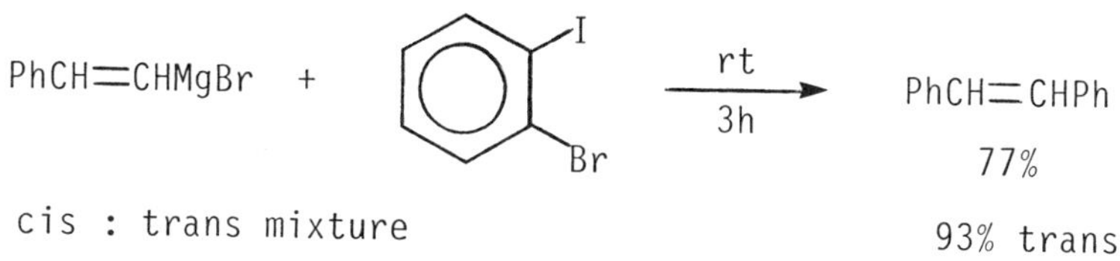

I.F.2-10 M.T. Rahman and S.K. Nahar, J. Organomet. Chem., 329, 133 (1987); M.W. Rathke and D. Vogiazoglou, J. Org. Chem., 52, 3697 (1987); H. Suzuki, T. Ogawa et al., Chem. Lett., 887 (1987).

$$2\ PhMgBr + CuI \xrightarrow[PhNO_2,\ THF]{RX} \underset{41\text{-}84\%}{Ph\text{-}R} + \underset{trace\text{-}14\%}{Ph\text{-}Ph}$$

I.F.2-11 K.S. Kyler et al., J. Org. Chem., 52, 4296 (1987); F.E. Ziegler et al., Org. Synth., 65, 108 (1987); J. Cornforth et al., J. Chem. Soc., Perkin Trans. 1, 859, 867 and 871 (1987); A.J. Bell and R.W. Read, Aust. J. Chem., 40, 1813 (1987).

$$ArSnBu_3 \xrightarrow[THF,\ rt]{Cu(NO_3)_2 \cdot 3H_2O} \underset{14\text{-}85\%}{Ar\text{—}Ar}$$

similar coupling of aryl halides with other copper salts

I.F.2-12 R.F. Heck et al., Organometallics, 6, 1941 (1987).

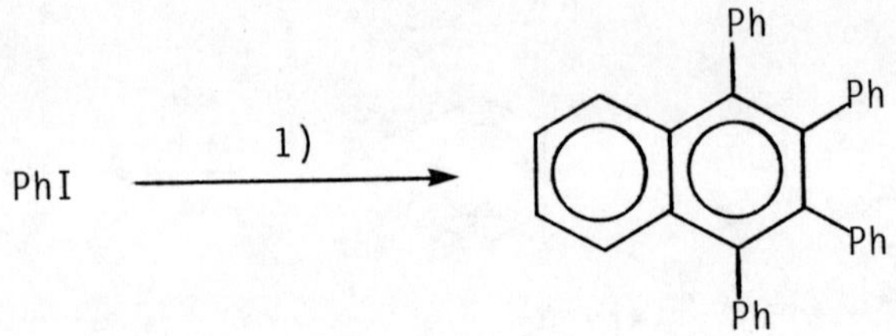

1) $Pd(OAc)_2$, Ph_3P, $MeNO_2$, 41h, 100°C

47%

I.F.2-13 D.H.R. Barton et al., Tetrahedron, 43, 323 (1987); idem, J. Chem. Soc., Perkin Trans. 1, 241 (1987).

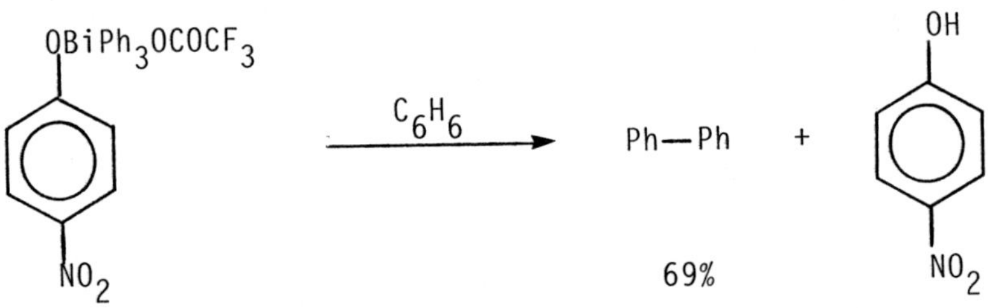

69% 34%

O-phenylation with other substituents and conditions

I.F.2-14 J.-P. Robin et al., Tetrahedron Lett., 28, 5161 (1987).

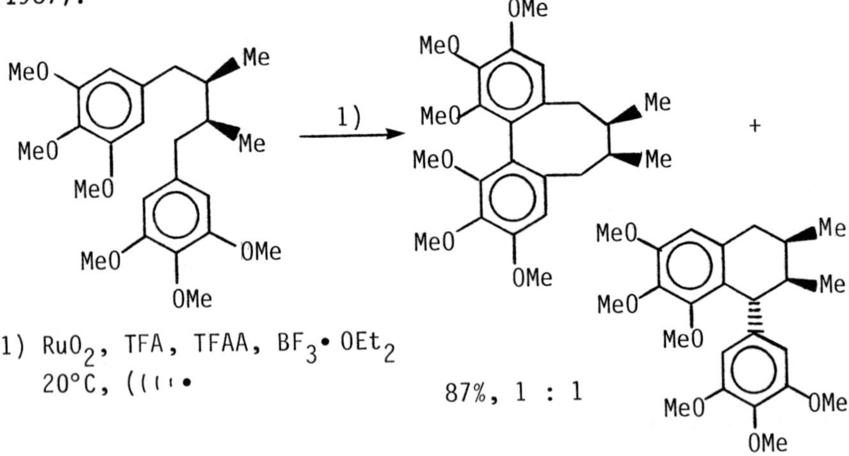

1) RuO_2, TFA, TFAA, $BF_3 \cdot OEt_2$
20°C, (((•

87%, 1 : 1

I.F.2-15 R.N. Butler et al., J. Chem. Soc., Perkin Trans. 1, 1039 (1987).

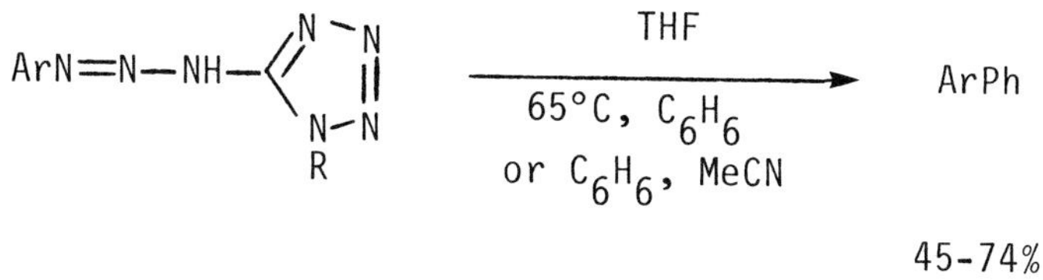

I.F.2-16 D. Stossel and T.H. Chan, J. Org. Chem., 52, 2105 (1987).

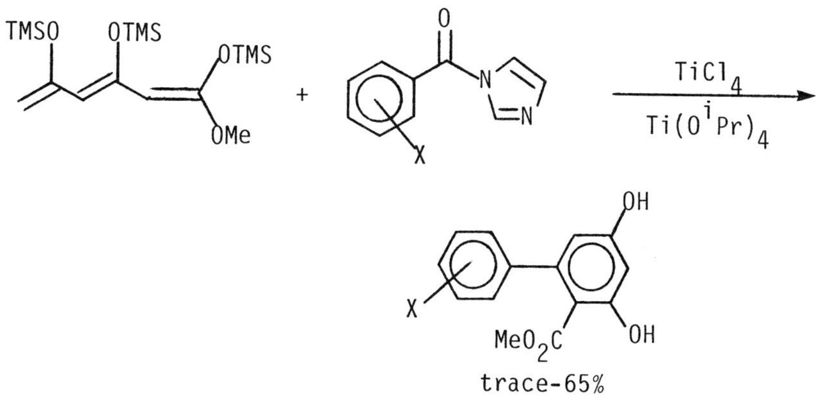

I.F.2-17 M. Tashiro et al., J. Org. Chem., 52, 3196 (1987); T. Yamato and M. Tashiro, J. Chem. Res. (S), 198 (1987).

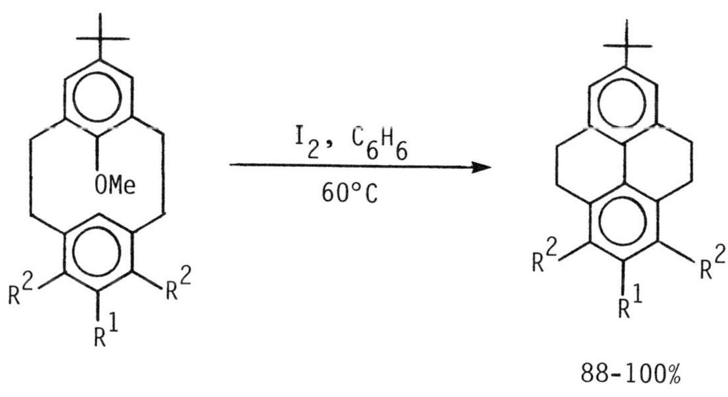

I.F.2-18 H.W. Moore et al., J. Org. Chem., 52, 2530 and 2537 (1987).

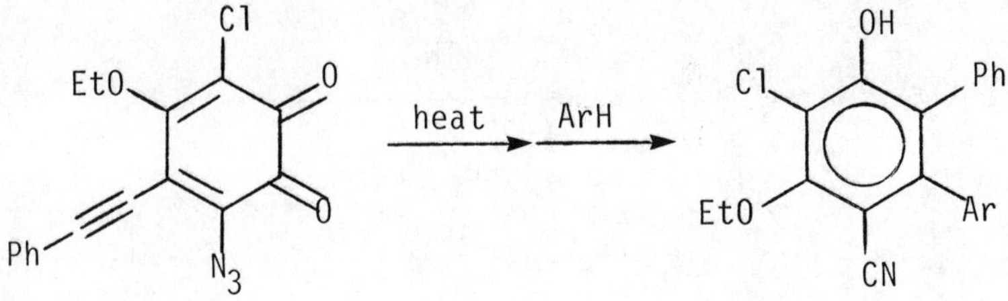

no yield given

I.F.2-19 D.J. Ackland and J.T. Pinhey, J. Chem. Soc., Perkin Trans. 1, 2689 and 2695 (1987).

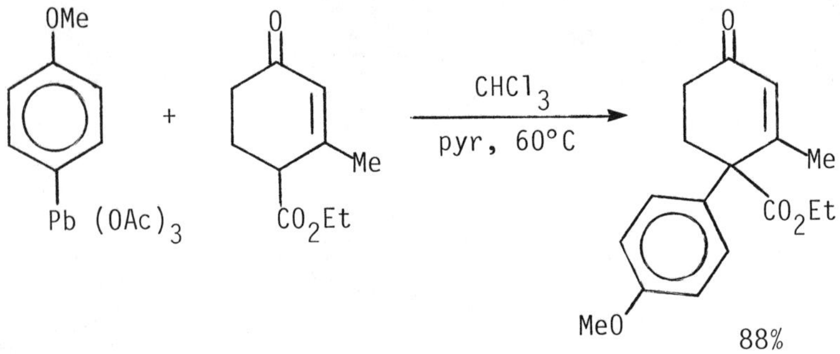

88%

I.F.2-20 M. Oda et al., Chem. Lett., 1263 (1987); K. Nakatani, Tetrahedron Lett., 28, 165 (1987); V. Barre and D. Uguen, ibid., 28, 6045 (1987); I. Murata et al., Angew. Chem., Int. Ed. Engl., 26, 1247 (1987).

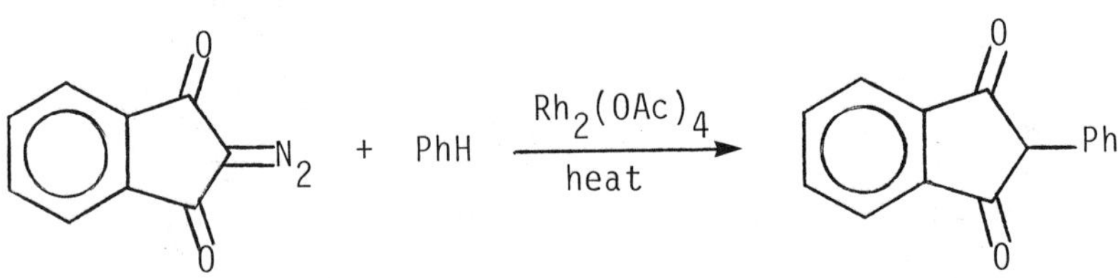

81%

in some cases rearrangement also observed
see also I.A.8.

I.F.2-21 W.R. Cullen et al., <u>Chem. Commun.</u>, 439 (1987).

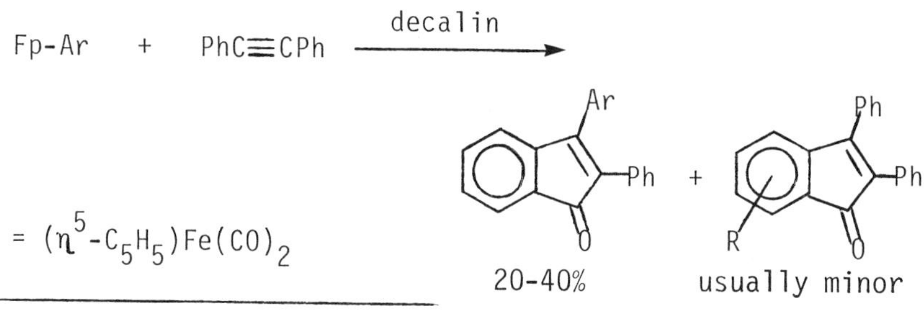

Fp = (η^5-C_5H_5)Fe$(CO)_2$

I.F.3. Other Aromatic Substitutions

I.F.3-1 G.P. Ellis and T.M. Romney-Alexander, <u>Chem. Rev.</u>, <u>87</u>, 779 (1987).

Review: "Cyanation of Aromatic Halides."

I.F.3-2 A. Collet, <u>Tetrahedron</u>, <u>43</u>, 5725 (1987).

Review: "Cyclotriveratrylenes and Cryptophanes."

I.F.3-3 T.C.W. Mak and H.N.C. Wong, <u>Top. Curr. Chem.</u>, <u>140</u>, 141 (1987).

Review: "Inclusion Properties of Tetraphenylene and Synthesis of its Derivatives."

I.F.3-4 C.K. Bradsher, <u>Chem. Rev.</u>, <u>87</u>, 1277 (1987).

Review: "Formation of Six-Membered Aromatic Rings by Cyclialkyation of Some Aldehydes and Ketones."

I.F.3-5 N.S. Narasimhan and R.S. Mali, Top. Curr. Chem., 138, 63 (1987).

Review: "Heteroatom Directed Aromatic Lithiation Reactions for the Synthesis of Condensed Heterocyclic Compounds."

I.F.3-6 I. Kubo, T. Kamikawa et al., Tetrahedron, 43, 2653 and 3387 (1987); R.S. Mali et al., Synthesis, 630 (1987); A.R. Katritzky and W.-Q. Fan, Org. Prep. Proced. Int., 19, 263 (1987); A.R. Katritzky et al., Heterocycles, 26, 3135 (1987); R.D. Bindal and J.A. Katzenellenbogen, J. Org. Chem., 52, 3181 (1987); K. Katsuura and V. Snieckus, Can. J. Chem., 65, 124 (1987); A.I. Meyers and D.A. Dickman, J. Am. Chem. Soc., 109, 1263 (1987); V. Snieckus et al., Tetrahedron Lett., 28, 6561 (1987); A. Costa and J.M. Saa, ibid., 28, 5551 (1987); J.M. Saa and A. Llobera, ibid., 28, 5045 (1987).

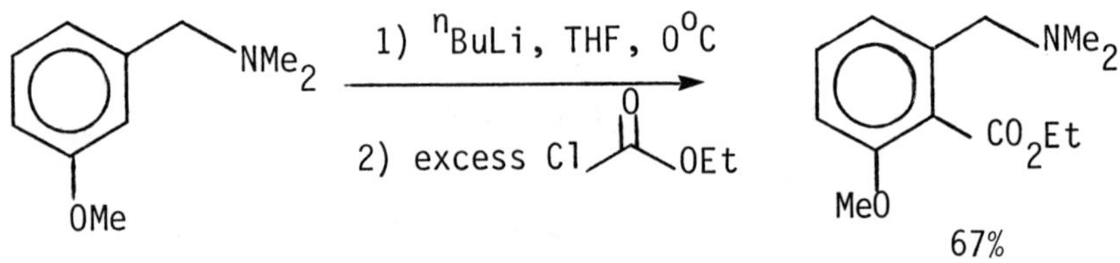

various other groups employed for directed lithiation

I.F.3-7 E.P. Kundig et al., Organometallics, 6, 1173 (1987); J.L. Davidson, H. Patel and P.N. Preston, J. Organomet. Chem., 336, C44 (1987).

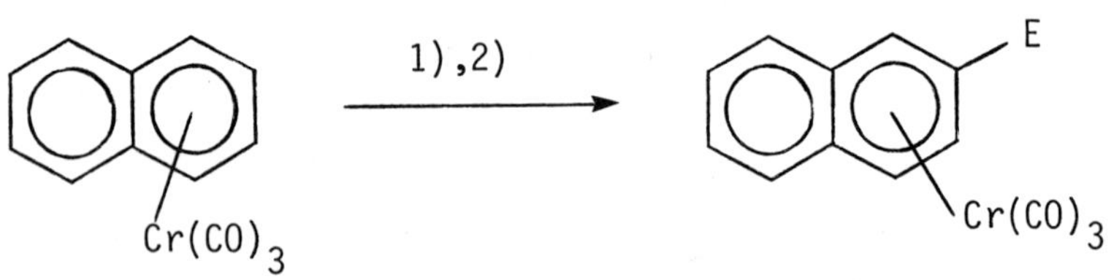

1) TMPLi, THF 2) E^+

I.F.3-8 F. Rose-Munch, E. Rose et al., Bull. Soc. Chim. Fr., 640 (1987); R.G. Sutherland, C.C. Lee et al., Synth. Commun., 17, 393 (1987).

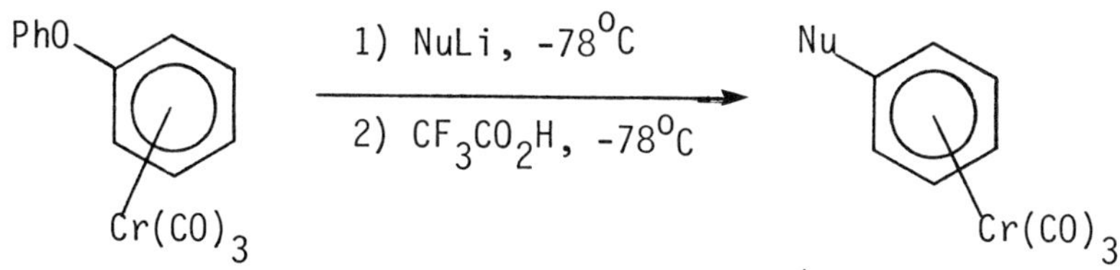

Nu = CH_2CN, CH(Me)CN, $C(Me)_2CN$, $CH(Me)CO_2{}^tBu$ 52-86%

similar results with halo or nitroarene $^+$FeCp PF_6^- complexes

---

I.F.3-9 M. Makosza et al., Liebigs Ann. Chem., 28, 711 (1987); M. Makosza and Z. Owczarczyk, Tetrahedron Lett., 28, 3021 (1987); M. Makosza and J. Winiarski, Acc. Chem. Res., 20, 282 (1987).

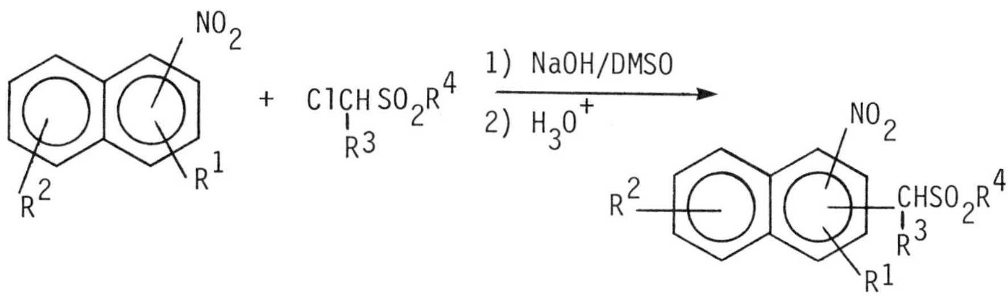

vicarious nucleophilic substitution of hydrogen 9-93%

I.F.3-10 T.C. McKenzie and W. Hassen, Tetrahedron Lett., 28, 2563 (1987); P. Sutter and C.D. Weis, J. Heterocycl. Chem., 24, 69 (1987).

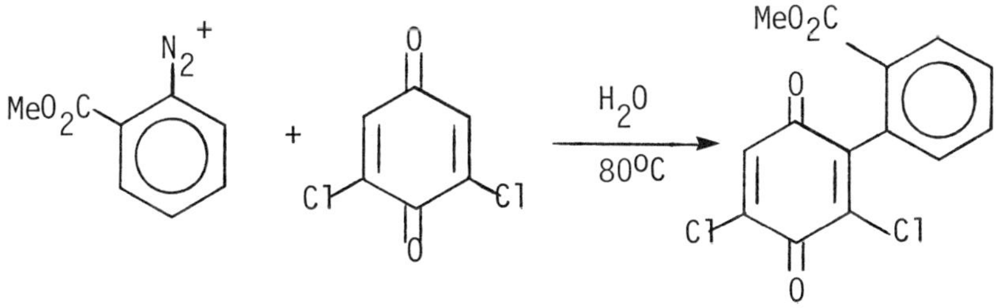

84%

I.F.3-11 M. Novi et al., Tetrahedron Lett., 28, 1345 (1987); G. Petrillo, Tetrahedron, 43, 4625 (1987).

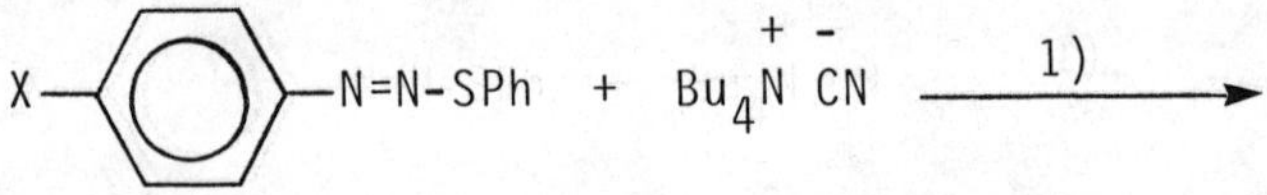

X = electron withdrawing

$4\text{-}XC_6H_4CN$ + $4\text{-}XC_6H_4SPh$

4-74% 3-40%

1) sunlamp, rt

I.F.3-12 E.R. Biehl et al., J. Org. Chem., 52, 5685 (1987); Y. Nishimura et al., Chem. Lett., 1655 (1987).

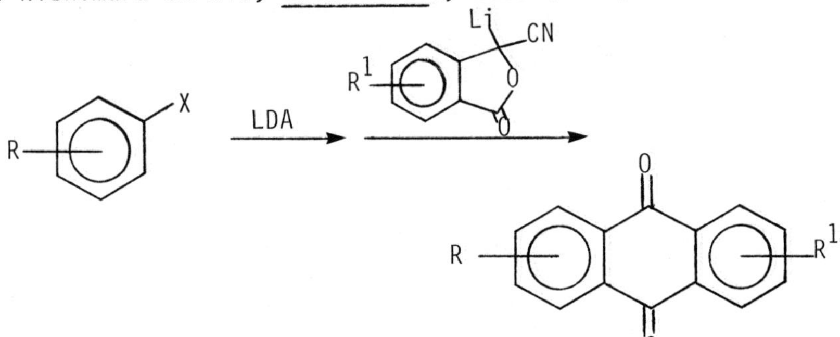

32-75%

I.F.3-13 K.M. Nicholas et al., Tetrahedron Lett., 28, 5415 (1987).

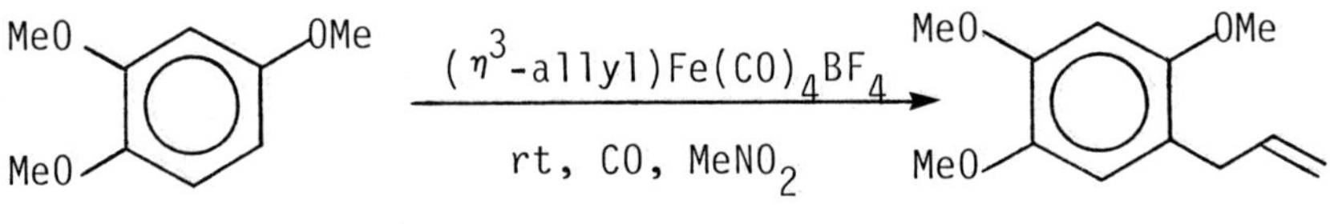

87%

other activated aryls also used

I.F.3-14 S.P. Khanapure and E.R. Biehl, J. Org. Chem., 52, 1333 (1987).

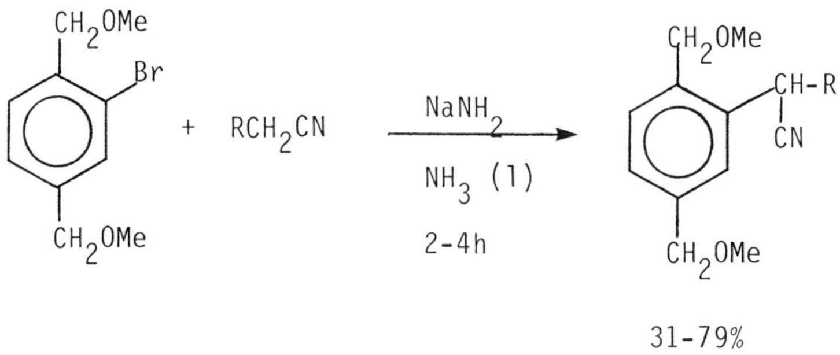

I.F.3-15 A.L.J. Beckwith et al., Chem. Commun., 666 (1987).

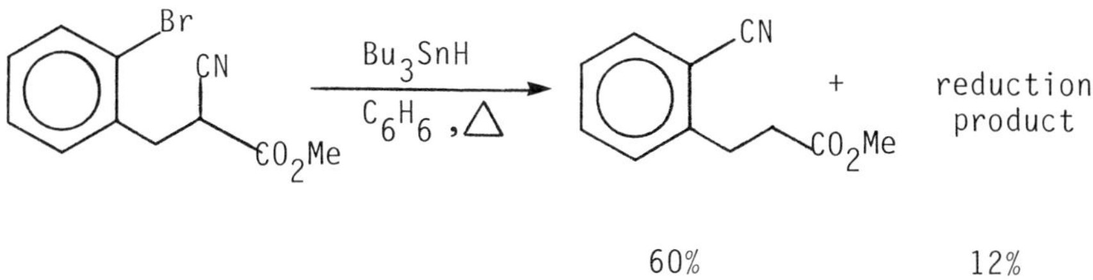

I.F.3-16 R. Beugelmans et al., J. Org. Chem., 52, 3880 (1987).

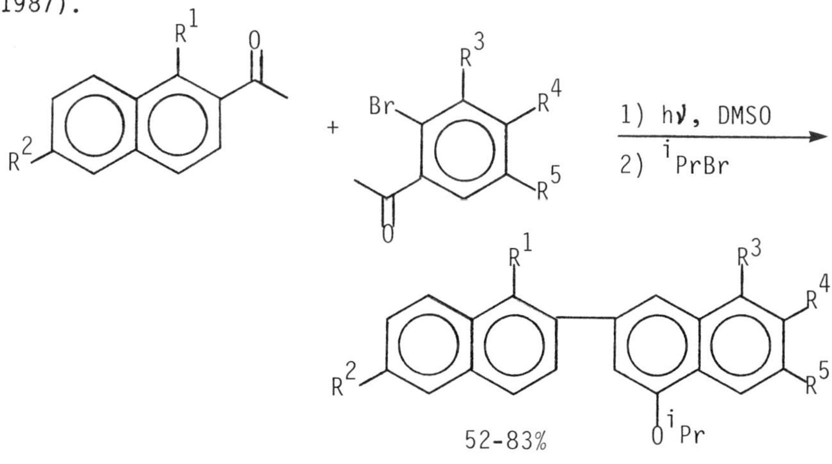

I.F.3-17 S.L. Buchwald et al., Tetrahedron Lett., 28, 3245 (1987); idem, J. Am. Chem. Soc., 109, 7137 (1987).

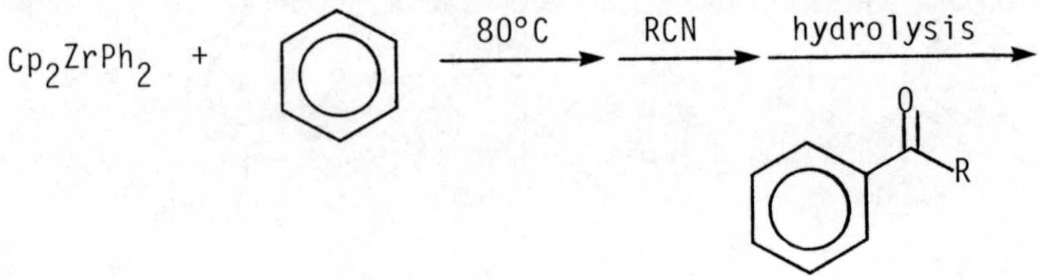

40-86%

I.F.3-18 F.M. Nongrum and B. Myrboh, Synthesis, 845 (1987).

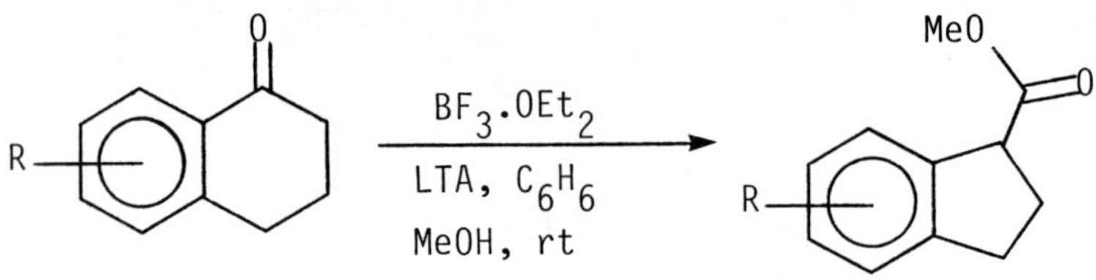

53-68%

I.F.3-19 G. Casiraghi, M. Cornia et al., Chem. Commun., 794 (1987); C. Monneret et al., J. Org. Chem., 52, 1051 (1987); V.I. Dyachenko et al., J. Org. Chem. (USSR), 23, 808 (1987); R.A.M.C. De Groote et al., Bull. Soc. Chim. Fr., 89 (1987); V. Snieckus et al., Tetrahedron Lett., 28, 2933 (1987).

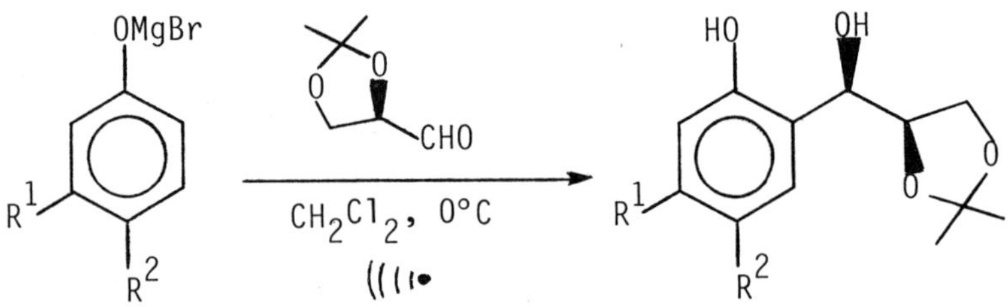

63-76%, 91-93% de

ANTI product favoured by $Ti(O^iPr)_3$ phenolate

I.F.3-20 M.H. Nantz and P.L. Fuchs, Synth. Commun., 17, 761 (1987); H. Ila, H. Junjappa et al., Synthesis, 284 and 873 (1987); idem, Tetrahedron Lett., 28, 1459 (1987); K. Eichinger et al., Synthesis, 1061 (1987); D. Murali and G.S.K. Rao, ibid., 254 (1987).

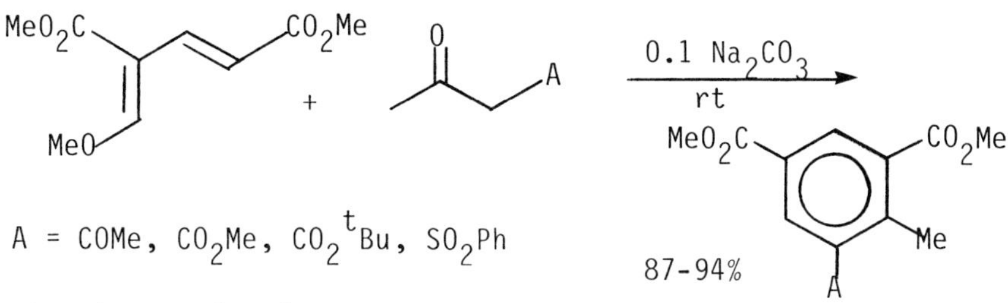

A = COMe, CO_2Me, $CO_2{}^tBu$, SO_2Ph

other bases, leaving groups and nucleophiles also used

I.F.3-21 G.S.K. Rao et al., Synthesis, 197 and 231 (1987); idem, Ind. J. Chem., 26B, 173, 175 and 177 (1987).

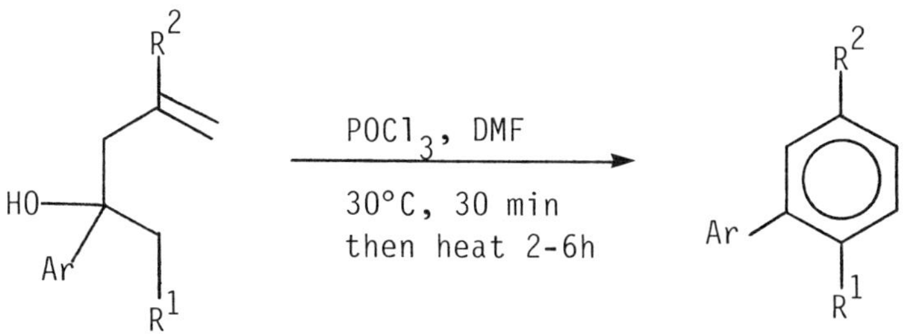

30-98%

I.F.3-22 F.M. Abdel-Galil and M.H. Elnagdi, Liebigs Ann. Chem., 477 (1987); M.H. Elnagdi et al., Coll. Czech. Chem. Commun., 52, 1561 (1987); K. Gewald et al., Z. Chem., 27, 32 (1987); A.M. El-Torgoman et al., Chem. Ind., 60 (1987).

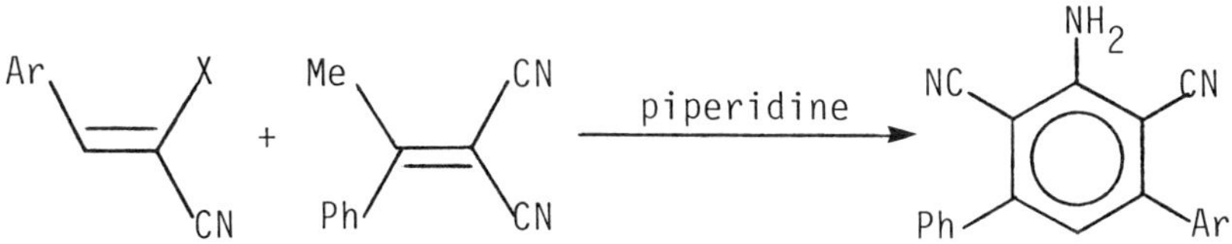

65-80%

Ar = 4-$MeOC_6H_4$, 4-ClC_6H_4, 2-furyl

I.F.3-23 M. Hidai et al., Chem. Commun., 575 (1987).

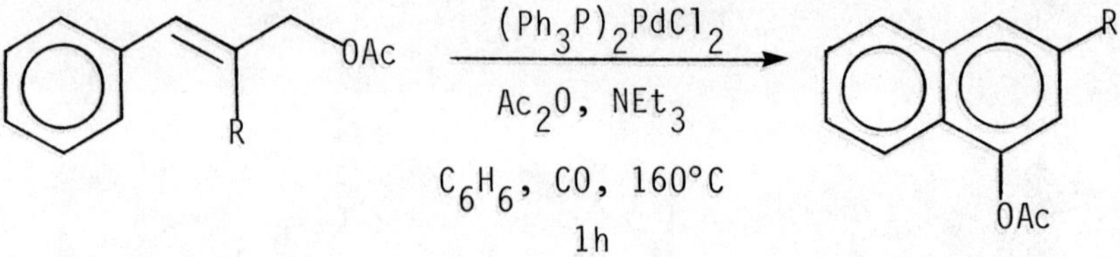

74-76%

76-92% conversion

I.F.3-24 J.R. Peterson et al., Tetrahedron Lett., 28, 6109 (1987).

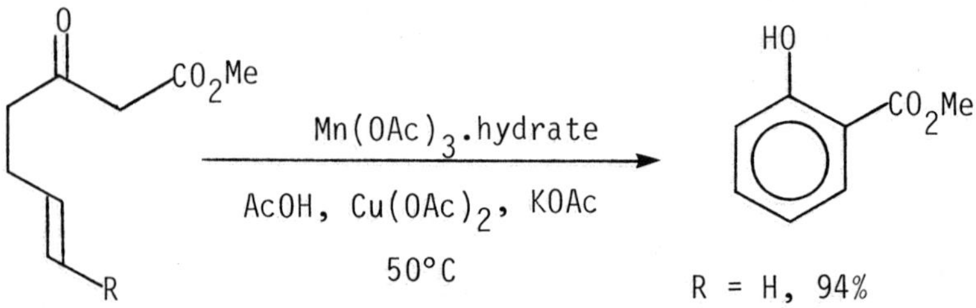

R = H, 94%

R = Ph - a cyclopentanone formed

I.F.3-25 M.E. Jung and D.D. Grove, Chem. Commun., 753 (1987).

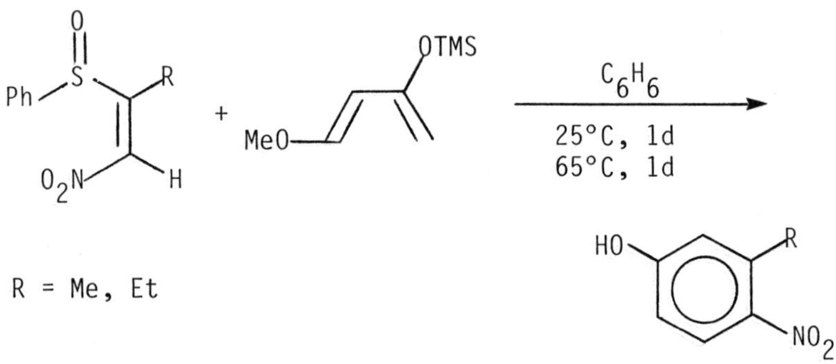

R = Me, Et

52-73%

I.F.3-26 W.D. Wulff et al., Tetrahedron Lett., 28, 1381 (1987); idem, J. Organomet. Chem., 334, 9 (1987); K.H. Dotz et al., ibid., 334, 57 (1987); K.H. Dotz and M. Popall, Angew. Chem., Int. Ed. Engl., 26, 1158 (1987).

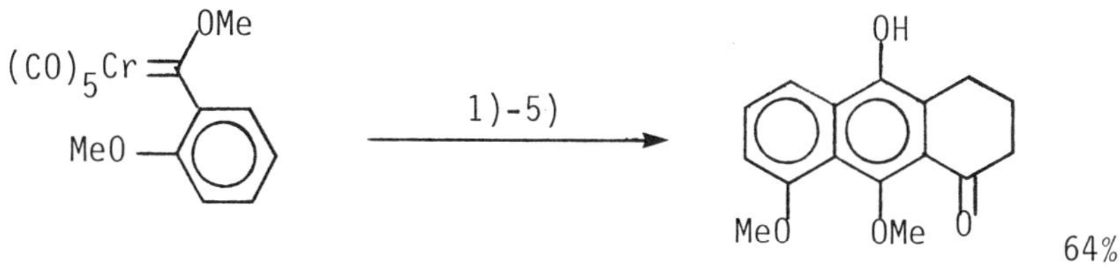

1) $\equiv$—CH$_2$CH$_2$CH$_2$—$CO_2{}^tBu$, CH_2Cl_2, 40°C, 36h 2) air, 25°C, 15min
3) TFAA, NaOAc, 10min 4) TFA, 45min 5) HO^-

I.F.3-27 W. Ding et al., Tetrahedron Lett., 28, 81 (1987).

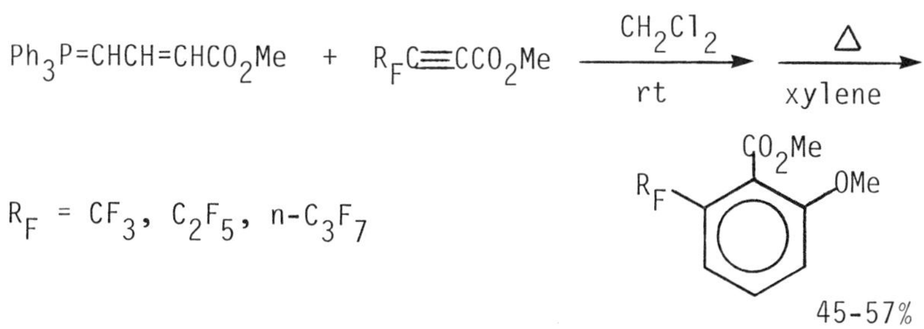

I.F.3-28 M. Unroe and B.A. Reinhardt, Synthesis, 981 (1987).

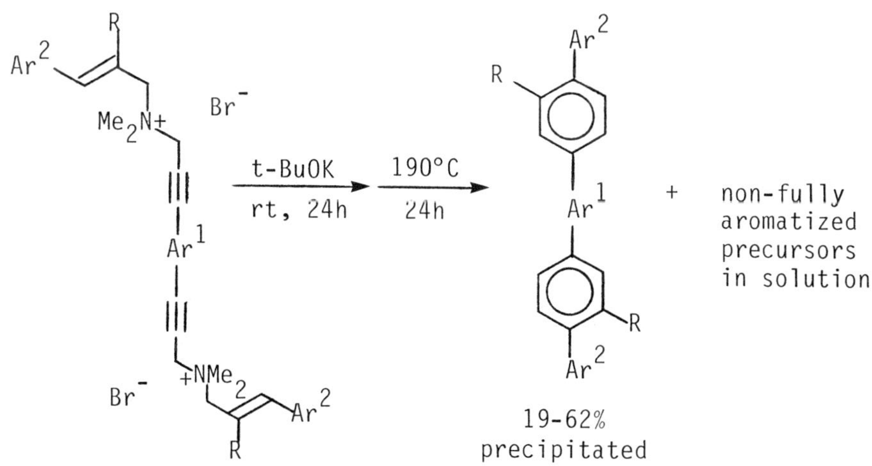

I.F.3-29 M. Ariga and E. Matsumura, Bull. Chem. Soc. Jpn., 60, 1198 (1987).

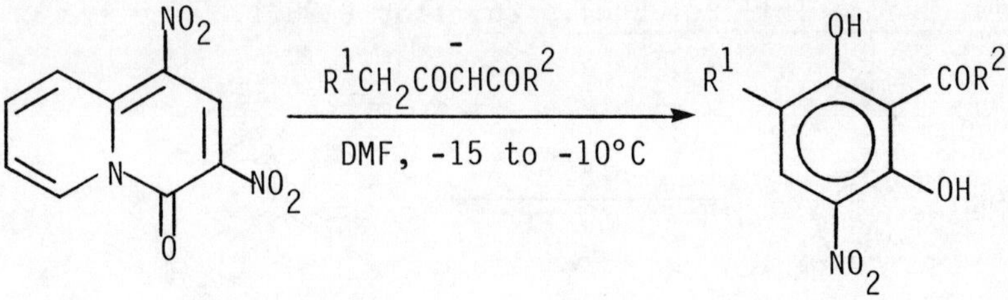

25-76%

I.F.3-30 T. Zimmermann and G.W. Fischer, J. Prakt. Chem., 329, 499 (1987); C. Roussel et al., Bull. Soc. Chim. Fr., 205 (1987).

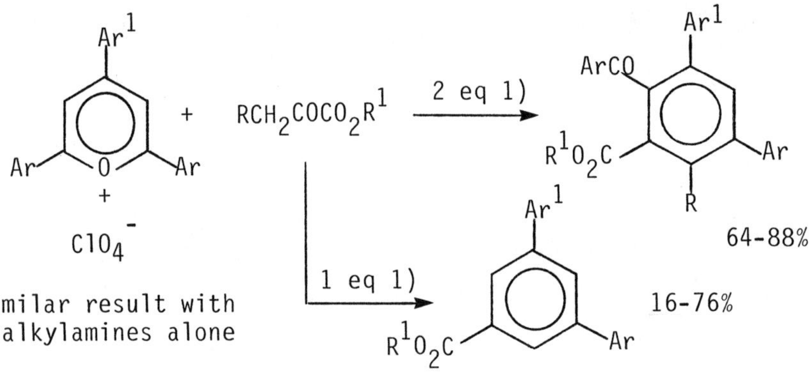

similar result with dialkylamines alone

1) dialkylamine salts of weak acids, NEt_3/AcOH or NaOAc

I.F.3-31 K. Fukumoto et al., Chem. Lett., 2113 and 2117 (1987); S.T. Perri and H.W. Moore, Tetrahedron Lett., 28, 4507 (1987).

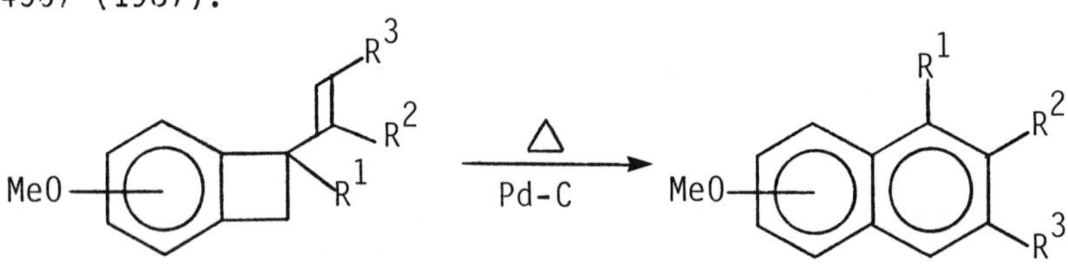

70-96%

I.F.3-32 A.G. Schultz et al., J. Org. Chem., 52, 5482 (1987); J.A. Valderrama et al., J. Chem. Soc., Perkin Trans. 1, 2855 (1987).

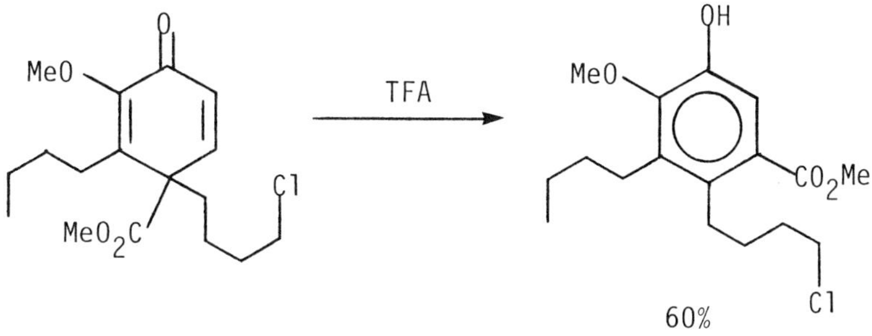

I.F.3-33 M.F. Semmelhack et al., J. Am. Chem. Soc., 109, 4397 (1987).

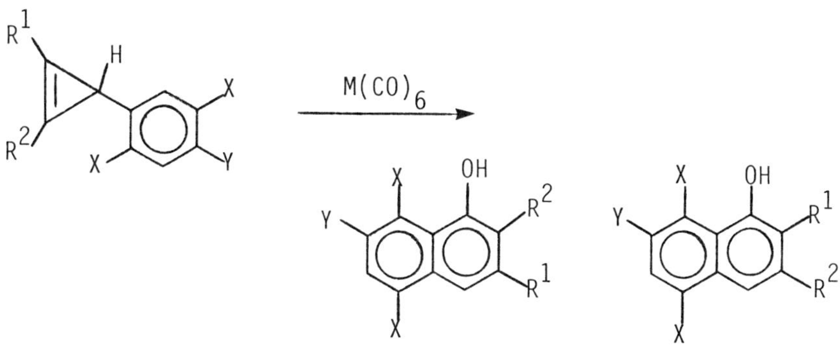

35-76%, 100 : 0 to 60 : 40

I.F.3-34 R.E. Lehr et al., J. Org. Chem., 52, 5574 (1987); R. Cosmo and S. Sternhell, Aust. J. Chem., 40, 1499 (1987).

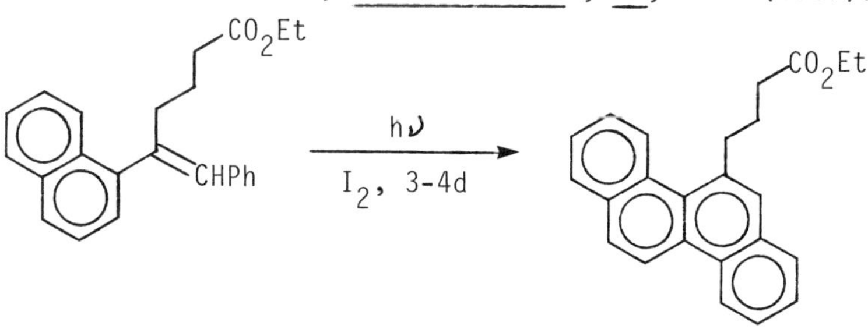

53%

I.F.3-35 A.K. Jhingan and W.F. Maier, J. Org. Chem., 52, 1161 (1987); K.P.C. Vollhardt et al., Angew. Chem., Int. Ed. Engl., 26, 1246 (1987).

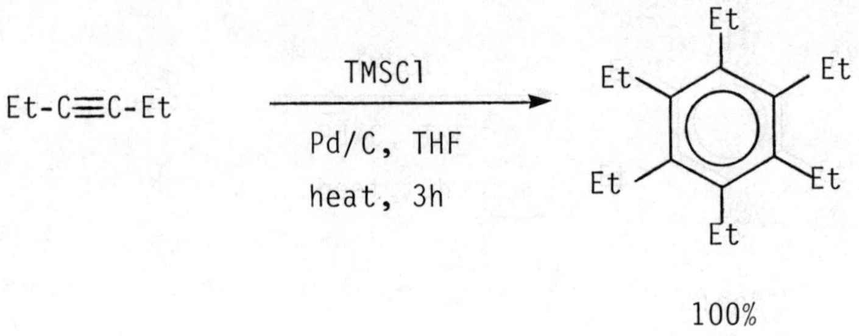

I.F.3-36 M. Hanack and R. Rieth, Chem. Ber., 120, 1659 (1987).

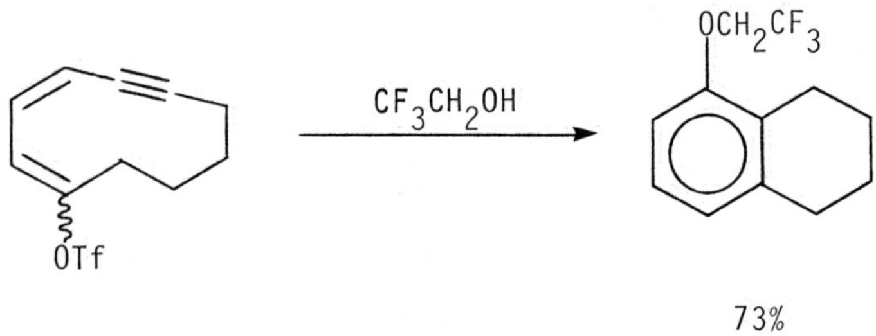

I.F.3-37 K. Chow and H.W. Moore, Tetrahedron Lett., 28, 5013 (1987).

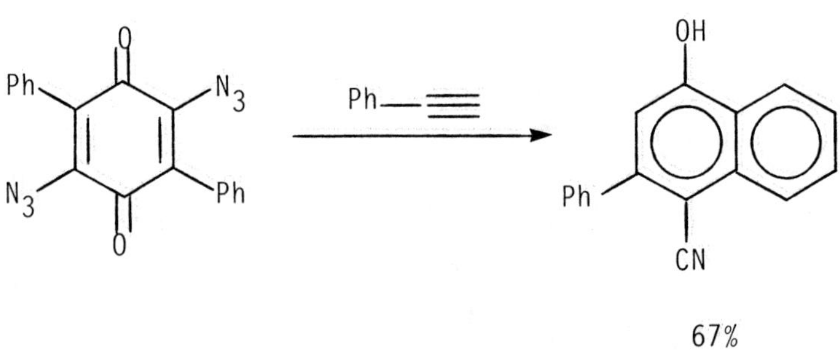

I.F.3-38 H. Prinzbach et al., J. Am. Chem. Soc., 109, 4626 (1987); A. de Meijere et al., Angew. Chem., Int. Ed. Engl., 26, 1277 (1987).

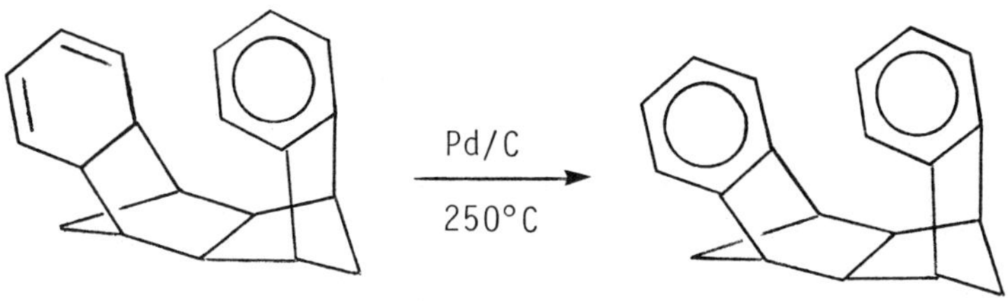

70% (major)

towards Pagodane

I.G. Synthesis via Organometallics

I.G.1. Synthesis via Organoboranes

I.G.1-1 K. Smith, Chem. Ind., 603 (1987).

Review: "Advances in Organoboron Chemistry - Prospects for Industry."

I.G.1-2 G.W. Kabalka, J. Organomet. Chem., 337, 169 (1987).

Review: "Boranes in Organic Synthesis. Annual Survey Covering the Year 1985."

I.G.1-3 Pure and Appl. Chem., 59, 837-914 (1987).

Reviews: "6th International Symposium on Boron Chemistry, Bechyne, Czechoslovakia, 22-26 June 1987."

I.G.1-4 M. Srebnik and P.V. Ramachandran, Aldrichimica Acta, 20, 9 (1987).

Review: "The Utility of Chiral Organoboranes in the Preparation of Optically Active Compounds."

I.G.1-5 A. Arase et al., Chem. Commun., 531 and 1629 (1987).

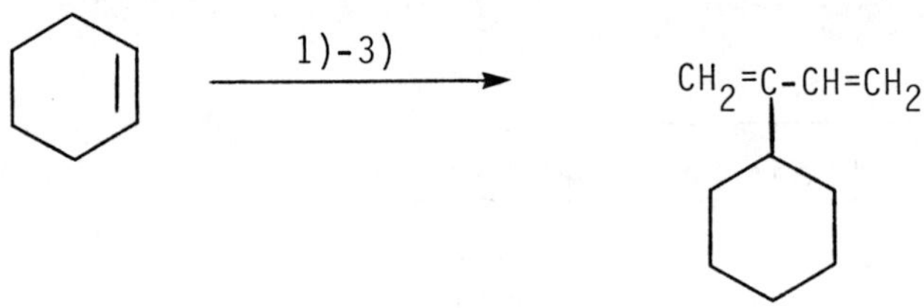

1) BH_3 , THF 2) $ClCH_2C{\equiv}CCH_2Cl$, 0°C, 2h 3) MeLi

I.G.1-6 F. Sato et al., Tetrahedron Lett., 28, 5849 (1987); Y. Kishi et al., J. Am. Chem. Soc., 109, 4756 (1987).

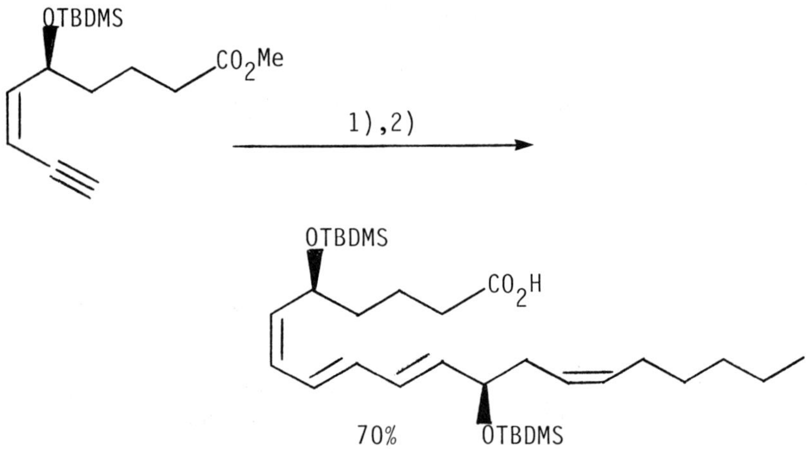

1) $(Sia)_2BH$, THF, 0°C 2) 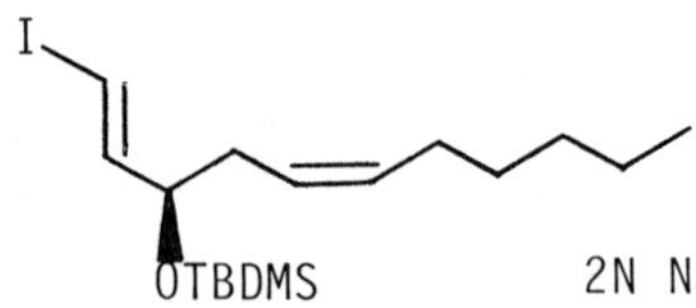

, $Pd(PPh_3)_4$ 2N NaOH, 50°C

I.G.1-7 J.Y. Gauthier and Y. Guindon, Tetrahedron Lett., 28, 5985 (1987).

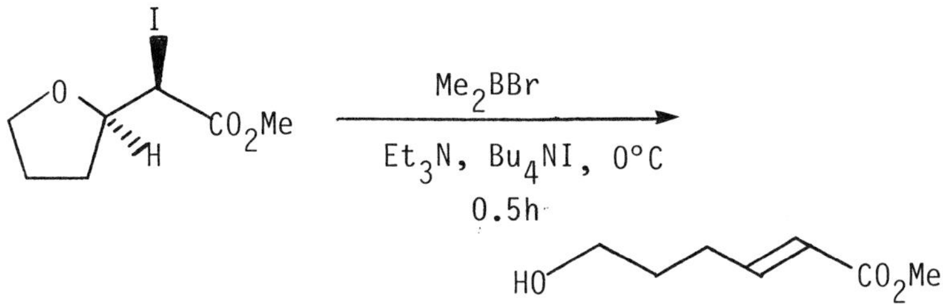

91%

I.G.1-8 H.C. Brown et al., J. Am. Chem. Soc., 109, 5420 (1987); idem, Tetrahedron Lett., 28, 3771 (1987).

91%

1) LiO^tBu, Et_2O 2) H_2O_2, phosphate buffer, pH8

also extended to chiral borinic esters

I.G.1-9 K.K. Wang et al., Tetrahedron Lett., 28, 1003 and 1007 (1987); G. Zweifel et al., J. Org. Chem., 52, 5484 (1987).

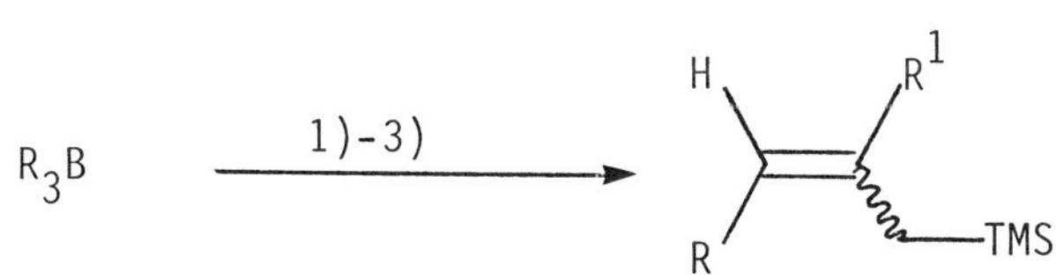

45-81%

1) LiC≡CR^1 2) $CF_3SO_3CH_2TMS$ 3) $MeCO_2H$

I.G.1-10 Z. Weike et al., Tetrahedron Lett., 28, 2599 (1987).

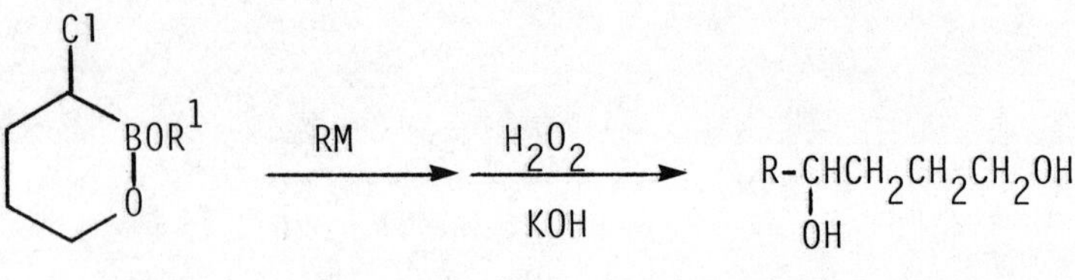

M = Li, MgBr

25-80%

I.G.1-11 D.S. Matteson and E.C. Beedle, Tetrahedron Lett., 28, 4499 (1987); D.S. Matteson et al., J. Org. Chem., 52, 5116 and 5121 (1987).

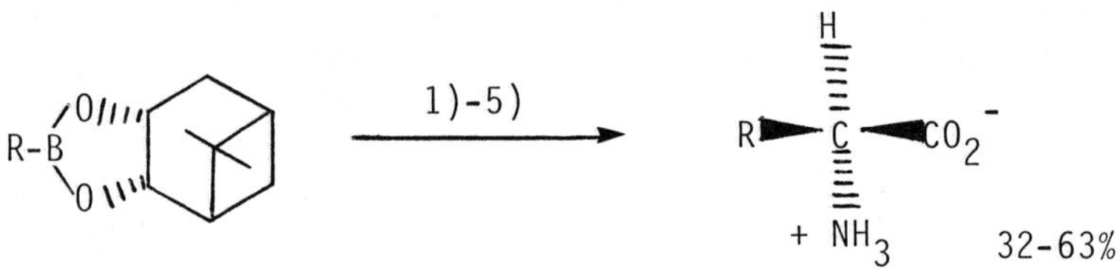

isomer ratio 25 : 1 to 50 : 1

1) $LiCHX_2$ 2) NaN_3 3) $LiCHCl_2$ 4) $NaClO_2$ 5) H_2/Pd

I.G.2. Carbonylation Reactions

I.G.2.1 T. Jintoku, H. Taniguchi and Y. Fujiwara, Chem. Lett., 1159 (1987); T. Sakakura and M. Tanaka, ibid., 249 and 1113 (1987); R. Ugo and A. Chiesa, J. Chem. Soc., Perkin Trans. 1, 2625 (1987).

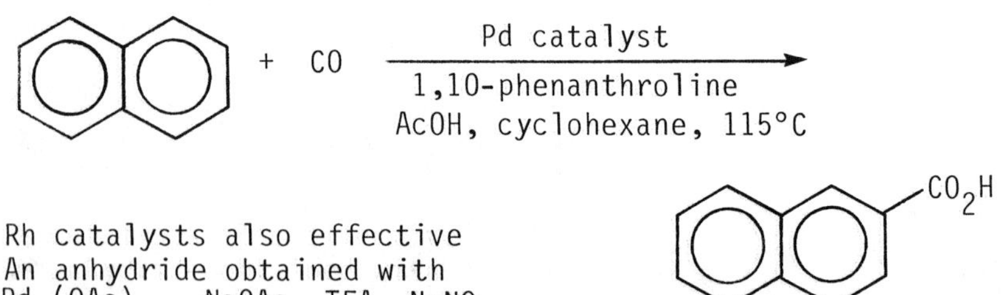

Rh catalysts also effective
An anhydride obtained with
$Pd_3(OAc)_6$, NaOAc, TFA, $NaNO_3$

421% [major]

I.G.2-2 M. Tanaka et al., *J. Org. Chem.*, *52*, 5733 (1987); M. Miura et al., *ibid.*, *52*, 2623 (1987); *idem*, *J. Chem. Soc., Perkin Trans. 1*, 1021 (1987); R.E. Dolle et al., *Chem. Commun.*, 904 (1987); A. Yamamoto et al., *Organometallics*, *6*, 1640 (1987); K. Kudo et al., *Chem. Lett.*, 577 (1987); M. Tanaka et al., *Tetrahedron Lett.*, *28*, 2721 (1987); idem, *J. Organomet. Chem.*, *334*, 205 (1987); C.J. Elsevier et al., *ibid.*, *325*, C23 (1987); N. Satyanarayana and M. Periasamy, *ibid.*, *333*, C33 (1987); idem, *Tetrahedron Lett.*, *28*, 2633 (1987).

$$\mathrm{PhI} + \mathrm{MeOH} \xrightarrow[\mathrm{NEt_3,\ C_6H_6,\ CO}\quad 80^\circ\mathrm{C,\ 14h}]{\mathrm{PdCl_2(PPh_3)_2}} \underset{67\%}{\mathrm{PhCOCO_2Me}} + \mathrm{PhCO_2Me}$$

similar results with different Pd or Co species.

Depending on catalyst, nucleophiles and conditions : carboxylic acids, esters, acyl fluorides or ketones can be obtained

---

I.G.2-3 K. Kikukawa et al., *J. Chem. Soc., Perkin Trans. 1*, 1511 (1987).

$$\mathrm{ArN_2X} + \mathrm{Me_4Sn} \xrightarrow[\mathrm{CO,\ MeCN}\quad \mathrm{rt}]{\mathrm{Pd(OAc)_2}} \underset{55\text{-}93\%}{\mathrm{ArCOMe}}$$

I.G.2-4 N.E. Schore et al., J. Org. Chem., 52, 569 and 3595 (1987); P. Magnus et al., ibid., 52, 1483 (1987); P. Magnus and D.P. Becker, J. Am. Chem. Soc., 109, 7495 (1987); S.L. Buchwald et al., ibid., 109, 2544 (1987); P. Hong et al., J. Organomet. Chem., 334, 129 (1987); R. Aumann and H.-J. Weidenhaupt, Chem. Ber., 120, 23 (1987); F. Camps et al., Tetrahedron Lett., 28, 4745 (1987).

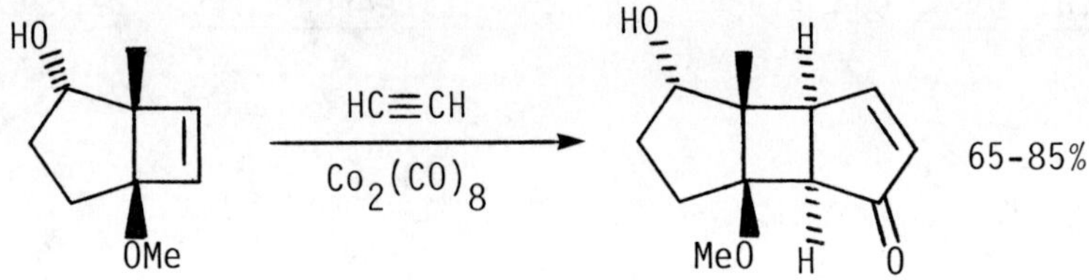

similar results with $Ni(CO)_4$, zirconocene-alkyne complexes/CO, $Rh_4(CO)_{12}$/CO and $Fe(CO)_5$/CO / hν

I.G.2-5 S. Uemura et al., J. Organomet. Chem., 326, 35 (1987); idem, J. Org. Chem., 52, 4859 (1987).

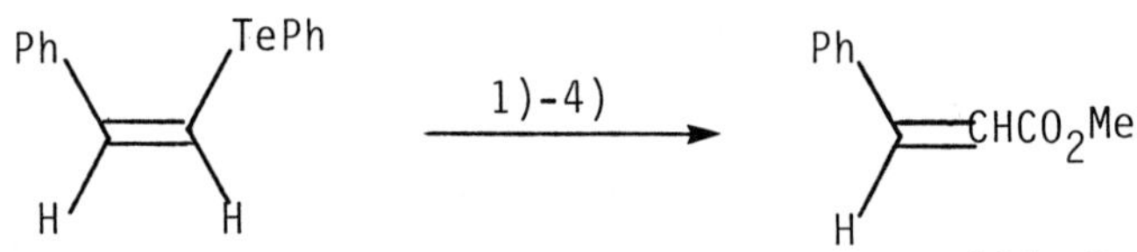

83%, Z:E 34:66

1) CO, Li_2PdCl_4, MeCN 2) ^-OH 3) H_3O^+ 4) CH_2N_2

I.G.2-6 L. Kollar et al., J. Organomet. Chem., 330, 305 (1987); R. Choukroun, P. Kalck et al., ibid., 335, C9 (1987); G. Parrinello and J.K. Stille, J. Am. Chem. Soc., 109, 7122 (1987); I. Ojima et al., ibid., 109, 7714 (1987); M. Ichikawa et al., Chem. Lett., 941 (1987); J.M. Brown and A.G. Kent, J. Chem. Soc., Perkin Trans. 2, 1597 (1987).

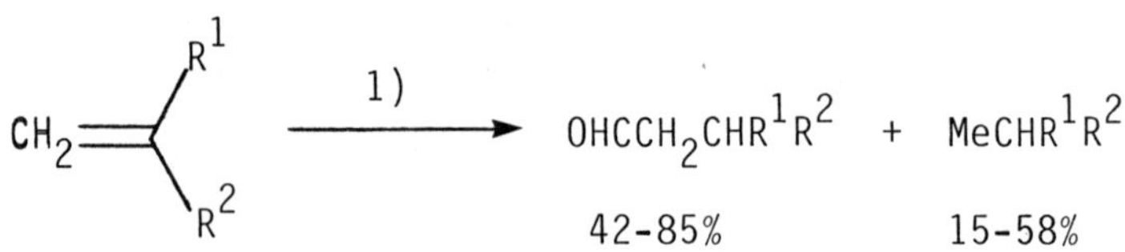

42-85% 15-58%
10-42.5% ee

1) $PtCl(SnCl_3)[(R,R)\text{-DIOP}]$, CO/H_2, 80bar, 100°C

other catalysts and conditions also used

I.G.2-7 S.E. Thomas, Chem. Commun., 226 (1987); D. Rakshit and S.E. Thomas, J. Organomet. Chem., 333, C3 (1987); P. Vogel et al., Helv. Chim. Acta, 70, 1791 (1987); F. Serratosa et al., J. Chem. Soc., Perkin Trans. 1, 2749 (1987).

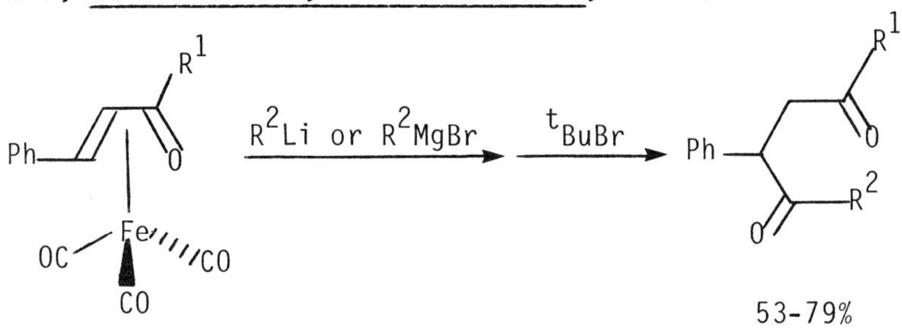

I.G.2-8 R.C. Larock et al., Tetrahedron, 43, 2891 (1987).

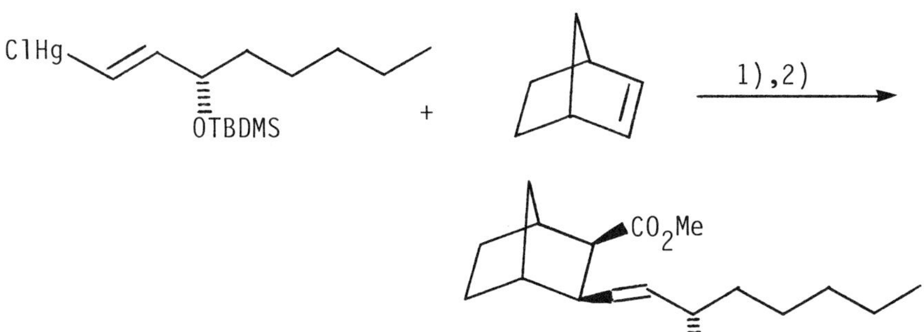

1) Li_2PdCl_4, THF 2) CO, MeOH, NEt_3

I.G.2-9 P. De Shong et al., Tetrahedron Lett., 28, 2229 and 2233 (1987); H. Alper et al., Organometallics, 6, 1591 (1987); P.P. Nicholas, J. Org. Chem., 52, 5266 (1987).

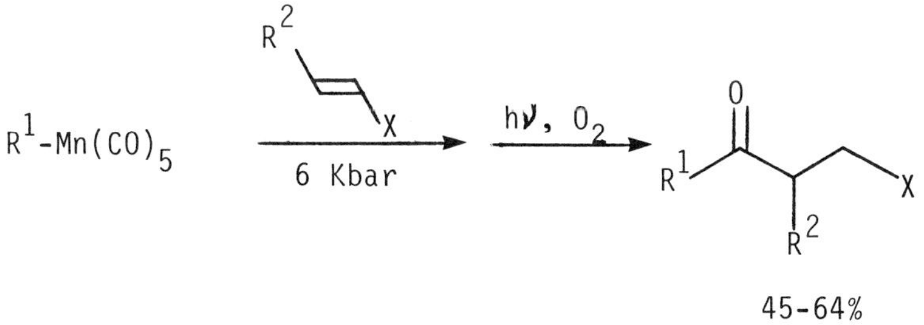

other catalysts with CO also used

I.G.2-10 N.S. Nudelman et al., J. Organomet. Chem., 332, 9 (1987).

$$\underset{1}{PhLi} + \underset{3}{BuBr} + CO \xrightarrow{THF} Ph_2C(OH)\text{-}Bu$$

1 : 3

80%

I.G.2-11 P.M. Maitlis et al., J. Organomet. Chem., 330, C31 (1987).

$$[C_5Me_5Ph(CO)(p\text{-}XC_6H_4)I] + MeI \xrightarrow{1)} p\text{-}XC_6H_4COMe$$

80-90%

1) sealed tube, 60°C

I.G.2-12 H. Alper et al., Organometallics, 6, 1625 (1987).

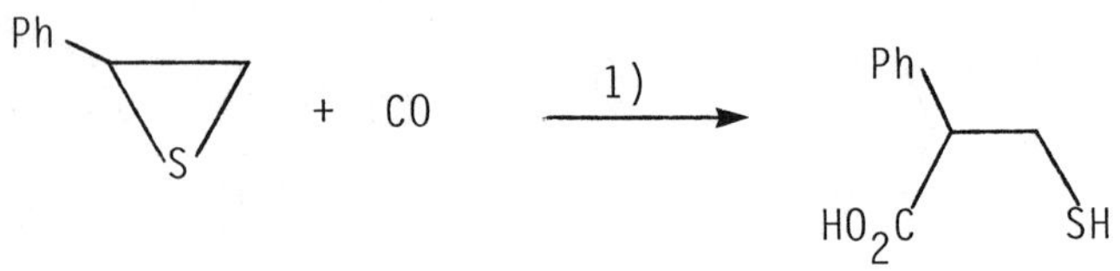

78%

1) MeI, 3N KOH, PEG-400, $Co_2(CO)_8$, C_6H_6, rt, 1 atm.

I.G.2-13 P. Magnus and M. Slater, Tetrahedron Lett., 28, 2829 (1987); V. Rautenstrauch, Helv. Chim. Acta, 70, 593 (1987); E. Lindner et al., Z. Naturforsch., 42b, 1527 (1987).

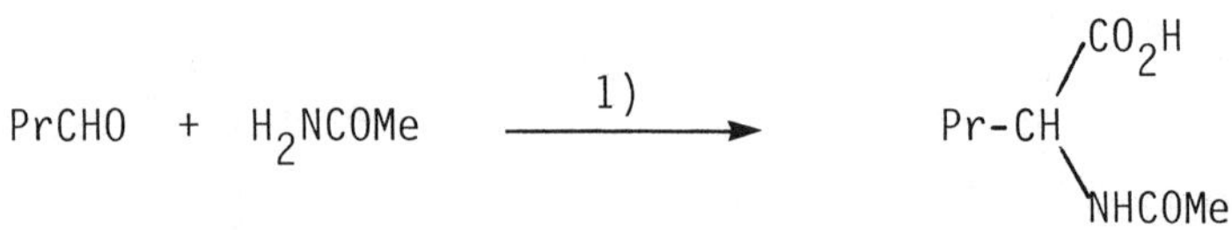

72%

1) $Co_2(CO)_8$, CO, H_2, 145-160°C, EtOAc

I.G.2-14 H. Hoberg et al., Angew. Chem., Int. Ed. Engl., 26, 153 (1987).

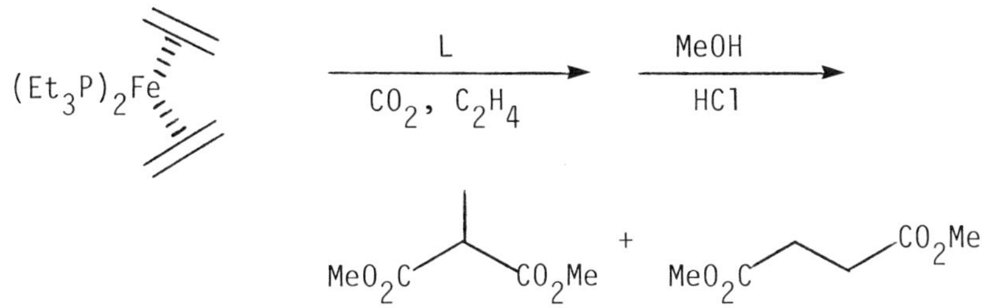

	L			
49-60%	L = PMe_3	400	:	1
	= DMPE	7	:	1
	= DCPE	1	:	10

DMPE = $Me_2PCH_2CH_2PMe_2$, DCPE = $(c\text{-}C_6H_{11})PCH_2CH_2P(c\text{-}C_6H_{11})$

I.G.2-15 H. Hoberg et al., Angew. Chem., Int Ed. Engl., 26, 571 (1987).

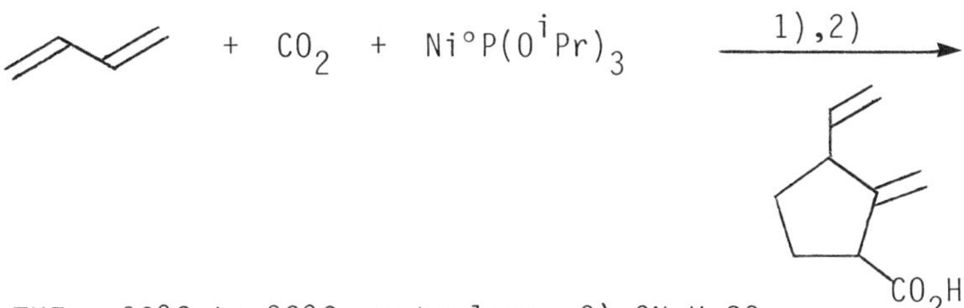

95%

1) THF, -20°C to 20°C, autoclave 2) 2N H_2SO_4

I.G.2-16 Y. Sasaki and P.H. Dixneuf, J. Org. Chem., 52, 314 and 4389 (1987).

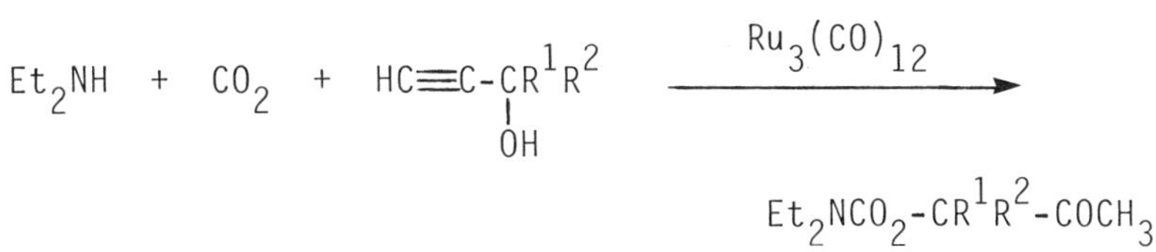

52-64%

I.G.3. Other Syntheses via Organometallics

I.G.3-1 T.M. Swager and R.H. Grubbs, J. Am. Chem. Soc., 109, 894 (1987).

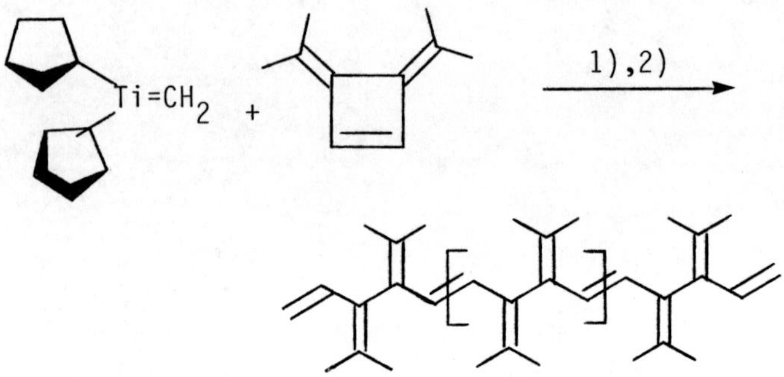

1) C_6H_6 2) MeOH

no yield given

I.G.3-2 Y. Ishii, M. Ogawa et al., J. Org. Chem., 52, 2239 (1987).

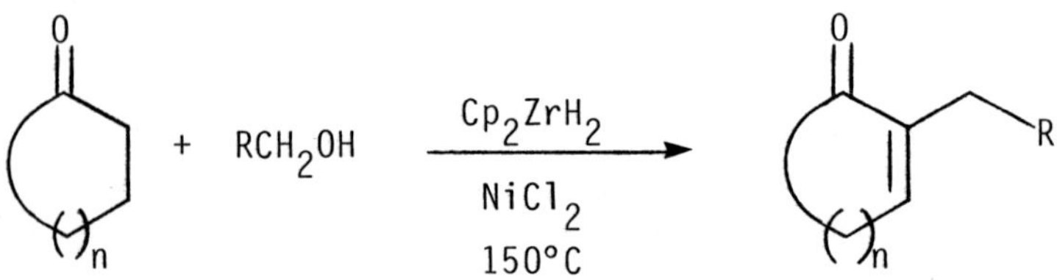

n = 2,1

25-48%

I.G.3-3 M. Franck-Neumann et al., J. Organomet. Chem., 326, 389 (1987).

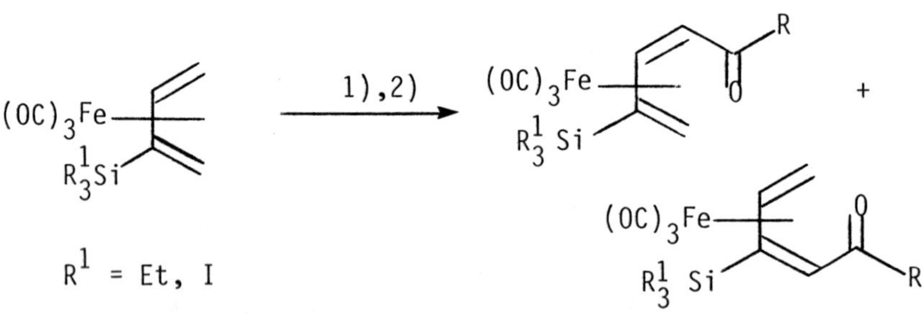

R^1 = Et, I

1) RCOCl, $AlCl_3$ 2) H_2O

89-92%, 2.7-8 : 1

I.G.3-4 M.F. Semmelhack and J. Park, J. Am. Chem. Soc., 109, 935 (1987).

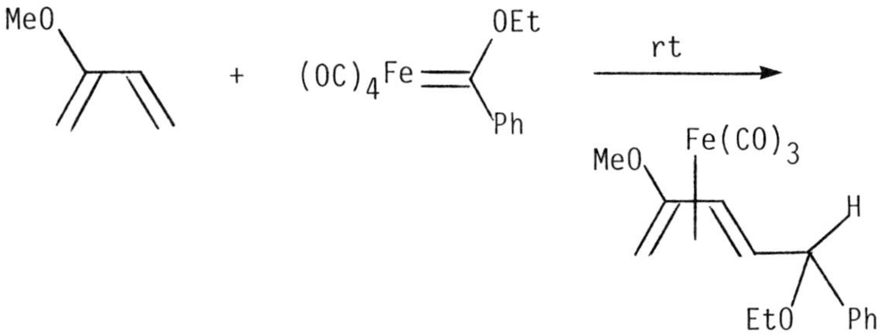

87%, 3 : 1

I.G.3-5 R. Aumann et al., Chem. Ber., 120, 1287 and 1293 (1987).

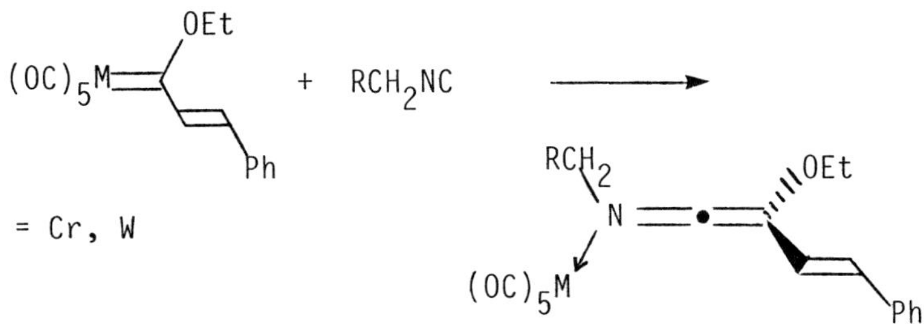

no yields given

I.G.3-6 L.A. Paquette and M.P. Trova, Tetrahedron Lett., 28, 2795 (1987).

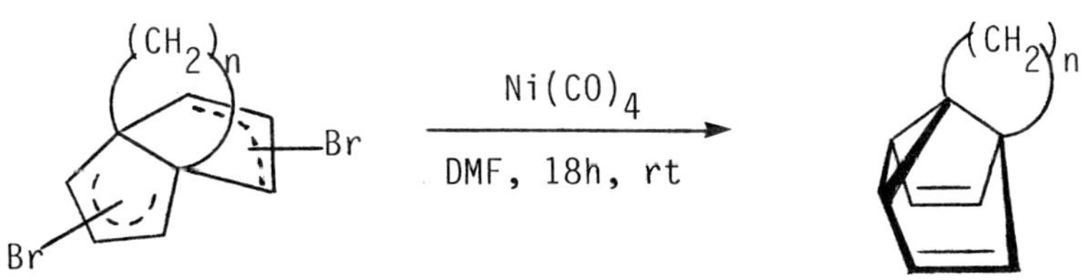

61-89%

I.G.3-7 R.F. Heck et al., Organometallics, 6, 2386 (1987).

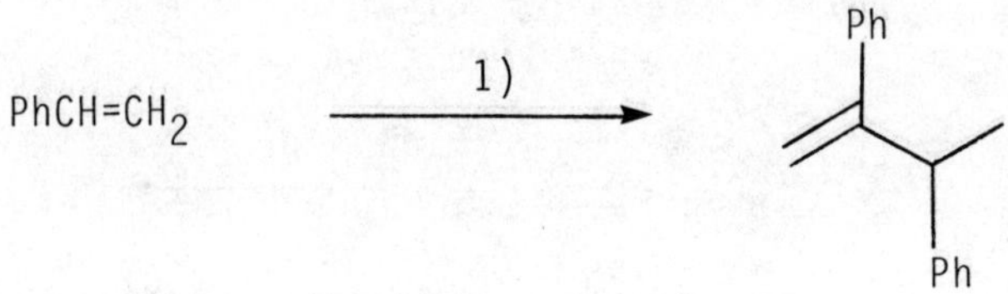

90%

1) (azobenzene-H)Pd$(MeNO_2)_2{}^+BF_4{}^-$, rt

I.G.3-8 E. Negishi et al., Chem. Lett., 623 (1987); W.A. Nugent et al., J. Am. Chem. Soc., 109, 2788 (1987); J.J. Eisch and M.P. Boleslawski, J. Organomet. Chem., 334, C1 (1987).

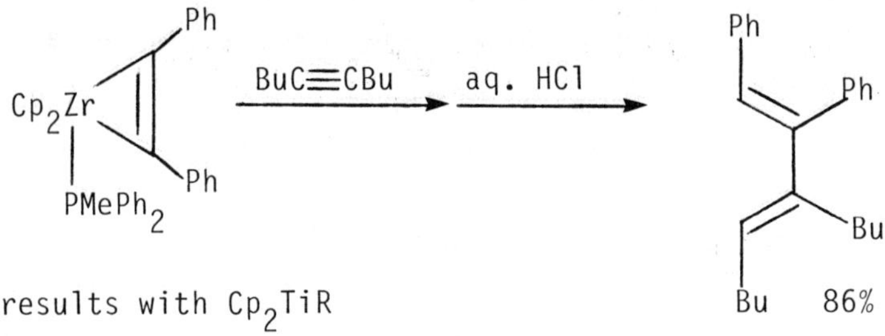

similar results with Cp_2TiR

86%

I.G.3-9 T. Takeichi et al., Chem. Lett., 1301 (1987).

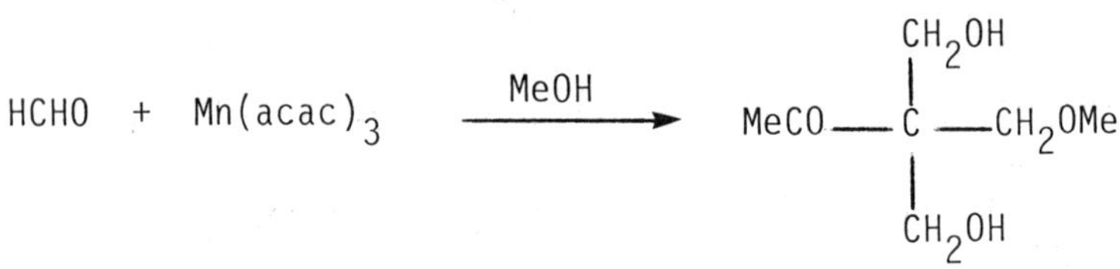

ca. 20%

I.G.3-10 Y. Watanabe et al., Tetrahedron Lett., 28, 6229 (1987).

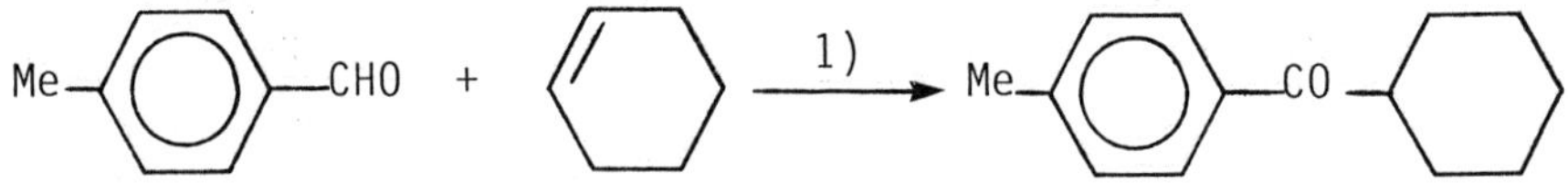

54%

1) $Ru_3(CO)_{12}$, 200°C, CO

I.G.3-11 Y. Watanabe et al., J. Org. Chem., 52, 3186 (1987).

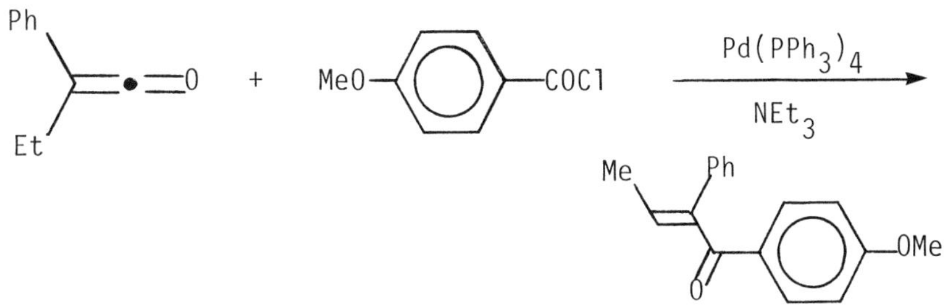

86%, Z : E = 2 : 3

I.G.3-12 L.S. Trifonov, A.S. Orahovats and H. Heimgartner, Helv. Chim. Acta, 70, 1070 (1987).

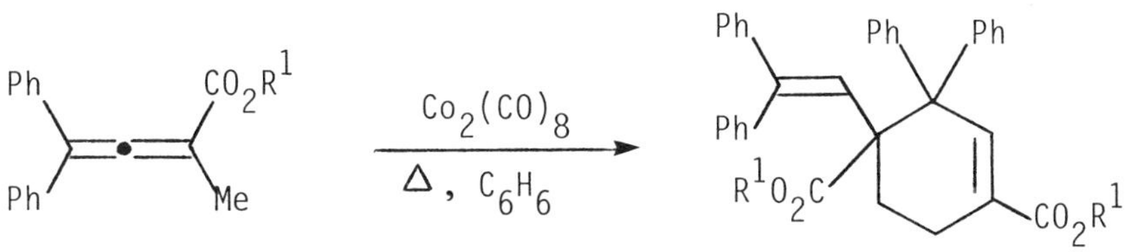

61-83%

I.G.3-13 T.S. Sorensen et al., Can. J. Chem., 65, 1499 (1987).

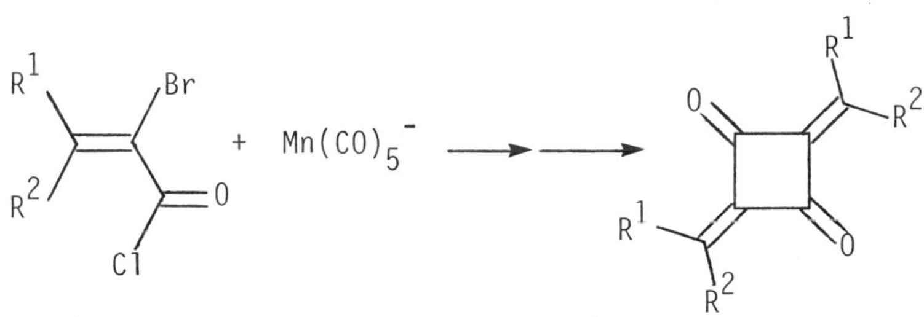

60-90%

I.G.3-14 S. Iyer and L.S. Liebeskind, J. Am. Chem. Soc., 109, 2759 (1987); L.S. Liebeskind and R. Chidambaram, ibid., 109, 5025 (1987).

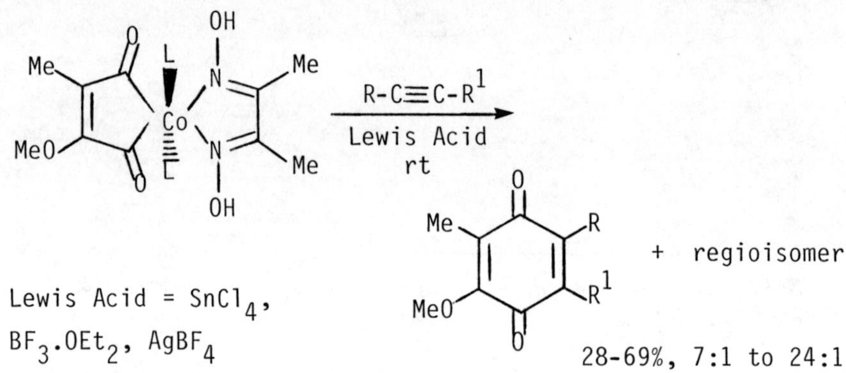

I.G.4. Organometallic Reviews

I.G.4-1 Y. Yamamoto, Aldrichimica Acta, 20, 45 (1987).

Review: "Allylic Tin Compounds in Organic Synthesis."

I.G.4-2 H. Kurosawa, J. Organomet. Chem., 334, 243 (1987).

Review: "Molecular Basis of Catalytic Reactions Involving η^3-Allyl Complexes of Group 10 Metals as Key Intermediates."

I.G.4-3 J. Tsuji and I. Minami, Acc. Chem. Res., 20, 140 (1987).

Review: "New Synthetic Reactions of Allyl Alkyl Carbonates, Allyl β-Keto Carboxylates, and Allyl Vinylic Carbonates Catalyzed by Palladium Complexes."

I.G.4-4 G.L. Larson, J. Organomet. Chem., 337, 195 (1987).

Review: "Organosilicon Survey 1985. The Silicon-Carbon Bond."

I.G.4-5 A.J. Fatiadi, Synthesis, 959 (1987).

Review: "New Applications of Tetracyanoethylene in Organometallic Chemistry."

I.G.4-6 H. Yasuda and A. Nakamura, Angew. Chem., Int. Ed. Engl., 26, 723 (1987).

Review: "Diene, Alkyne, Alkene and Alkyl Complexes of Early Transition Metals: Structures and Synthetic Applications in Organic and Polymer Chemistry."

I.G.4-7 H.M. Colquhoun, Chem. Ind., 612 (1987).

Review: "Homogeneous, Palladium-Catalyzed, Carbon-Carbon Bond Formation."

I.G.4-8 P.A. Chaloner, J. Organomet. Chem., 324, 283 (1987); idem, ibid., 337, 431 (1987).

Reviews: "Nickel, Palladium and Platinum. Annual Survey Covering the Years 1980 and 1981."

I.G.4-9 K.M. Nicholas, Acc. Chem. Res., 20, 207 (1987).

Review: "Chemistry and Synthetic Utility of Cobalt-Complexed Propargyl Cations."

I.G.4-10 T. Mitsudo et al., J. Organomet. Chem., 334, 157 (1987).

Review: "Novel Organic Syntheses Catalyzed by (Cyclooctadiene) (Cyclooctatriene) Ruthenium and its Derivatives."

I.G.4-11 I.D. Sadekov et al., Russ. Chem. Rev., 56, 343 (1987).

Review: "Organotellurium Compounds in Organic Synthesis."

I.G.4-12 B.H. Lipshutz, Synthesis, 325 (1987).

Review: "Applications of Higher-Order Mixed Organocuprates to Organic Synthesis."

I.G.4-13 W.H. Ilsley, J. Organomet. Chem., 337, 1 (1987).

Review: "Mercury: Annual Survey Covering the Year 1980."

I.H. Rearrangements

I.H.1. Claisen, Cope and Similar Processes

I.H.1-1 J.M. Cassady et al., J. Org. Chem., 52, 342 (1987); L.M. Harwood et al., Chem. Commun., 1615 (1987).

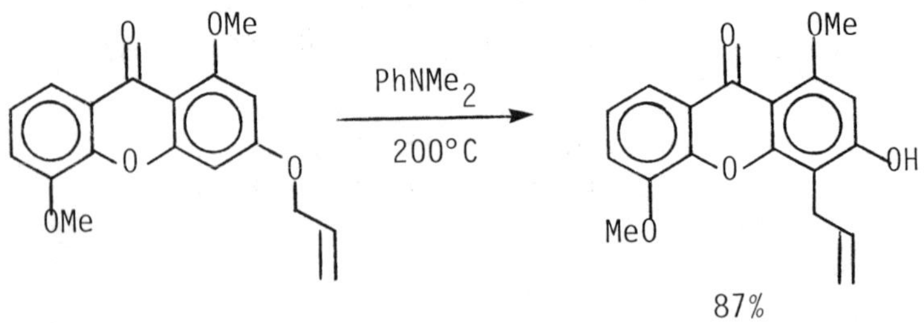

I.H.1-2 H. Takeshita et al., Bull. Chem. Soc. Jpn., 60, 1429, 1747 and 2497 (1987); J.A. Miller, J. Org. Chem., 52, 322 (1987).

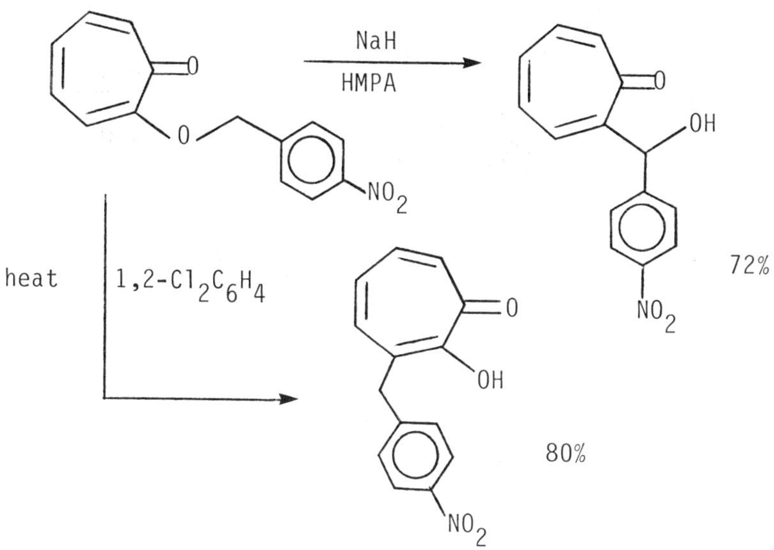

I.H.1-3 T. Katsuki et al., Chem. Lett., 351 (1987); T. Fujisawa et al., ibid., 1979 (1987); S. Takano et al., ibid., 1379 (1987); J. Kallmerten et al., J. Org. Chem., 52, 3889 (1987); J.E. Baldwin et al., Chem. Commun., 250 (1987); W.A. Smit et al., Tetrahedron Lett., 28, 4893 (1987).

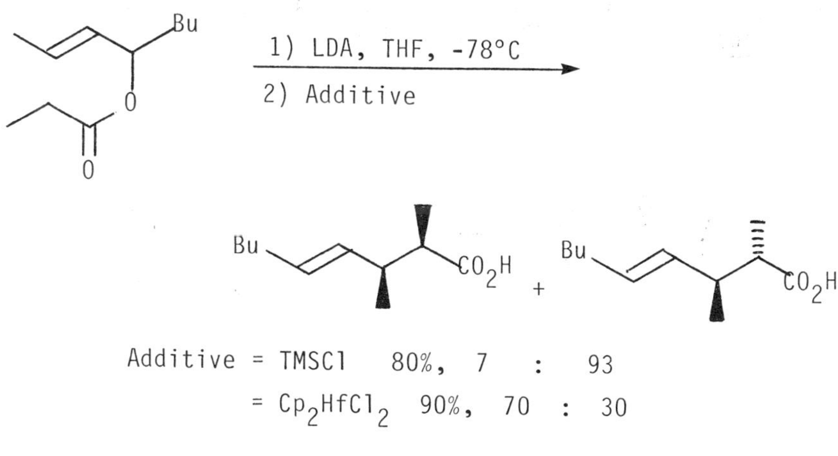

Additive = TMSCl 80%, 7 : 93

= Cp_2HfCl_2 90%, 70 : 30

I.H.1-4 J.A. Marshall et al., J. Org. Chem., 52, 3860 and 3883 (1987); G.A. Kraus et al., ibid., 52, 1273 (1987); J. Barluenga et al., ibid., 52, 5190 (1987); J.H. Rigby and C. Senanayake, ibid., 52, 4635 (1987); J.A. Marshall and J. Lebreton, Tetrahedron Lett., 28, 3323 (1987); S.L. Schreiber and M.T. Goulet, ibid., 28, 1043 (1987); L. Castedo et al., ibid., 28, 2099 (1987); T. Takahashi et al., Tetrahedron, 43, 5499 (1987); J.J. Gajewski, B. Ganem, B.K. Carpenter et al., J. Am. Chem. Soc., 109, 1170 (1987); R.M. Coates, D.P. Curran et al., ibid., 109, 1160 (1987).

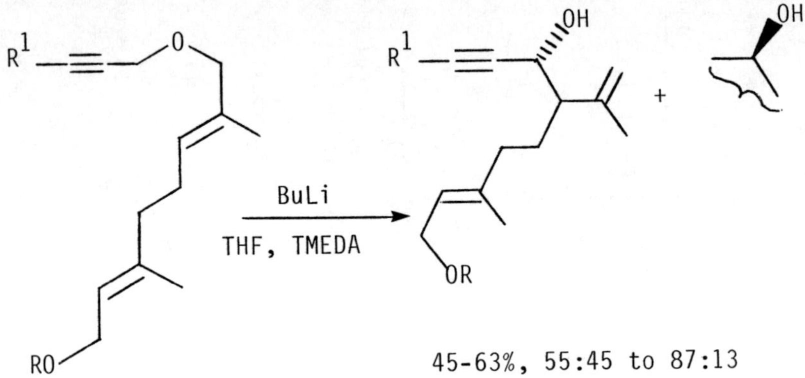

45-63%, 55:45 to 87:13

I.H.1-5 G. Mehta and N. Krishnamurthy, Tetrahedron Lett., 28, 5945 (1987); D.V. Pratt and P.B. Hopkins, ibid., 28, 3065 (1987); M.J. Kurth and C.J. Soares, ibid., 28, 1031 (1987); T. Tsunoda et al., ibid., 28, 2537 (1987); P.S. Rutledge et al., Aust. J. Chem., 40, 1063 (1987).

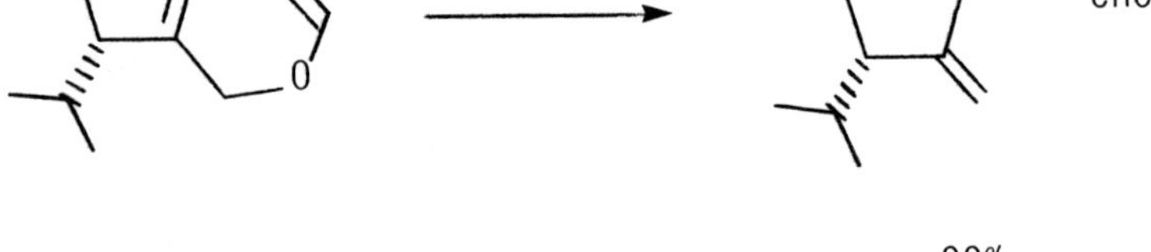

90%

I.H.1-6 T. Nakai et al., Chem. Lett., 69 (1987); T. Katsuki et al., Tetrahedron Lett., 28, 803 (1987); S.E. Denmark and J.E. Marlin, J. Org. Chem., 52, 5742 (1987).

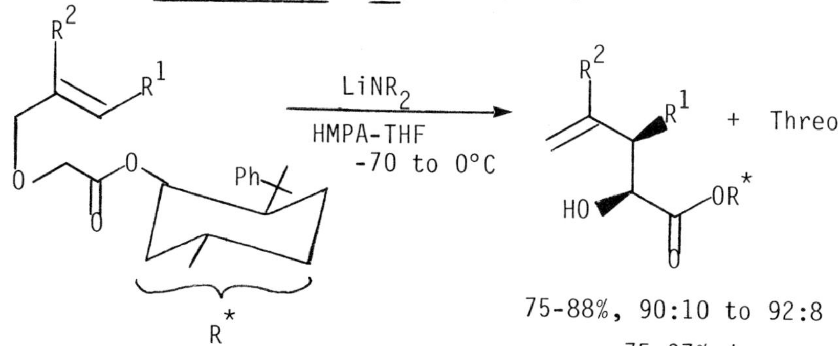

75-88%, 90:10 to 92:8

75-97% de

I.H.1-7 P. Gosselin, Synthesis, 267 (1987).

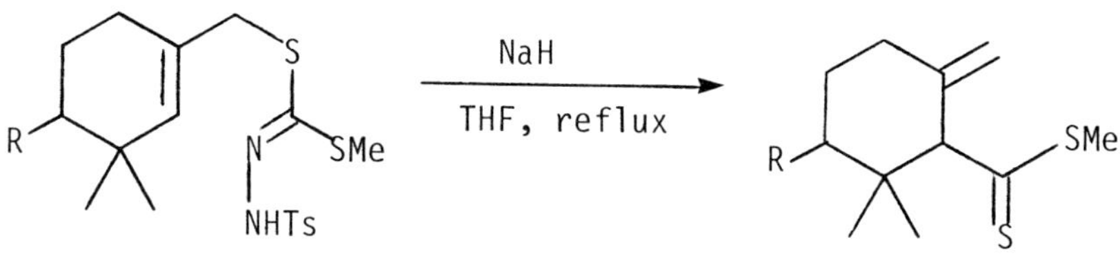

78-83%

I.H.1-8 L.A. Paquette and W.H. Ham, J. Am. Chem. Soc., 109, 3025 (1987); M. Sakakibara et al., Tetrahedron Lett., 28, 2597 (1987); A. Chattopadhyay and V.R. Mamdapur, Ind. J. Chem., 26B, 161 (1987); K. Tadano, T. Suami et al., J. Org. Chem., 52, 1201 (1987); M. Sworin and K.-C Lin, ibid., 52, 5640 (1987); P.F. Schuda et al. Tetrahedron, 43, 463 (1987).

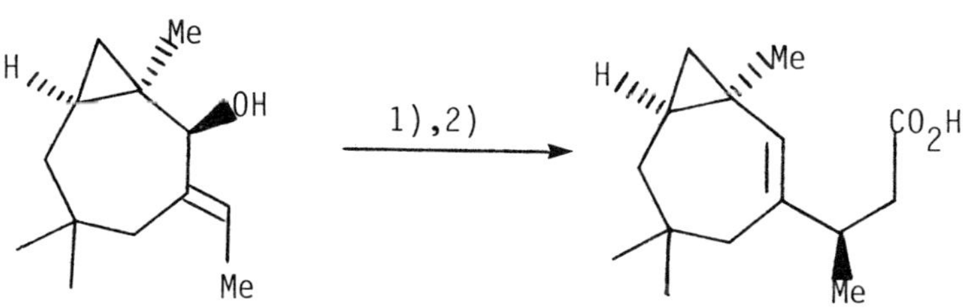

1) $MeC(OEt)_3$, $EtCO_2H$, Δ 2) KOH

no yield given

towards Africanol

I.H.1-9 J.T. Welch and S. Eswarakrishnan, J. Am. Chem. Soc., 109, 6716 (1987).

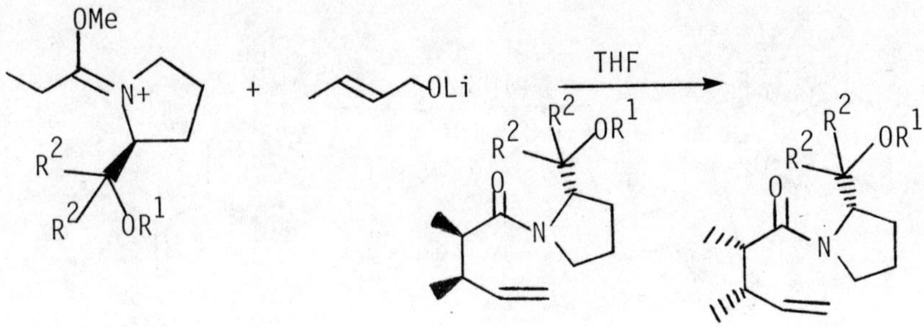

4 : 1 to 6.4 : 1

I.H.1-10 T.C. McKenzie, Org. Prep. Proced. Int., 19, 435 (1987).

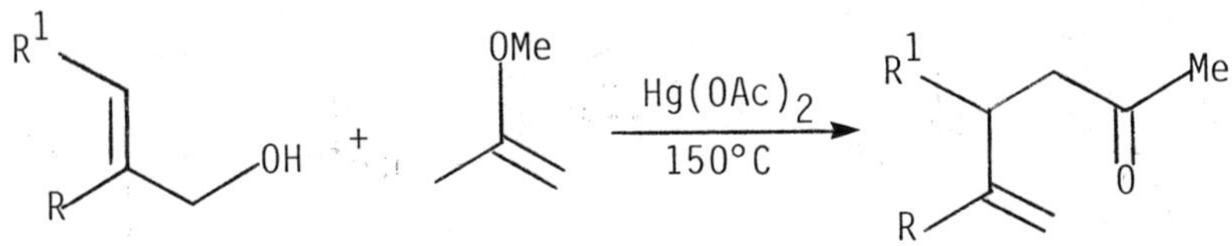

53-89%

I.H.1-11 H.G. Viehe et al., Tetrahedron Lett., 28, 2591 (1987).

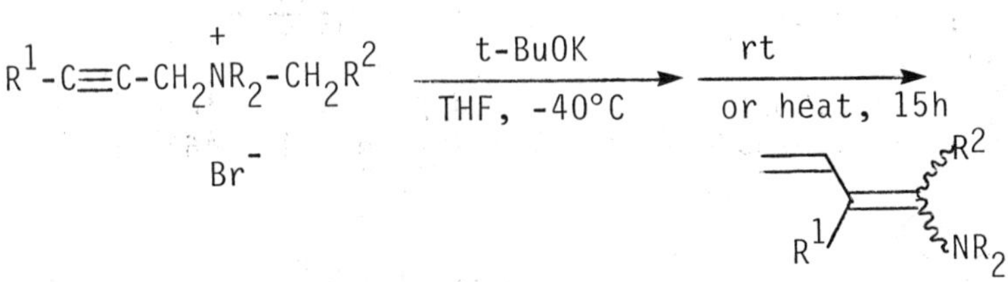

48-87% E:Z 55:45 to 0:100

I.H.1-12 J.H. Bowie et al., J. Org. Chem., 52, 5224 (1987).

Sigmatropic Rearrangements of Deprotonated Allyl Phenylacetates in the Gas Phase

I.H.1-13 L.A. Paquette et al., Tetrahedron Lett., 28, 31 (1987); J.A. Oplinger and L.A. Paquette, ibid., 28, 5441 (1987); L.A. Paquette et al., J. Am. Chem. Soc., 109, 5731 (1987).

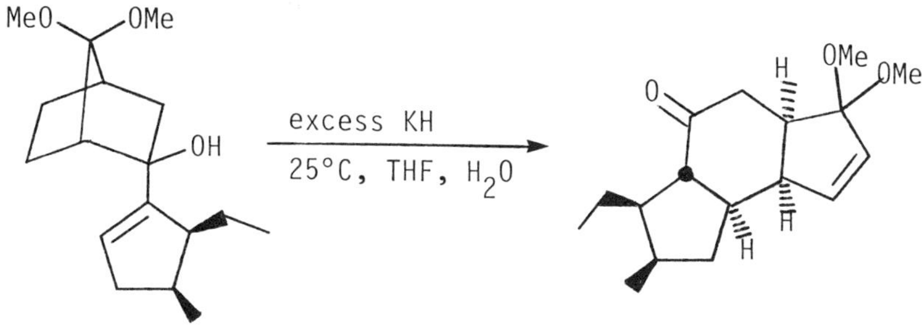

72%

I.H.1-14 W.R. Dolbier, Jr. et al., J. Am. Chem. Soc., 109, 3046 (1987).

Kinetic and Stereochemical Effect of a Fluorine Substituent on the Cope and the Homodienyl 1,5-Hydrogen Shift Rearrangements

---

I.H.1-15 N.M. Przheval'skii and I.I. Grandberg, Russ. Chem. Rev., 56, 477 (1987).

Review: "The Cope Aza-rearrangement in Organic Synthesis."

I.H.1-16 E. Piers et al., Can. J. Chem., 65, 670 (1987).

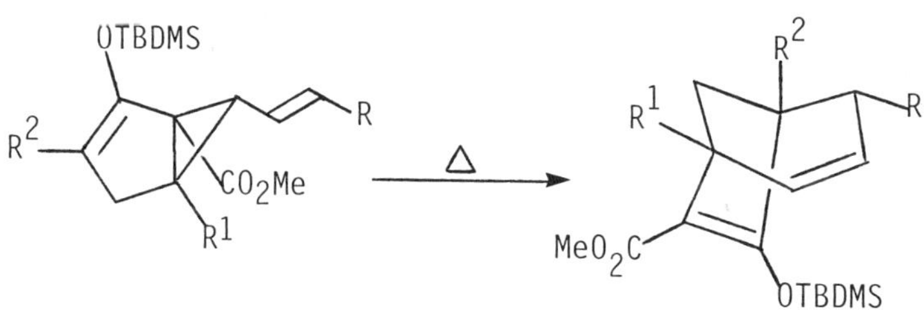

93-99%

I.H.1-17 L.A. Paquette et al., Tetrahedron, 43, 4989 (1987).

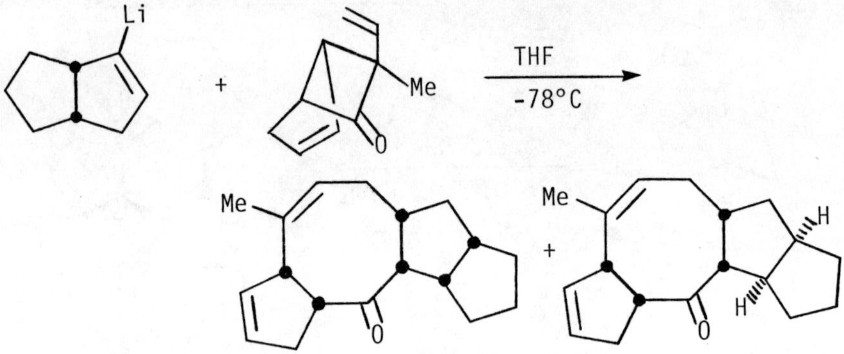

88%, 1.43 : 1

I.H.2. Other Rearrangements

I.H.2-1 K. Suzuki, G. Tsuchihashi et al., Tetrahedron Lett., 28, 3515 and 5891 (1987); idem, Chem. Lett., 113 (1987); B.M. Trost and G.K. Mikhail, J. Am. Chem. Soc., 109, 4124 (1987); A. Krief et al., Tetrahedron Lett., 28, 1545 and 1549 (1987); T. Cohen et al., ibid., 28, 2203 (1987); G. Seitz and R. van Gemmern, Synthesis, 953 (1987); W.H. Bunnelle and W.R. Shangraw, Tetrahedron, 43, 2005 (1987); R. Baudouy et al., ibid., 43, 1099 (1987); D.H. Hua et al., J. Org. Chem., 52, 2509 (1987); C.S. Swindell et al., ibid., 52, 2346 (1987); D.H.G. Crout and D.L. Rathbone, Chem. Commun., 290 (1987); T.L. Gilchrist and J.E. Stanford, J. Chem. Soc., Perkin Trans. 1, 225 (1987); B. Halton, P.J. Stang et al., Aust. J. Chem., 40, 1375 (1987).

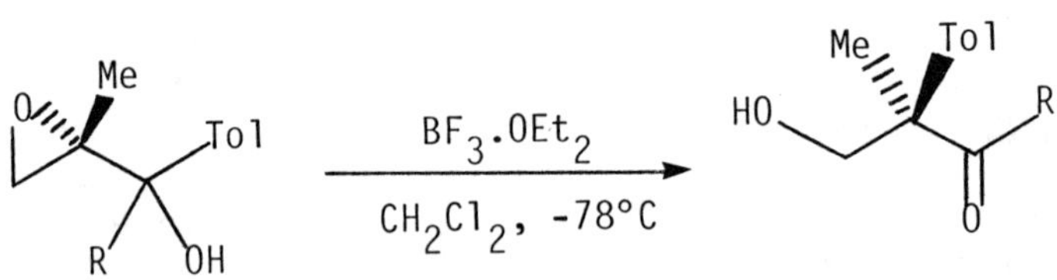

100%, 87 ± 1% ee

various other leaving groups, acids or bases also used

I.H.2-2 L.S. Liebeskind et al., J. Am. Chem. Soc., 109, 7908 (1987); H. Suginome et al., Chem. Lett., 1527 (1987).

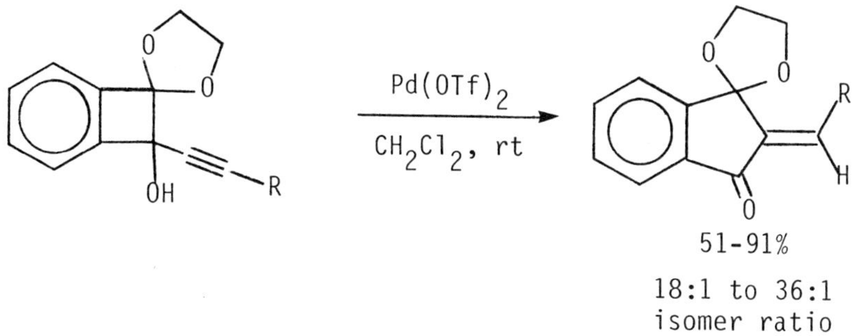

I.H.2-3 D. Liotta et al., Tetrahedron Lett., 28, 1069 (1987).

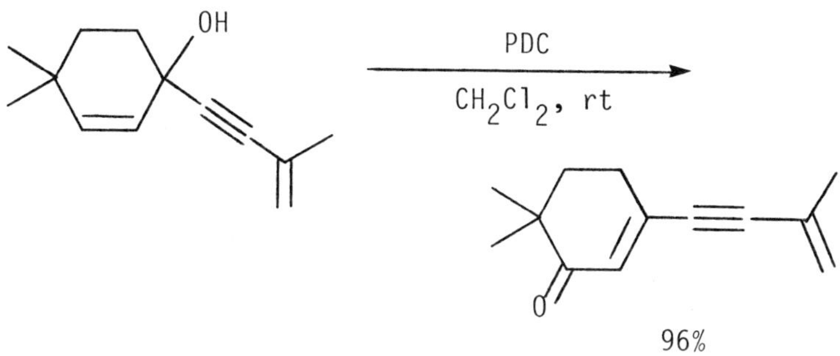

I.H.2-4 K. Okamoto et al., Synthesis, 612 (1987).

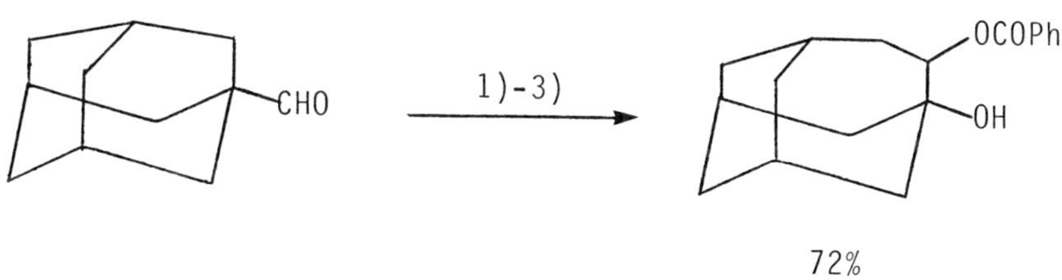

1) $PhCO_2Tf$, CCl_4 2) TfOH 3) H_2O

I.H.2-5 S. Kanemoto et al., Tetrahedron Lett., 28, 6313 (1987); M. Shimizu and H. Yoshioka, ibid., 28, 3119 (1987).

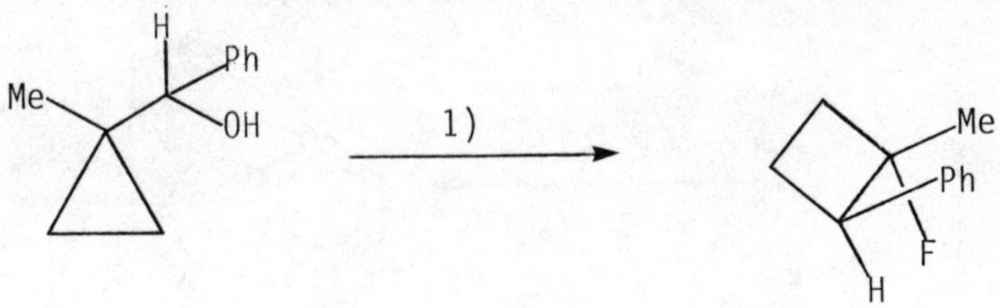

71%, trans:cis 95:5

1) $^{i}Pr_2NH\text{-}KHF_2\text{-}(HF)_n$pyr complex

I.H.2-6 T. Uyehara et al., Chem. Lett., 137 (1987); idem, Tetrahedron, 43, 5605 (1987); A. Hofland and Th.J. de Boer, Rec. Trav. Chim., 106, 558 (1987); Y. Honda, A. Ori and G. Tsuchihashi, Bull. Chem. Soc. Jpn., 60, 1027 (1987).

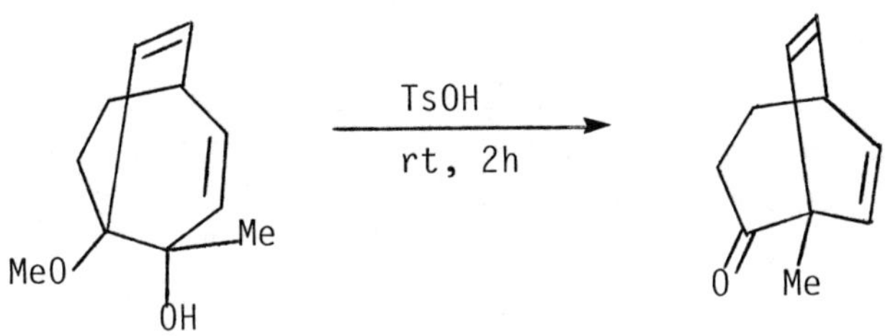

74%

other leaving groups and conditions also used

I.H.2-7 G. Castaldi, C. Giordano et al., J. Org. Chem., 52, 3018 (1987); H. Mayr et al., Tetrahedron Lett., 28, 387 (1987); A.Chatterjee et al., Synthesis, 806 (1987).

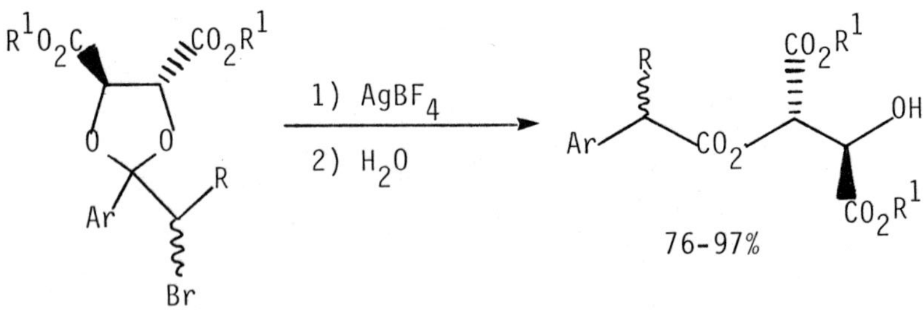

76-97%

$SnCl_4$ and $ZnCl_2$ used similarly

I.H.2-8 T. Sakai et al., Bull. Chem. Soc. Jpn., 60, 2295 (1987); A. Takeda et al., ibid., 60, 836 (1987).

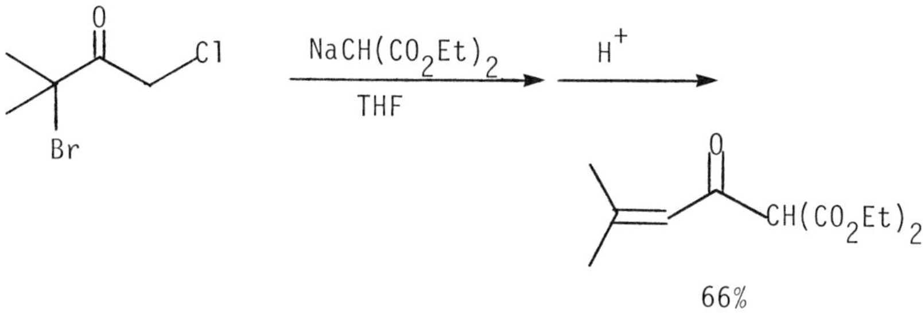

I.H.2-9 R. Menicagli et al., J. Org. Chem., 52, 5700 (1987).

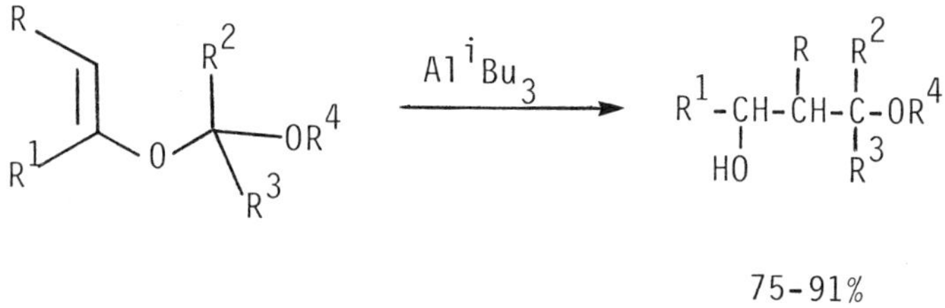

I.H.2-10 M.S. Baird and H.H. Hussain, Tetrahedron, 43, 215 (1987).

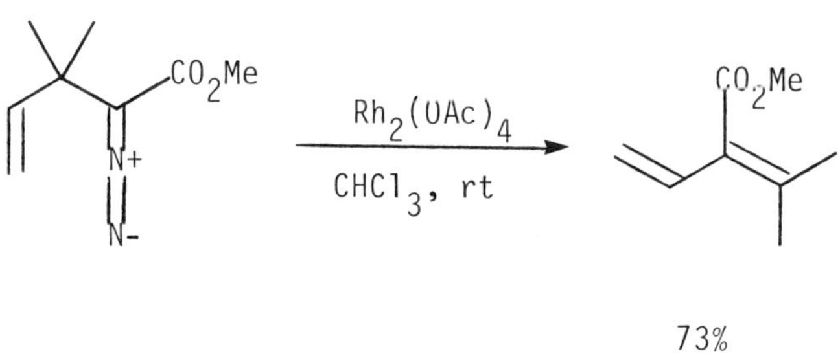

I.H.2-11 L.A. Paquette et al., J. Org. Chem., 52, 4732 (1987).

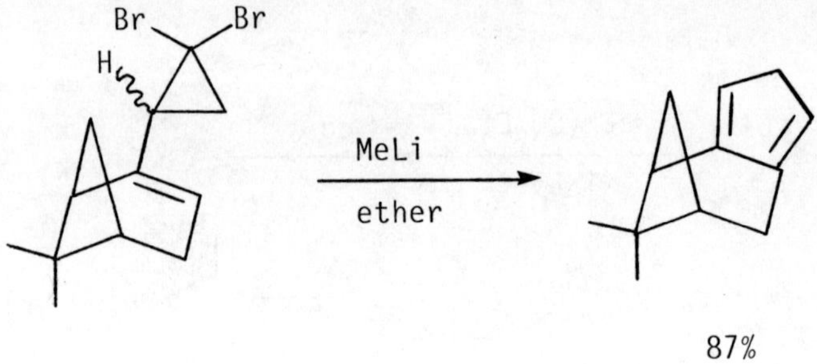

I.H.2-12 K. Okamoto et al., J. Org. Chem., 52, 2183 (1987).

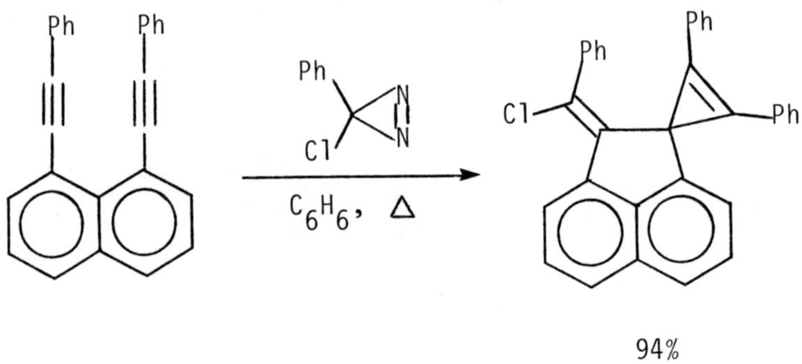

I.H.2-13 D.W. Jones et al., J. Chem. Soc., Perkin Trans. 1, 1471 and 1479 (1987).

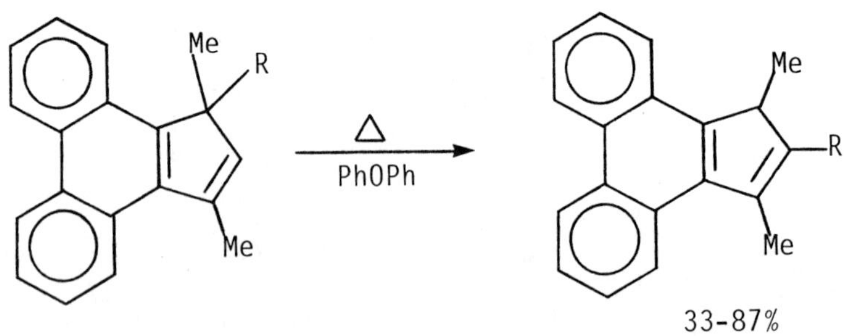

R = CHO, CO_2Me, CN or other unsaturated groups

I.H.2-14 L. Fitjer and U. Quabeck, Angew. Chem., Int. Ed. Engl., 26, 1023 (1987); M. Ue et al., Tetrahedron Lett., 28, 3979 (1987).

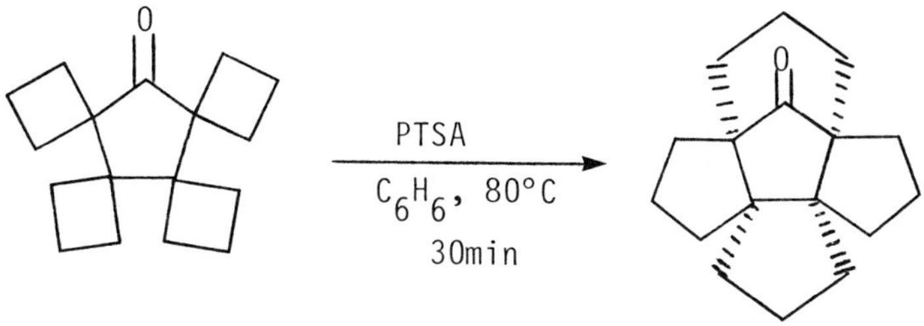

100%

I.H.2-15 P.C. Page et al., Tetrahedron Lett., 28, 4455 (1987).

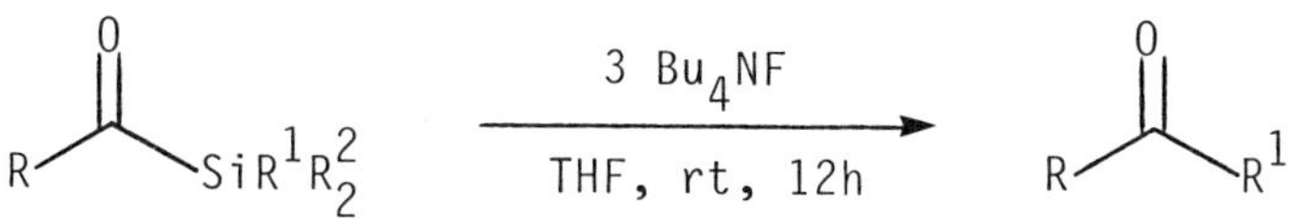

52-68%

I.H.2-16 T. Yamauchi et al., J. Chem. Soc., Perkin Trans. 1, 1433 (1987).

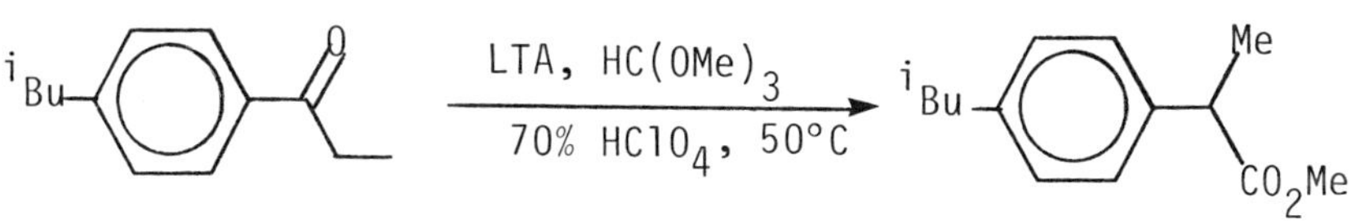

85%

I.H.2-17 R.M. Moriarty et al., Tetrahedron Lett., 28, 2845 (1987).

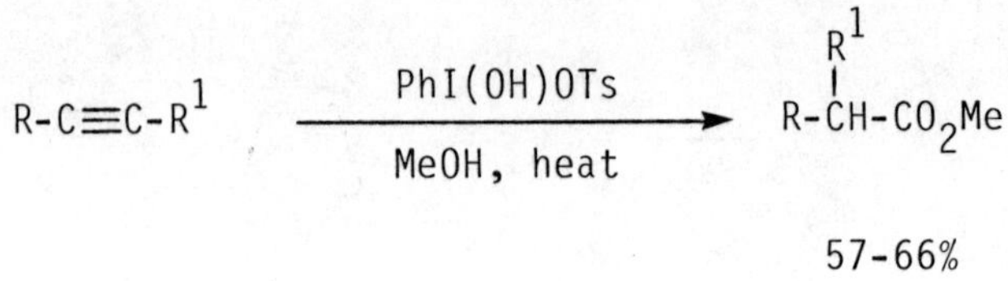

I.H.2-18 B.B. Snider et al., J. Org. Chem., 52, 5419 (1987); H. Ishibashi et al., J. Chem. Res. (S), 296 (1987); M. Asaoka and H. Takei, Tetrahedron Lett., 28, 6343 (1987); V.K. Aggarwal and S. Warren, ibid., 28, 1925 (1987); J.B.P.A. Wijnberg and Ae. de Groot, ibid., 28, 3007 (1987); W.H. Okamura et al., J. Am. Chem. Soc., 109, 7499 (1987); F.E. Ziegler et al., ibid., 109, 3987 (1987); I. Kuwajima et al., Chem. Lett., 385 (1987); V.N. Plotkin et al., J. Org. Chem. (USSR), 23, 1007 (1987); M. Sakamoto et al., J. Chem. Soc., Perkin Trans. 1, 1937 (1987).

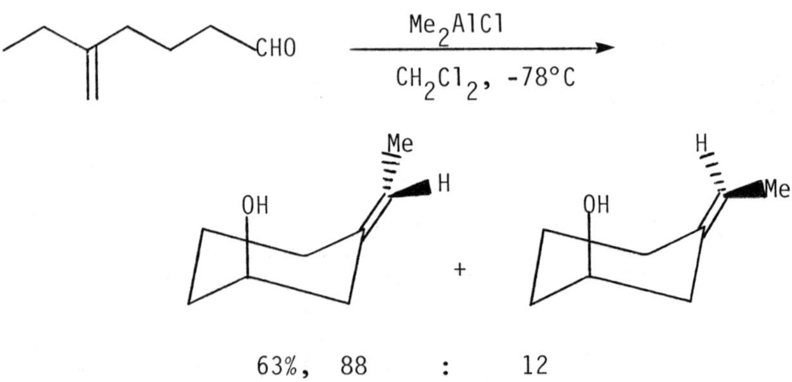

other acids, bases or leaving groups also employed

I.H.2-19 J.M. Takacs et al., J. Am. Chem. Soc., 109, 2542 (1987); idem, Angew. Chem., Int. Ed. Engl., 26, 1013 (1987).

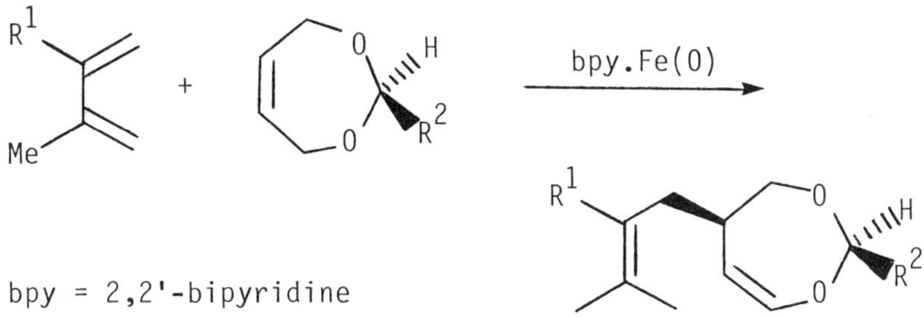

bpy = 2,2'-bipyridine

70-80%, >95:5 ds

I.H.2-20 B. Ben. Hassine et al., Bull. Soc. Chim. Belg., 96, 801 (1987).

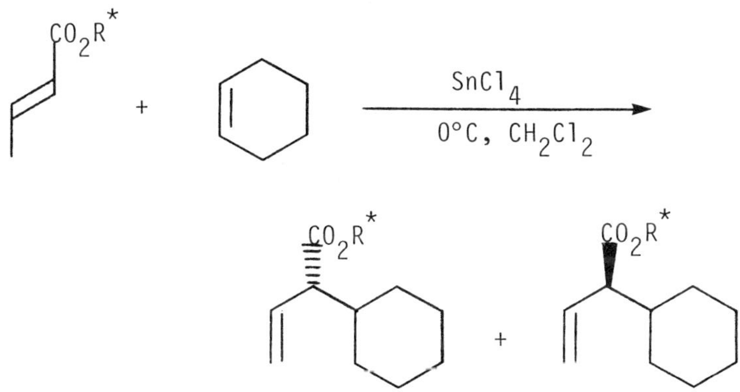

76-83%, 74:26 to 100:0

R^* = 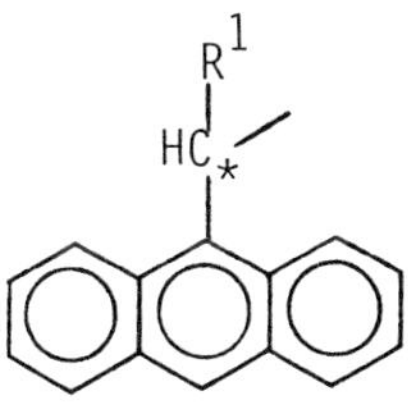, R^1 = CH_3, CF_3, CCl_3

II.A. C—O Oxidations

1. Alcohol→Ketone, Aldehyde

II.A.1-1 Y. Lin, D. Ma and X. Lu, Tetrahedron Lett., 28, 3054 (1987); B. Byrne and M. Karras, Tetrahedron Lett., 28, 769 (1987).

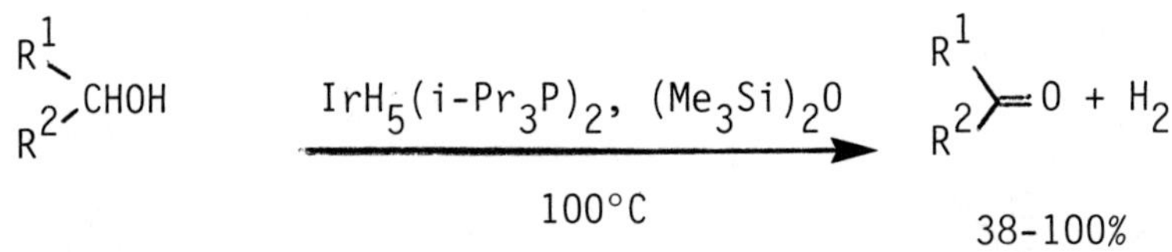

II.A.1-2 R.J. Linderman and D.M. Graves, Tetrahedron Lett., 28, 4259 (1987).

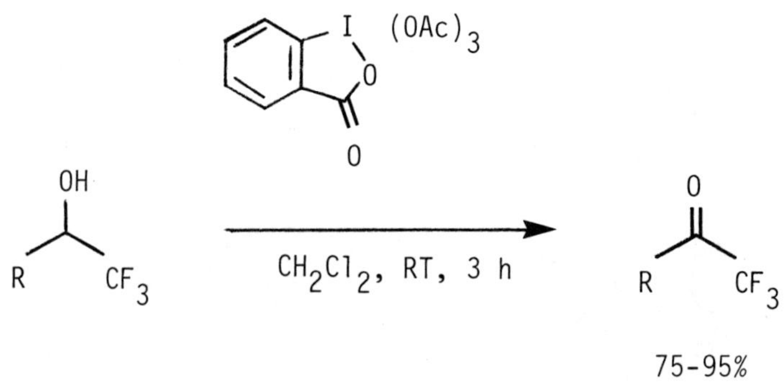

II.A.1-3 K.S. Kim , S.J. Kim, Y.H. Song and C.S. Hahn, Synthesis, 1017 (1987); H. Cano-Yelo and A. Deronzier, Nouv. J. Chem., 11, 479 (1987).

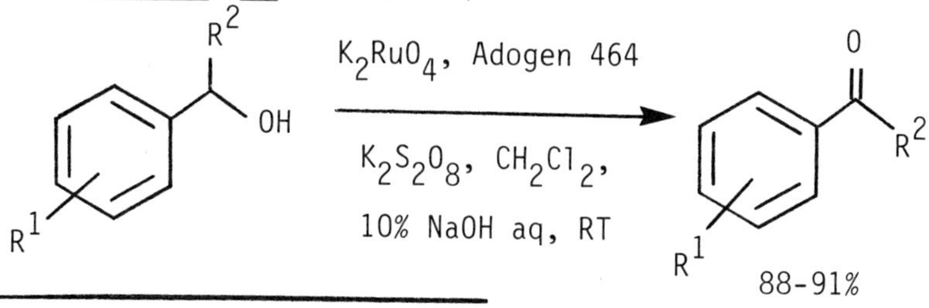

II.A.1-4 D.H.R. Barton, J.P. Finet, W. B. Motherwell and C. Pichon, J. Chem. Soc.,Perkin I, 251 (1987).

BTMG = N-tert-butyl-N', N', N", N"-tetramethylguanidine

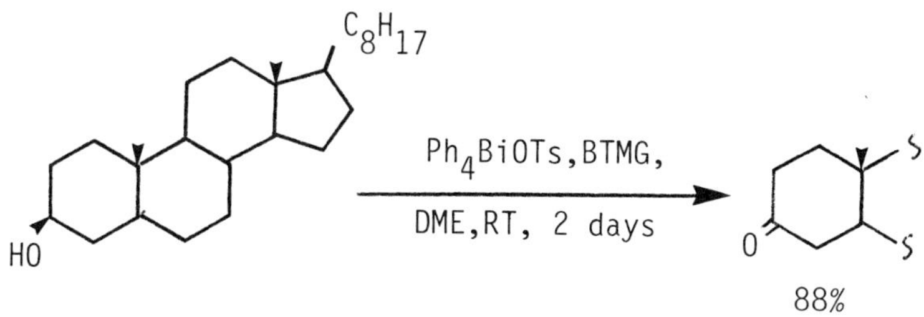

II.A.1-5 P.E. Morris, Jr. and D.E. Kiely , J. Org. Chem., 52, 1149 (1987).

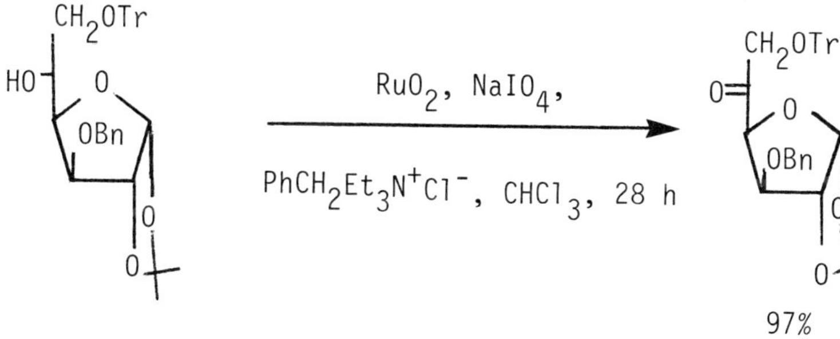

II.A.1-7 T. Nakano, Y. Ishii and M. Ogawa, J. Org. Chem., 52, 4855 (1987).

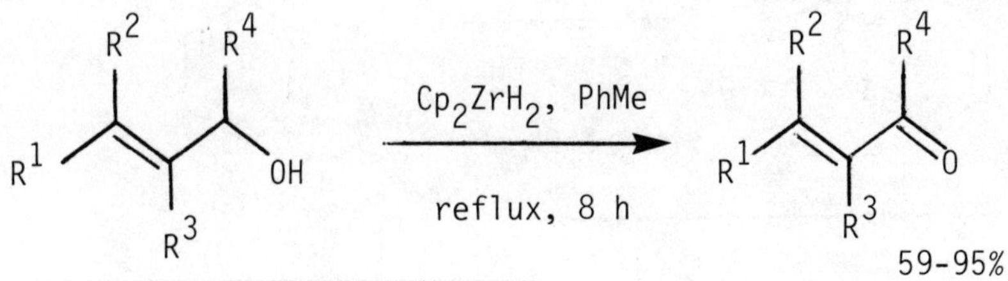

II.A.1-8 C.M. Amon, M.G. Banwell and G.L. Gravatt, J. Org. Chem., 52, 4851 (1987); R. Schobert, Synthesis, 741 (1987).

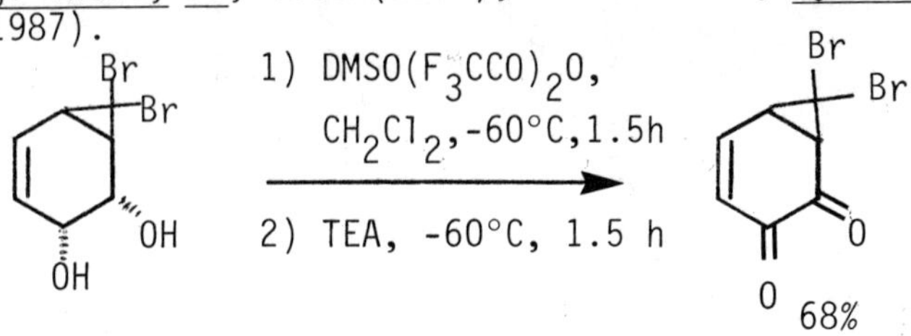

II.A.1-9 Y. Sasson, H. Wiener and S. Bashir, J. Chem. Soc. Chem. Commun., 1574 (1987).

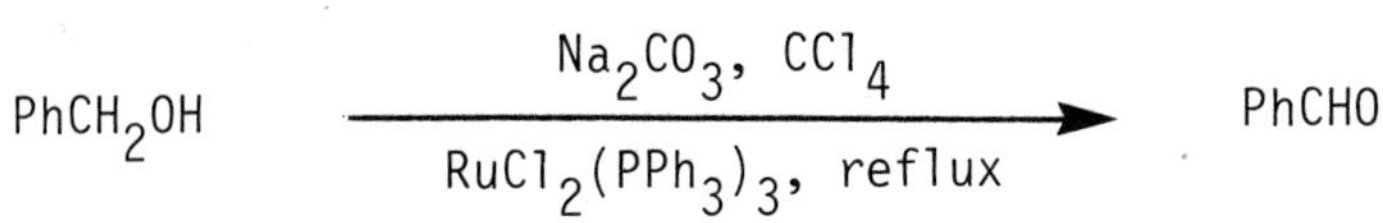

II.A.1-10 P.L. Anelli, C. Biffi, F. Montanari and S. Quici, J. Org. Chem., 52, 2559 (1987).

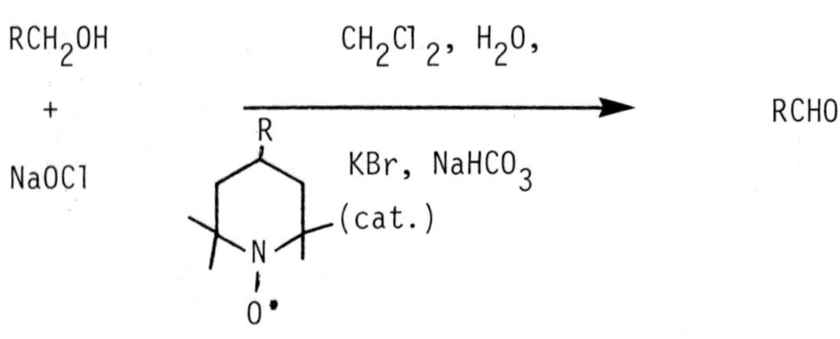

II.A.1-11 A. McKillop and L.S. Mills, Synth. Commun., 17, 647 (1987).

$$ArCH_2OH \xrightarrow[CH_2Cl_2,\ RT]{KMnO_4,\ TDA\text{-}1} ArCHO \quad 54\text{-}91\%$$

II.A.1-12 F.P. Cossio, M. Concepcion Lopez and C. Palomo, Tetrahedron, 43, 3963 (1987).

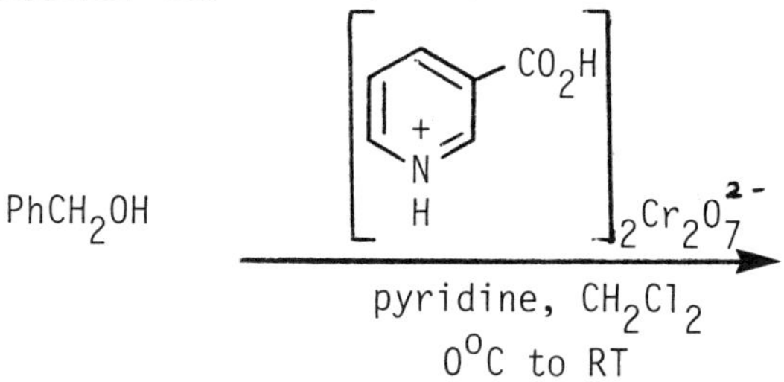

II.A.1-13 J.D. Lou and Y. Y. Wu, Synth. Commun., 17, 1717 (1987); J. Lou and Y.Y. Wu, Chem. Ind. (London), 531 (1987); H. Schick, et al., J. Prakt. Chem., 328 777 (1986).

$$RCH_2OH \xrightarrow{1)} RCHO \quad 74\text{-}90\%$$

1) Chromic acid, aluminum silicate, CCl_4.

II.A.1-14 D.F. Taber, J.C. Amedio, Jr. and K. Jung, J. Org. Chem., 52, 5621 (1987).

$$Me(CH_2)_{15}OH \xrightarrow[2)\ TEA]{1)\ P_2O_5,\ DMSO,\ RT} Me(CH_2)_{14}CHO \quad 83\%$$

II.A.1-15 W.P. Griffith, S.V. Ley, G.P. Whitcombe and A.D. White, J. Chem. Soc. Chem. Commun., 1625 (1987).

$$\text{n-BuOH} \xrightarrow[\text{4 A sieves, RT; N-methylmorpholine N-oxide}]{Bu_4NRuO_4,\ CH_2Cl_2} \text{PrCHO}\quad 94\%$$

II.B. C—H Oxidations

1. C—H ⟶ C—O

II.B.1-1 R.P. Kreh, R.M. Spotnitz and J.T. Lundquist, Tetrahedron Lett., 28, 1067 (1987); E. Vismara, F. Fontana and F. Minisci, Gazz. Chim. Ital., 117, 135 (1987).

$$ArCH_3 \xrightarrow{1)} ArCHO\quad 75\text{-}91\%$$

1) $Ce(OMs)_4$ or $Ce(OTf)_4$ generated electrochemically from carbonate in MsOH or TfOH, H_2O, 25-80°C.

II.B.1-2 S. Murata, M. Miura and M. Nomura, J. Chem. Soc., Perkin I, 1259 (1987).

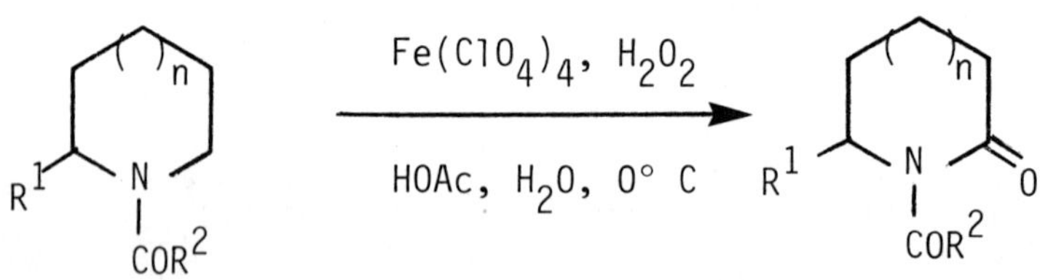

40-75%

II.B.1-3 J. Muzart, Tetrahedron Lett., 28, 2131, 2133 (1987).

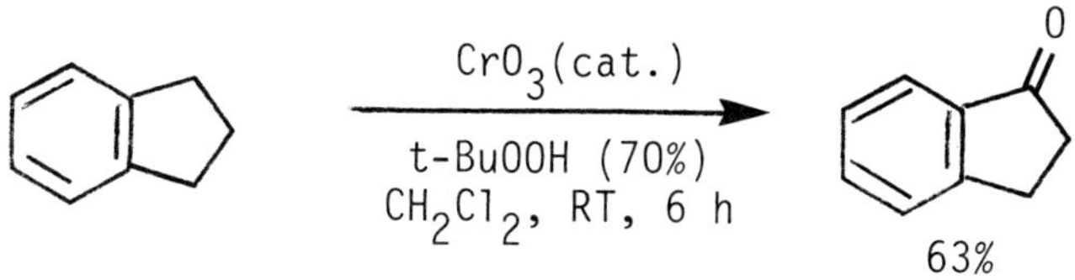

II.B.1-4 F.M. Hauser and S.R. Ellenberger, Synthesis, 723 (1987).

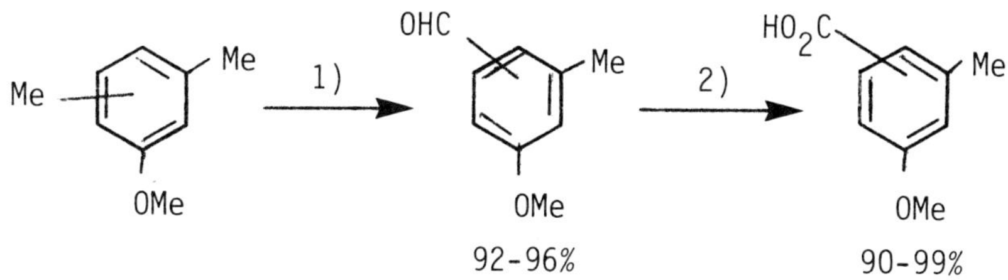

1) Cu^{2+}, $K_2S_2O_8$, MeCN aq, reflux;

2) $NaClO_3$, H_2O, THF, H_2NSO_3H, RT.

II.B.1-5 N. Chidaabaram and S. Chandrasekaran, J. Org. Chem., 52, 5048 (1987).

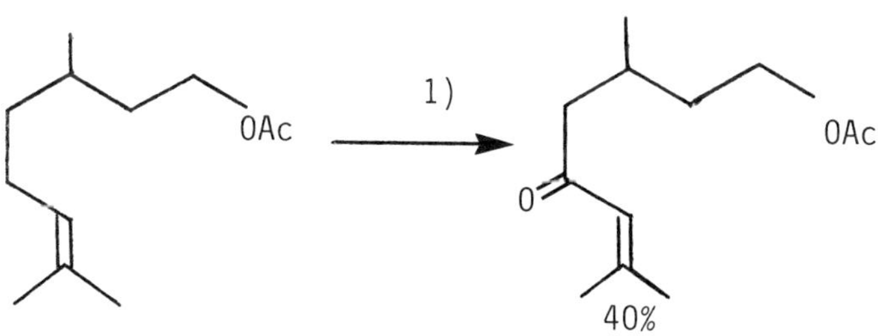

1) t-BuOOH, PDC, PhH, 10°C, 15 min, then 25°C, 11 h.

II.B.1-6 P.H.J. Carlsen, Synth. Commun., 17, 19 (1987).

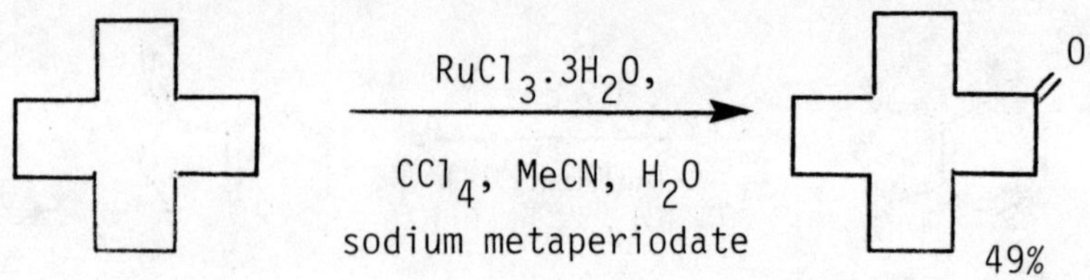

II.B.1-7 H.L. Holland et al., Can. J. Chem., 65, 502 (1987).

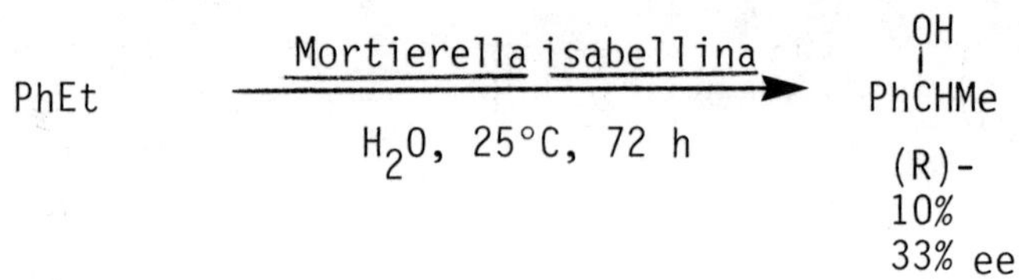

II.B.1-8 D. Pletcher and E.M. Valdes, J. Chem. Res., 386 (1987).

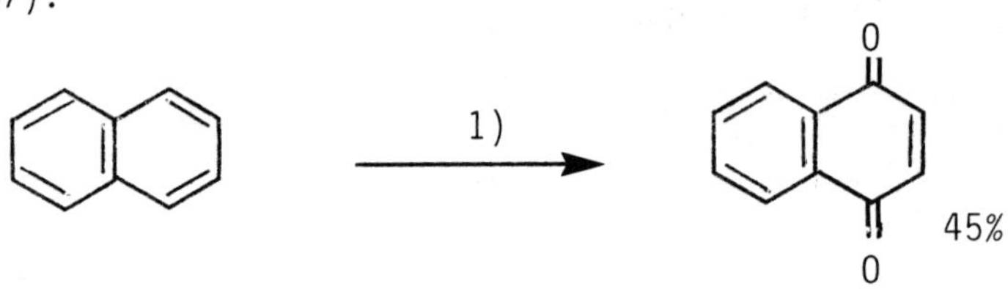

1) Bu_4NNO_3, $(NH_4)_2Cr(NO_3)_6$, HNO_3, H_2O, $ClCH_2CH_2Cl$, 45°C.

II.B.1-9 K.A. Parker and K.A. Koziski, J. Org. Chem., 52, 674 (1987).

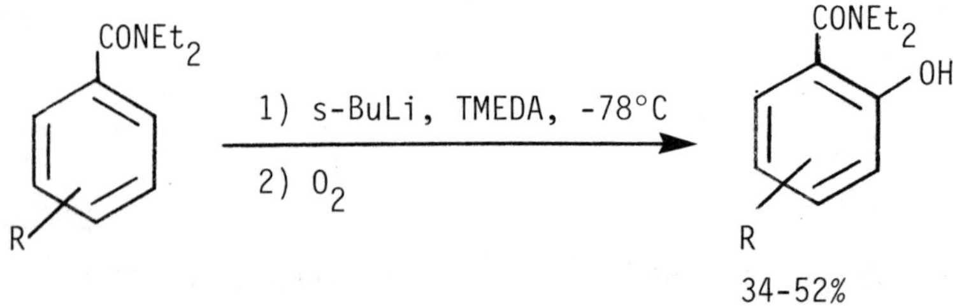

II.B.1-10 C.P. Rao and G. Srimannarayana, Synth. Commun., 17, 1507 (1987).

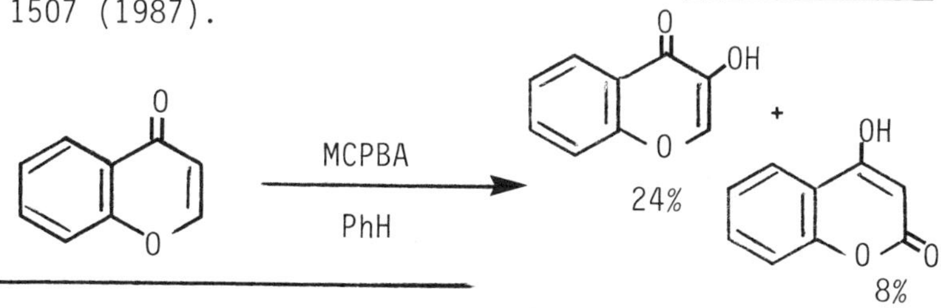

II.B.1-11 F.A. Davis, T.G. Ulatowski and M.S. Haoue, J. Org. Chem., 52, 5288 (1987).

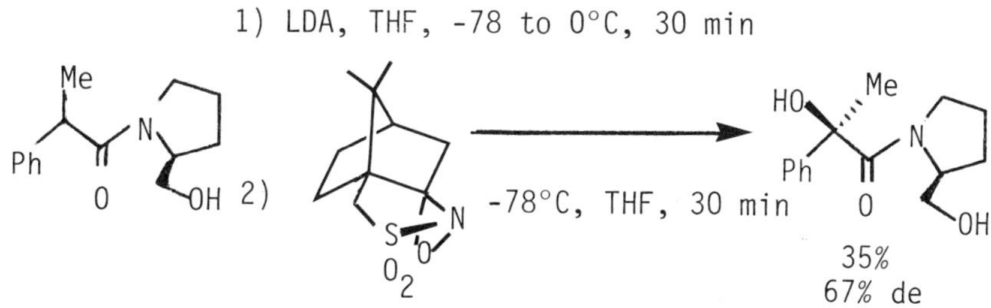

II.B.1-12 S. Rozen, D. Hebel and D. Zamir, J. Am. Chem. Soc., 109, 3789 (1987).

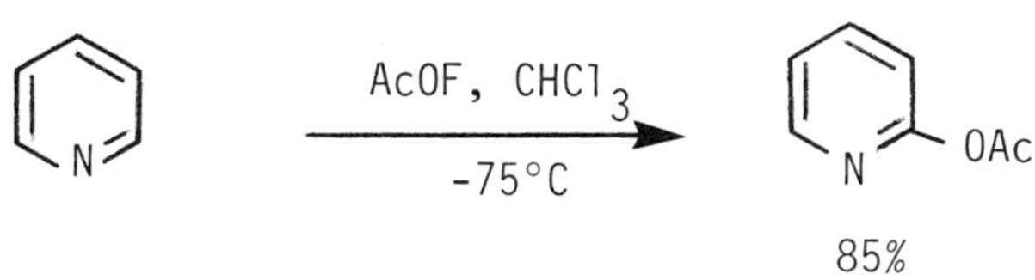

II.B.1-13 T. Sugiyama, Chem. Lett., 1013 (1987).

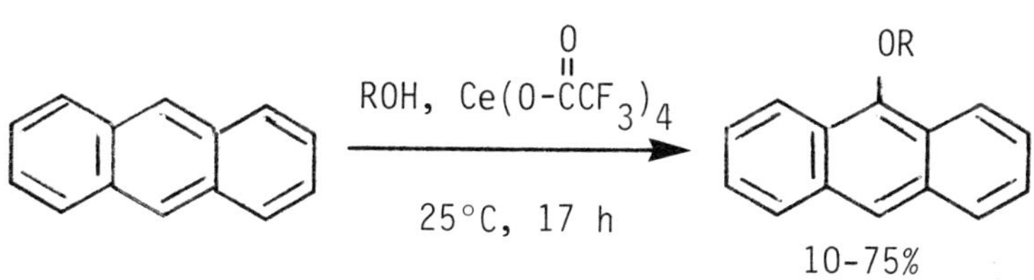

II.B.1-14 M. Kimura, K. Koie, S. Matsubara, Y. Sawaki and H. Iwamura, J. Chem. Soc. Chem. Commun., 122 (1987).

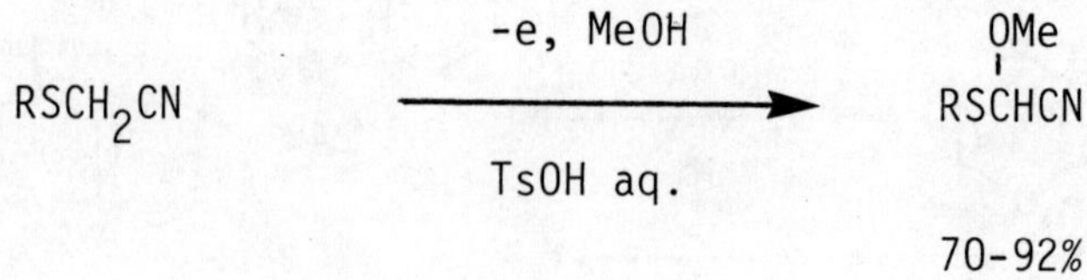

II.B.1-15 T. Shono, Y. Matsumura, O. Onomura and Y. Yamada, Synthesis, 1099 (1987).

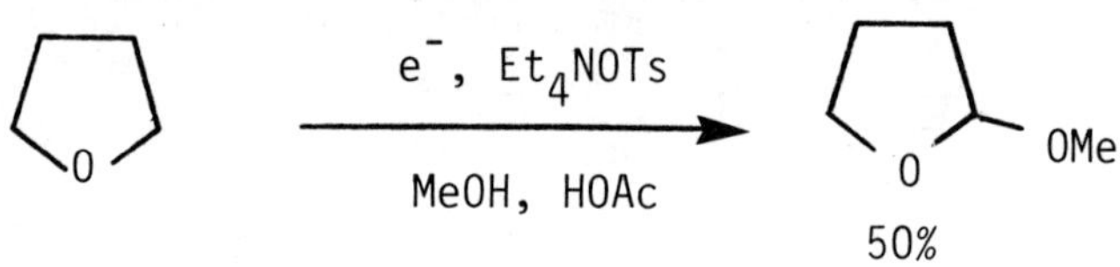

II.B.1-16 A.M. Maione and A. Romeo, Synthesis, 250 (1987).

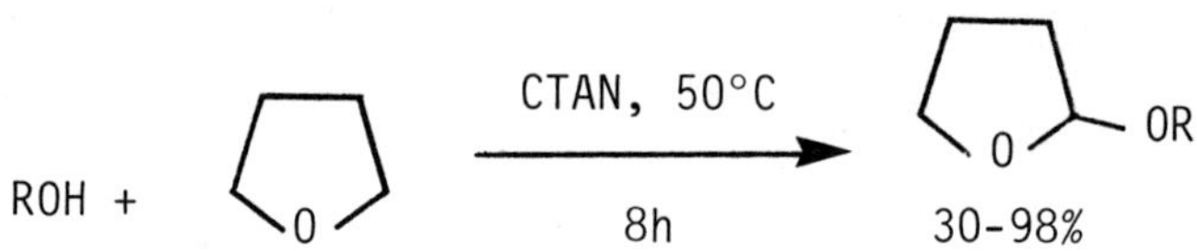

II.B.2. C — H ⟶ C — Hal

II.B.2-1 S. Rozen and C. Gal, J. Org. Chem., 52, 4928 (1988).

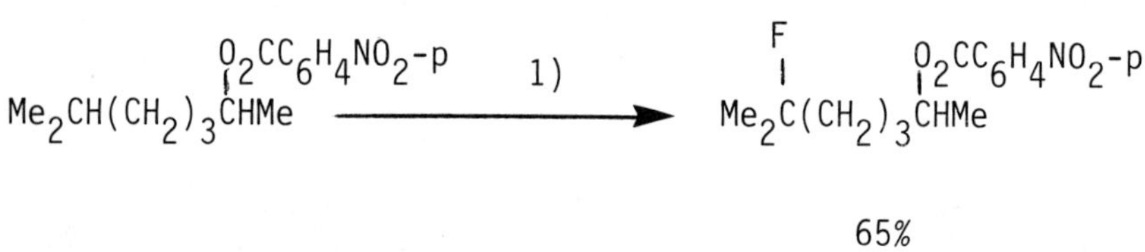

1) 4% F_2, N_2, $CFCl_3$, $CHCl_3$, -75°C

II.B.2-2 S. Singh, D.D. DesMarteau, S.S. Zuberi, M. Witz and H. Huang, J. Am. Chem. Soc., 109, 7194 (1987).

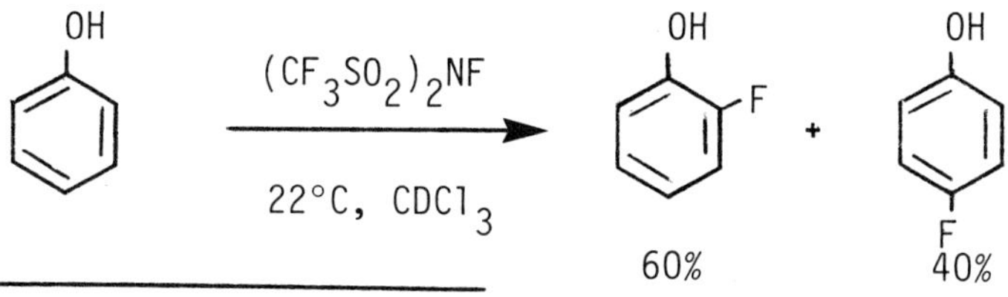

II.B.2-3 M. Van Der Puy, Tetrahedron Lett., 28, 255 (1987).

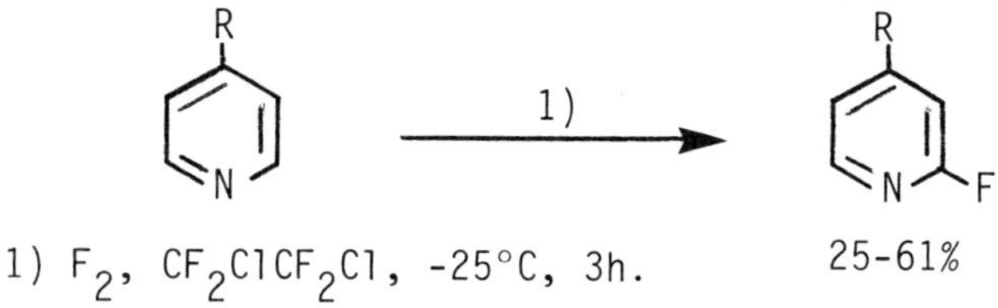

II.B.2-4 M. Kodomari, S. Takahashi and S. Yoshitomi, Chem. Lett., 1901 (1987).

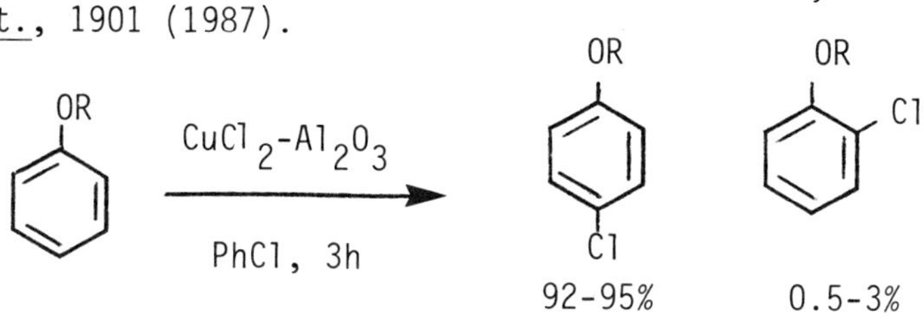

II.B.2-5 D.A. Jaeger and M.W. Clennan, Tetrahedron Lett., 28, 4805 (1987).

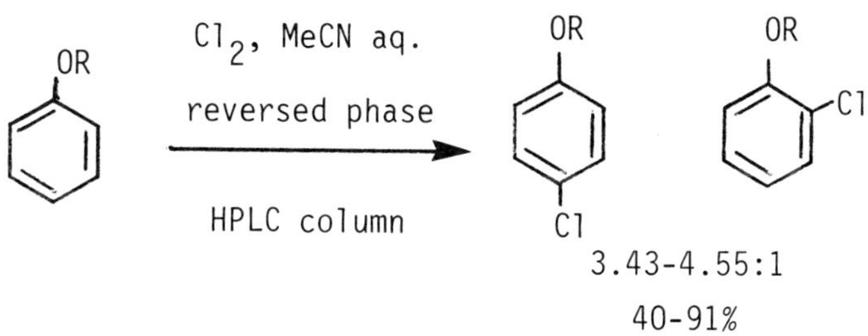

II.B.2-6 G.E. Jeromin, W. Orth, B. Rapp and W. Weiss, Chem. Ber., 120, 649 (1987).

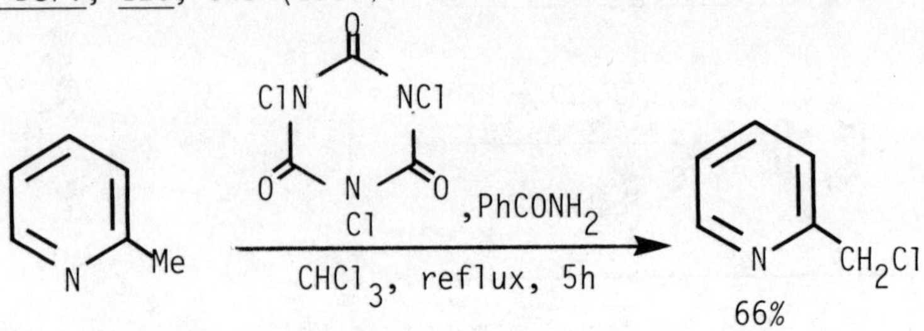

II.B.2-7 M.C.R. Franssen, H.G. van Boven and H.C. van der Plas, J. Heterocycl. Chem., 24, 1313 (1987).

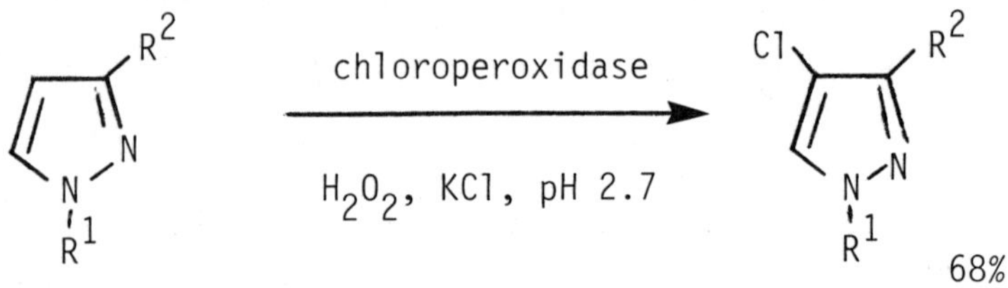

II.B.2-8 S. Kajigueshi, T. Kakinami, H. Tokiyama, T. Hirakuwa and T. Okamoto, Bull. Chem. Soc. Jpn., 60, 2667 (1987); P. Mencarelli and F. Stegel, Gazz. Chim. Ital., 117, 83 (1987).

$$\mathrm{RCOMe} \xrightarrow[\mathrm{CH_2Cl_2,\ MeOH,\ RT}]{\mathrm{PhCH_2\overset{+}{N}Me_3Br_3^-}} \mathrm{RCOCHBr_2}$$

75-89%

II.B.2-9 M. Srebnik, R. Mechoulam and I. Yona, J. Chem. Soc. Perkin I, 1423 (1987); S. Rozen and M. Brand, J. Chem. Soc., Chem. Commun., 752 (1987); S. Kajigaeshi, T. Kakinami, H. Tokiyama and T. Hirakawa, Chem.Lett., 627 (1987).

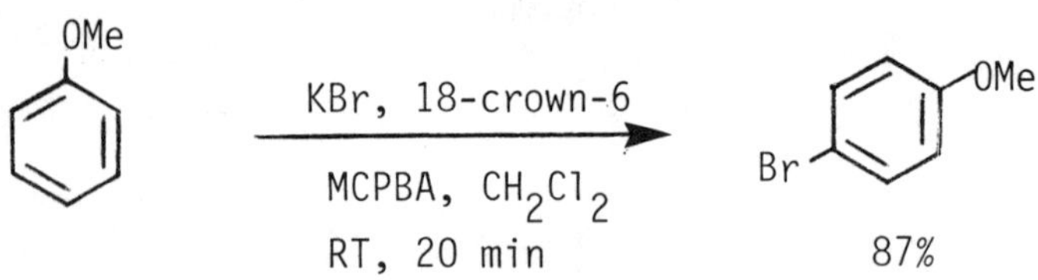

II.B.2-10 M. Tashiro, T. Makishima and S. Horie, J. Chem. Res. Synop., 342 (1987).

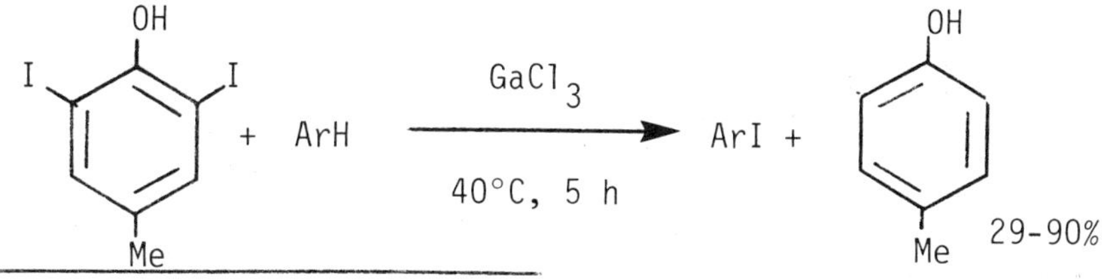

II.B.3. Other C—H Oxidations

II.B.3-1 M. Lemaire, A. Guy, J. Roussel and J.P. Guette, Tetrahedron, 43, 835 (1987).

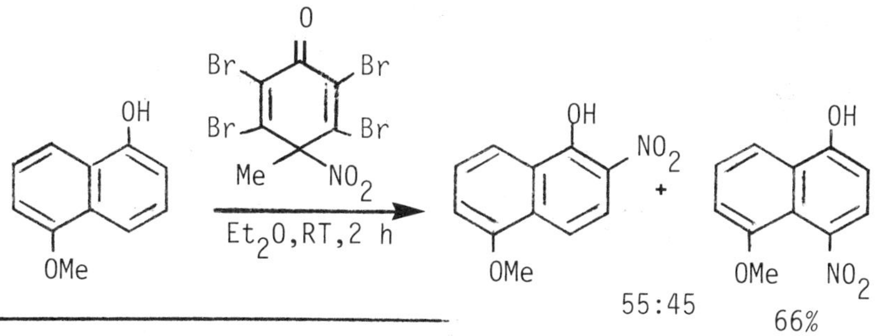

II.C. C—N Oxidations

II.C-1 S.B. Shim, K. Kim and Y.H. Kim, Tetrahedron Lett., 28, 645 (1987).

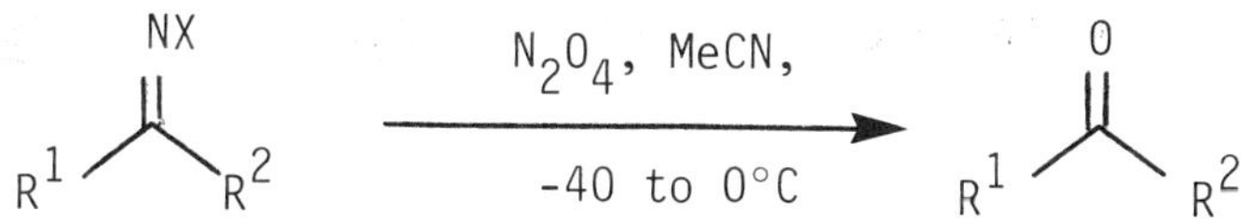

II.C-2 C. Alvarez, A.C. Cano, V. Rivera and C. Marquez, Synth. Commun., 17, 279 (1987).

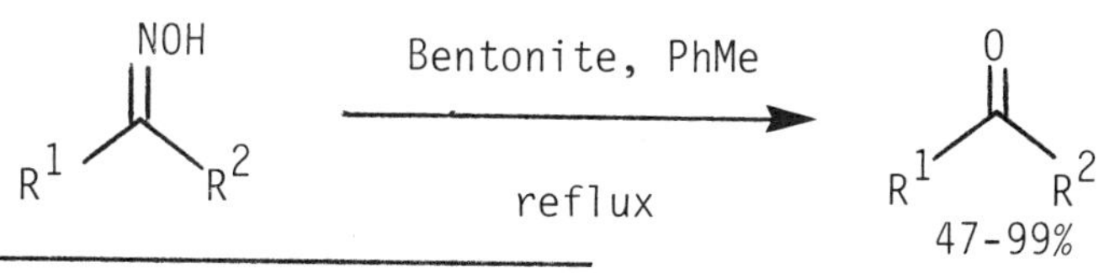

II.C-3 Y.H. Kim, H.K. Lee and H.S. Chang, Tetrahedron Lett., 28, 4285 (1987).

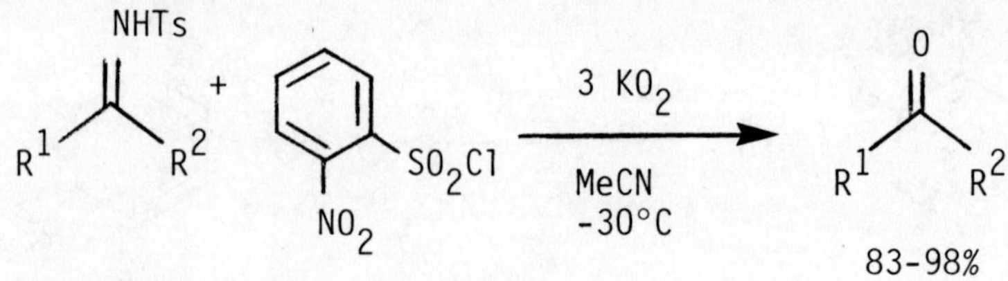

II.C-4 A. Kotali and P.G. Tsoungas, Tetrahedron Lett., 28, 4321 (1987).

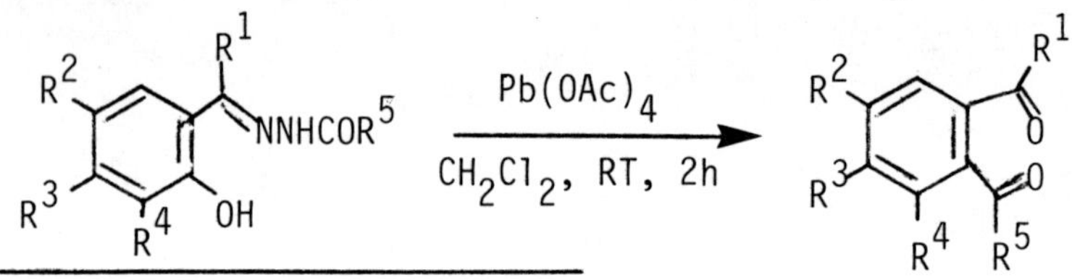

II.C-5 P.S. Vankar, R. Rathmore and S. Chandrasekaran, Synth. Commun., 17, 195 (1987).

$$PhCH_2NO_2 \xrightarrow[CH_2Cl_2,\ 3h]{Me(CH_2)_{15}NMe^+_3MnO^-_4} PhCHO \quad 89\%$$

II.D. Amine Oxidations

II.D-1 S. Murahashi and T. Shiota, Tetrahedron Lett., 28, 2383 (1987).

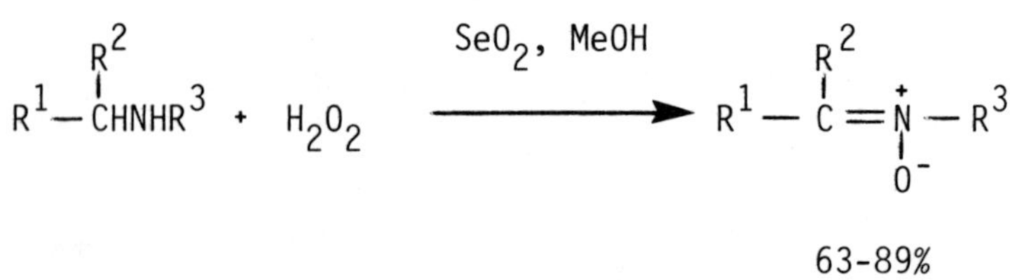

II.E. Sulfur Oxidations

II.E-1 J. Youn and R. Herrmann, Synthesis, 72 (1987); R. Davis, Synth. Commun., 17, 823 (1987).

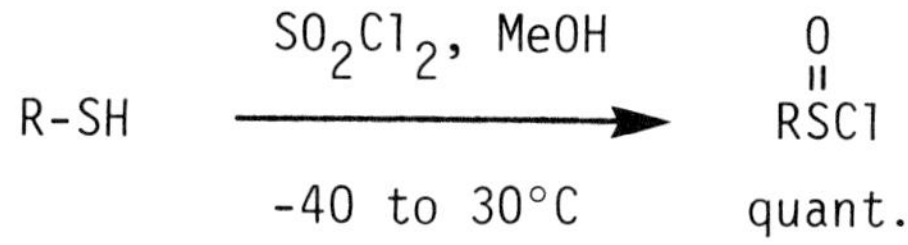

II.E-2 I. Torrini, M.P. Paradisi, G.P. Zecchini and F. Agrosi, Synth. Commun., 17, 515 (1987); S. Colonna, et al., J. Chem. Soc., Perkin I, 71 (1987); C. Srinivasan, P. Subramanian and S. Radha, Ind. J. Chem.B, 26, 193 (1987).

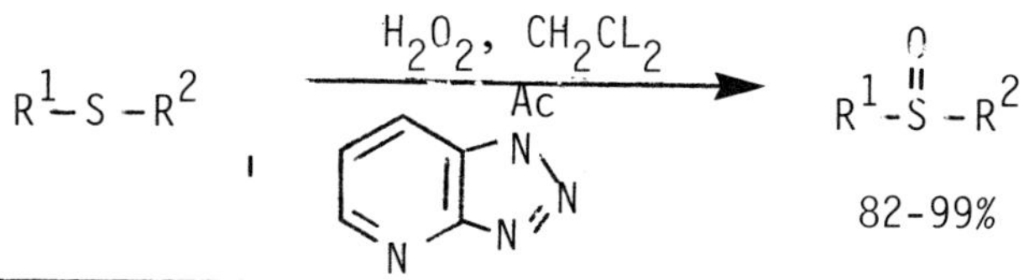

II.E-3 Y.H. Kim and H.K. Lee, Chem.Lett., 1499 (1987).

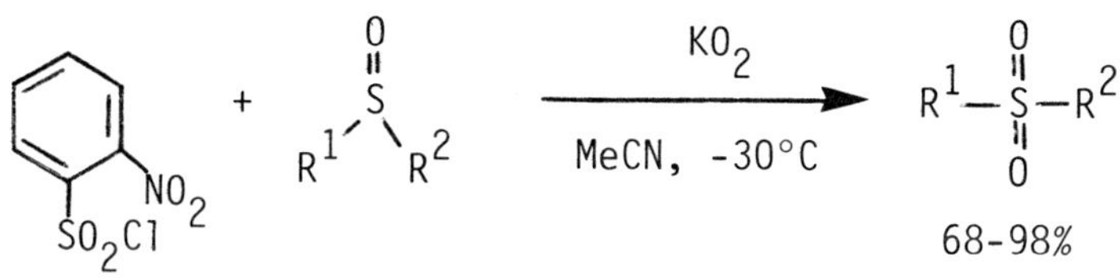

II.E-4 H.J. Kim and Y.H. Kim, Tetrahedron Lett., 28, 1669 (1987); Y. Tominaga, S. Kohra and A. Hosomi, Tetrahedron Lett., 28, 1529 (1987).

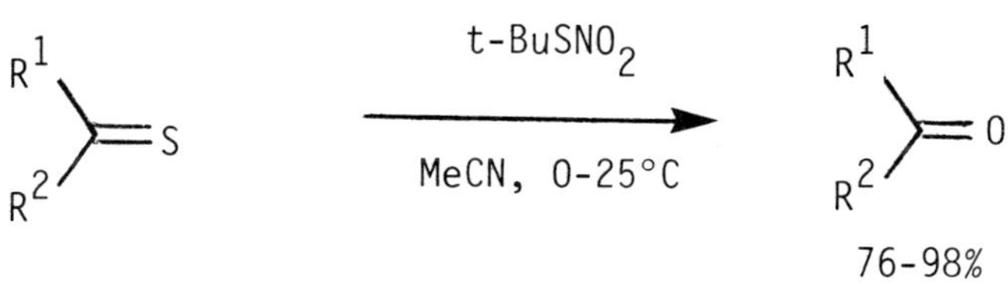

II.E-5 R.M. Moriarty, I. Prakash, D.E. Clarisse, R. Penmasta and A.K. Awasthi, J. Chem. Soc., Chem. Commun., 1209 (1987).

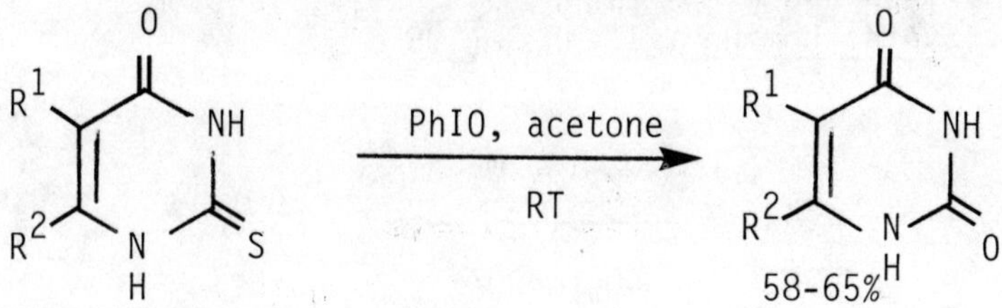

II.E-6 K. Harano, I. Shinohara, M. Murase and T. Hisano, Heterocycles, 26, 2583 (1987).

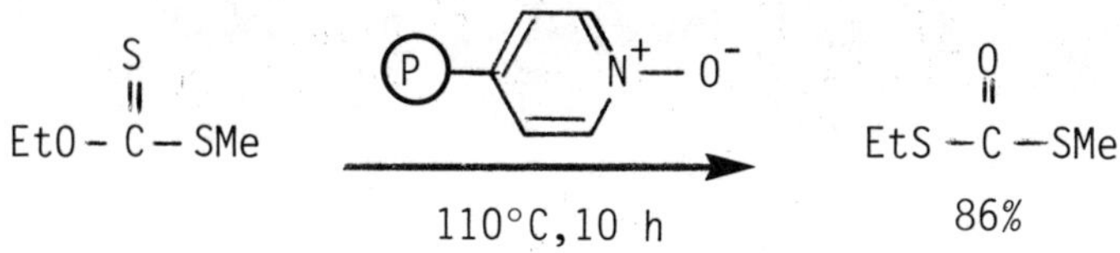

II.F. Phosphorus and Selenium Oxidation

II.F-1 Y. Okamoto, T. Kusano and S. Takamuku, Chem. Ind. (London), 529 (1987).

$$(RO)_2\overset{O}{\overset{\|}{P}}P + R^1OH \xrightarrow[50°C,\ 3h]{O_2,\ CuCl_2} (RO)_2\overset{O}{\overset{\|}{P}}OR^1 + H_2O$$

63-95%

II.F-2 L. Syper and J. Mkochowski, Tetrahedron, 43, 207 (1987).

$$RSeO_2H \xrightarrow[0°C\ to\ RT]{H_2O_2,\ H_2O} R\overset{O}{\overset{\|}{Se}}OOH + H_2O$$

II.F-3 M. Tiecco, M. Tingoli, L. Testaferri and D. Bartoli, Tetrahedron Lett., 28, 3849 (1987).

(+) - DIPT	(S-)	72-82% 18-40% ee
(-) - DIPT	(R-)	70-75% 20-28% ee

II.G. Oxidative Additions to C — C Multiple Bonds

II.G.1. Epoxidations

II.G.1-1 Z. Wang and W. Zhou, Tetrahedron, 43, 2935 (1987).

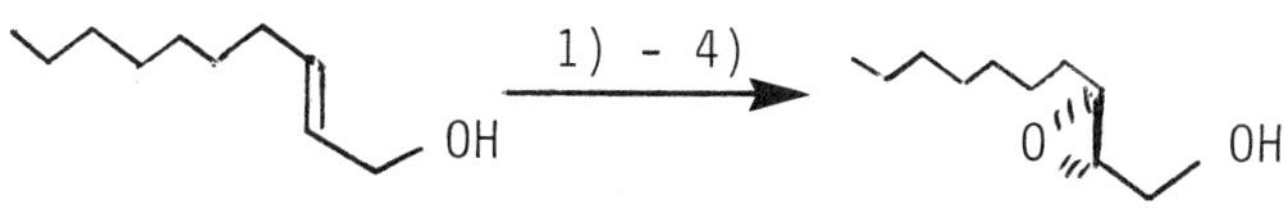

71%
95% ee

1) Ti(OiPr)$_4$, L-(+)-diethyl tartrate, CaH_2, silica gel, CH_2Cl_2, -20°C, 10 min ;
2) t-BuOOH, -40°C, 1 h;
3) tartaric acid, H_2O, -20°C, then RT;
4) NaOH aq, Et_2O, 0°C, 30 min

II.G.1-2 J.M. Chong and S. Wong, J. Org. Chem., 52, 2597 (1987).

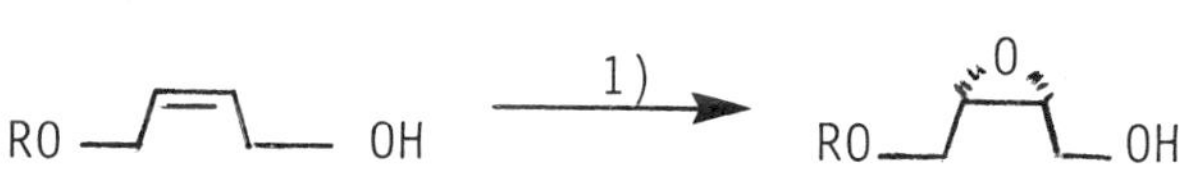

65%

1) 10-20 mol % Ti(O-i-Pr)$_4$,(-)-diisopropyl tartrate (12-28 mol %), TBHP (2-3 Eq), 4 A mol. sieves.

II.G.1-3 H. Kogen and T. Nishi, J. Chem. Soc., Chem. Commun., 311 (1987).

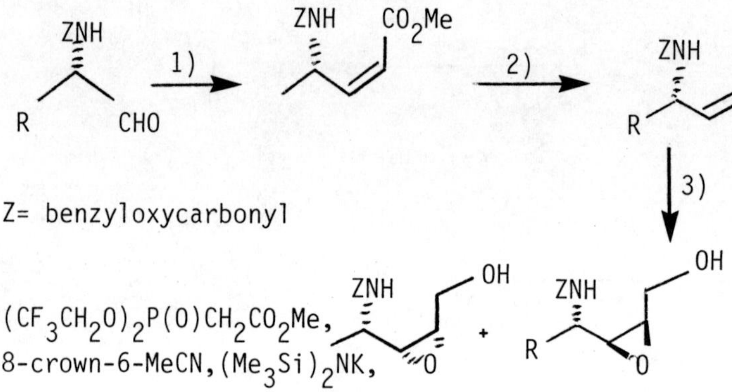

Z= benzyloxycarbonyl

1) $(CF_3CH_2O)_2P(O)CH_2CO_2Me$, 18-crown-6-MeCN, $(Me_3Si)_2NK$, THF, -78°C; 2) DIBAL; 3) MCPBA

II.G.1-4 M. Miyashita, T. Suzuki and A. Yoshikosh, Chem. Lett., 285 (1987); J. Muzart, Tetrahedron Lett., 28, 4665 (1987).

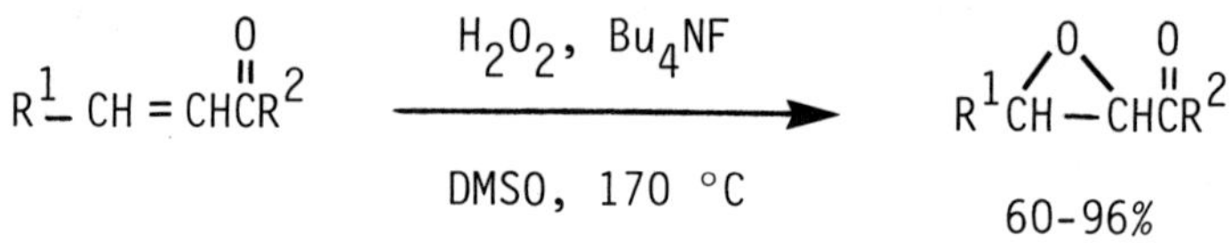

II.G.1-5 A. Foucaud and M. Bakouetila, Synthesis, 854 (1987).

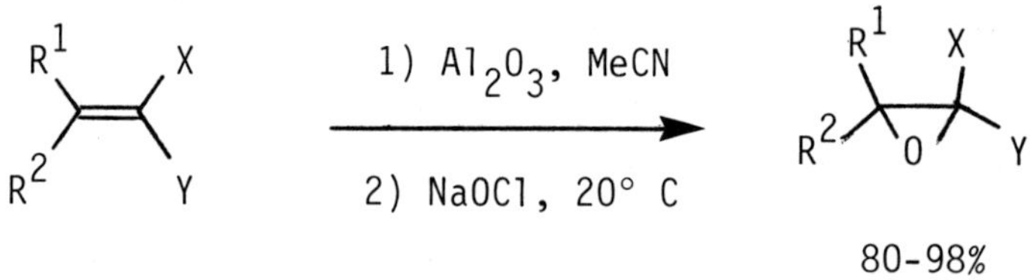

II.G.1-6 C.J.C. DeCock, J.-L. DeKeyser, J.H. Poupaert and P. Dumont, Bull. Soc. Chim. Belg., 96, 783 (1987).

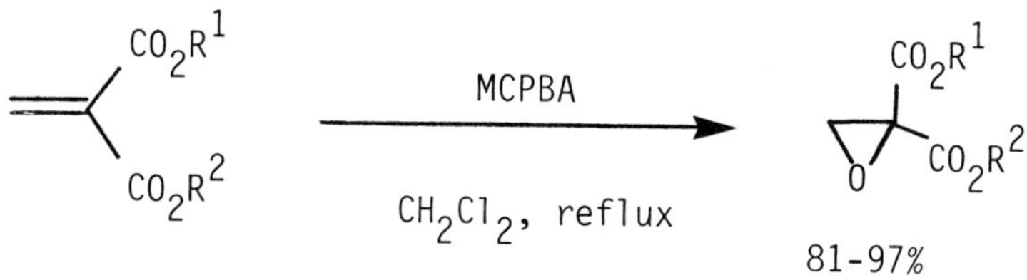

II.G.1-7 Y. Ishii et al., J. Org. Chem., 52, 1868 (1987).

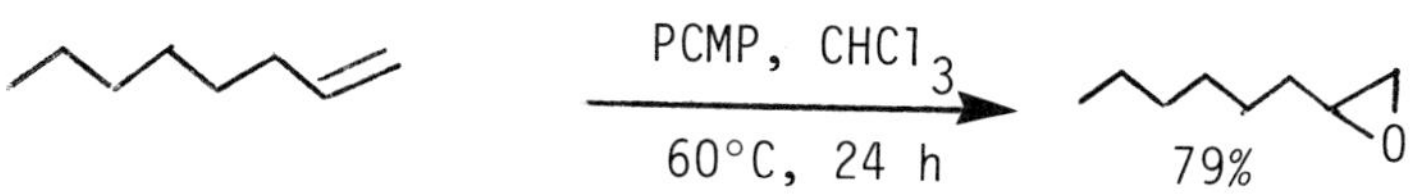

PCMP = peroxy-molybdeuum complex of tris(cetylpyridinium) 12-molybdophosphate

II.G.1-8 M. Quenard, V. Bonmarin and G. Gelbard, Tetrahedron Lett., 28, 2237 (1987).

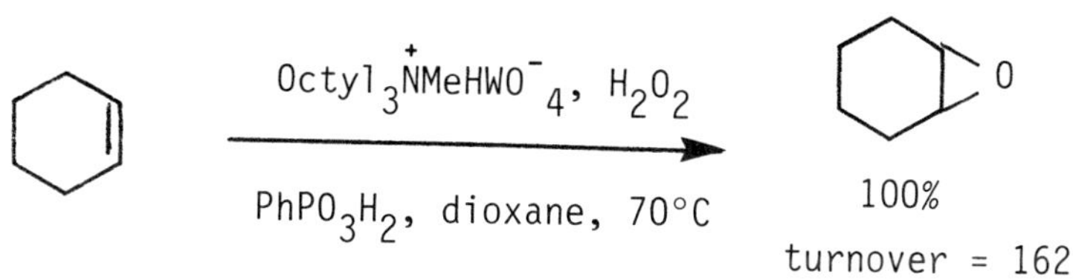

II.G.2. Hydroxylation

II.G.2-1 Y. Tamura, T. Yakura, H. Terashi, J. Haruta and Y. Kita, Chem. Pharm. Bull., 35, 570 (1987).

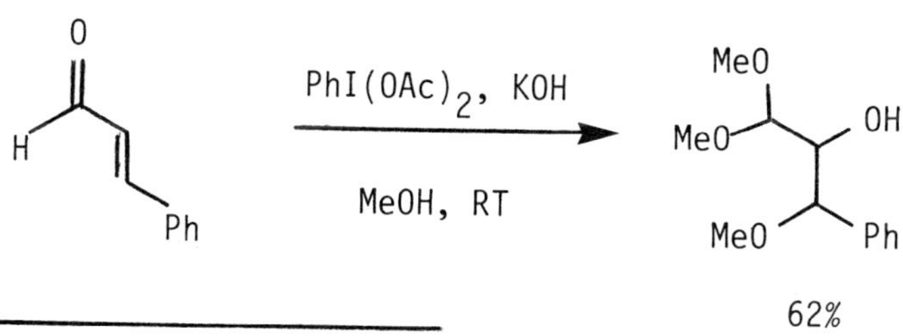

II.G.2-2 C.T. Goralski, B. Singaram and H.C. Brown, J. Org. Chem., 52, 4014 (1987); K. Maruoka, H. Sano, K. Shinoda and H. Yamamoto, Chem. Lett., 73 (1987).

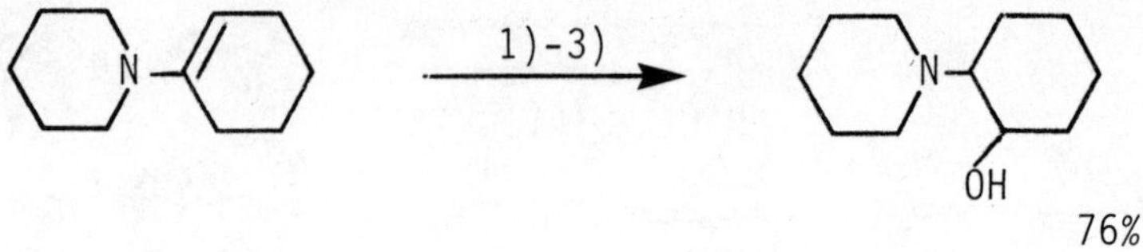

1) $BH_3.Me_2S$, THF, 0°C; 2) MeOH, 25°C; 3) NaOH, H_2O_2, THF MeOH, H_2O, 25°C; 4) HCl, 25°C.

II.G.2-3 G.D. Sriyani Ananda, P.J. Cremins and R.J. Stoodley, J.C.S. Chem. Commun., 882 (1987).

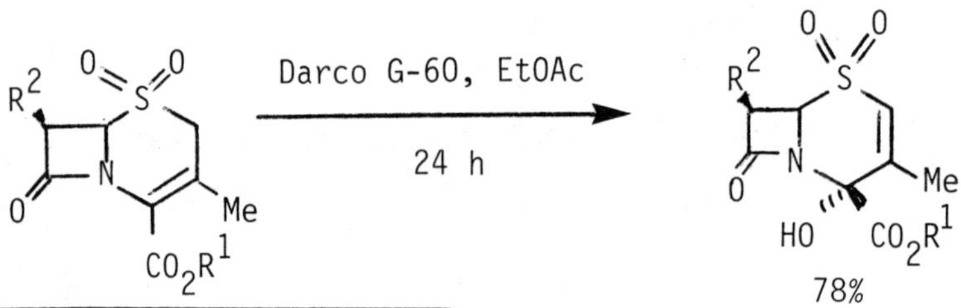

II.G.2-4 H. Kim and M.F. Schlecht, Tetrahedron Lett., 28, 5229 (1987).

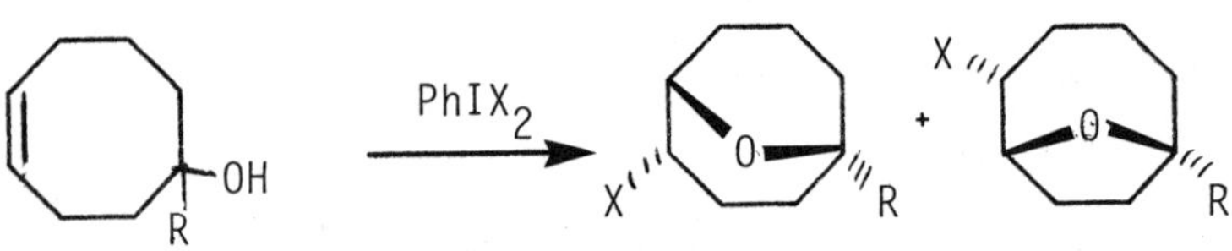

II.G.2-5 R.M. Moriarty, M.P. Duncan and O. Prakash, J. Chem. Soc., Perkin I, 1781 (1987).

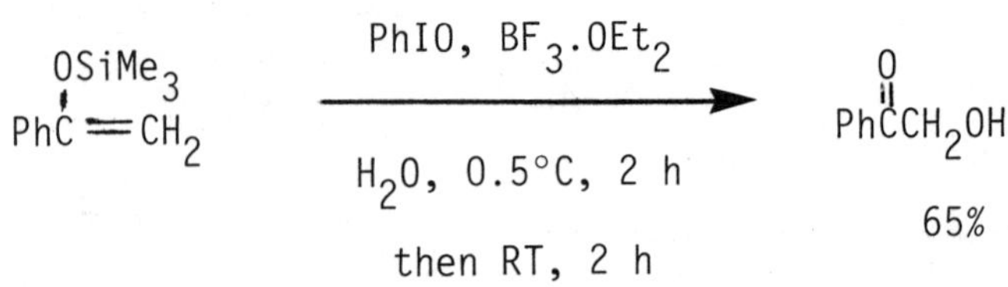

II.G.2-6 Y. Tamura et al., J. Org. Chem., 52, 3927 (1987); N. Lewis and P. Wallbank, Synthesis, 1103 (1987).

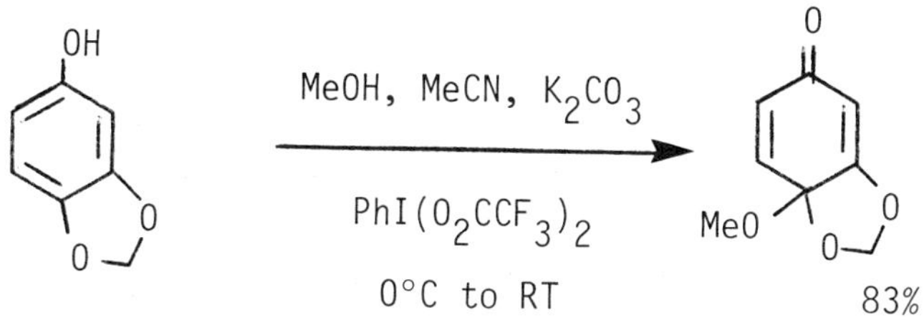

II.G.2-7 M.P. Capparelli, R.E. DeSchepper and J.S. Swenton, J. Org. Chem., 52, 4953 (1987).

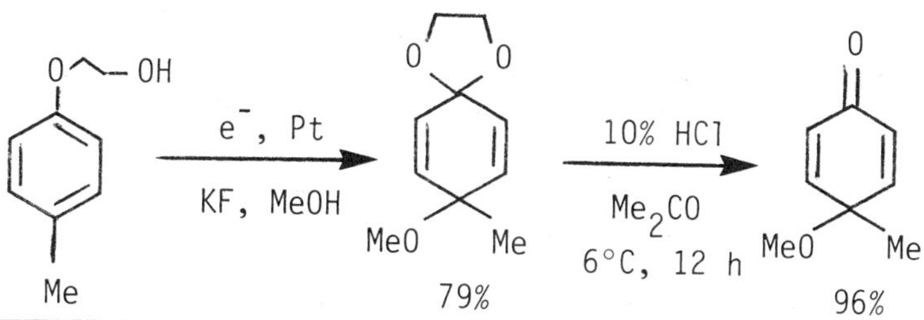

II.G.3. Other Oxidative Additions to C — C Multiple Bonds

II.G.3-1 K. Kloc, E. Kubicz, J. Mlochowski and L. Syper, Synthesis, 1084 (1987).

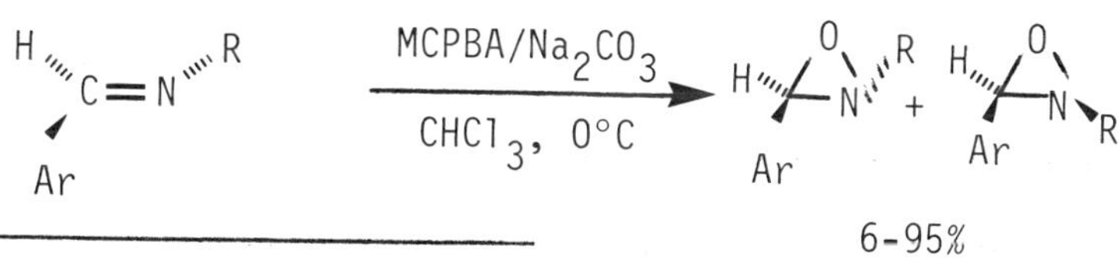

II.G.3-2 H.C. Brown and M.M. Midland, Tetrahedron, 43, 4059 (1987).

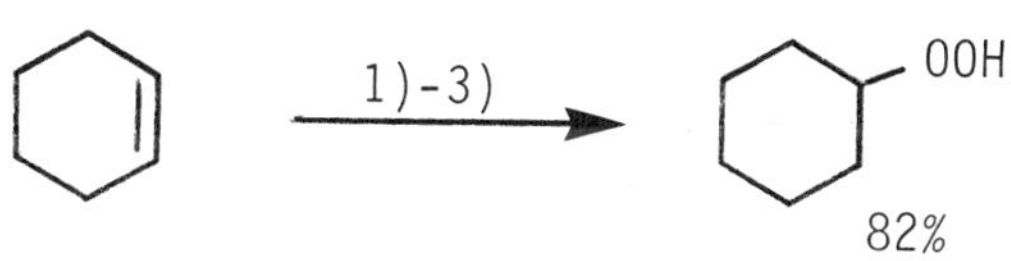

1) BH_3, THF, 0°, then 50°C, 3 h; 2) H_2O_2, H_2O, MeOH, O_2 -78°C, 1 h; 3) H_2O_2, H_2O, 30 min, 0°C.

II.G.3-3 T. Hosokawa, T. Takahashi, T. Ohta and S. Murahashi, J. Organomet. Chem., 334, C5 (1987).

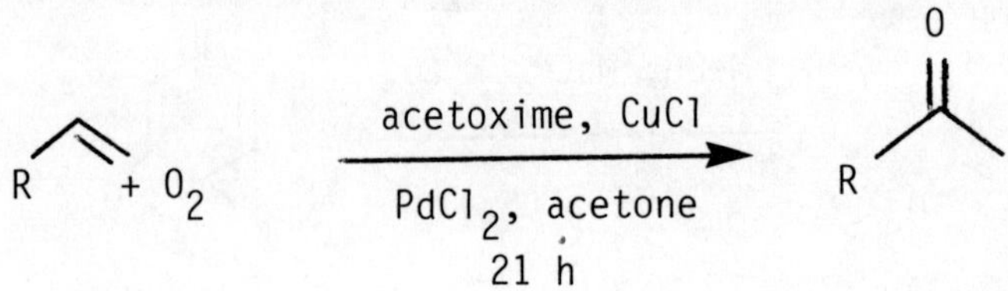

II.G.3-4 S. Torii, T. Inokuchi and Y. Hirata, Synthesis, 377 (1987).

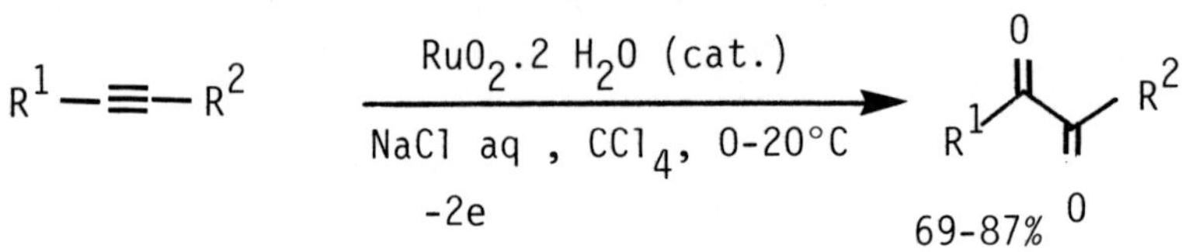

II.G.3-5 A. Stern and J.S. Swenton, J. Org. Chem., 52, 2763 (1987).

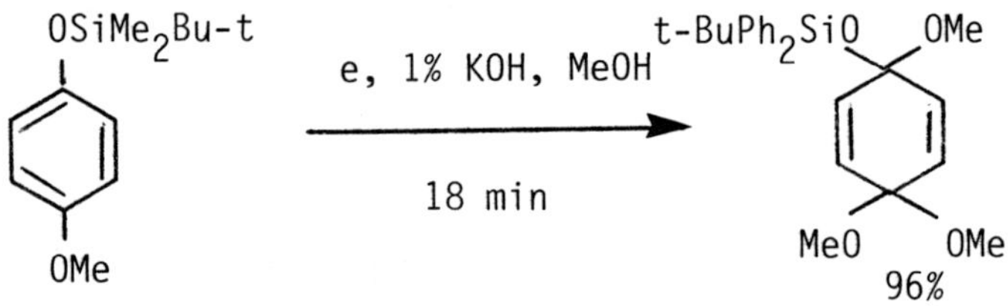

II.G.3-6 D. Hebel and S. Rozen, J. Org. Chem., 52, 2588 (1987).

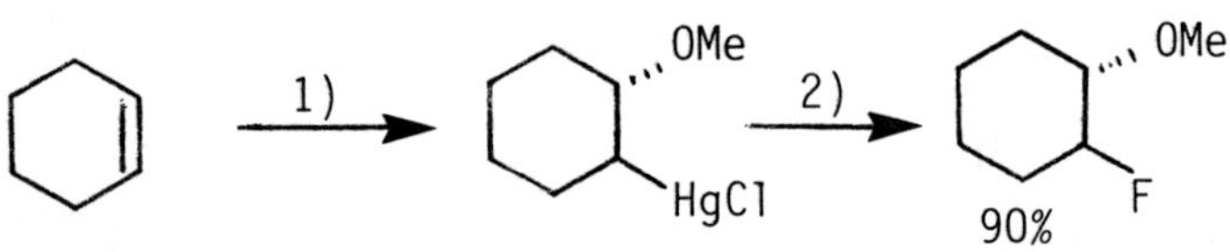

1) $Hg(OAc)_2$, MeOH, RT; 2) AcOF, $CHCl_3$, -78°C, 5 min.

II.G.3-7 F. Ghelfi *et al.*, *J. Chem. Res.*, (S), 24, 238 (1987).

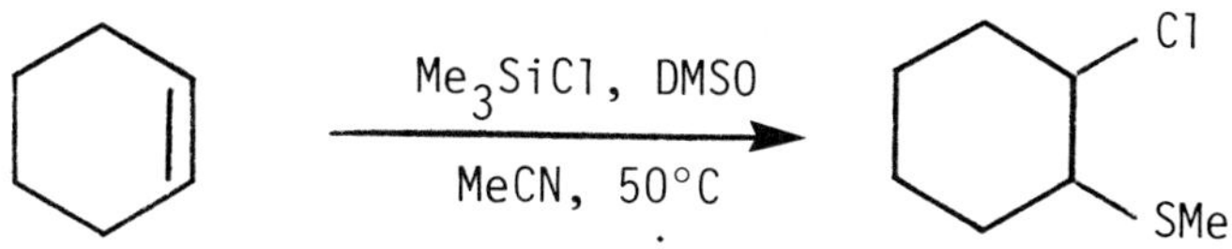

96%

II.G.3-8 L.J. Stewart, D. Gray, R.M. Pagni and G.W. Kabalka, *Tetrahedron Lett.*, 28, 4497 (1987).

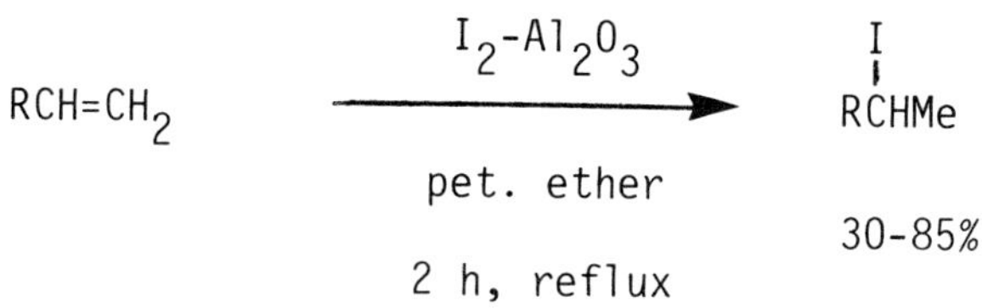

30-85%

II.H. Phenol-Quinone Oxidation

II.H-1 D.V. Pratt, F. Ruan and P.B. Hopkins, *J. Org. Chem.*, 52, 5053 (1987); J. Tsuji and M. Minato, *Tetrahedron Lett.*, 28, 3683 (1987).

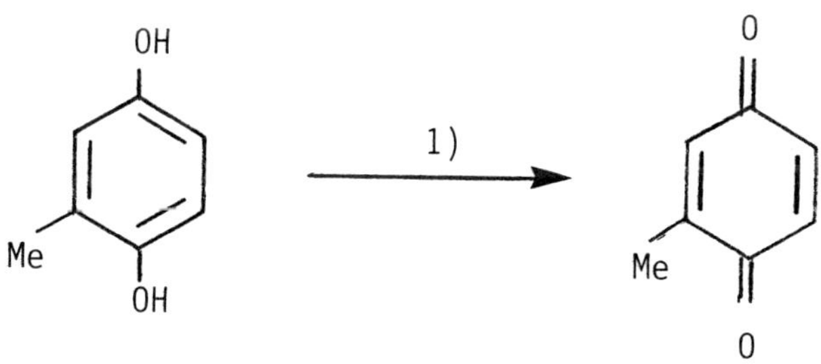

82%

1) H_2O_2, Ph_2Se_2, $Bu_4N^+HSO_4^-$, CH_2Cl_2, reflux, 1 h.

II.H-2 R. Araya, R. Tapia and J.A. Valderrama, J. Chem. Res. (S), 84 (1987); J.M. Saa, A. Llobera and P.M. Deya, Chem. Lett., 771 (1987).

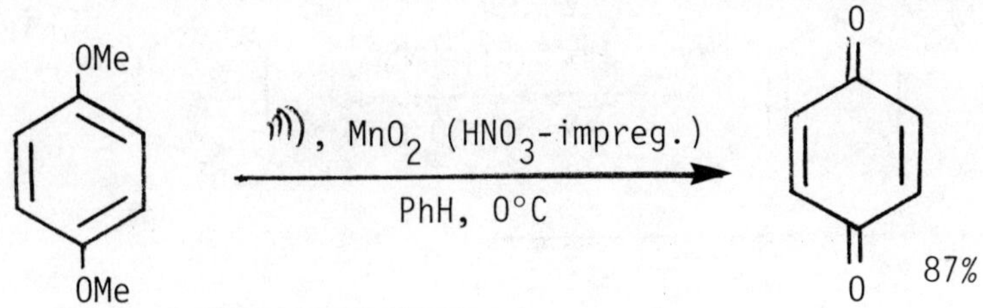

II.I. Dehydrogenation

II.I-1 T.L. Ho, Chem. Ind. (London), 295 (1987).

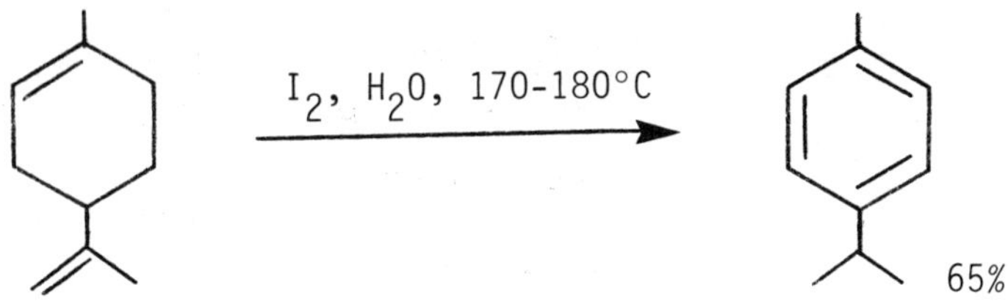

II.I-2 R.M. Moriarty, I. Prakash and R. Penmasta, Synth. Commun., 17, 409 (1987).

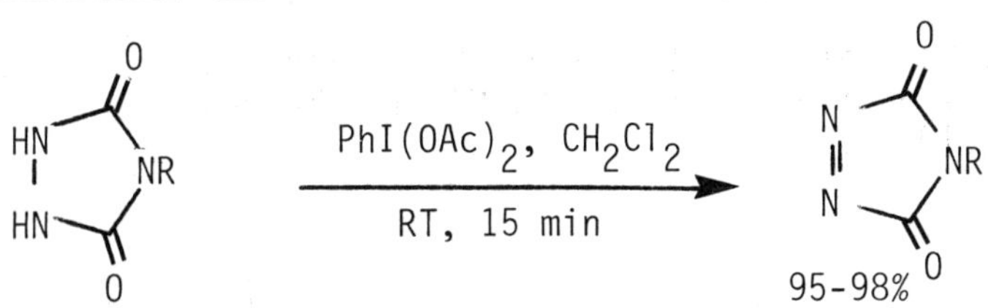

II.J. Other Oxidations

II.J-1 A. Lebouc, J. Simonet, J. Gelas and A. Dehbi, Synthesis, 320 (1987).

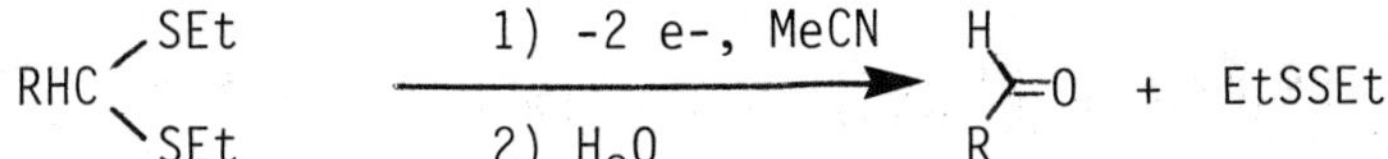

II.J-2 C.M. Alfonso, M.T. Barros and C.D. Maycock, J. Chem. Soc., Perkin I, 1221 (1987).

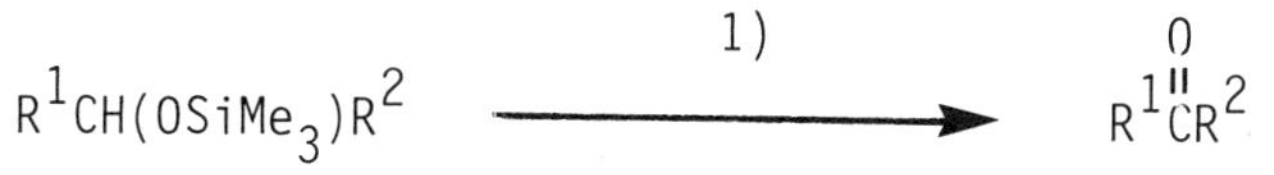

40-94%

1) $(COCl)_2$, DMSO, -60°C, then -30 to -40°C.

II.J-3 T.R. Beebe et al., J. Org. Chem., 52, 3165 (1987).

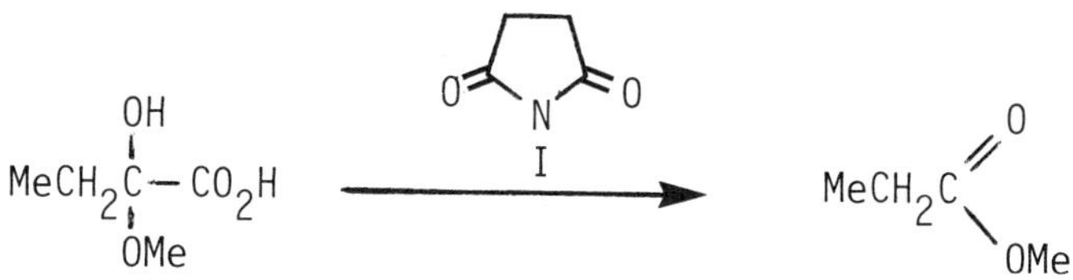

76-81%

II.K. Reviews

II.K-1 A.J. Fatiadi, Synthesis, 85 (1987).

Review: "The Classical Permanganate Ion: Still a Novel Oxidant in Organic Chemistry"

II.K-2 N. Mizuno and M. Misono, J. Synth.Org.Chem.Jpn., 45, 258 (1987.

Review: "Selective Oxidation of Lower Alkanes Over Oxide Catalysts"

II.K-3 P. Kovacic and M.B. Jones, Chem. Rev., 87, 357 (1987).

Review: "Dehydro Coupling of Aromatic Nuclei by Catalyst-Oxidant Systems: Poly(p-phenylene)"

II.K-4 S.F. Nelsen, Acc. Chem. Res., 20, 269 (1987).

Review: "Cation Radical Catalyzed Chain Oxygenation of Alkylated Olefins and Dienes"

II.K-5 R.V. Hoffman, Org. Prep. Proced. Int., 18, 179 (1986).

Review: "The Oxidation of Electron Donors With Sulphonyl Peroxides"

III.A. C═O Reductions

III.A-1 M. Kira, K. Sato and H. Sakurai, J. Org. Chem., 52, 948 (1987).

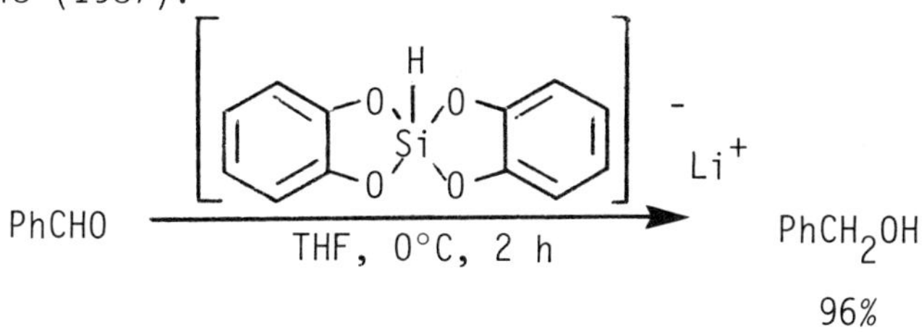

III.A-2 M. Borbaruah, N.C. Barua and R.P. Sharma, Tetrahedron Lett., 28, 5741 (1987); C. Francese, M. Tordeux and C. Wakselman, J. Chem. Soc., Chem. Commun., 642 (1987); A. Ono and H. Hayakawa, Chem. Lett., 853 (1987); W. Boland and U. Niedermeyer, Synthesis, 28 (1987).

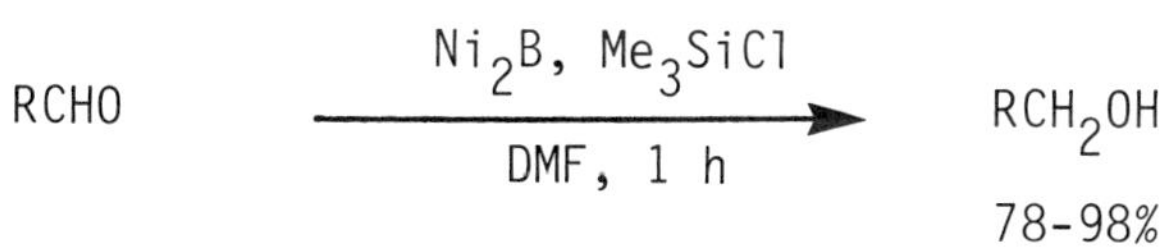

III.A 3 S. Torii, S. Takagishi, T. Inokuchi and H. Okumoto, Bull. Chem. Soc. Jpn., 60, 775 (1987).

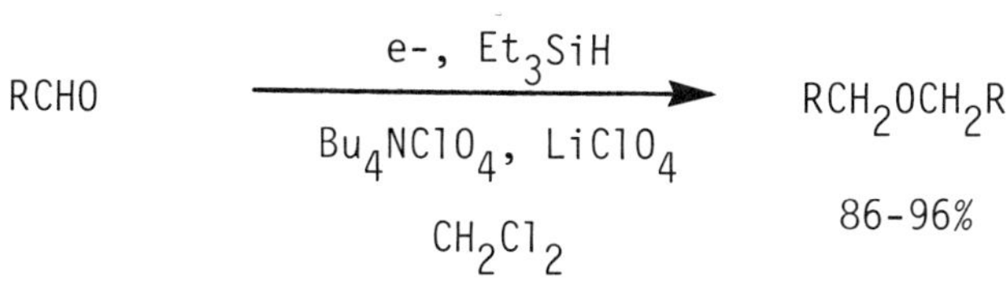

III.A-4 E.J. Corey, R.R. Bakshi, S. Shibata, C.P. Chen and V.K. Singh, J. Am. Chem. Soc., 109, 7925 (1987).

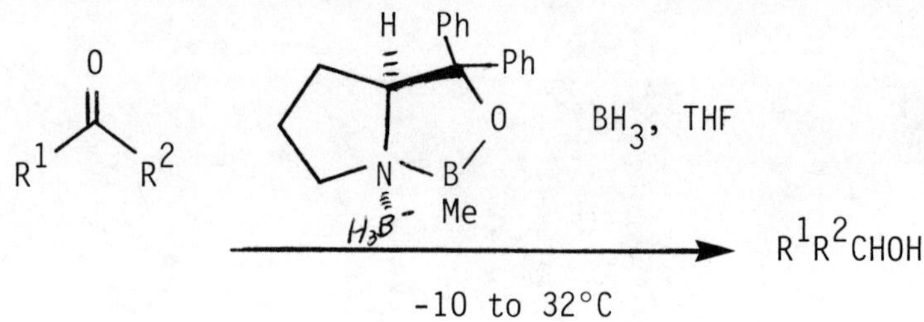

III.A-5 H.C. Brown, B.T. Cho and W.S. Park, J. Org. Chem., 52, 4020 (1987).

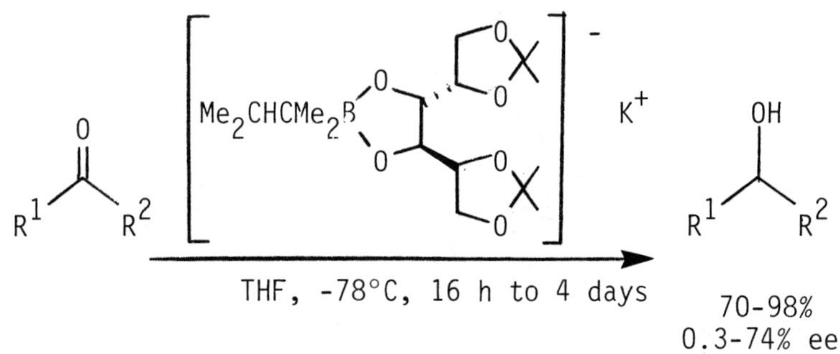

III.A-6 H. Takahashi, T. Morimoto and K. Achiwa, Chem. Lett., 855 (1987).

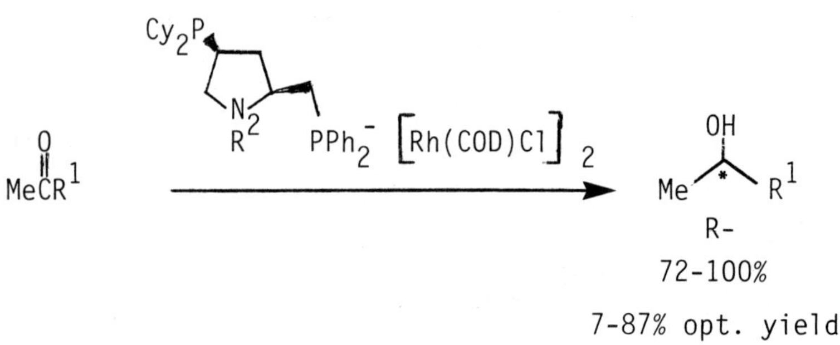

III.A-7 M. Kira, K. Sato and H. Sakurai, <u>Chem. Lett.</u>, 2243 (1987).

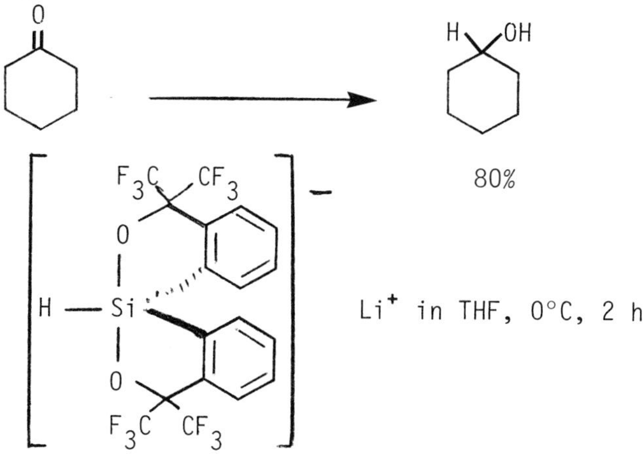

III.A-8 T. Okano, M. Matsuoka, H. Konishi and J. Kiji, <u>Chem. Lett.</u>, 181 (1987).

$$R^1C(=O)R^2 \xrightarrow[\text{i-PrOH}]{\text{Gd(i-PrO)}_3} R^1CH(OH)R^2 + MeC(=O)Me$$

21-99%

III.A-9 T. Fujisawa, H. Hayashi and Y. Kishioka, <u>Chem. Lett.</u>, 129 (1987); T.E.A. Nieminen and T.A. Hase, <u>Tetrahedron Lett.</u>, <u>28</u>, 4725 (1987); N.M. Yoon and K.E. Kim, <u>J. Org. Chem.</u>, <u>52</u>, 5564 (1987).

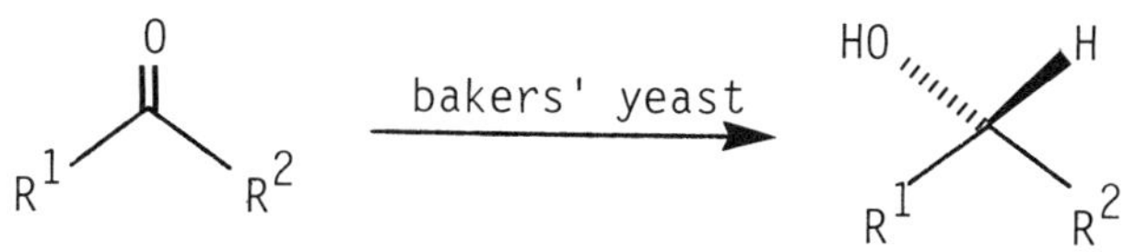

41-94%

45->96% ee

III.A-10 Y. Hanzawa, K. Kawagoe and Y. Kobayashi, Chem. Pharm. Bull., 35, 2609 (1987).

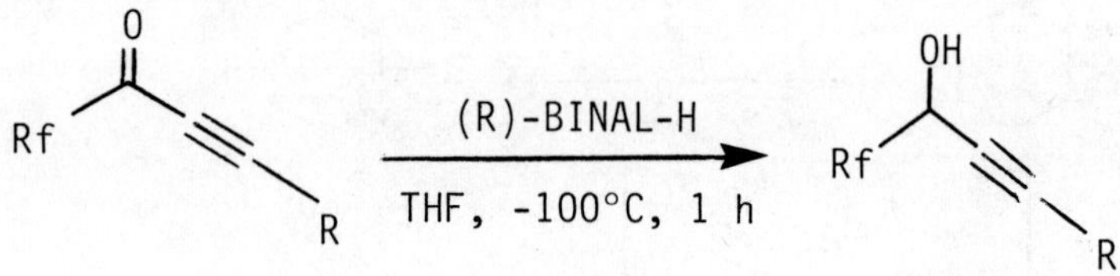

79-99%
19-78% ee

III.A-11 K. Soai, S. Niwa and T. Kobayashi, J. Chem. Soc., Chem. Commun., 801 (1987); M. Takeshita et al., Heterocycles, 26, 3051 (1987).

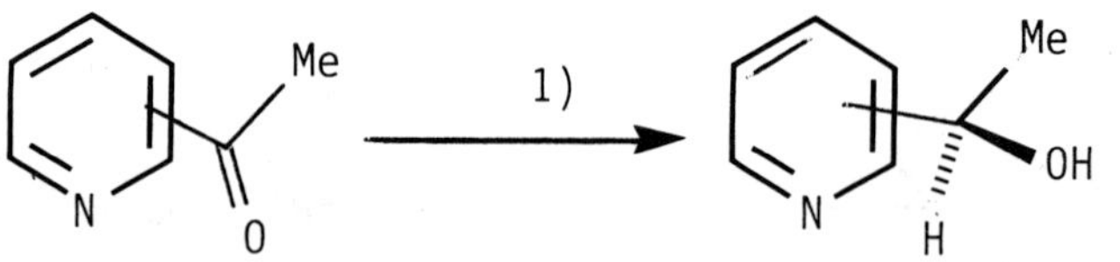

72%

1) $LiBH_4$, EtOH, THF, 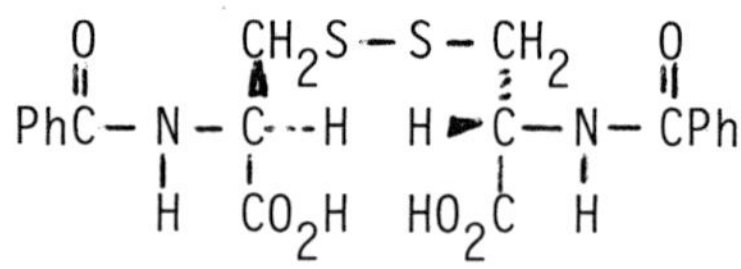

(R,R[1])-

III.A-12 A. Manzocchi, A. Fiecchi and E. Santaniello, Synthesis, 1007 (1987).

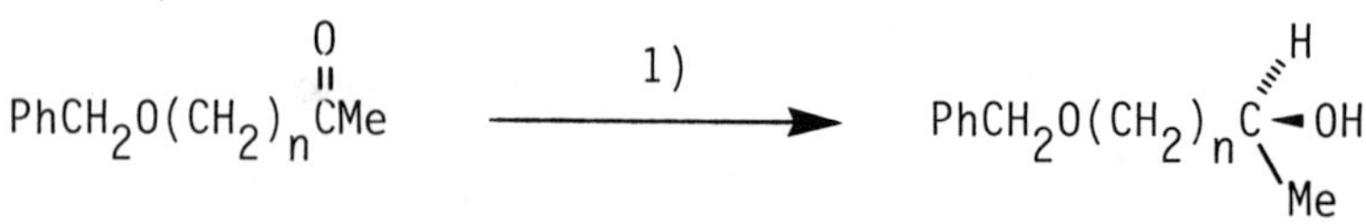

57-76%

1) bakers' yeast, EtOH, H_2O, 30°C.

III.A-13 R. Noyori, T. Ohkuma and M. Kitamura, J. Am. Chem. Soc., 109, 5856 (1987).

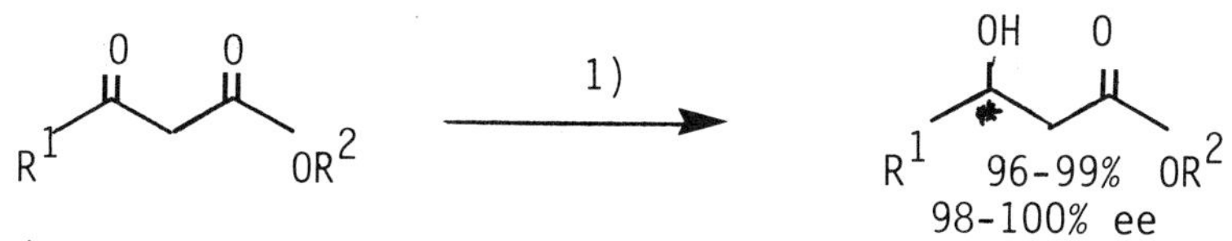

1) BINAP-Ru complex, H_2, 73-100 atmos, 34-106 h, MeOH, RT

III.A-14 J. Barluenga, J. G. Resa, B. Olani and S. Fustero, J. Org. Chem., 52, 1425 (1987); Y. Kawanami, et al., Chem. Lett., 2021 (1987).

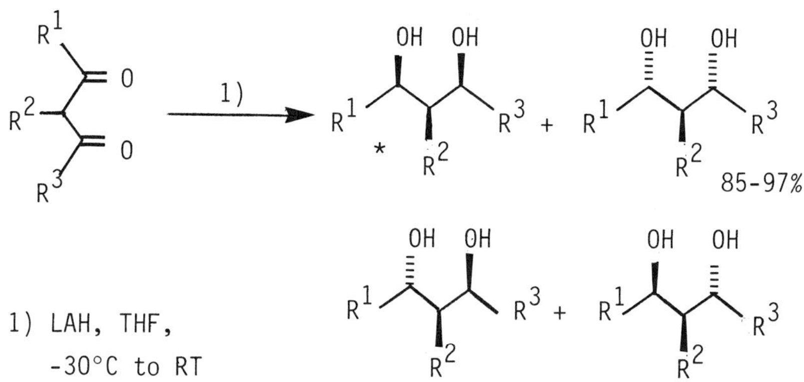

Predominant products with $LiAlH(Ot\text{-}Bu)_3$

*Predominant isomer using LAH, $TiCl_4$.

III.A-15 D.W. Brooks, H. Mazdiyasni and P.G. Grothaus, J. Org. Chem., 52, 3223 (1987).

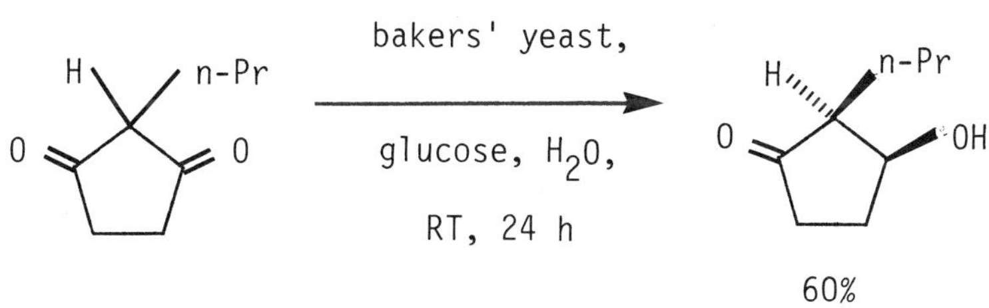

III.A-16 K. Chen, G.E. Hardtmann, K. Prasad, O. Repič and M.J. Shapiro, Tetrahedron Lett., 28, 155 (1987).

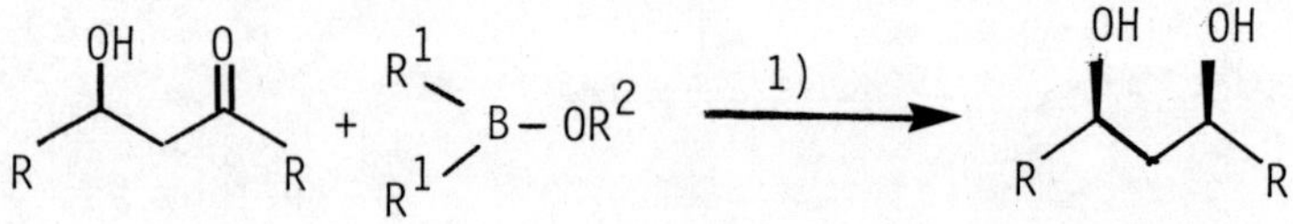

syn:anti>98:2

68-99%

1) THF, ahydrous MeOH, -70°C, Ar

III.A-17 Y. Tamura, H. Annoura and H. Fujioka, Tetrahedron Lett., 28, 5681 (1987).

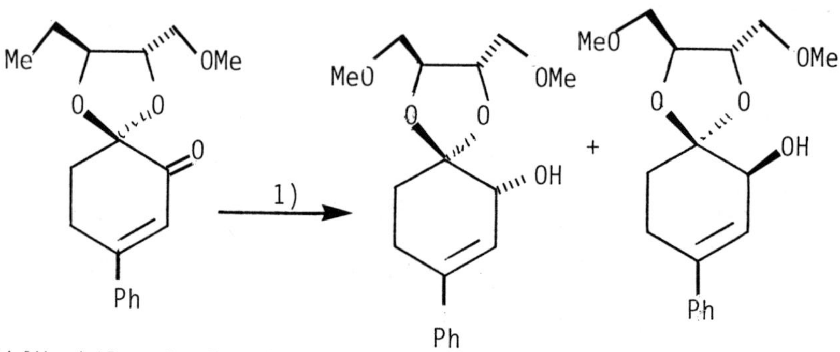

1)LAH, LiBr, Et_2O, -100°C.

96:4

III.A-18 K. Hasegawa, F. Matsuda, M. Yanagiya and T. Matsumoto, Tetrahedron Lett., 28, 1671 (1987).

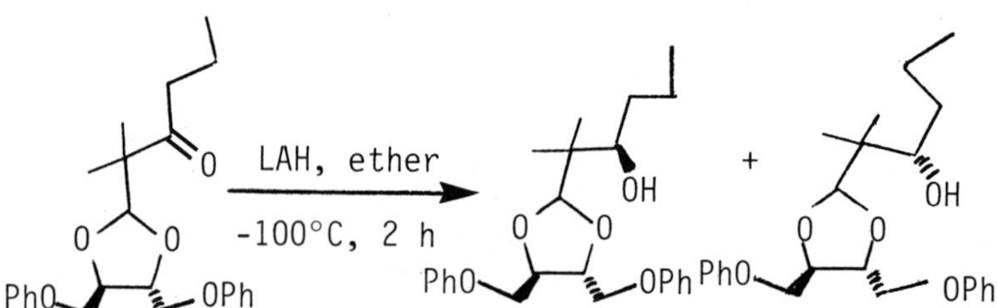

95-98%

+LiBr	72	:	28
+$MgBr_2$	9	:	91
No salt	67	:	33

III.A-19 D. Seebach <u>et al.</u>, <u>Helv. Chim. Acta</u>, <u>70</u>, 1605 (1987).

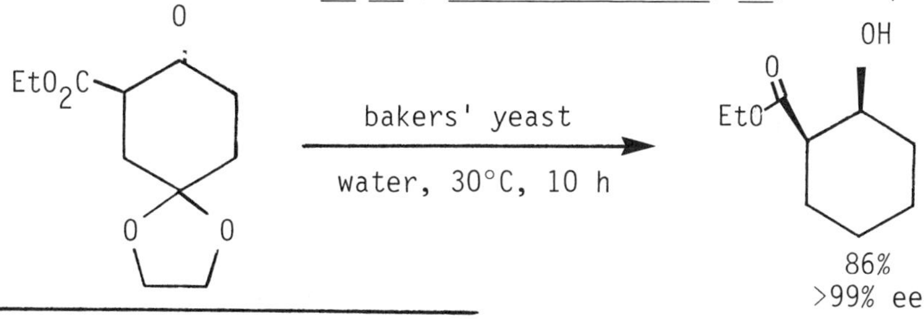

III.A-20 T. Kikukawa, Y. Iizuka, T. Sugimura, T. Harad and A. Tai, <u>Chem. Lett.</u>, 1267 (1987).

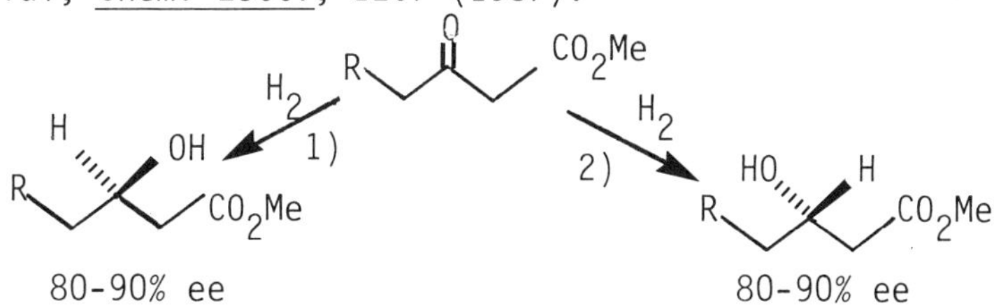

1) (R,R)-TA-NaBr-MRNi 2) (S,S)-TA-NaBr-MRNi

III.A-21 M. Utaka, H. Higashi and A. Takeda, <u>J. Chem. Soc., Chem. Commun.</u>, 1368 (1987).

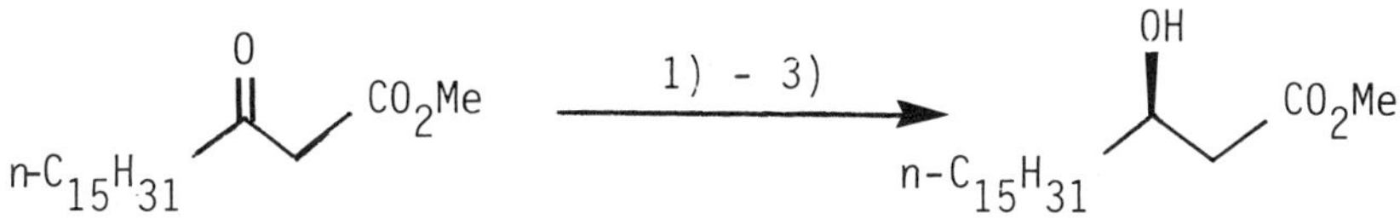

1) aq. KOH, EtOH; 2) bakers' yeast, 48 h; 3) CH_2N_2

III.A-22 K. Nakamura, K. Inoue, K. Ushio, S. Oka and A. Ohno, <u>Chem. Lett.</u>, 679 (1987).

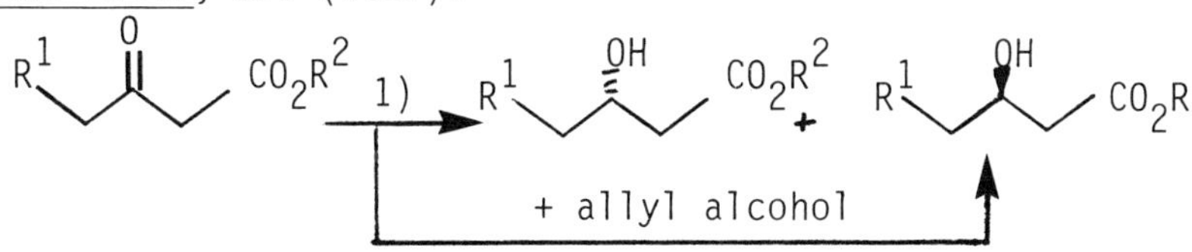

1) bakers' yeast

III.A-23 H.J. Dyke, J.A. Elix, S.M. Marcuccio and A.A. Whitton, Aust. J. Chem., 40, 431 (1987).

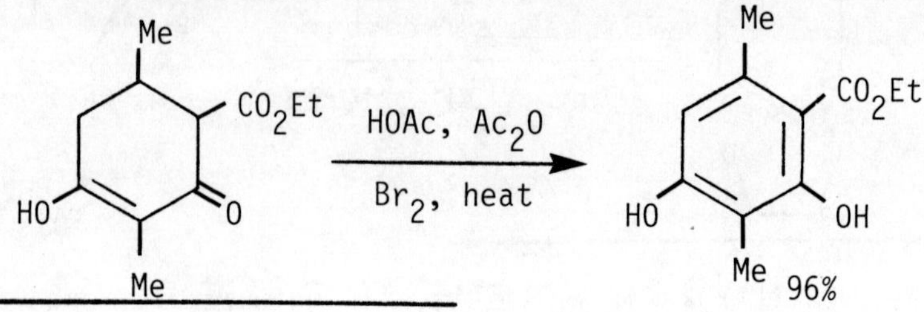

III.A-24 C. Kashima, K. Harada and Y. Omote, J. Chem. Soc., Perkin I, 1495 (1987).

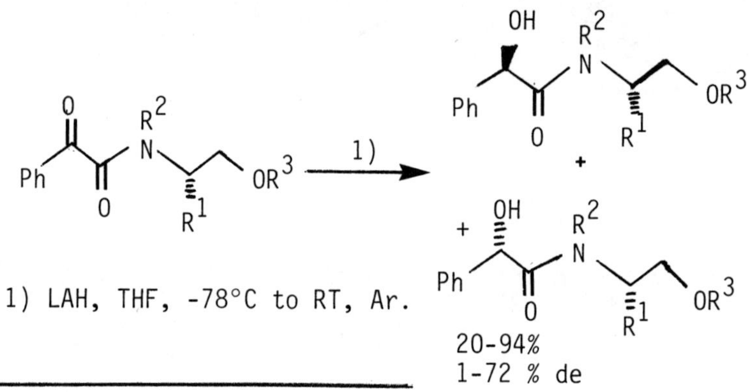

III.B. C—N Multiple Bond Reductions

1. Imine Reduction

III.B.1-1 S. Itsuno et al., Bull. Chem. Soc. Jpn., 60, 395 (1987).

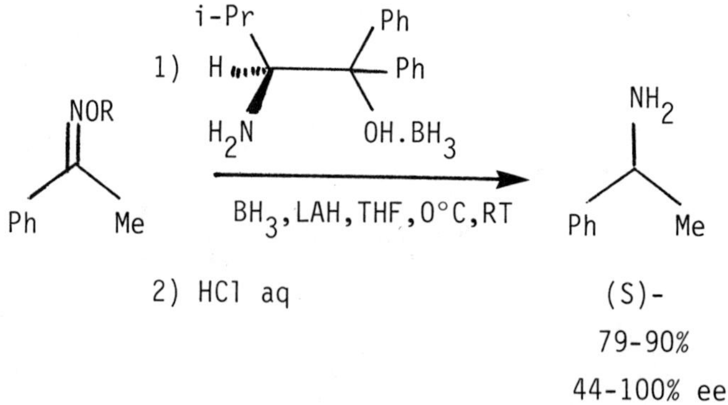

III.B.1-2 R.O. Hutchins and M.C. Rutledge, Tetrahedron Lett., 28, 5619 (1987).

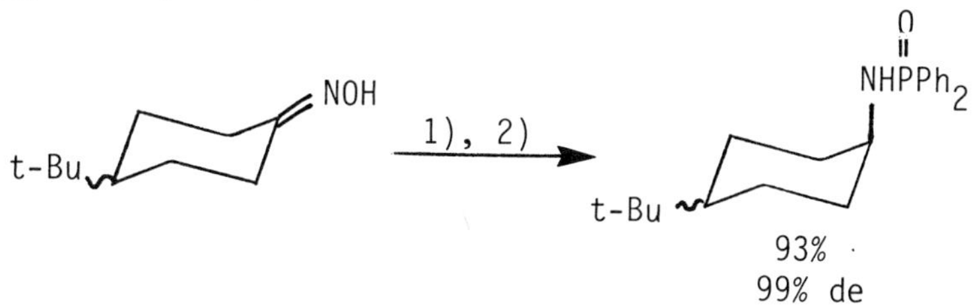

1) Ph_2PCl, TEA, CH_2Cl_2, pet. ether, -40° to -10°C, 20 min ;

2) $LiBH(BuS)_3$, RT.

III.B.1-3 F. Clerici, A. DiMare, M.L. Gelmi and D. Pocar , Synthesis, 719 (1987).

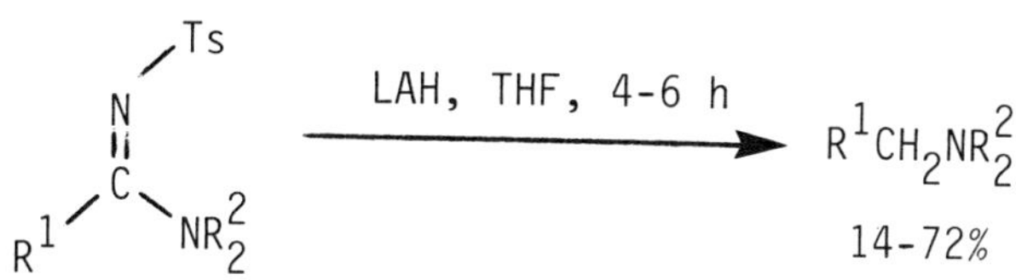

---

2. Reduction of Heterocycles

III.B.2-1 G. Lunn, J. Org. Chem., 52, 1043 (1987).

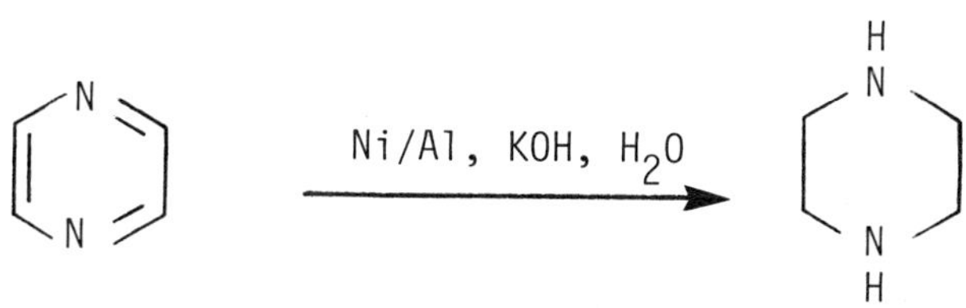

III.B.2-2 H. Kotsuki, U. Ushio and M. Ochi, Heterocycles, 26, 1771 (1987).

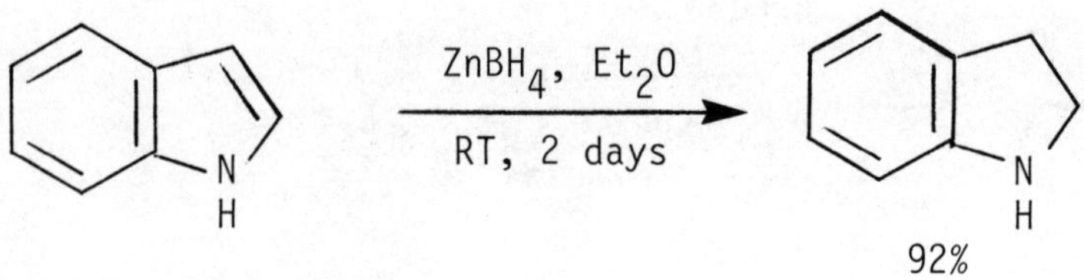

III.B.2-3 S. Murahashi, Y. Imada and Y. Hirai, Tetrahedron Lett., 28, 77 (1987); J.E. Shaw and P.R. Stapp, J. Heterocycl. Chem., 24, 1477 (1987).

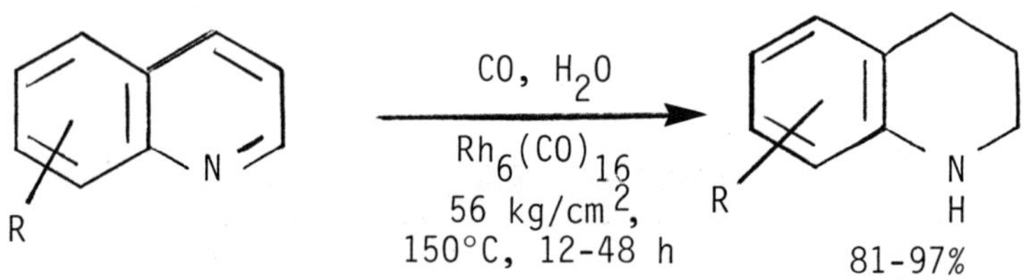

III.C. Reduction of Sulfur Compounds

III.C.1-1 A. Ogawa, Y. Nishiyama, N. Kambe, S. Murai and N. Sonada, Tetrahedron Lett., 28, 3271 (1987).

$$(PhS)_2 \xrightarrow[\text{N-methylpyrrolidine}]{CO,\ H_2O,\ Se,\ 70°C} PhSH$$

62-72%

III.C.1-2 K. Nagasawa, A. Yoneta, T. Umezawa and K. Ito, Heterocycles, 26, 2607 (1987); R. Lin and Y. Zhang, Synth. Commun., 17, 1403 (1987).

$$R^1\overset{O}{\overset{\|}{S}}R^2 \xrightarrow[0°C]{Zn,\ Cl_2SiMe_2,\ Me_2CO} R^1\text{—}S\text{—}R^2$$

70-100%

III.C.1-3 A. Nose and T. Kudo, Chem. Pharm. Bull., 35, 1770 (1987).

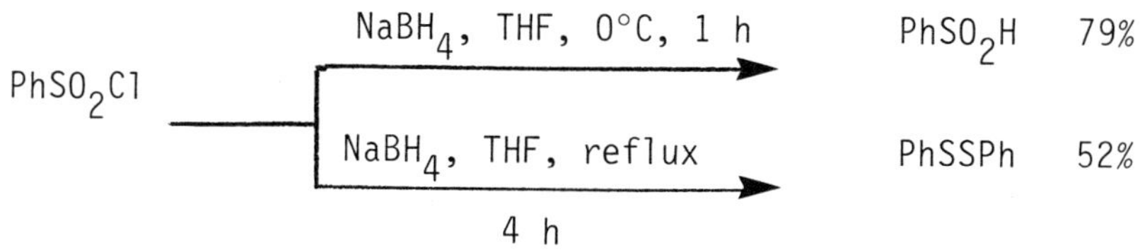

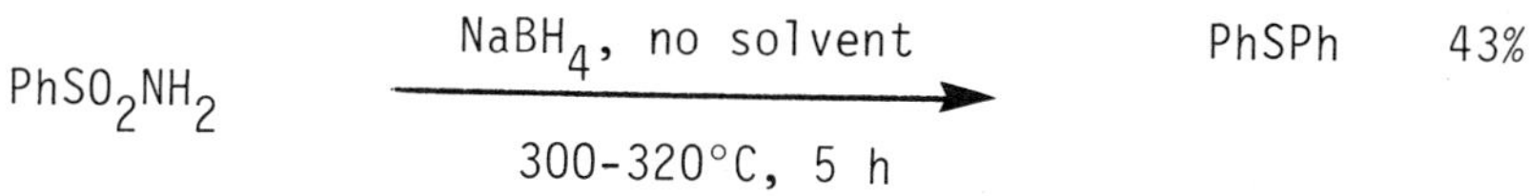

III.C.1-4 K. Langner, S. Tesch-Schmidtke and J. Voss, Chem. Ber., 120, 67 (1987).

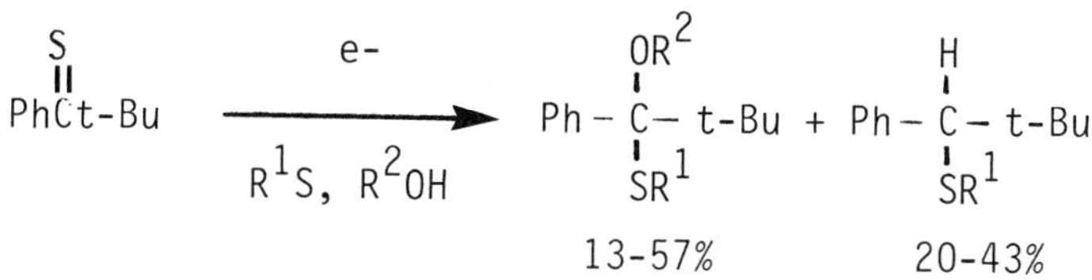

III.D. N — O Reductions

III.D.1-1 H. Suzuki, H. Manabe, T. Kawaguchi and M. Inouye, Bull. Chem. Soc. Jpn., 60, 771 (1987).

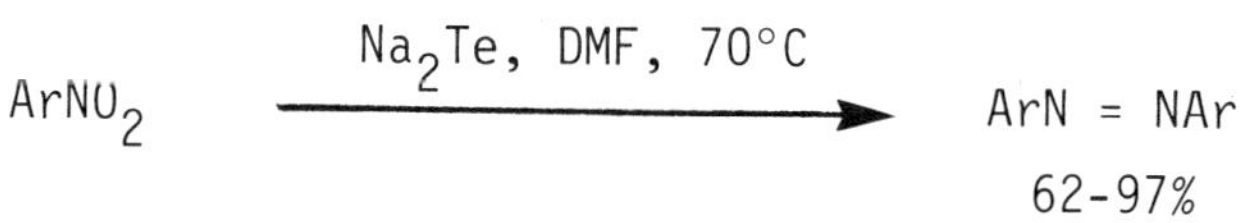

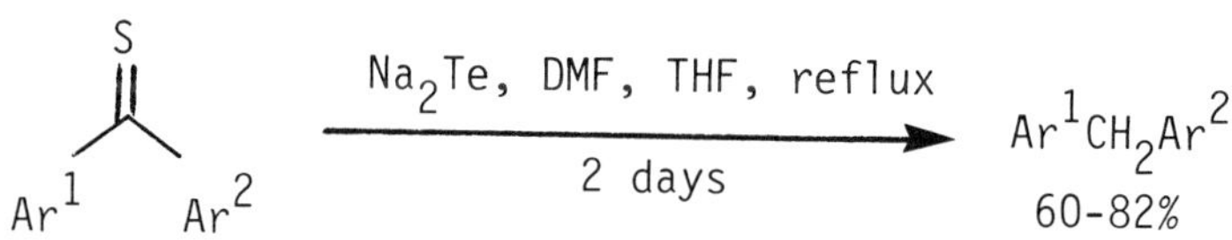

III.D.1-2 D.H.R. Barton, L. Fernandez, C.S. Richard and S.Z. Zard, Tetrahedron, 43, 551 (1987); M.E. Jung, D. D. Grove and S.I. Khan, J. Org. Chem., 52, 4570 (1987).

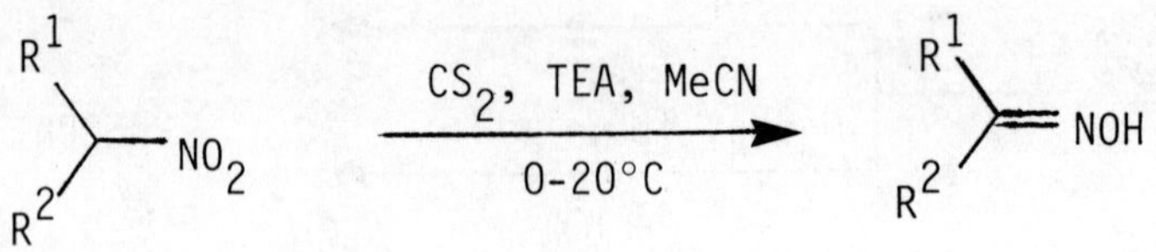

III.D.1-3 A. Bose and C.R. Saha, Chem. Ind., 199 (1987); M. Petrini, R. Ballini and G. Rosini, Synthesis, 713 (1987).

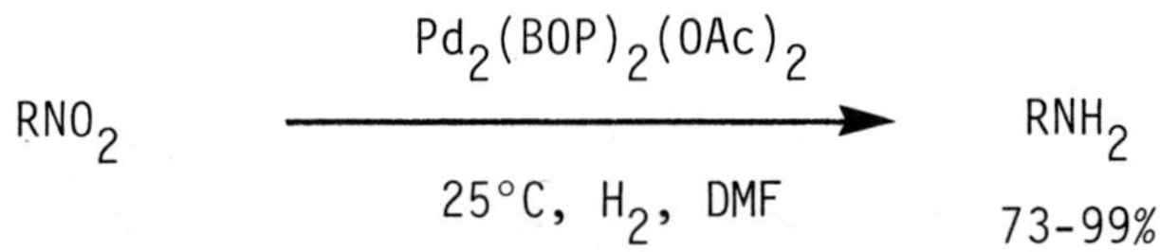

III.D.1-4 K. Klier, G. Kresze, O. Werbitzky and H. Simon, Tetrahedron Lett., 28, 2677 (1987).

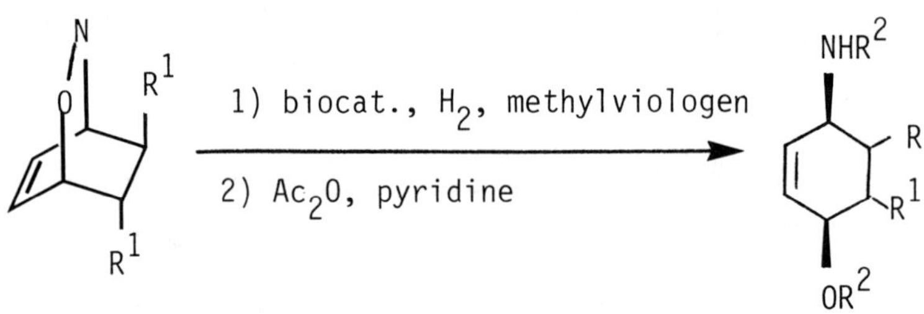

biocat. - Methanosarcina barkeri DSM800

Methanobacterium thermoautotrophicum DSM2133

Clostridium kluyveri DSM555

III.E. C—C Multiple Bond Reductions

1. C=C Reductions

III.E.1-1 Z. Hou, H. Taniguchi and Y. Fujiwara, Chem. Lett., 305 (1987); T. Negoro and Y. Ikeda, Org. Prep. Proced. Intl., 19, 71 (1987).

$$Ph_2C{=}CH_2 \xrightarrow[\text{THF, RT}]{\text{PhYbI, MeOH}} Ph_2CHMe$$

94%

III.E.1-2 I. Amer, T. Bravdo, J. Blum and K.P.C. Vollhardt, Tetrahedron Lett., 28, 1321 (1987); C. Larpent, R. Dabard and H. Patin, 28, 2507 (1987).

$$3\text{-}NO_2C_6H_4CH{=}CH_2 \xrightarrow[\text{Aliquat}^{R}\ 336,\ MeNO_2,\ 30°C,\ 7\ h]{H_2,\ RhCl_3.3H_2O} 3\text{-}NO_2C_6H_4CH_2CH_3$$

70%

III.E.1-3 H. Chikashita, et al., Bull. Chem. Soc. Jpn., 60, 737 (1987).

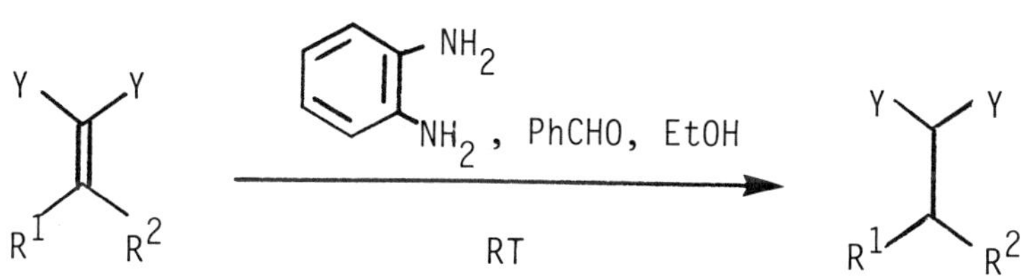

50-93%

III.E.1-4 P.W. Rabideau and G.L. Karrick, Tetrahedron Lett., 28, 2481 (1987).

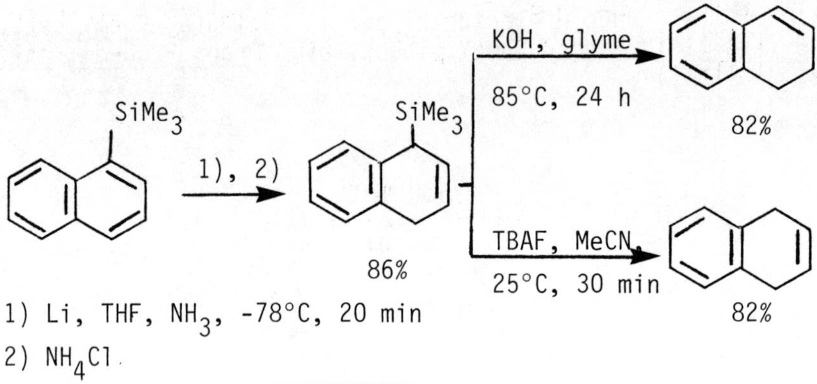

II.E.1-5 J. Chaussard, C. Combellas and A. Thiebault, Tetrahedron Lett., 28, 1173 (1987).

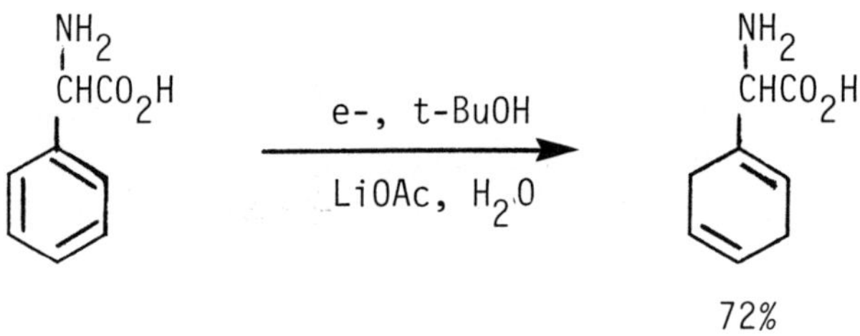

III.E.1-6 H. Takaya et al., J. Am. Chem. Soc., 105, 1596 (1987).

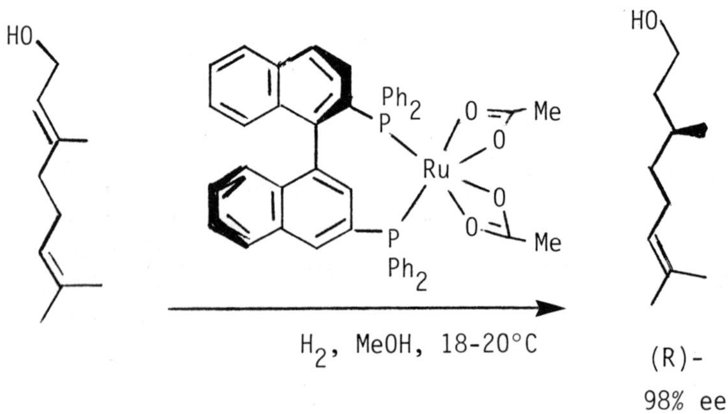

III.E.1-7 T. Sakai, K. Miyata, K. Hondo, M. Utaka and A. Takeda, Bull. Chem. Soc. Jpn., 60, 3421 (1987); Y.D. Vankar, G. Kumaravel, N. Mukherjee and C.T. Rao, Synth. Commun., 17, 181 (1987); T. Sakai, K. Miyata, M. Utaka and A. Takeda, Bull. Chem. Soc. Jpn., 60, 1063 (1987).

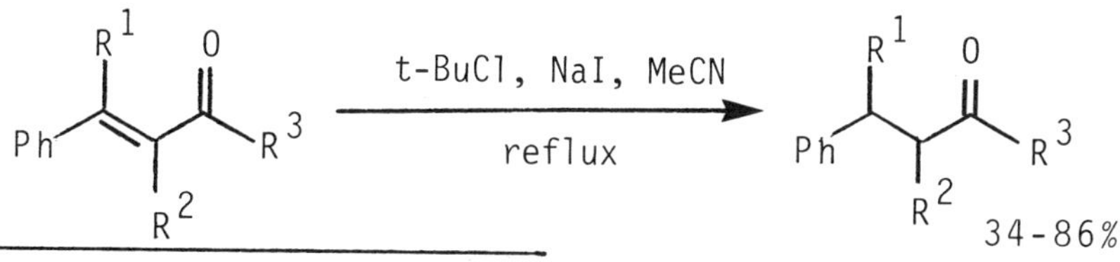

III.E.1-8 M. Utaka, S. Onoue and A. Takeda, Chem. Lett., 971 (1987).

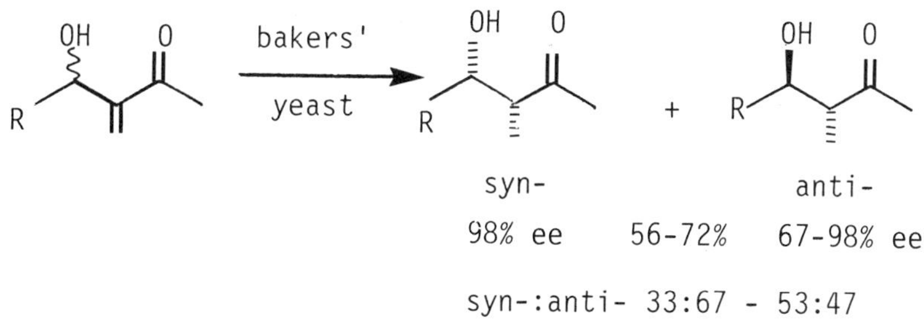

III.E.1-9 A. Ohno et al., Bull. Chem. Soc. Jpn., 60, 2423 (1987); V. Massonneau, P. Le Maux and G. Simonneaux, J. Organomet. Chem., 327, 269 (1987).

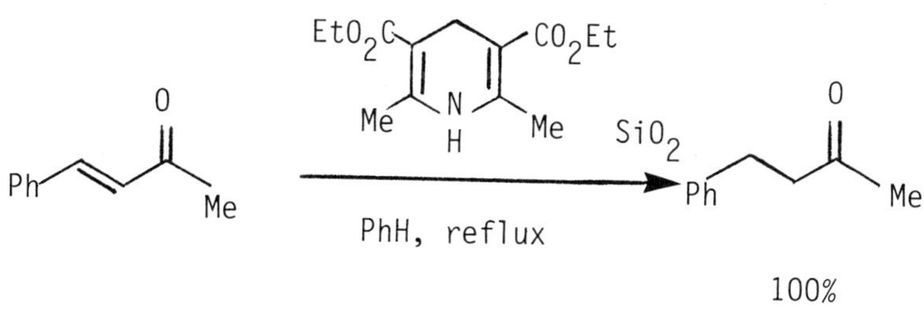

III.E.1-10 C. Petrier and J. Luche, Tetrahedron Lett., 28, 2347 (1987).

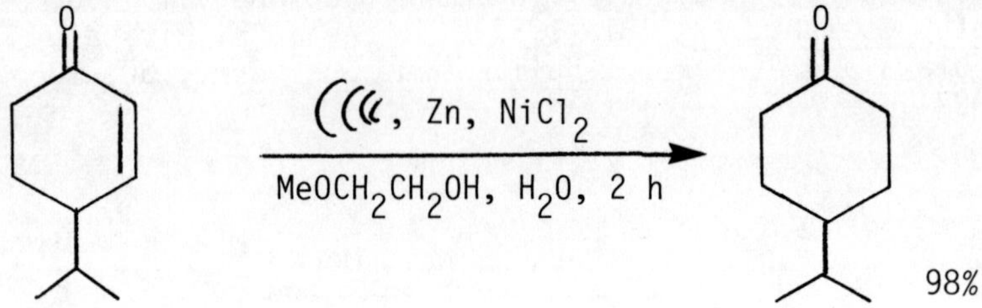

III.E.1-11 E. Keinan and D. Perez, J. Org. Chem., 52, 2576 (1987).

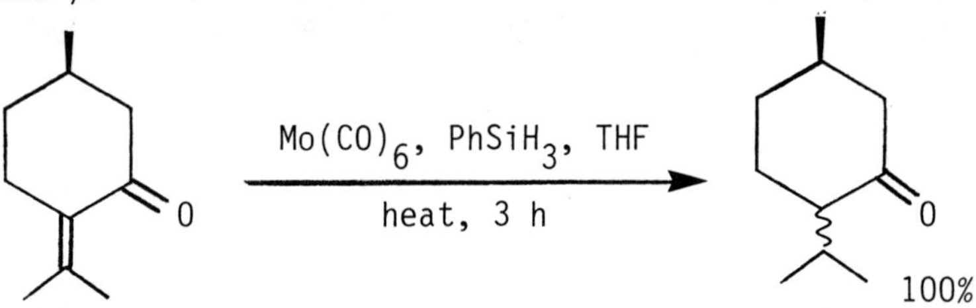

III.E.1-12 T. Hayashi, N. Kawamura and Y. Ito, J. Am. Chem. Soc., 109, 7876 (1987); H. Takahashi and K. Achiwa, Chem. Lett., 1921 (1987).

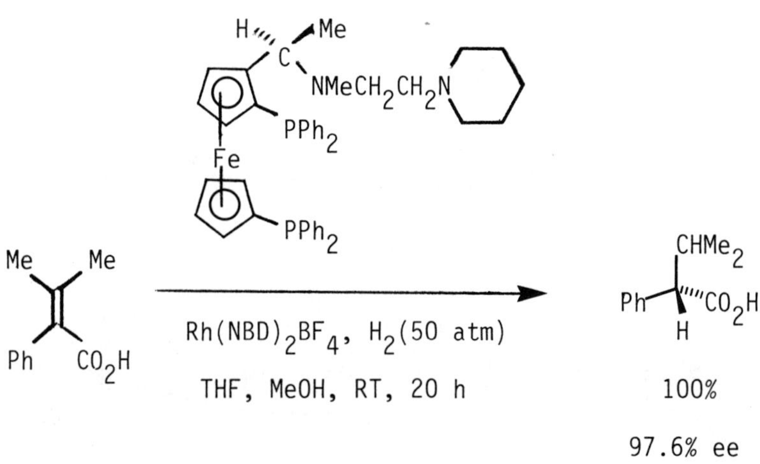

III.E.1-13 H. Kawano, et al., Tetrahedron Lett., 28, 1905 (1987); T. Ohta, H. Takaya, M. Kitamura, K. Nagai and R. Noyori, J. Org. Chem., 52, 3174 (1987).

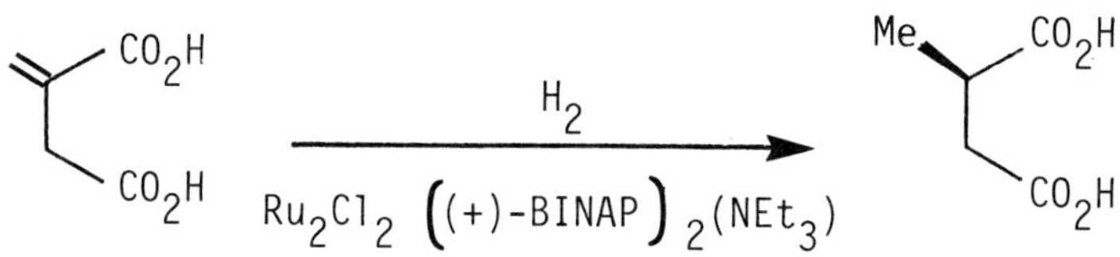

(S)- 100% yield

88% o.p.

III.E.1-14 J.M. Brown and A.P. Jones, J. Chem. Soc., Chem. Commun., 181 (1987); C.W. Hung and H.N.C. Wong, Tetrahedron Lett., 28, 2393 (1987); R.M. Moriarty, R.K. Vaid and M.P. Duncan, Synth. Commun., 17, 703 (1987); T. Hudlicky, G. Sinai-Zingde and M.G. Natchus, Tetrahedron Lett., 28, 5287 (1987).

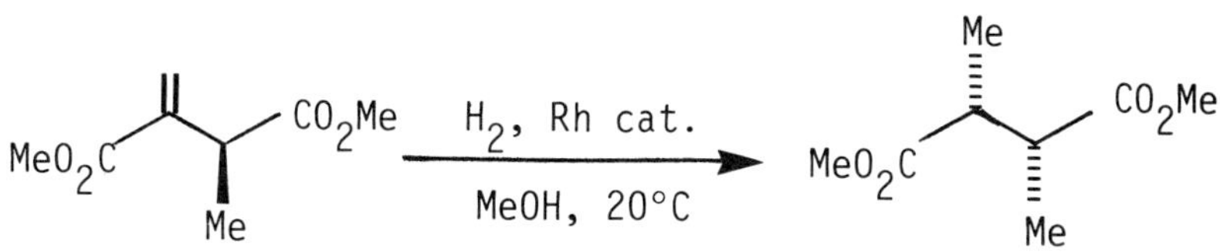

65%

100% ee

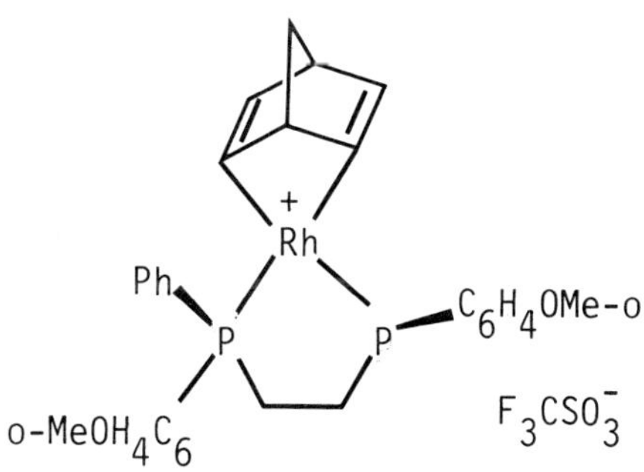

III.E.1-15 B.D. Zwick, A.M. Arif, A.T. Patton and J.A. Gladysz, Angew. Chem. Int. Ed. Engl., 26, 910 (1987); D.C. Snyder, J. Organomet. Chem., 320, 163 (1987).

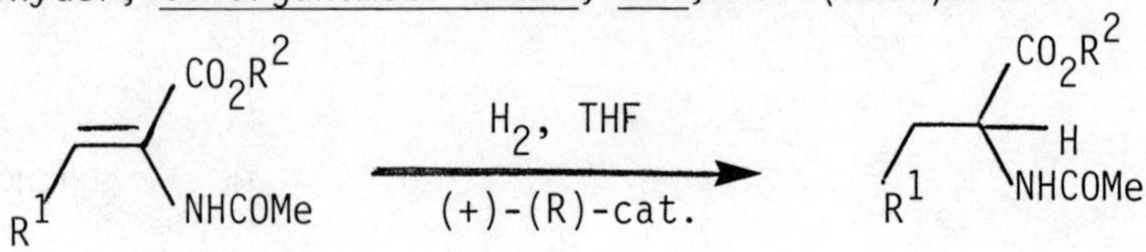

(R)-

70-94%

82-98% ee

cat.=

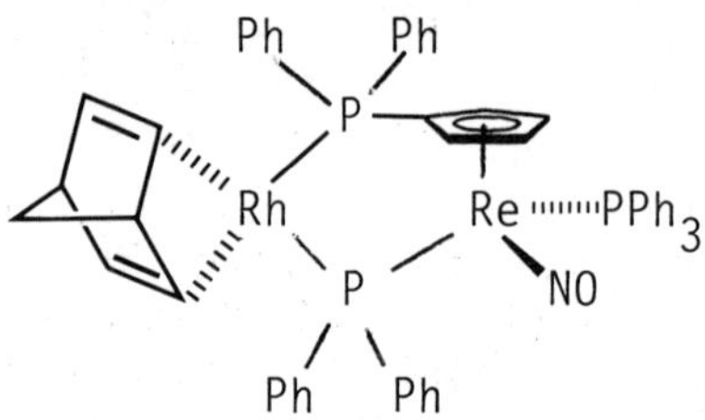

III.E.1-16 J. Aizpurua, M. Oiarbide and C. Palomo, Tetrahedron Lett., 28, 5365 (1987); H. Ohta, K. Ozaki and G. Tsuchihashi, Chem. Lett., 191 (1987).

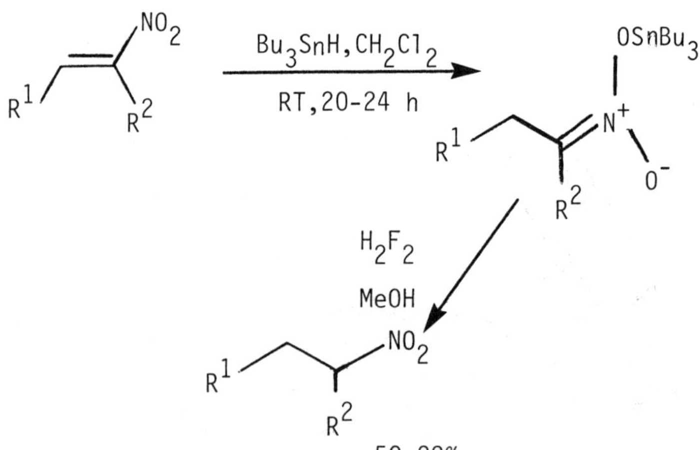

50-99%

III.E.1-17 A. Sera, H. Tani, I. Nishiguchi and T. Hirashima, <u>Synthseis</u>, 631 (1987).

$$\mathrm{RCH{=}CHNO_2} \xrightarrow[\mathrm{e}]{\mathrm{TiCl_4,\ DMF,\ N^{+}Et_4OTs^{-}}} \mathrm{RCH_2CN}$$

---

2. C ≡ C Reductions

III.E.2-1 T. Tsuda, T. Yoshida, T. Kawamoto and T. Saegusa, <u>J. Org. Chem.</u>, <u>52</u>, 1624 (1987).

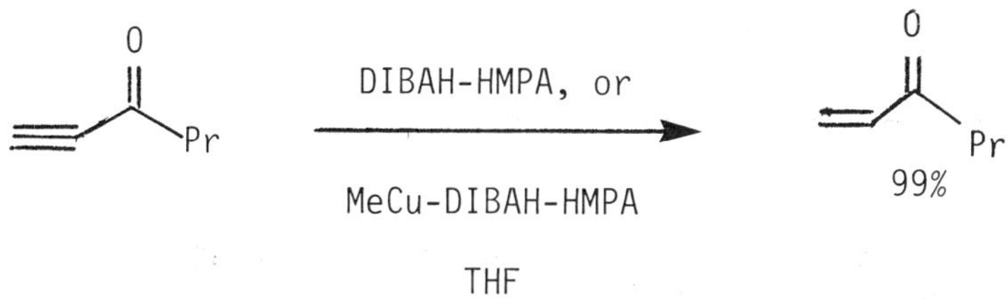

III.E.2-2 T-L. Ho and S. Liu, <u>Synth. Commun.</u>, <u>17</u>, 969 (1987).

$$\text{1-Heptyne} \xrightarrow[\text{Lindlar cat., 1-dodecene}]{\mathrm{H_2}\text{ (low pres.)}} \text{1-heptene} \quad 98.3\%$$

III. F Hetero-Bond Reductions

1. C — O ⟶ C — H

III.F.1-1 Nazar-Ul-Islam, D.W. Sopher and J.H.P. Utley, Tetrahedron, 43, 2741 (1987); A. Ono, N. Suzuki and J. Kamimura, Synthesis, 736 (1987); A. Ono, T. Maruyama and N. Suzuki, Synth. Commun., 17, 1001 (1987).

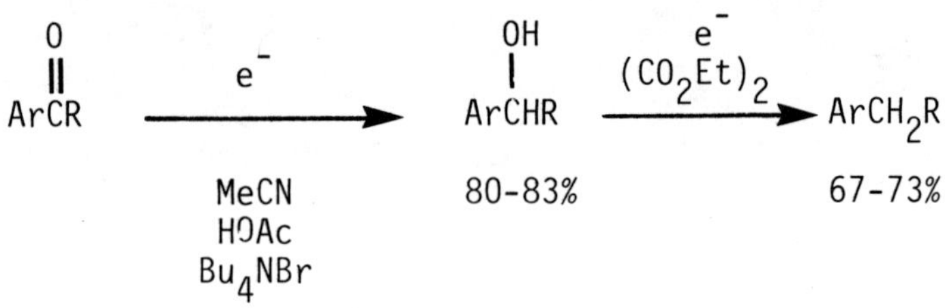

III.F.1-2 C. Soucy, D. Faureau and M.M. Kayser, J. Org. Chem., 52, 129 (1987).

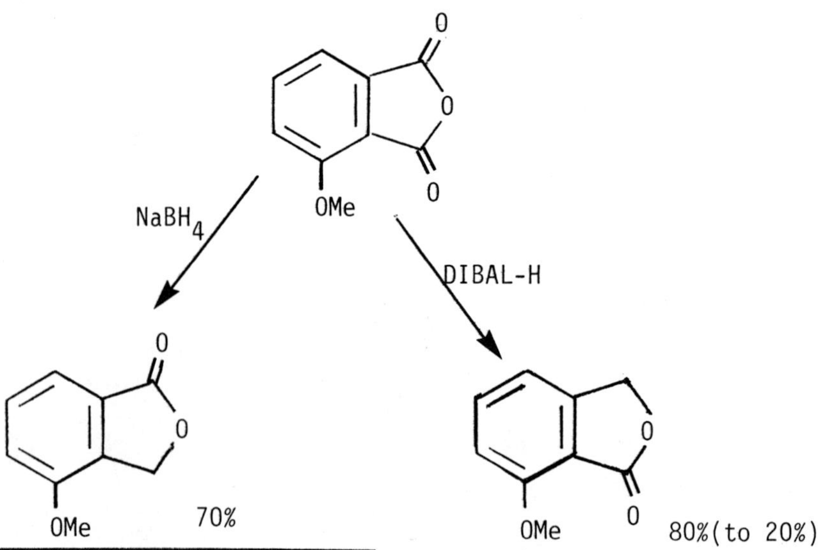

III.F.1-3 G.A. Peterson, F. Kunng, J.S. McCallum and W.D. Wulff, Tetrahedron Lett., 28, 1381 (1987).

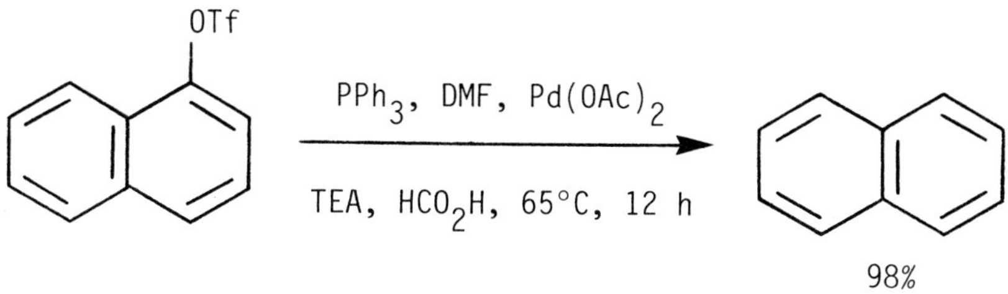

III.F.1-4 T. Sakai, K. Miyata, M. Utaka and A. Takeda, Tetrahedron Lett., 28, 3817 (1987).

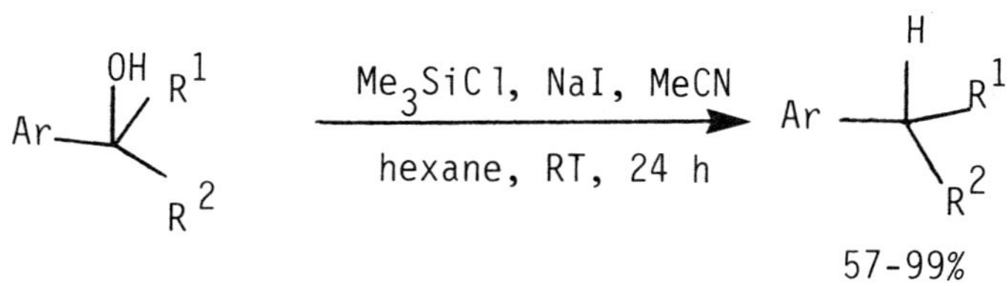

III.F.1-5 T. Ishihara, T. Maekawa, Y. Yamasaki and T. Ando, J. Org. Chem., 52, 300 (1987).

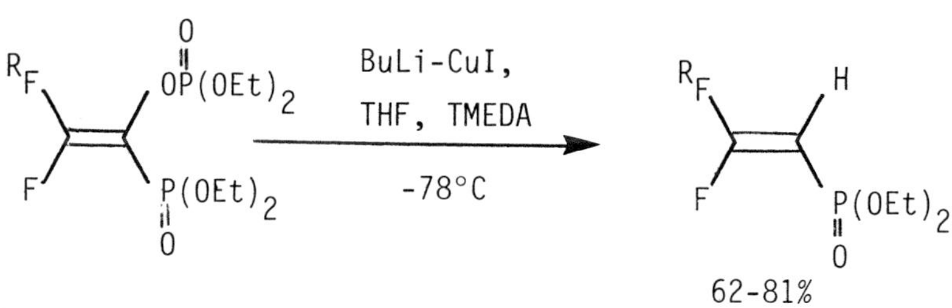

III.F.1-6 T.J.P. Corriu, G.F. Lanneau and M. Perrot, Tetrahedron Lett., 28, 3941 (1987).

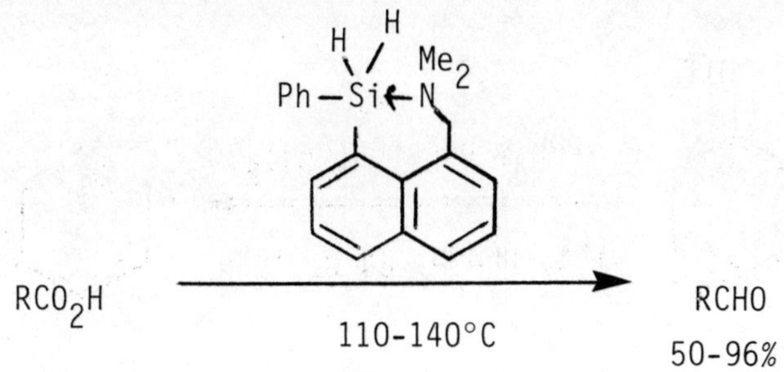

III.F.1-7 J.S. Cha, J.E. Kim and K. W. Lee, J. Org. Chem., 52, 5030 (1987); J.S. Cha, J.E. Kim, S.Y. Oh, J.C. Lee and K.W. Lee, Tetrahedron Lett., 28, 2389 (1987); J.S. Cha, J.E. Kim, S.E. Oh and J.D. Kim, Tetrahedron Lett., 28, 4575 (1987).

$RCO_2M \xrightarrow{1)} RCHO$ 85-99%

1) Me_2CHCMe_2BBrH, Me_2S, CS_2, CH_2Cl_2, RT, 1 to 9 h

III.F.1-8 J.S. Cha and S.S. Kwon, J. Org. Chem., 52, 5486 (1987).

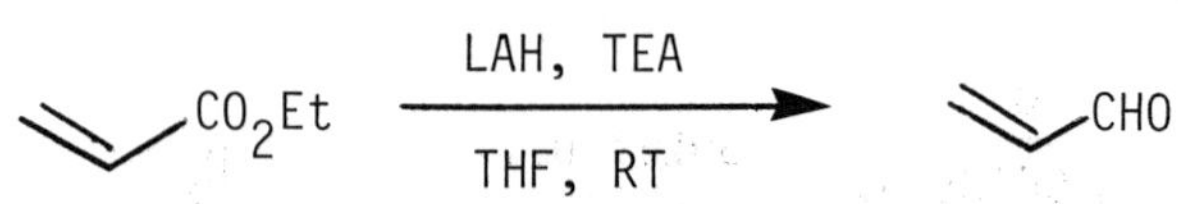

94%

2. C—Hal → C—H

III.F.2-1 J. Inanaga, M. Ishikawa and M. Yamaguchi, Chem. Lett., 1485 (1987); A.R. Harris, Synth. Commun., 17, 1587 (1987).

$$\mathrm{R{-}X} \xrightarrow[\text{HMPA, i-PrOH}]{\mathrm{SmI_2},\ \text{THF}} \mathrm{R{-}H}$$

N_2, 5 min to 8 h   82-99%

III.F.2-2 D.E. Bergbreiter and J.R. Blanton, J. Org. Chem., 52, 472 (1987).

$$\mathrm{RX} \xrightarrow{1)} \mathrm{RH}$$

1) n-Bu_3SnCl, excess $NaBH_4$, PhMe, 110°C, with benzo-15-crown-5, polyethylene-bound crown ether or poly(ethylene glycol) dimethyl ether

---

III.F.2-3 R.L. Danheiser and S. Savariar, Tetrahedron Lett., 28, 3299 (1987).

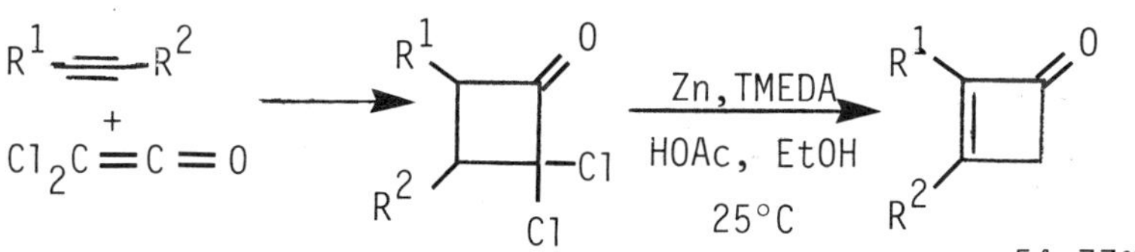

54-77%

III.F.2-4 D. Perez, N. Greenspoon and E. Keinan, J. Org. Chem., 52, 5570 (1987); A. Ono, T. Maruyama and S. Kamimura, Synthesis, 1093 (1987); A. Ono, J. Kamimura and N. Suzuki, Synthesis, 406 (1987).

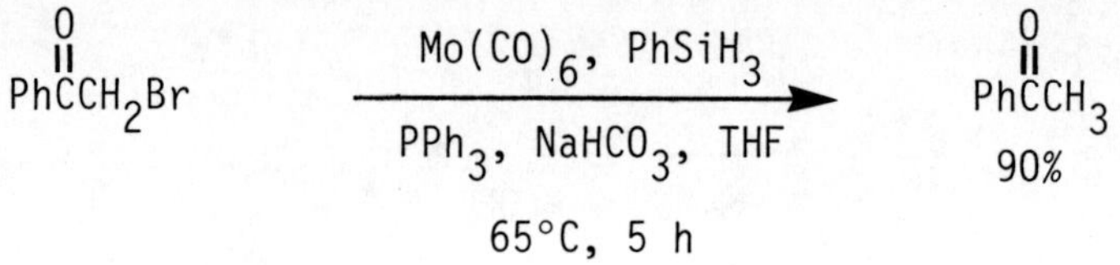

III.F.2-5 G.F. Meijs, J. Org. Chem., 52, 3923 (1987).

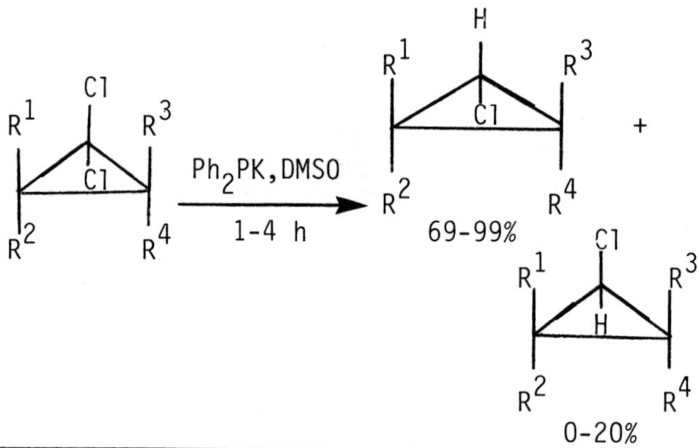

---

3. C—S ⟶ C—H

III.F.3-1 P.G. Theobald and W.H. Okamura, Tetrahedron Lett., 28, 6565 (1987).

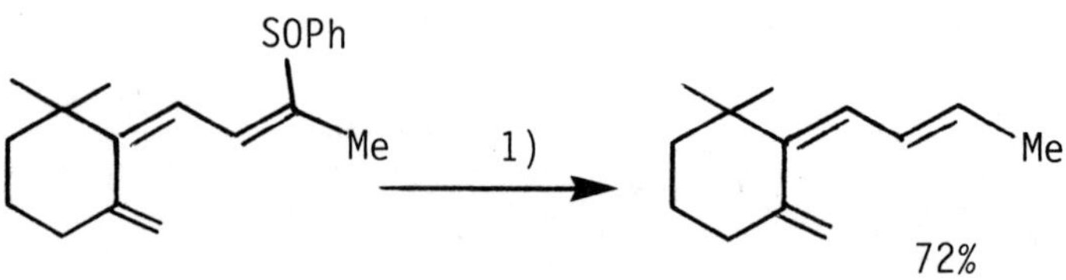

1) t-BuLi, MeOH, Et_2O, -78°C

III.F.3-2 R.A. Holton, D.J. Crouse, A.D. Williams and R.M. Kennedy, J. Org. Chem., 52, 2317 (1987).

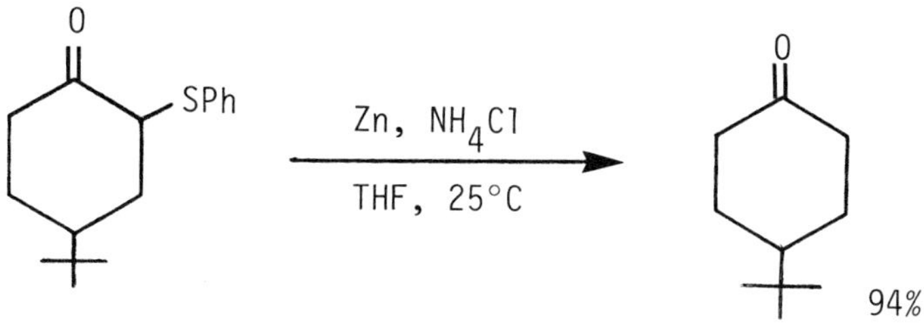

---

4. C — N ⟶ C — H

III.F.4-1 N. Ono, T. Hashimoto, T. Jun and A. Kajii, Tetrahedron Lett., 28, 2277 (1986); N. Ono, M. Jujii and A. Kaji, Synthesis, 532 (1987).

$$RNO_2 + Et_3SiH \xrightarrow[CH_2Cl_2,\ 0\text{-}20°C]{SnCl_4} R{-}H + Et_3SiONO$$

50-94%

III.F.4-2 N. Suzuki et al., J. Chem. Soc., Perkin I, 1951 (1987).

$$ArNH_2 \xrightarrow{1)} ArN_2^{+}Cl^{-} \xrightarrow{2)} ArH$$

51-100%

1) $NaNO_2$, HCl, PEG, CH_2Cl_2, 0°C;

2) Sn powder, PEG, CH_2Cl_2, 0°C, RT

III.G. Reductive Cleavages

1. Oxiranes

III.G.1-1 K. Otsubo, J. Inanaga and M. Yamaguchi, Tetrahedron Lett., 28, 4437 (1987).

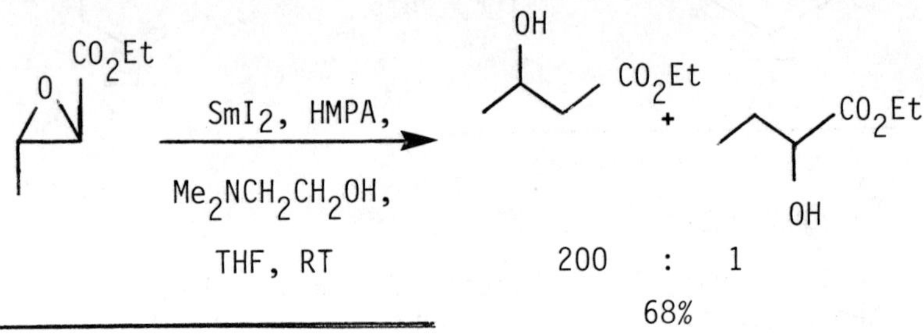

III.G.1-2 M. Miyashita, T. Suzuki and A. Yoshikoshi, Tetrahedron Lett., 28, 4293 (1987).

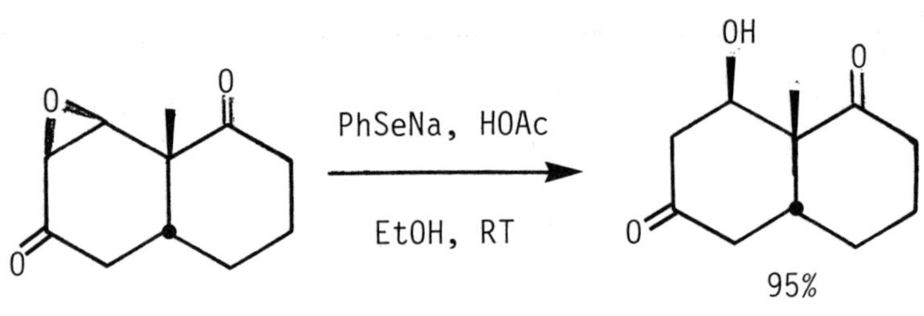

III.G.1-3 K. Otsubo, J. Inanaga and M. Yamaguchi, Tetrahedron Lett., 28, 4435 (1987); L.D. Corey, S.M. Singh and A.C. Oehlschlager, Can. J. Chem., 65, 1821 (1987).

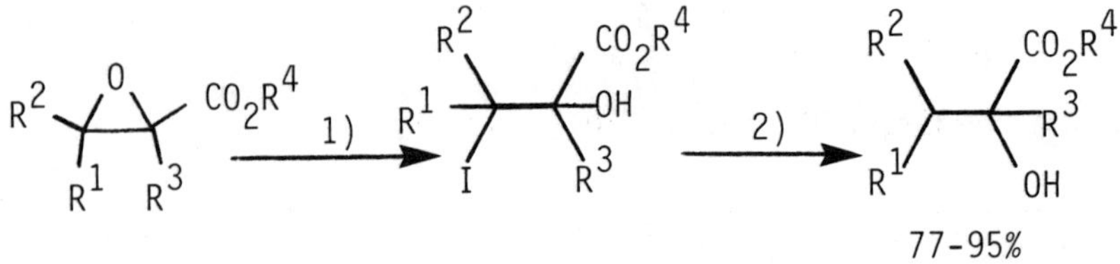

1)MgI_2Et_2O,RT; 2)Bu_3SnH, hexane.

III.G.1-4 E. Santaniello, *et al.*, *J. Org. Chem.*, *52*, 671 (1987).

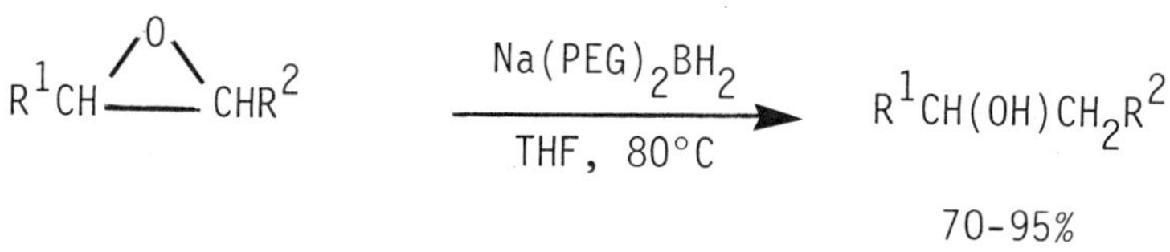

PEG = polyethyleneglycol

---

2. N—O Cleavage

III.F.2-1 Y. Zhang and R. Lin, *Synth. Commun.*, *17*, 329 (1987).

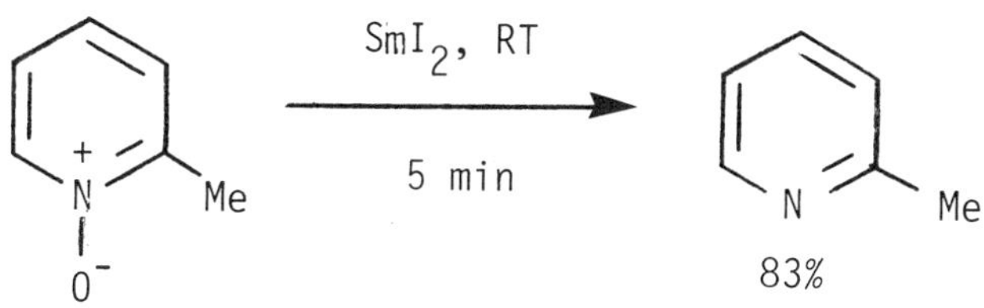

III.G.2-2 M. Malinowski and L. Kaczmarek, *Synthesis*, 1013 (1987); M. Malinowski, *Synthesis*, 732 (1987).

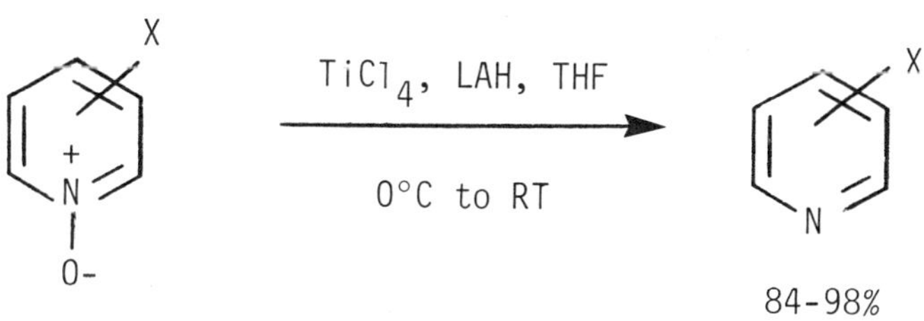

III.G.2-3 B. Jousseaume and E. Chanson, Synthesis, 55 (1987).

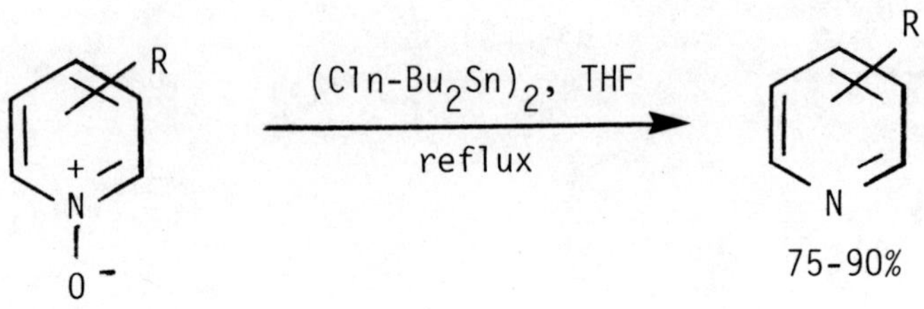

3. Others

III.G.3-1 M. Hasebe and T. Tsuchiya, Tetrahedron Lett., 28, 6207 (1987).

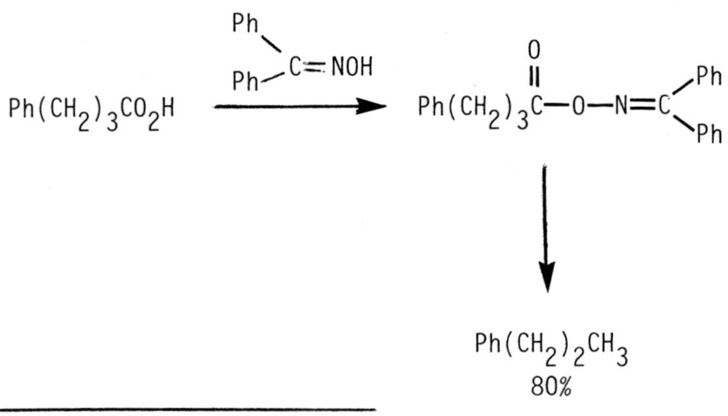

III.G.3-2 B.M. Adger, C.O'Farrell, N.J. Lewis and M.B. Mitchell, Synthesis, 53 (1987).

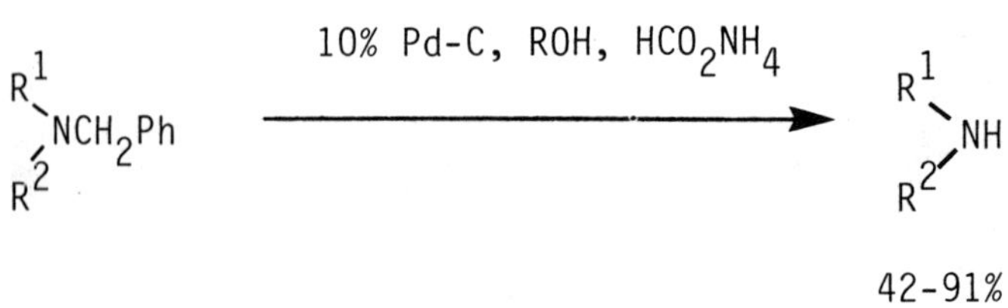

III.G.3-3 J.W. Misner, J. Org. Chem., 52, 3166 (1987).

$$Ph_2NCN \xrightarrow[\text{glycol, 130°C, 30 min}]{\text{NaOH, ethylene}} Ph_2NH$$

80%

III.G.3-4 K. Soai, S. Yokoyama and K. Mochida, Synthesis, 647 (1987).

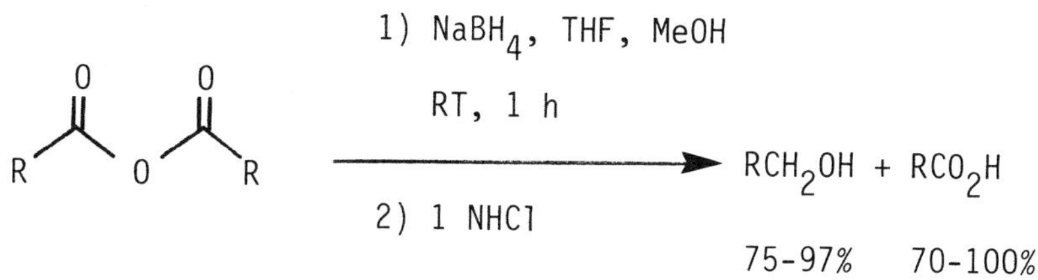

III.G.3-4 H. Kotsuki, et al., J. Org. Chem., 52, 2594 (1987).

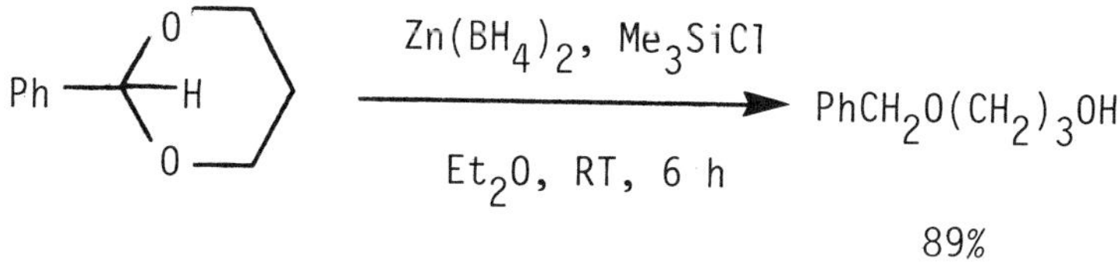

III.H. Reduction of Azides

III.H.1-1 M. Bartra, F. Urpí and J. Vilarrasa, Tetrahedron Lett., 28, 5941 (1987).

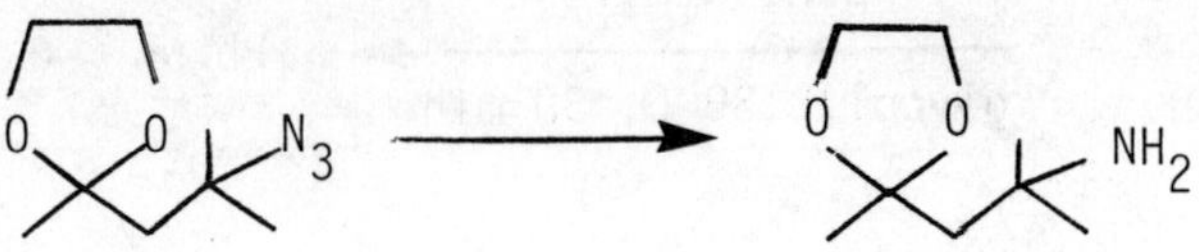

$Sn(SPh)_3^-$, PhH, 40°C, 40 min 100%

or

Bu_2SnH_2, PhH, 15°C, 7 h 92%

III.H.1-2 K. Soai, S. Yokoyama and A. Ookawa, Synthesis, 48 (1987).

$$R\text{—}N_3 \xrightarrow[\text{MeOH added dropwise}]{NaBH_4,\ \text{THF, heat}} R\text{—}NH_2$$

31-98%

III.H.1-3 J. Sarma and R.P. Sharma, Chem. Ind. (London), 764 (1987).

$$Me(CH_2)_{14}CH_2N_3 \xrightarrow[\text{MeOH, 0°C}]{\text{1) NiCl.6 } H_2O} Me(CH_2)_{14}CH_2NH_2$$

2) $NaBH_4$

83%

III.I. Reviews

III.I-1 J.M. Brown, Angew. Chem. Int. Ed. Engl., 26, 190 (1987).

Review: "Directed Homogenous Hydrogenation"

III.I-2 W.P. Neumann, Synthesis, 665 (1987).

Review: "Tri-n-butyltin Hydride as a Reagent in Organic Synthesis"

III.I-3 K. Müllen, Angew. Chem. Int. Ed. Engl., 26, 204 (1987).

Review: "New Reductive Transformations of Unsaturated Cyclic Hydrocarbons"

III.I-4 H.C. Brown and J.V.N. Vara Prasad, Heterocycles, 25, 641 (1987).

Review: "Hydroboration of Heterocyclic Olefins"

III.I-5 K. Smith, Chem. Ind.(London), 603 (1987).

Review: "Advances in Organoboron Chemistry -- Prospects for Industry"

III.I-6 D.S. Matteson, Synthesis, 973 (1986).

Review: "The Use of Chiral Organoboranes in Organic Synthesis"

III.I-7 G.W. Kabalka, J. Organomet. Chem., 337, 169 (1987).

Review: "Boron: Boranes in Organic Synthesis. Annual Survey Covering the Year 1985"

III.I-8 G.W. Kabalka, J. Organomet. Chem., 318, 1 (1987).

Review: "Boron: Boranes in Organic Synthesis. Annual Survey Covering the Year 1984"

III.I-9 T. Hirao and Y. Ohshiro, J. Synth. Org. Chem. Jpn., 45, 784 (1987).

Review: "Selective Transformations of Functional Groups by Use of Dialkyl Phosphonates"

IV.A. Oxiranes and Aziridines

IV.A-1 I. Saito, T. Mano, R. Nagata and T. Matsuura, Tetrahedron Lett., 28, 1909 (1987); P. Lescure and F. Huet, Synthesis, 404 (1987).

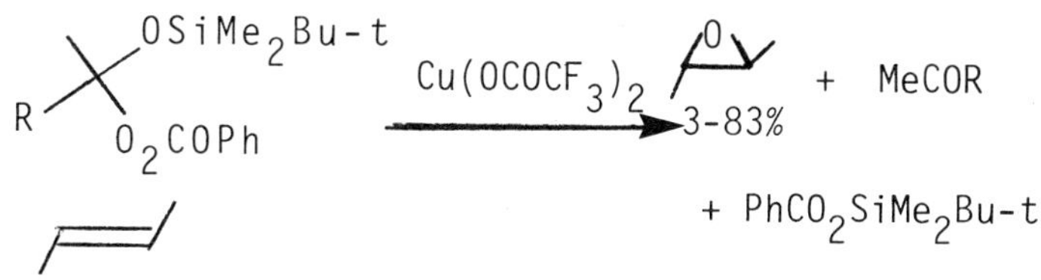

IV.A-2 Y. Honda, Y. Kataoka, M. Unno and G. Tsuchihashi, Chem. Lett., 2133 (1987); A. Hosomi, S. Kohra, Y. Tominaga, M. Ando and H. Sakura, Chem. Pharm. Bull., 35, 3058 (1987); S. Tsuboi, H. Furutani, M. Utaka and A. Takeda, Tetrahedron Lett., 28, 2709 (1987).

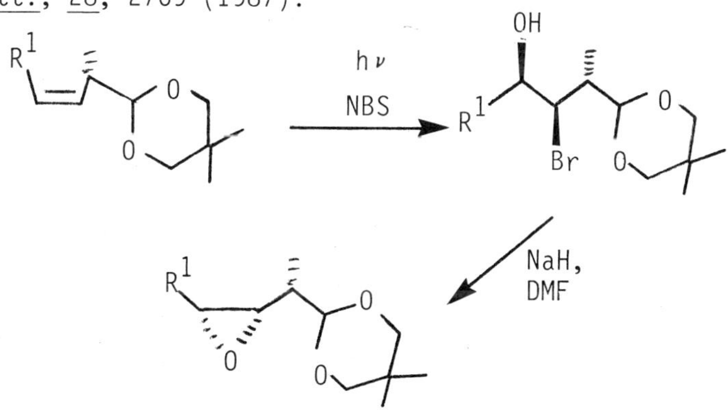

IV.A-3 H. Bouda, M.E. Borredon, M. Delmas and A. Gaset, Synth. Commun., 17, 503 (1987); K.O. Koua, M.E. Borredon, M. Delmas and A. Gaset, Synth. Commun., 17, 1593 (1987).

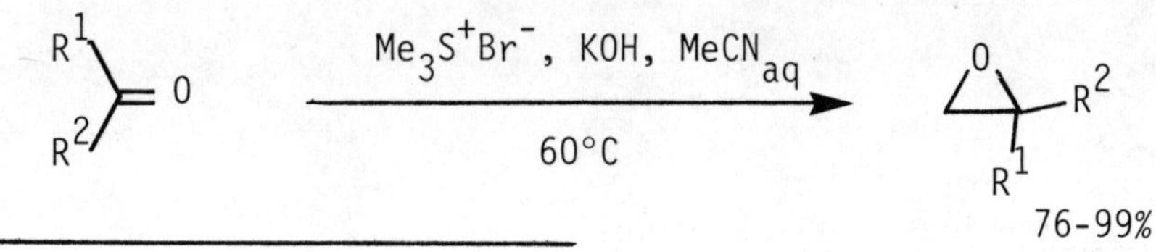

IV.A-4 R.S. Atkinson and G. Tughan, J. Chem. Soc., Perkin I, 2803 (1987).

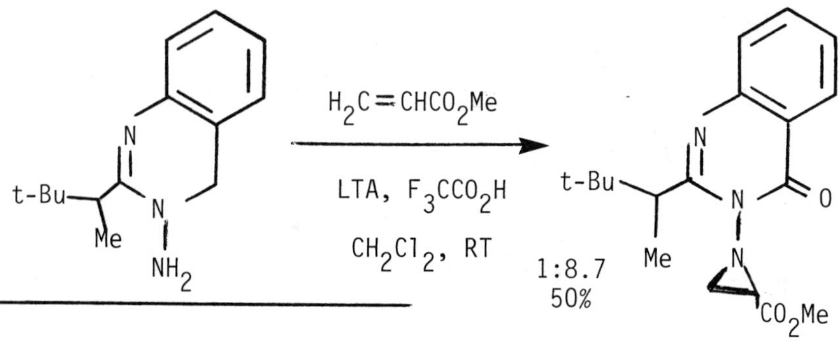

IV.B. Oxetanes

IV.B-1 L.M. Harwood and J. Robertson, Tetrahedron Lett., 28, 5175 (1987).

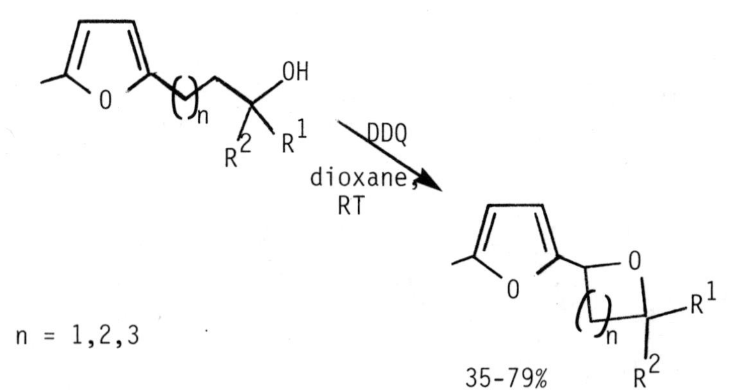

IV.B-2 T.H. Morris, E.H. Smith and R. Walsh, J. Chem. Soc., Chem. Commun., 964 (1987).

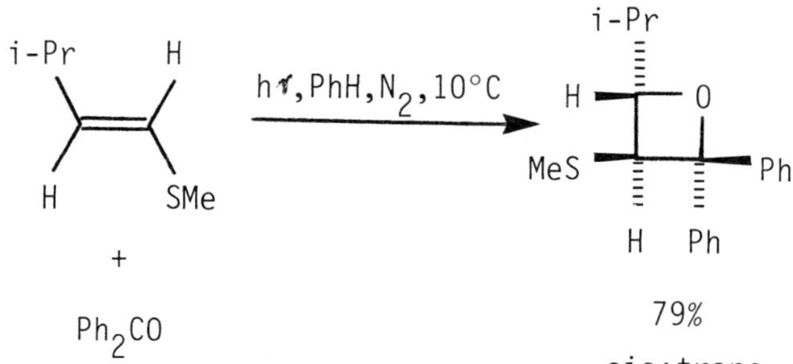

79%

cis:trans, 1:9

IV.C. Lactams

IV.C-1 F.P. Cossio et al., Tetrahedron Lett., 28, 1945 (1987); J.T.B.H. Jastrzebski et al., Rec. Trav. Chim., 106, 516 (1987); M. Wada et al., Tetrahedron Lett., 28, 3377 (1987); G. Cainelli et al., Tetrahedron Lett., 28, 5369 (1987); S.D. Sharma et al., Synthesis, 990 (1987); T. Yamada et al., Chem. Lett., 293(1987).

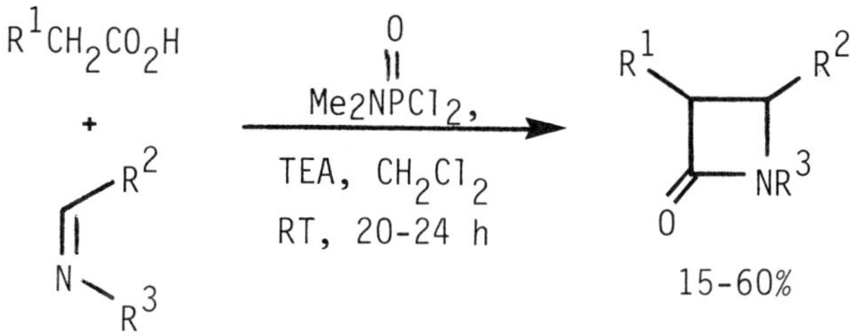

15-60%

IV.C-2 G. Rajendra and M.J. Miller, J. Org. Chem., 52, 4471 (1987); S. Kim, S.B. Chang and P.H. Lee, Tetrahedron Lett., 28, 2735 (1987); G. Rajendra and M.J. Miller, Tetrahedron Lett., 28, 6257 (1987);

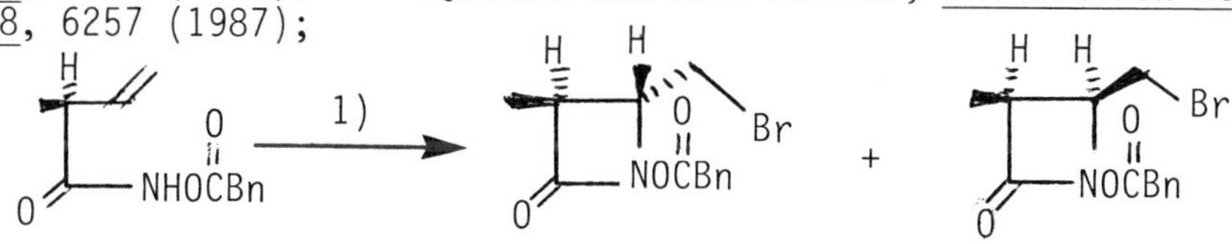

77:23 trans:cis
80.2%

1) Br_2, K_2CO_3, MeCN aq

IV.C-3 M.R.P. van Vliet, J.T.B.H. Jastrzebski, W.J. Klaver, K. Goubitz and G. van Koten, Rec. Trav. Chim., 106, 132 (1987).

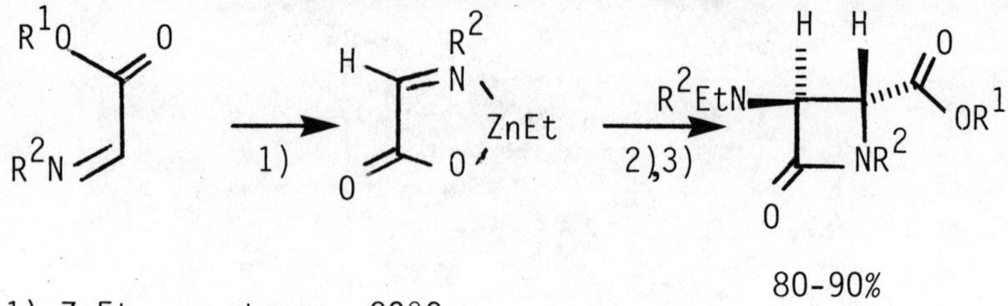

1) $ZnEt_2$, pentane, -80°C;
2) -80°C to RT; 3) H_2O.

IV.C-4 H. Alper and N. Hamel, Tetrahedron Lett., 28, 3237 (1987).

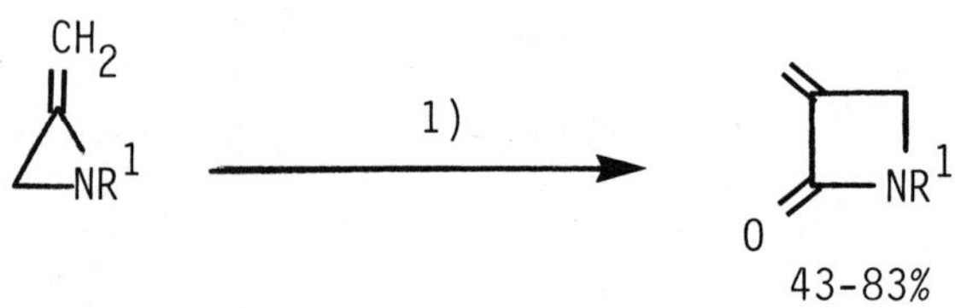

1) CO, $Pd(PPh_3)_4$, CH_2Cl_2, RT, 1 atmos.

IV.C-5 A. Battaglia et al., Tetrahedron Lett., 28, 4347 (1987).

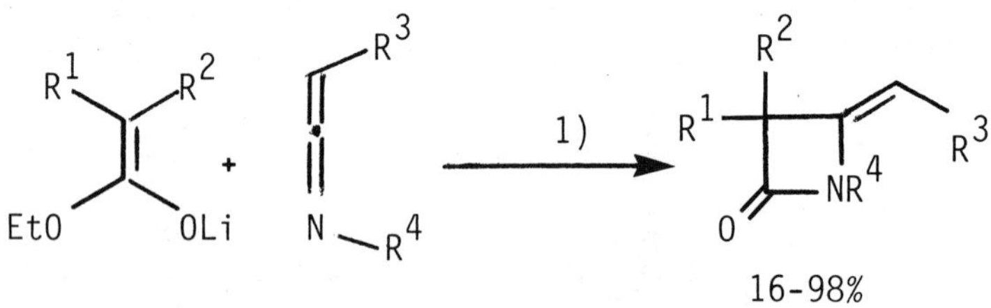

1) THF, -78°C to RT.

IV.C-6 G. Cainelli _et al._, _Tetrahedron Lett._, _28_, 3593 (1987).

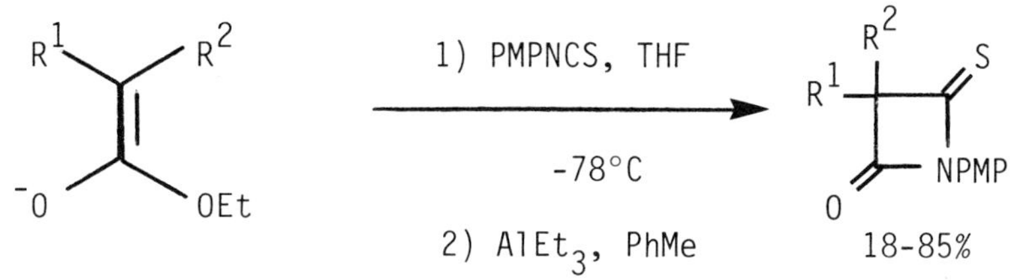

IV.C-7 S.W. Baldwin and J. Aubé, _Tetrahedron Lett._, _28_, 179 (1987).

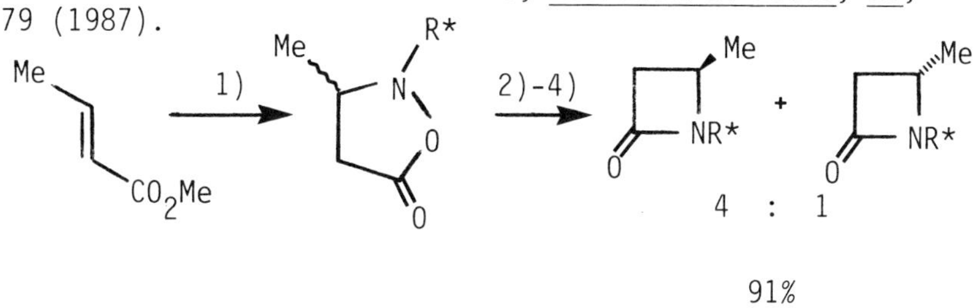

1) *RNHOH, PhH, K_2CO_3, 80°C; 2) H_2, Pd/C, EtOH;

3) n-Bu_4NHSO_4, $KHCO_3$aq, $CHCl_3$, MsCl; 4) Na, NH_3

IV.C-8 F. Toda, M. Yagi and S. Soda, _J. Chem. Soc., Chem. Commun._, 1413 (1987).

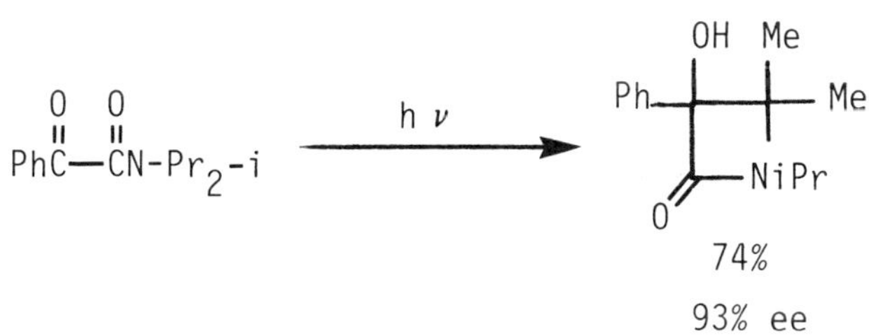

IV.C-9 H. Ishibashi, T. Sato, M. Irie, S. Harada and M. Ikeda, Chem. Lett., 795 (1987).

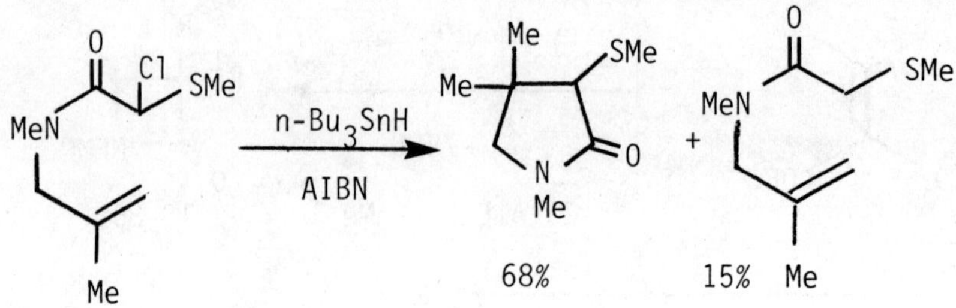

IV.C-10 J. Plumet et al., Heterocycles, 26, 1461 (1987); P.R. Babu and T.R. Balasubramanian, Ind. J. Chem.(B), 26, 63 (1987).

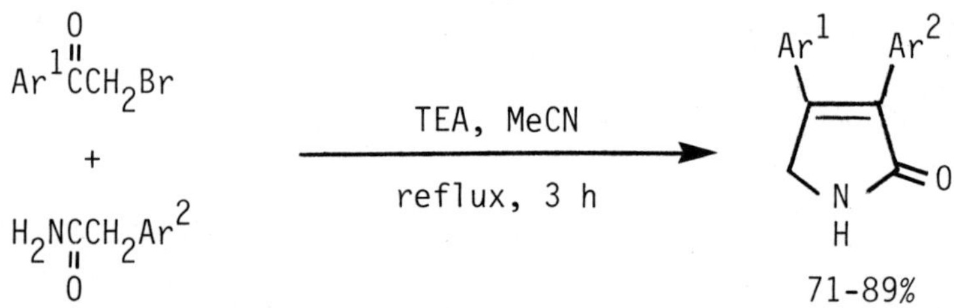

IV.C-11 F. Closs and R. Gompper, Angew. Chem. Int. Ed. Engl., 26, 552 (1987).

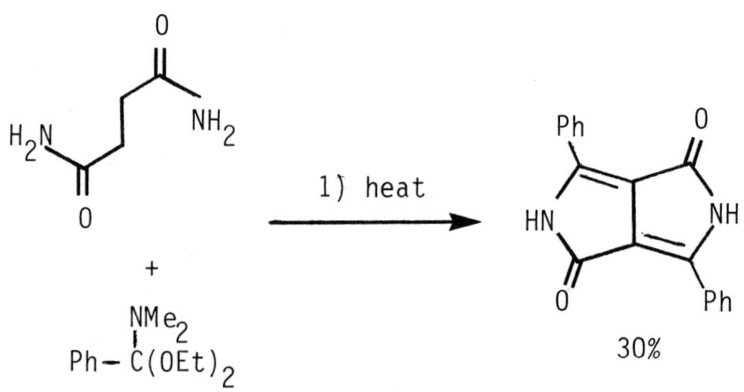

IV.C-12 Y. Kikugawa and M. Shimada, Chem. Lett., 1771 (1987); A. Marfat and M.P. Carta, Tetrahedron Lett., 28, 4027 (1987).

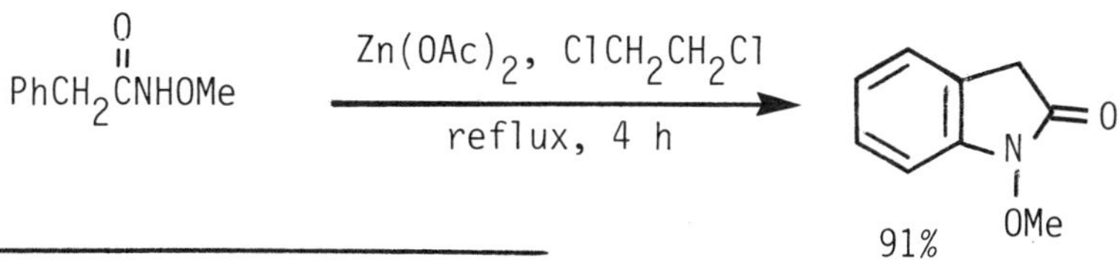

IV.C-13 D. Enders and A.S. Demir, Tetrahedron Lett., 28, 3795 (1987).

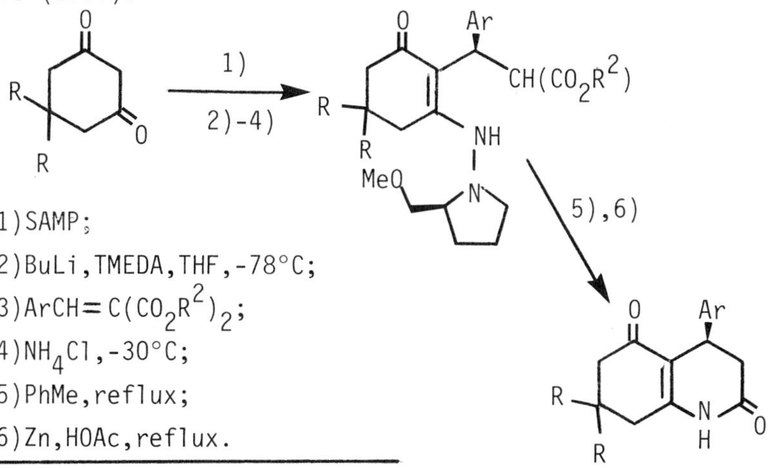

1) SAMP;
2) BuLi, TMEDA, THF, -78°C;
3) $ArCH{=}C(CO_2R^2)_2$;
4) NH_4Cl, -30°C;
5) PhMe, reflux;
6) Zn, HOAc, reflux.

IV.C-14 S. Murahashi, T. Oda, T. Sugahara and Y. Masui, J. Chem. Soc., Chem. Commun., 1471 (1987).

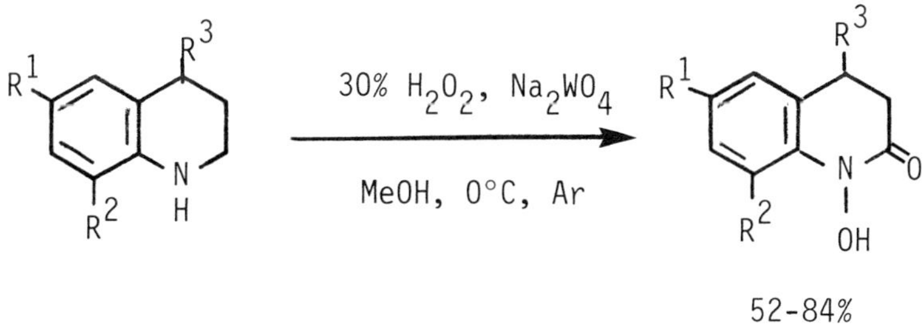

IV.C-15 R.D. Clark and Jahangir, J. Org. Chem., 52, 5378 (1987).

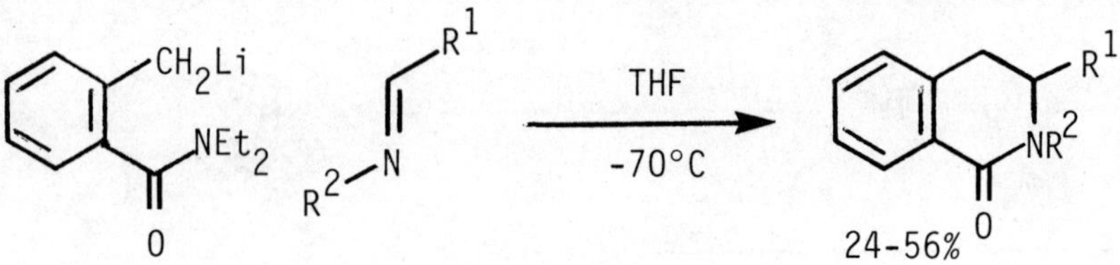

IV.C-16 W. Stadlbauer, Monatsh. Chem., 118, 1297 (1987).

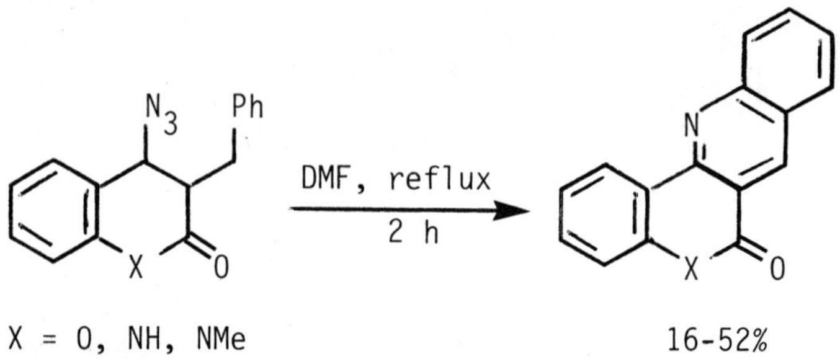

IV.C-17 A. Alberola, C. Andrés, A.G. Ortega, R. Pedrosa and M. Vicente, J. Heterocycl. Chem., 24, 709 (1987); M. Mittelbach, Monatsh. Chem., 118, 617 (1987).

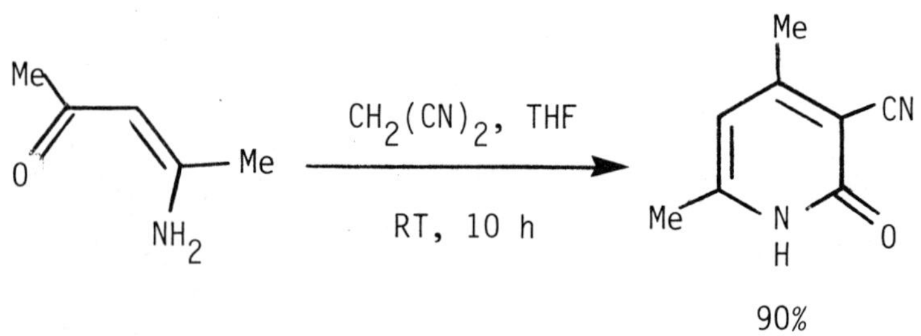

IV.C-18 M. Takahashi, C. Nozaki and Y. Shibazaki, Chem. Lett., 1229 (1987).

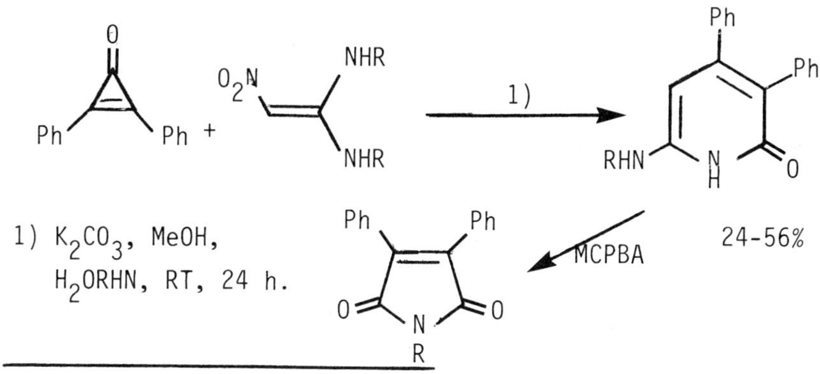

IV.C-19 B. Haag-Zeino, M.E. Maier and R.R. Schmidt, Chem. Ber., 120, 1505 (1987).

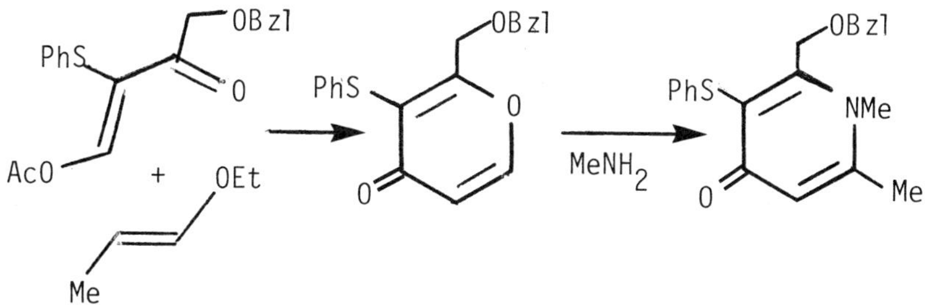

IV.C-20 B. Chen, X. Huang and J. Wang, Synthesis, 482 (1987).

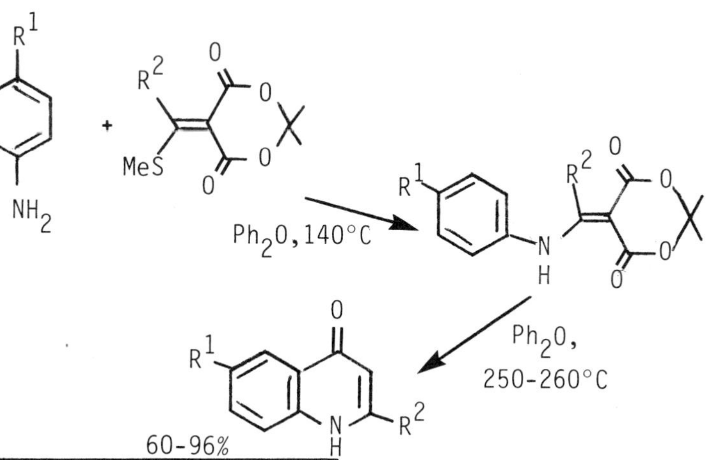

IV.C-21 A. McKillop and S.P. Brown, Synth. Commun., 17, 657 (1987).

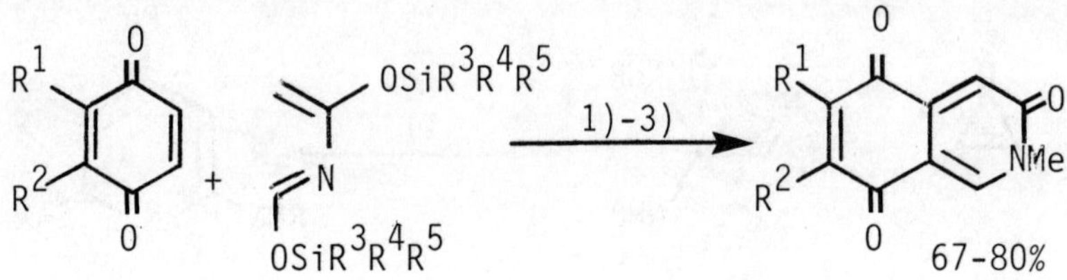

1) PhH, reflux; 2) HCl; 3) MeI, Na_2CO_3, DMF, TDA-1

IV.D. Lactones

IV.D-1 P.G. Gassman and K.J. Bottorff, J. Am. Chem. Soc., 109, 7547 (1987); K.F. Podraza, J. Heterocycl. Chem., 24, 293 (1987); T. Morikawa, T. Nishiwaki, Y. Iitaka and Y. Kobayashi, Tetrahedron Lett., 28, 671 (1987).

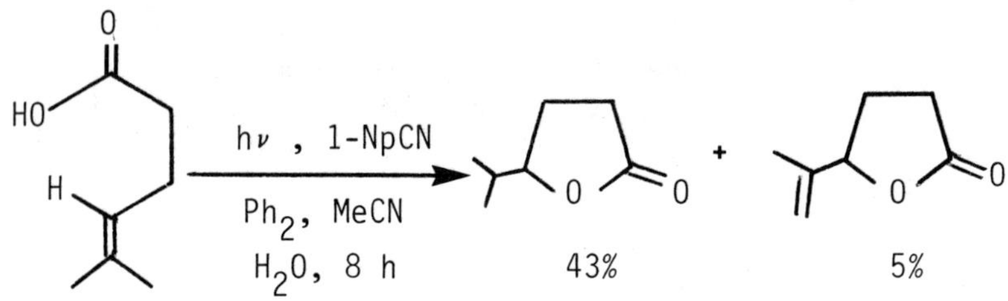

IV.D-2 J. Barluenga, J.R. Fernandez and M. Yus, J. Chem. Soc.,Chem. Commun., 1534 (1987); T. Nakano, M. Kayama and Y. Nagai, Bull. Chem. Soc. Jpn., 60, 1049 (1987); T.H. Black and W.J. Dubay III, Tetrahedron Lett., 28, 4787 (1987); T. Sakai, A. Yamawaki, T. Katayama, H. Okada, M. Utaka and A. Takeda, Bull. Chem. Soc. Jpn., 60, 1067 (1987); K. Otsubo, K. Kawamura, J. Inanaga and M. Yamaguchi, Chem. Lett., 1487 (1987).

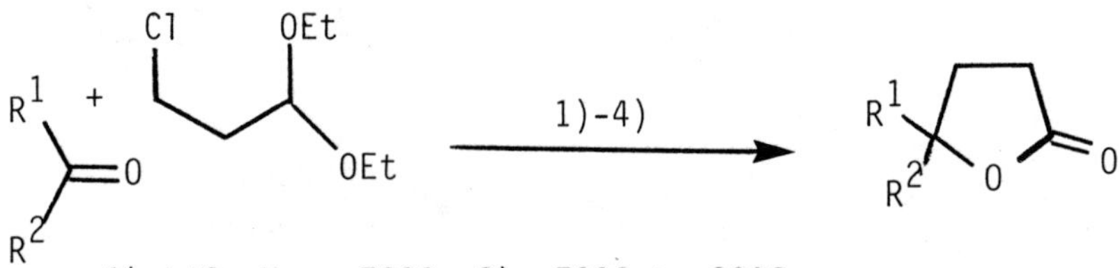

1) $LiC_{10}H_8$, -78°C; 2) -78°C to 20°C;
3) NH_4Cl; 4) MCPBA, $Et_2O.BF_3$.

IV.D-3 A.L. Gutman, K. Zuobi and A. Boltansky, Tetrahedron Lett., 28, 3861 (1987).

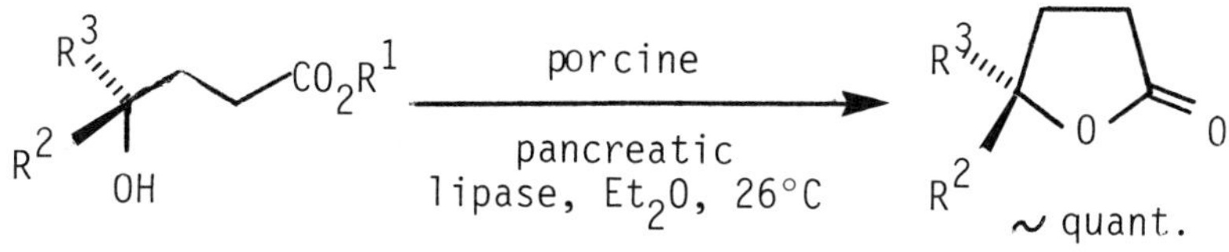

IV.D-4 G. Sabbioni and J.B. Jones, J. Org. Chem., 52, 4565 (1987).

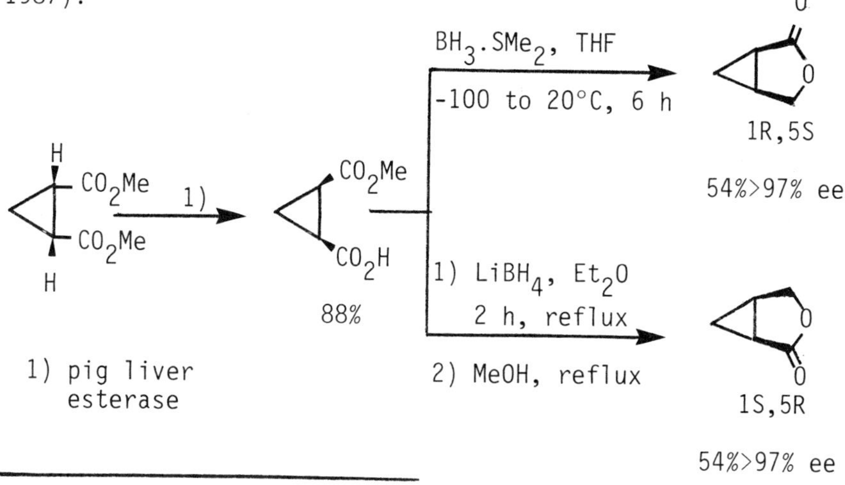

1) pig liver esterase

IV.D-5 M. Utaka, H. Watabu and A. Takeda, J. Org. Chem., 52, 4363 (1987).

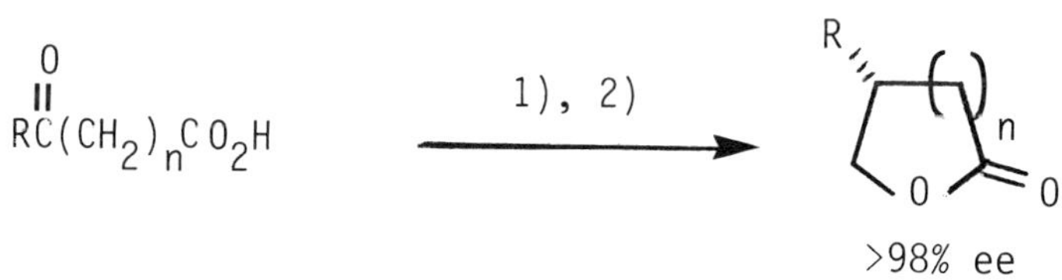

1) bakers' yeast, glucose; 2) TsOH, PhH, reflux

IV.D-6 J.P. Marino and M.W. Kim, Tetrahedron Lett., 28, 4925 (1987).

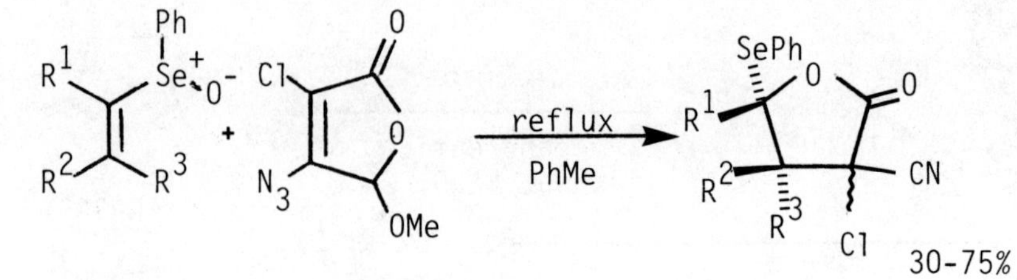

IV.D-7 S. Murata and T. Suzuki, Chem. Lett., 849 (1987).

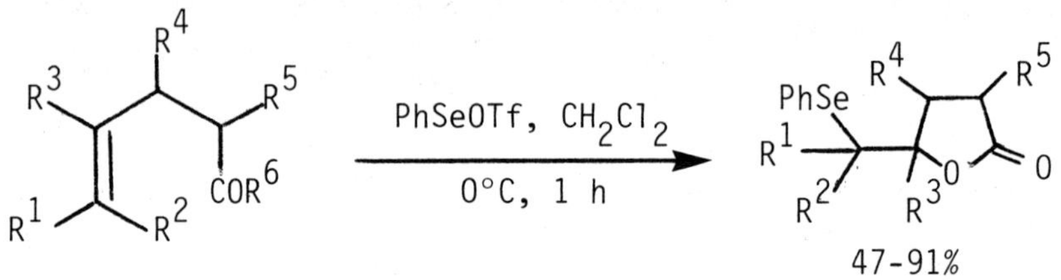

IV.D-8 S. Murahashi, et al., J. Org. Chem., 52, 4319 (1987).

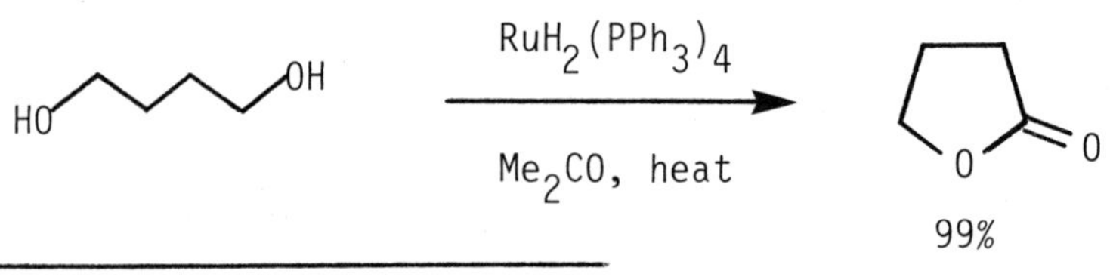

IV.D-9 Y. Tamura, M. Hojo and Z. Yoshida, Tetrahedron Lett., 28, 325 (1987).

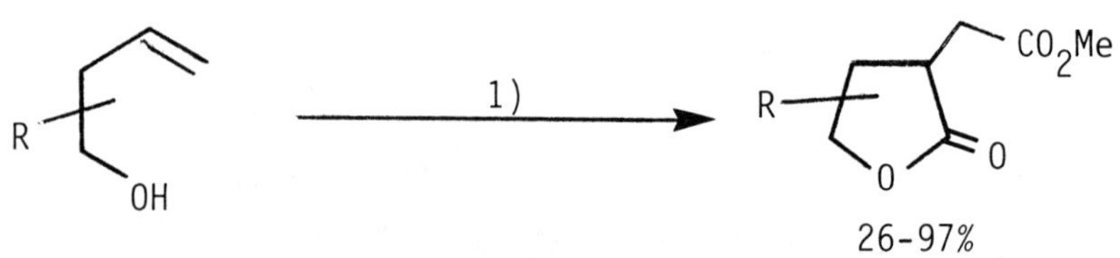

1) $PdCl_2$, $CuCl_2$, propylene oxide, $MeC(OEt)_3$, CO (1 atmos), CH_2Cl_2, MeOH, RT.

IV.D-10 R. Frenette, M. Kakushima, R. Zamboni, R.N. Young and T.R. Verhoeven, J. Org. Chem., 52, 304 (1987).

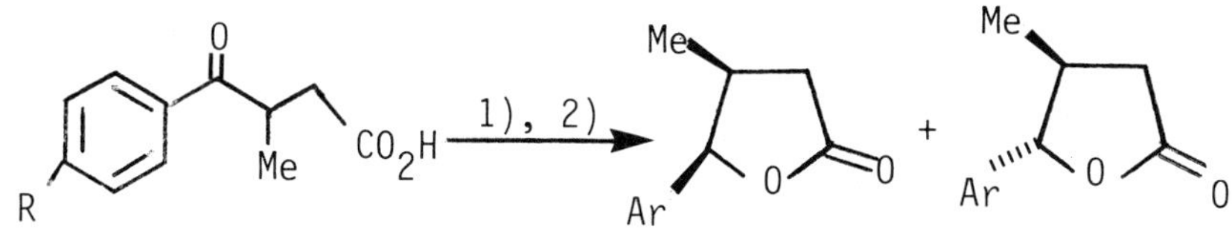

1) DIBAL-H, PhMe, $ZnCl_2$, THF; 2) TFA

60-97%

99:1 to 93:7

IV.D-11 R.C. Larock, D.J. Leuck and L.W. Harrison, Tetrahedron Lett., 28, 4977 (1987); C.W. Jefford and Y. Wang, J. Chem. Soc., Chem. Commun., 1513 (1987).

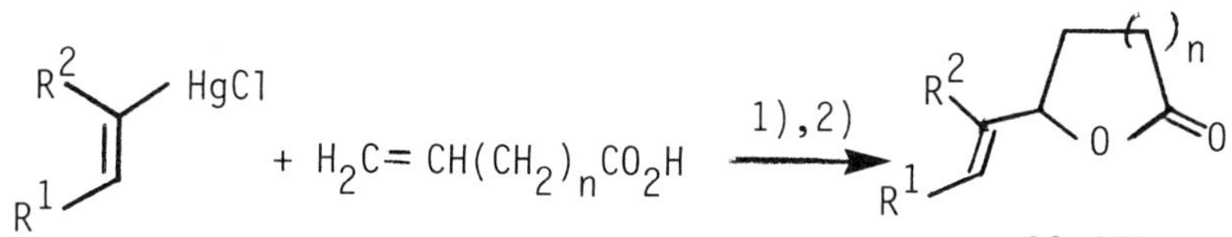

38-85%

1) $(MeCN)_2PdCl_2$, MeCN, -20 → 25°C, 20 h;
2) K_2CO_3, 80°C, 5 h.

IV.D-12 S. Fukuzawa, M. Iida, A. Nakanishi, T. Fujinami and S. Sakai, J. Chem. Soc., Chem. Commun., 920 (1987); N. Gnonlonfoun and H. Zamarlik, Tetrahedron Lett., 28, 4053 (1987).

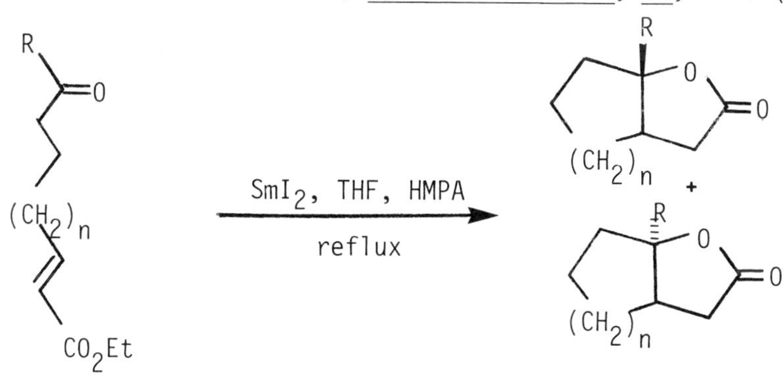

30-92%

trans:cis mixture

IV.D-13 M. Figuerido, J. Font and A. Virgili, Tetrahedron, 43, 1881 (1987); T. Hiyama et al., Bull. Chem. Soc. Jpn., 60, 2139 (1987); J.C. Carretero, S. deLombaert and L. Ghosez, Tetrahedron Lett., 28, 2135 (1987); P.M. Booth, C.M.J. Fox and S.V. Ley, J. Chem. Soc., Perkin I, 121 (1987); C.M. Cain and N.S. Simpkins, Tetrahedron Lett., 28, 3723 (1987).

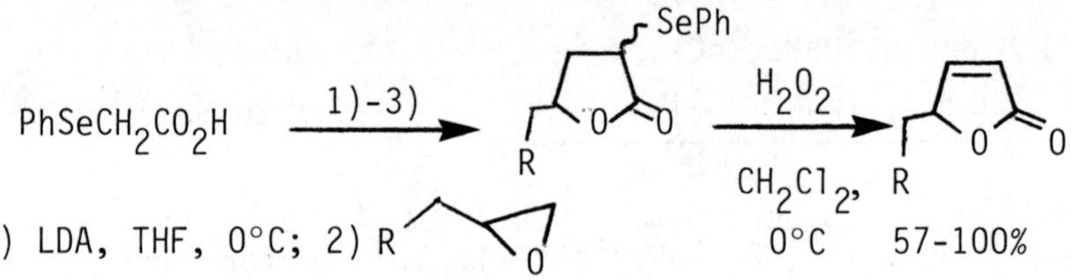

1) LDA, THF, 0°C; 2) R–(epoxide) RT, 3 h; 3) HOAc, reflux.

IV.D-14 S. Takahashi et al., J. Chem. Soc., Chem. Commun., 649 (1987); P. de Shong, D.R. Sidler and G.A. Slough, Tetrahedron Lett., 28, 2233 (1987).

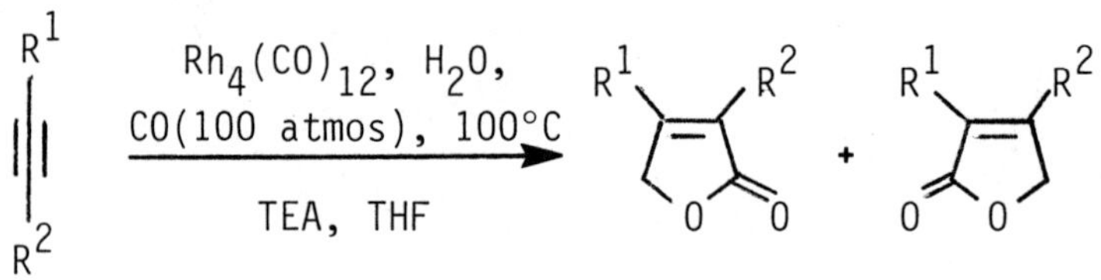

27-78:73-22

50-90%

IV.D-15 S. Tsuboi, H. Wada, S. Mimura and A. Takeda, Chem. Lett., 937 (1987).

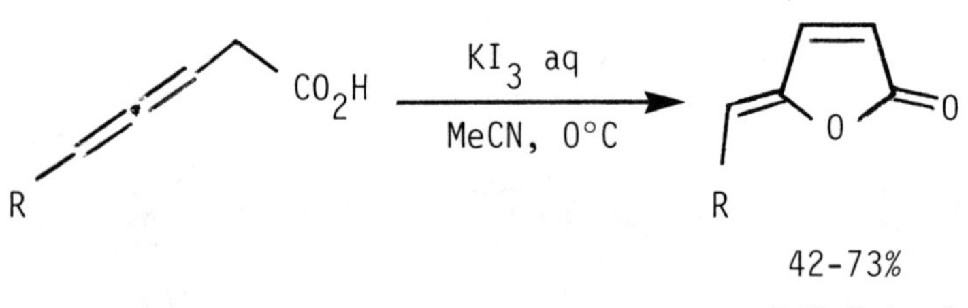

42-73%

Z/E=5 to 8:1

IV.D-16 T.R. Beebe *et al.*, *J. Org. Chem.*, *52*, 5472 (1987).

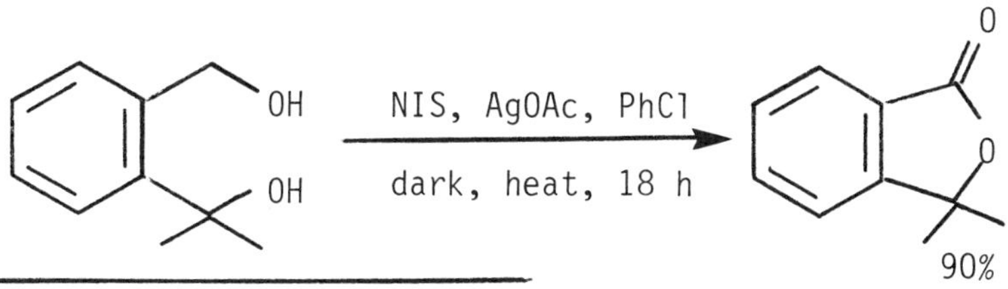

IV.D-17 K. Kobayashi, M. Itoh and H. Suginome, *Tetrahedron Lett.*, *28*, 3369 (1987).

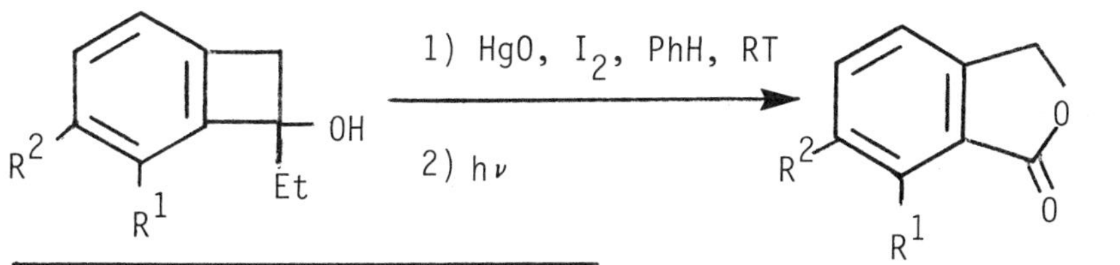

IV.D-18 K. Okazaki, K. Nomura and E. Yoshii, *Synth. Commun.*, *17*, 1021 (1987).

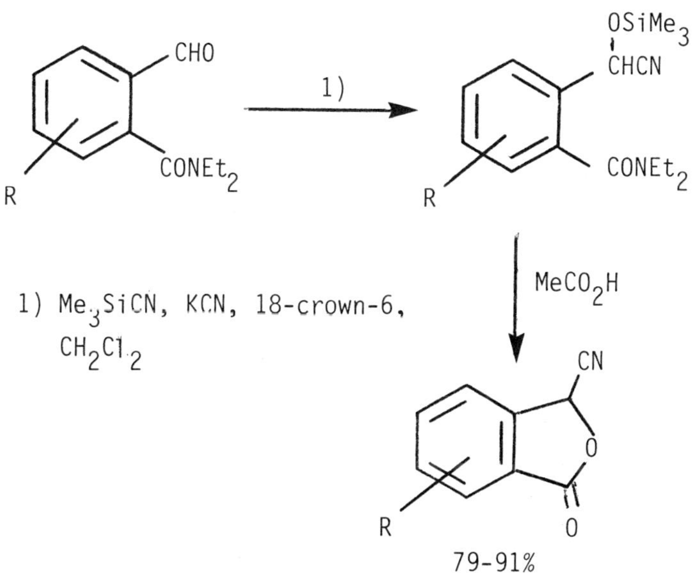

IV.D-19 D.M.T. Chan, T.B. Marder, D. Milstein and N.J. Taylor, J. Am. Chem. Soc., 109, 6385 (1987); A. Srikrishna, J. Chem. Soc., Chem. Commun., 587 (1987); P. Pale and J. Chuche, Tetrahedron Lett., 28. 6447 (1987).

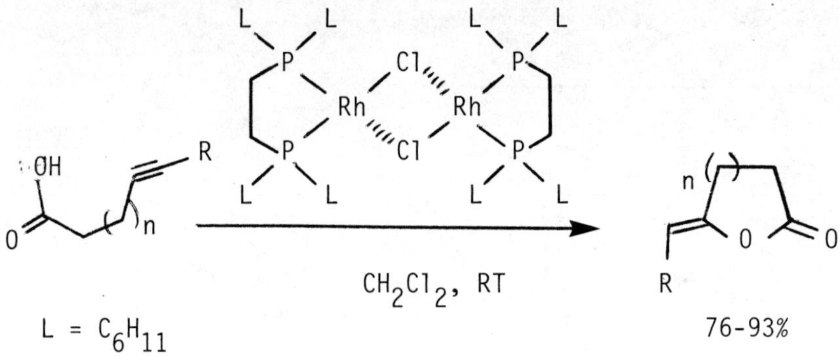

IV.D-20 N.E. Alami, C. Beland and J. Viliéras, Tetrahedron Lett., 28, 59 (1987).

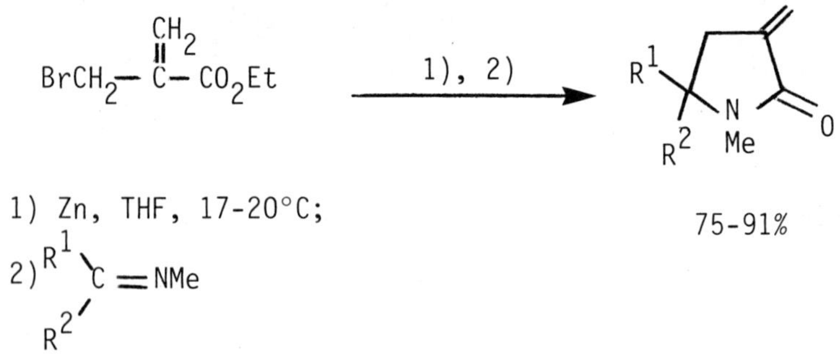

IV.D-21 K. Nishitani and K. Yamakawa, Tetrahedron Lett., 28, 655 (1987); K. Bock, I.M. Castilla, I. Lundt and C. Pedersen, Acta Chem. Scand.,Pt.B, 41, 13 (1987).

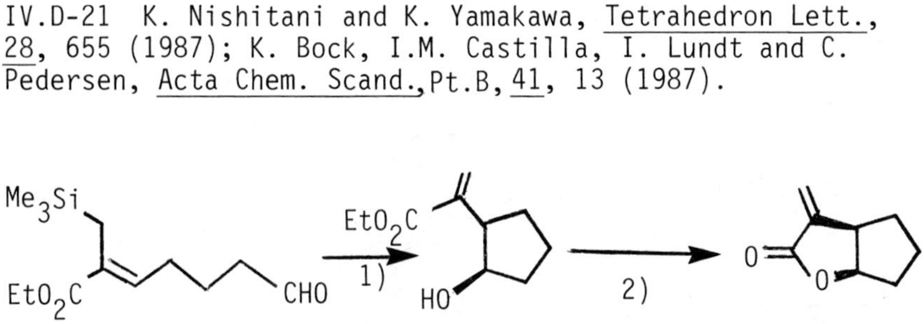

1)$TiCl_4$,CH_2Cl_2,-20°C; 2)p-TsOH,PhH,reflux.

IV.D-22 R.D. Miller and W. Theis, Tetrahedron Lett., 28, 1039 (1987).

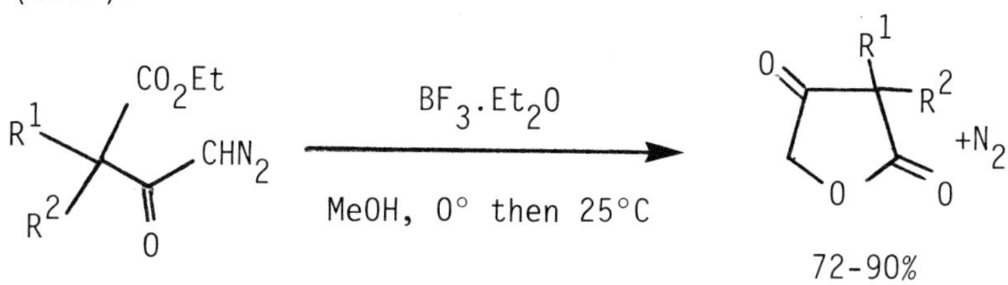

72-90%

IV.D-23 A. Sakamoto, Y. Yamamoto and J. Oda, J. Am. Chem. Soc., 109, 7188 (1987).

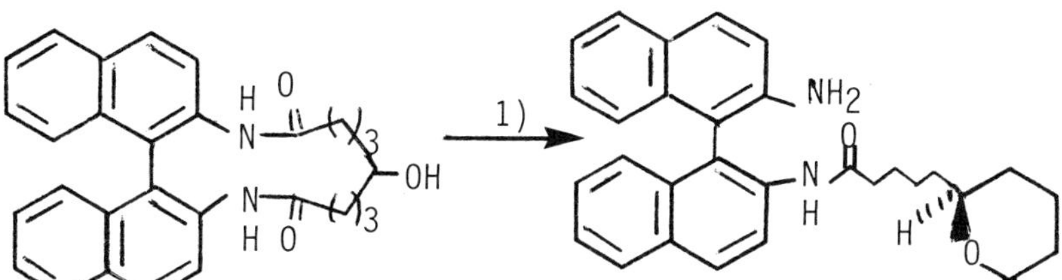

100%

97% de

1) $(F_3CCO)_2O$, CH_2Cl_2, -20°C, 12 h.

IV.D-24 A.T. Russell and G. Procter, Tetrahedron Lett., 28, 2041, 2045 (1987).

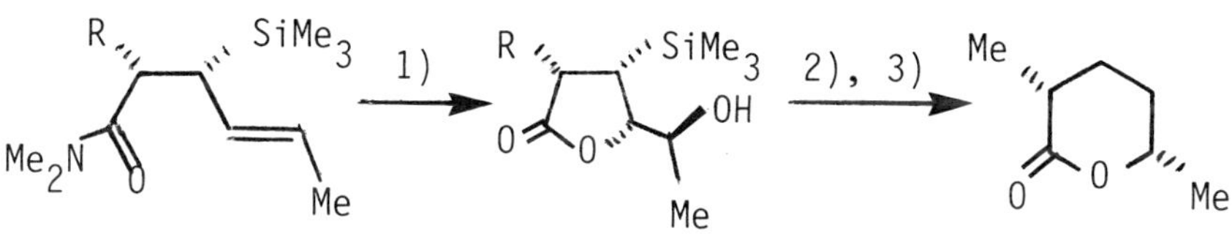

1) M-CPBA; 2) $BF_3.OEt_2$; 3) H_2, Pd/c

IV.D-25 F. Theil, B. Costisella, H. Gross, H. Schick and S. Schwarz, J. Chem. Soc., Perkin I, 2469 (1987).

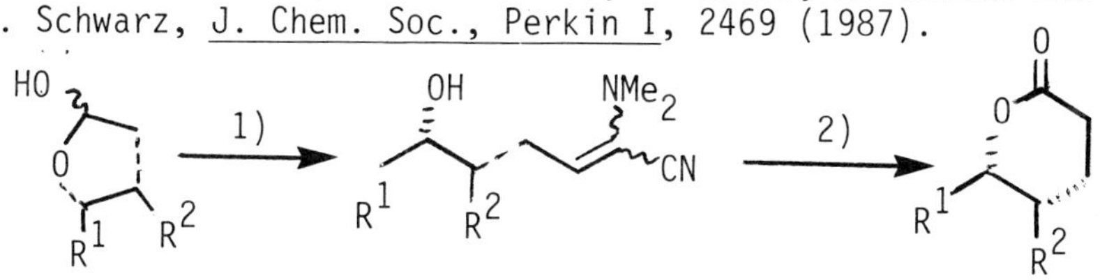

1) $(EtO)_2P(O)CH(CN)NMe_2$,BuLi, THF, 65°C, 2 h;
2) HOAc, THF, H_2O, 65°C

20-53%

IV.D-26 D. Enders and B.E.M. Rendenbach, Chem. Ber., 120, 1223 (1987).

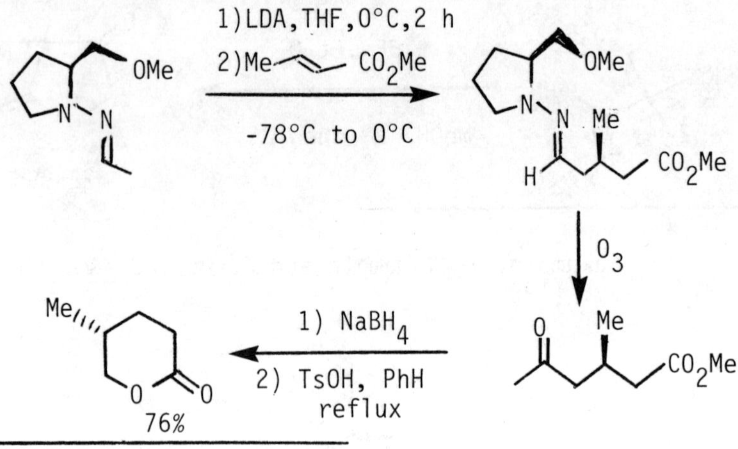

IV.D-27 C.M. Thompson, Tetrahedron Lett., 28, 4243 (1987).

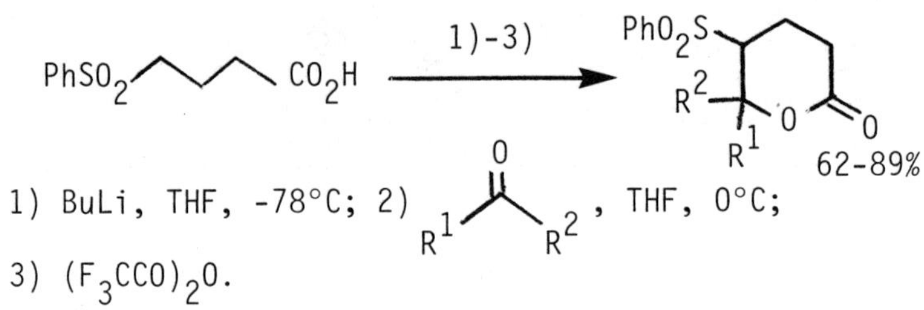

IV.D-28 T. Kubota, R. Aoyagi, H. Sando, M. Kawasumi and T. Tanaka, Chem. Lett., 1435 (1987).

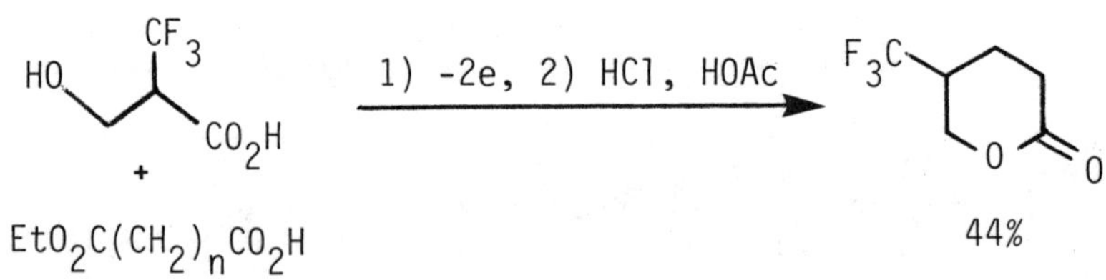

IV.D-29 P.C. Bulman Page, P.H. Williams, E.W. Collington and H. Finch, J. Chem. Soc., Chem. Commun., 756 (1987).

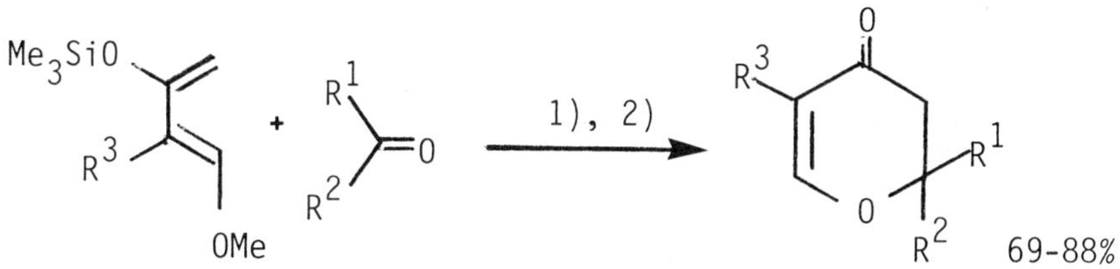

1) $ZnCl_2$, PhH, RT, 12 h; 2) HCl aq, Et_2O, 15 min.

IV.D-30 S. Valverde, S. Garcia-Ochoa and M. Martin-Lomas, J. Chem. Soc., Chem. Commun., 1714 (1987).

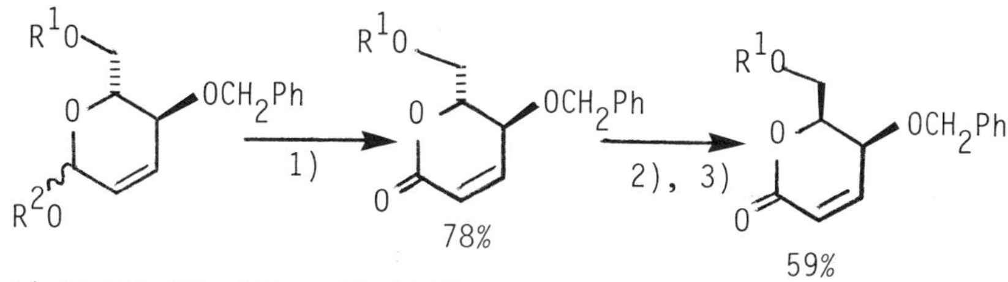

1) MCPBA, $BF_3.OEt_2$; 2) LiOH

3) Ph_3P, $EtO_2CN = NCO_2Et$

$R^1 = CH_2CH = CH_2$

IV.D-31 Y. Ito, E. Nakato, K. Sho and K. Tamao, Tetrahedron Lett., 28, 2247 (1987).

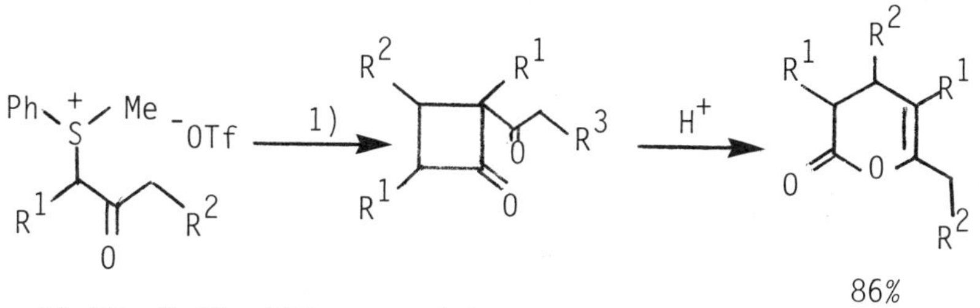

1) KF, MeCN, 0°C, overnight.

IV.D-32 S.P. Khanapure and E.R. Biehl, J. Org. Chem., 52, 1333 (1987).

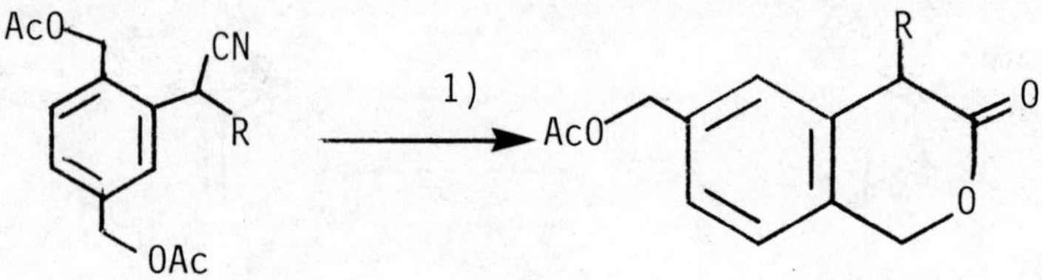

1) HOAc, H_2SO_4aq, reflux, 1 h 36-85%

IV.D-33 L. Bonsignore, S. Cabiddu, G. Loy and D. Secci, Heterocycles, 26, 1619 (1987).

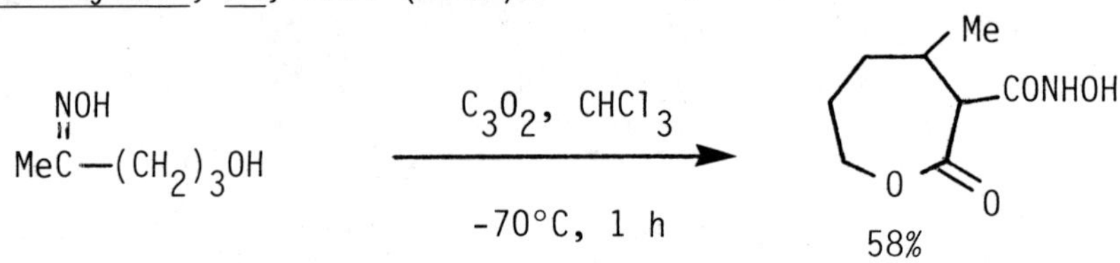

IV.D-34 J.K. Stille and M. Tanaka, J. Am. Chem. Soc., 109, 3785 (1987).

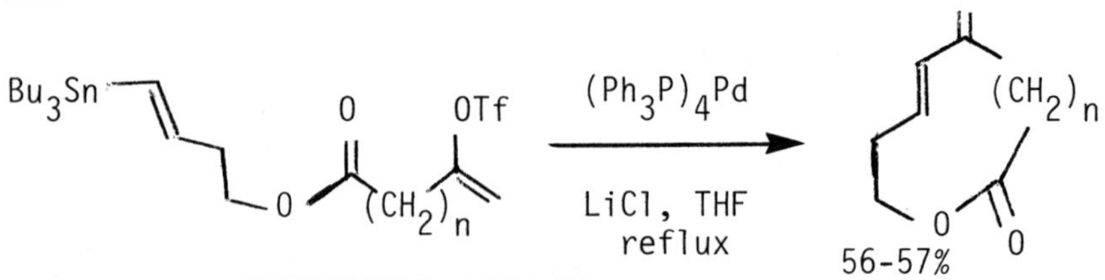

IV.D-35 G.H. Posner, E. Asirvatham, K.S. Webb and S. Jew, Tetrahedron Lett., 28, 5071 (1987).

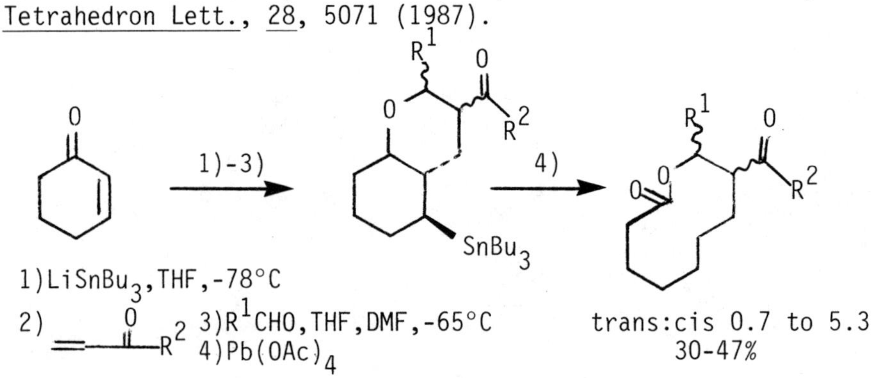

IV.D-36 G. Quinkert *et al.*, *Helv. Chim. Acta*, **70**, 771 (1987).

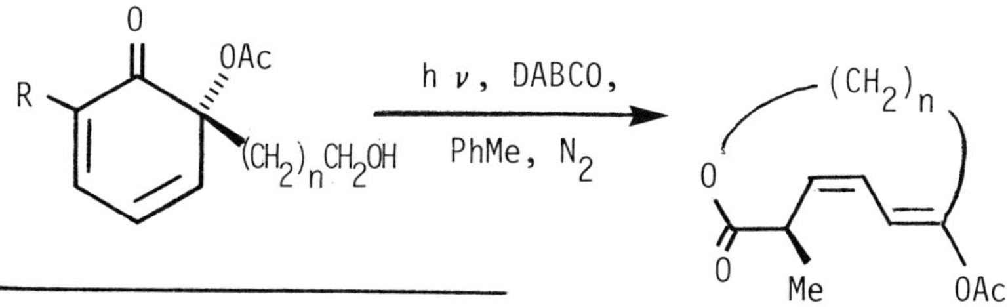

IV.D-37 H. Suginome and S. Yamada, *Tetrahedron*, **43**, 3371 (1987).

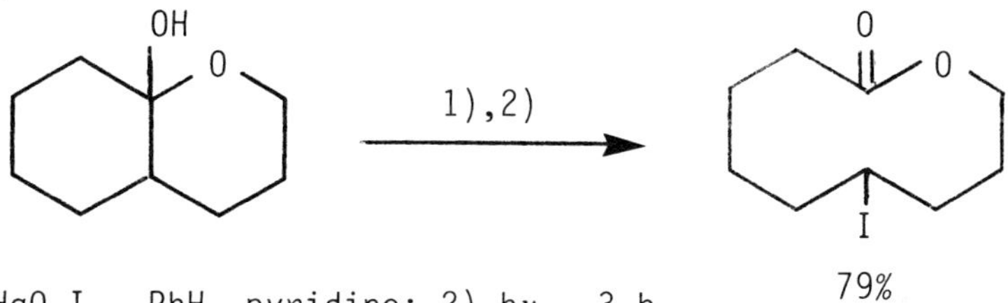

1) HgO,I_2, PhH, pyridine; 2) hν , 3 h

IV.D-38 N.E. Schore and S.D. Najdi, *J. Org. Chem.*, **52**, 5296 (1987).

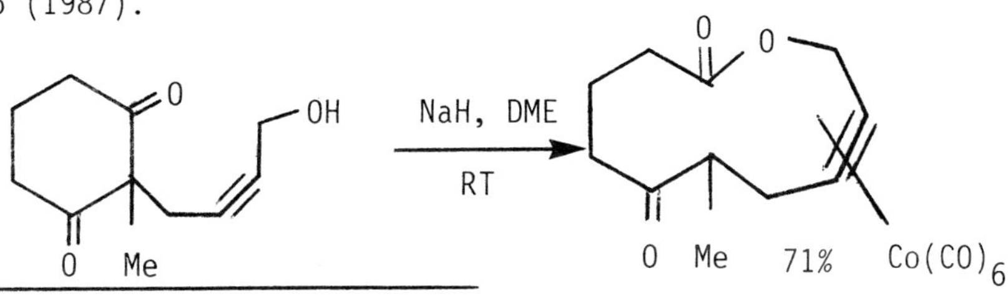

IV.D-39 M. Ochiai, S. Iwaki, T. Ukita and Y. Nagao, *Chem. Lett.*, 133 (1987).

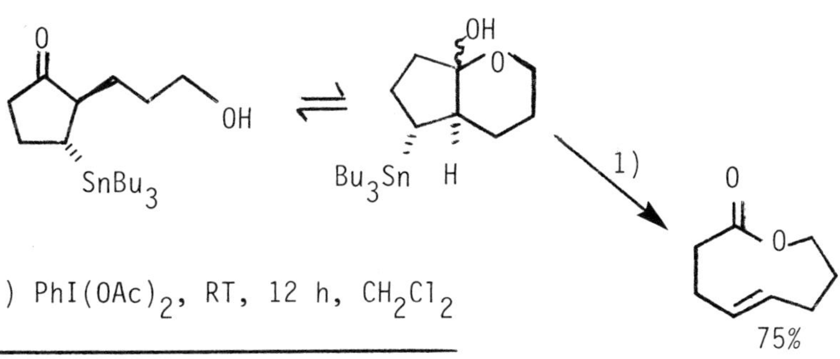

1) $PhI(OAc)_2$, RT, 12 h, CH_2Cl_2

IV.E. Furans, Thiophenes, Etc.

IV.E-1 M.J. Kurth and M.J. Rodriguez, J. Am. Chem. Soc., 109, 7577 (1987).

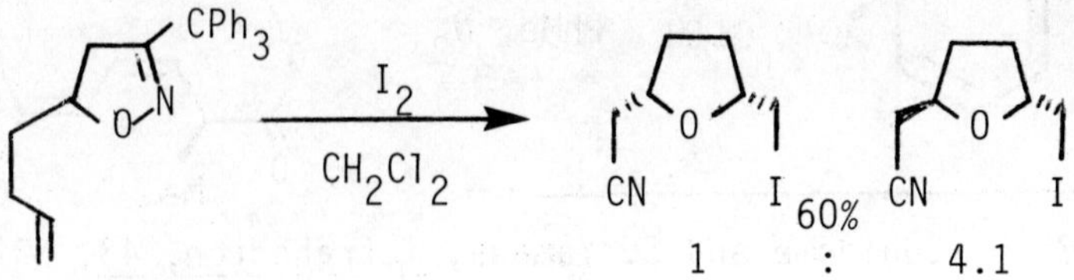

IV.E-2 M.H. Hopkins and L.E. Overman, J. Am. Chem. Soc., 109, 4748 (1987).

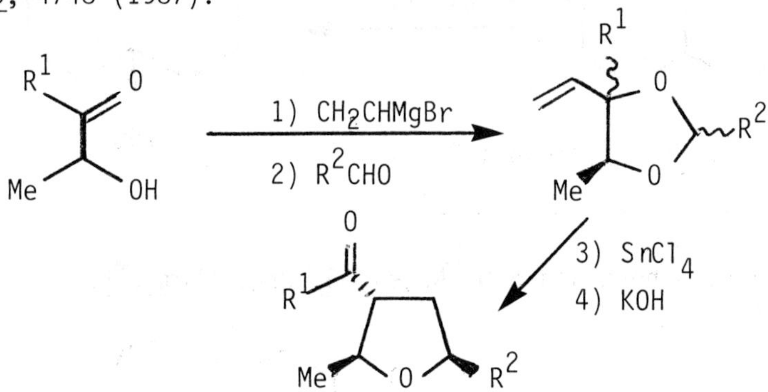

IV.E-3 K. Barlos et al., J. Chem. Soc., Chem. Commun., 474 (1987); A. Costa and J.M. Riego, Synth. Commun., 17, 1373 (1987); S. Murata and T. Suzuki, Tetrahedron Lett., 28, 4297 (1987); E. Magnol, J. Gore and M. Malacria, Bull. Soc. Chim. Fr., 455 (1987); S.M. Tuladhar and A.G. Fallis, Tetrahedron Lett., 28, 523 (1987).

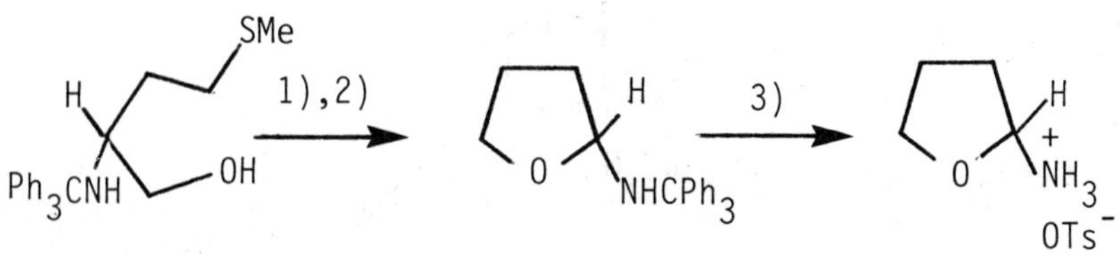

1) MeI, EtOAc; 2) NaH, DMF, -10°C;

3) TsOH aq, MeOH, reflux.

IV.E-4 K. Fugami, K. Oshima and K. Utimoto, Tetrahedron Lett., 28, 809 (1987).

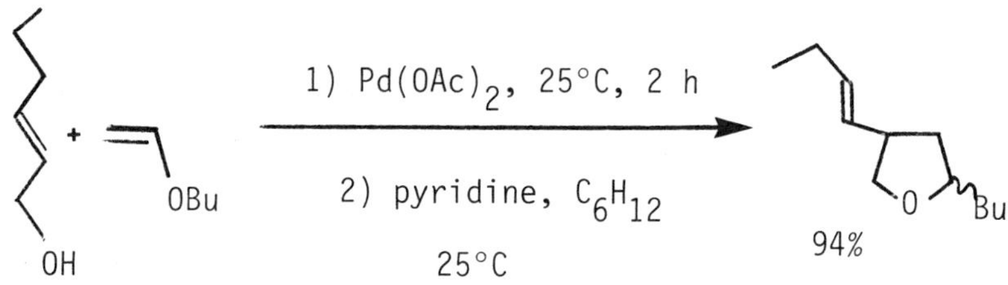

IV.E-5 G.A. Kraus and J. Thurston, Tetrahedron Lett., 28, 4011 (1987).

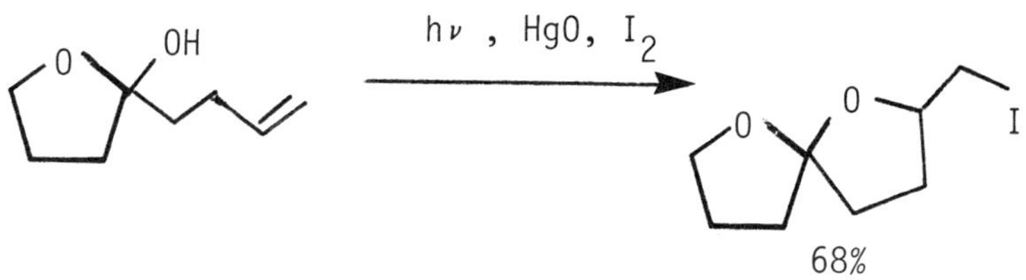

IV.E-6 M.C. Pirrung and C.V. DeAmicis, Heterocycles, 25, 189 (1987).

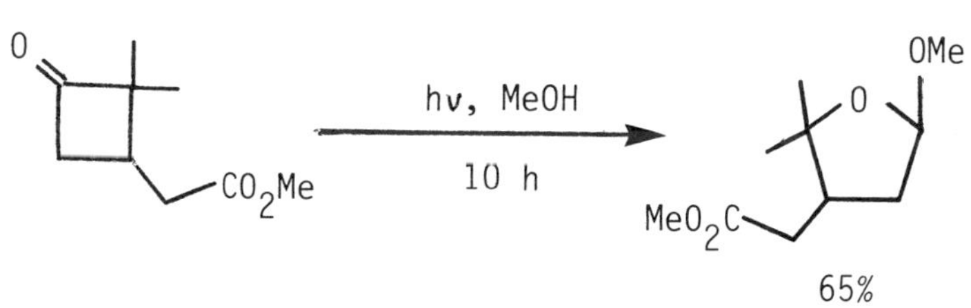

IV.E-7 B.J. Fitzsimmons, Y. LeBlanc and J. Rokach, J. Am. Chem. Soc., 109, 285 (1987).

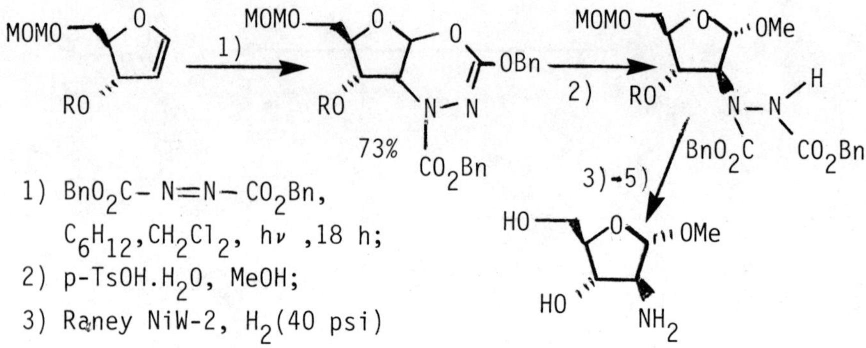

1) $BnO_2C-N{=}N-CO_2Bn$, C_6H_{12}, CH_2Cl_2, hν, 18 h;
2) p-TsOH.H_2O, MeOH;
3) Raney NiW-2, H_2(40 psi) MeOH-HOAc; 4) n-Bu_4NF;
5) AcCl, MeOH (4NHCl).

IV.E-8 P.M. Herrington, M.H. Hopkins, P. Mishra, M.J. Brown and L.E. Overman, J. Org. Chem., 52, 3711 (1987); S. Murata and T. Suzuki, Tetrahedron Lett., 28, 4415 (1987); G.N. Austin, G.W.J. Fleet, J.M. Peach, K. Prout and J.C. Son, Tetrahedron Lett., 28, 4741 (1987).

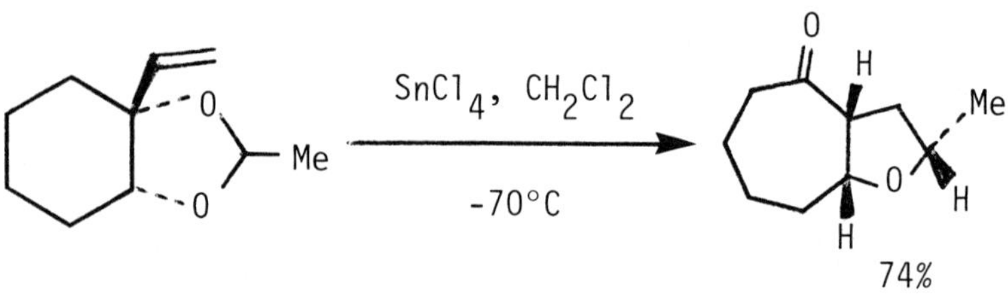

IV.E-9 T. Mukaiyama, Y. Sagawa and S. Kobayashi, Chem. Lett., 2169 (1987); R.S. Mali, M.I. Talale and M.A. Koshy, Synthesis, 630 (1987).

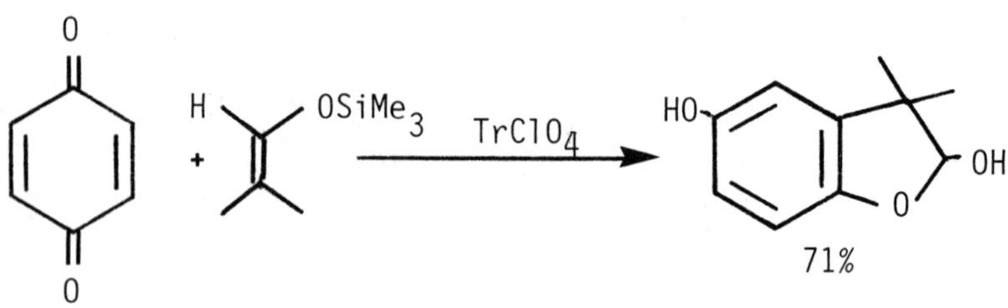

IV.E-10 M. Hrytsak and T. Durst, J. Chem. Soc., Chem. Commun., 1150 (1987).

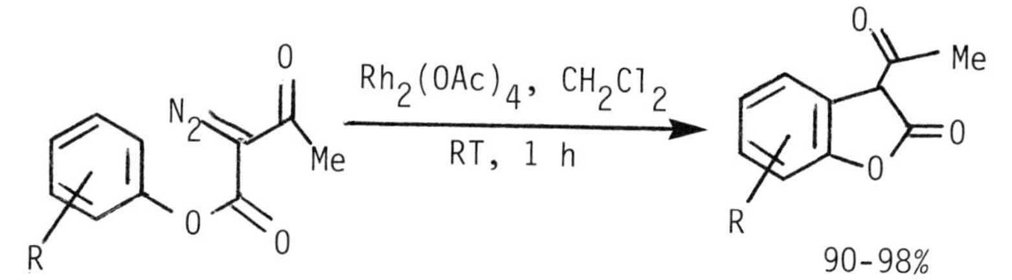

IV.E-11 E.J. Corey and A.K. Ghosh, Chem. Lett., 223 (1987).

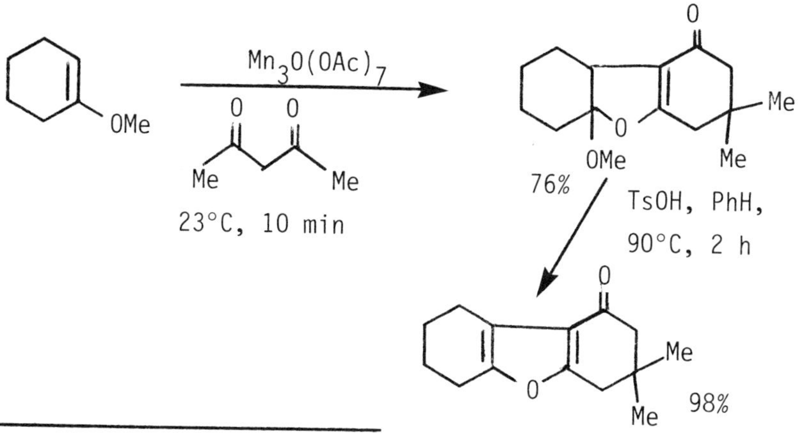

IV.E-12 S. Fukuzawa, T. Fujinami and S. Sakai, J. Chem. Soc., Chem. Commun., 919 (1987).

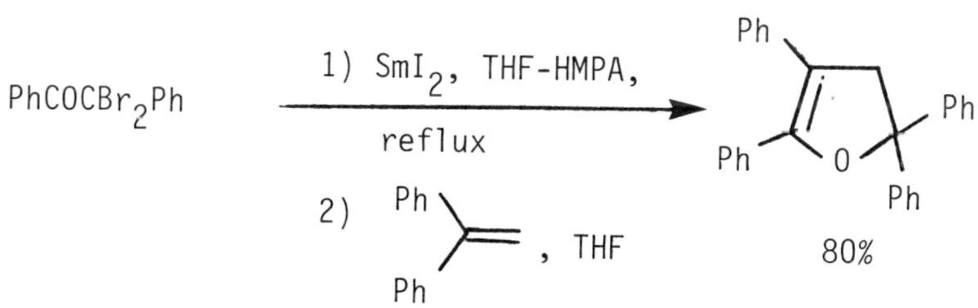

IV.E-13 I. Minami, M. Yuhara and J. Tsuji, Tetrahedron Lett., 28, 629 (1987).

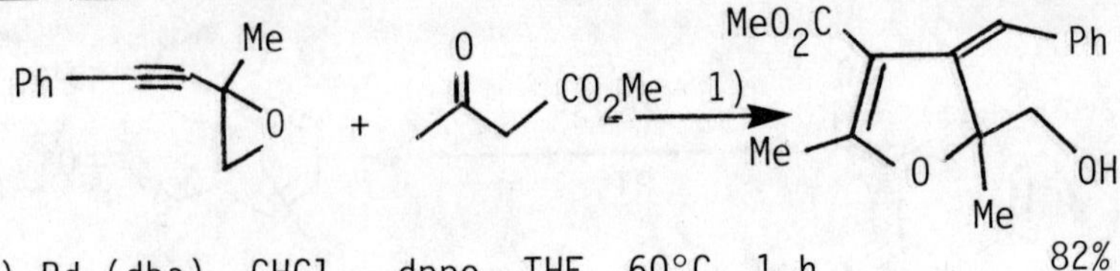

1) $Pd_2(dba)_3.CHCl_3$, dppe, THF, 60°C, 1 h.

IV.E-14 T.V. Lee, A.A. Galan and C.B. Chapleo, Tetrahedron Lett., 28, 2301 (1987).

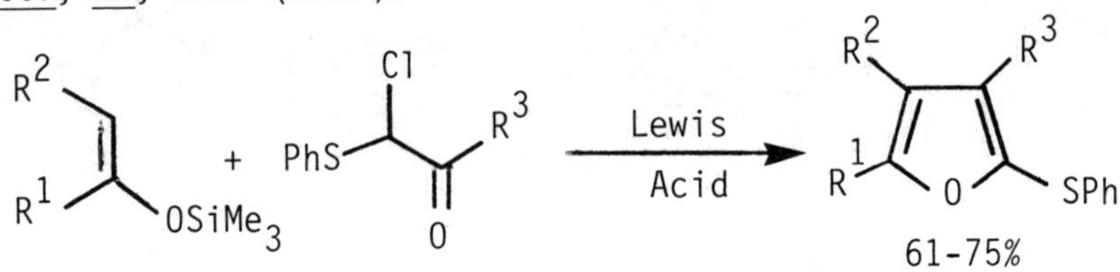

IV.E-15 N. Chatani and T. Hanafusa, J. Org. Chem., 52, 4408 (1987); E. Bartmann and R. Gompper, Synth. Commun., 17, 263 (1987).

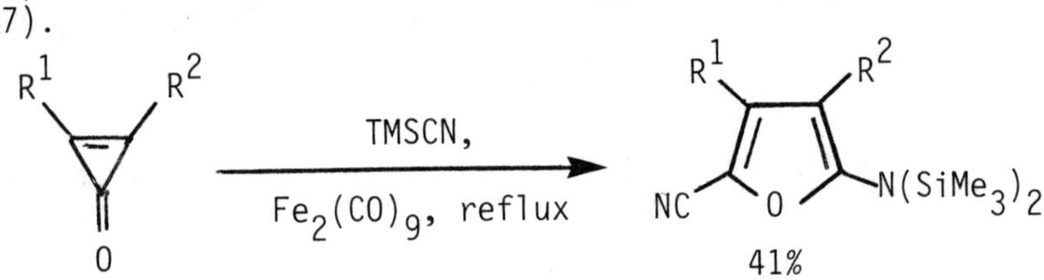

IV.E-16 S. Kosack and G. Himbert, Chem. Ber., 120, 71 (1988); S.W. McCombie et al., Tetrahedron Lett., 28, 4123 (1987).

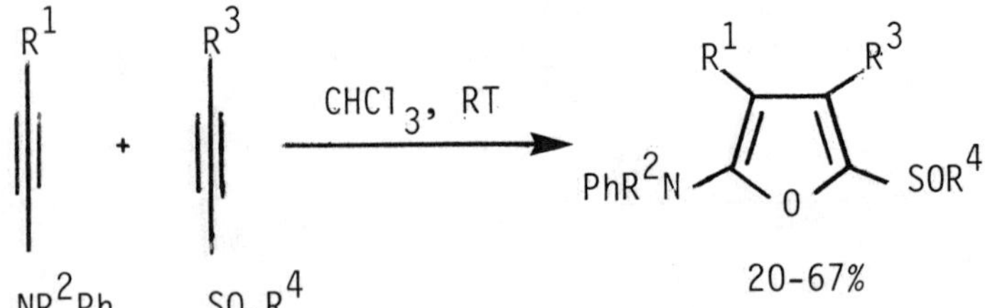

IV.E-17 H. Sheng, S. Lin and Y. Huang, Synthesis, 1022 (1987); A. Srikrishna and G. Sunderbabu, Tetrahedron Lett., 28, 6393 (1987); I. Minami, M. Yuhara, H. Watanabe and J. Tsuji, J. Organomet. Chem., 334, 225 (1987).

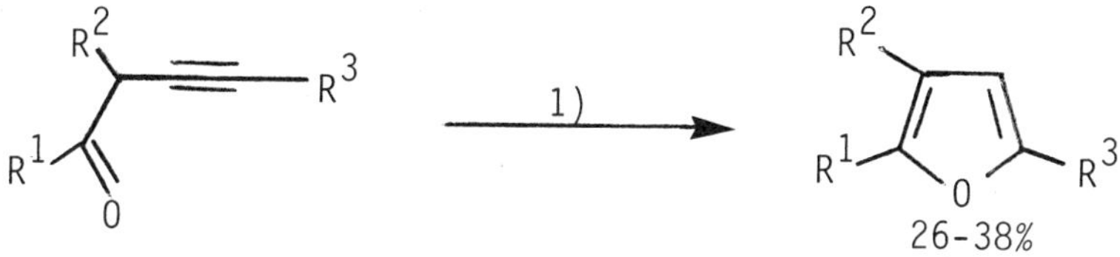

26-38%

1) $Pd(dba)_2$, PPh_3, PhMe, 70-100°C, 5-10 h.

IV.E-18 A. Srikrishna and K.C. Pulliah, Tetrahedron Lett., 28, 5203 (1987); B.J.M. Jansen, R.M. Peperzak and Ae. de Grout, Rec. Trav. Chim., 106, 549 (1987).

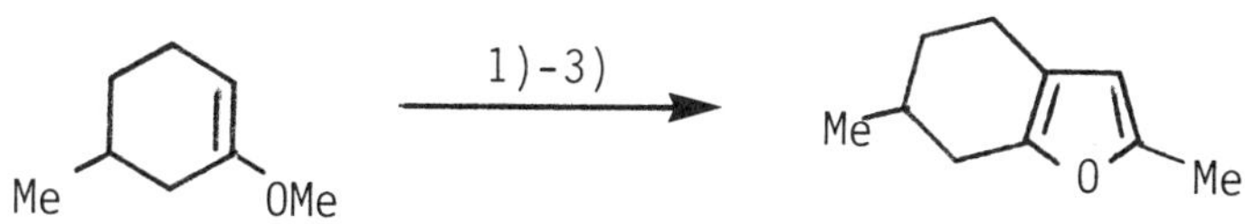

45%

1) NBS, $HC{\equiv}CCH_2OH$, CH_2Cl_2, -40°C, 4 h;

2) n-Bu_3SnCl, $NaCNBH_3$, AIBN (cat.), t-BuOH, 80°C, 1.5 h;

3) p-TsOH (cat.), PhH, RT.

IV.E-19 W.T. Brady, Y.F. Giang, A.P. Marchand and A. Wu, J. Org. Chem., 52, 3457 (1987).

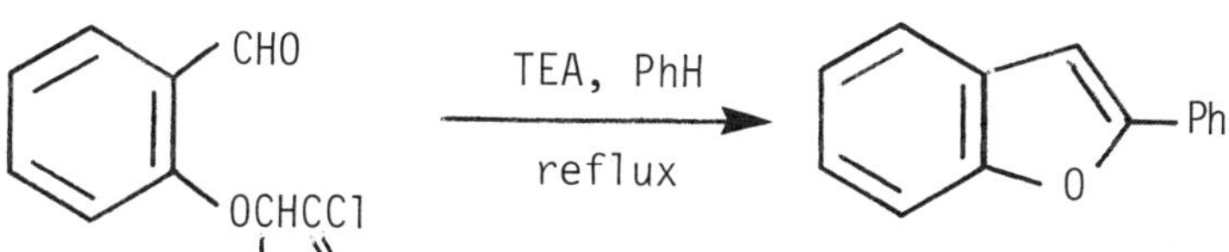

75%

IV.E-20 E.H.M. Abd Elall, M.I. Al Ashmawy and J.M. Mellor, J. Chem. Soc., Chem. Commun., 1577 (1987); Y. Terao, M. Aono, N. Imai and K. Achiwa, Chem. Pharm. Bull., 35, 1734 (1987).

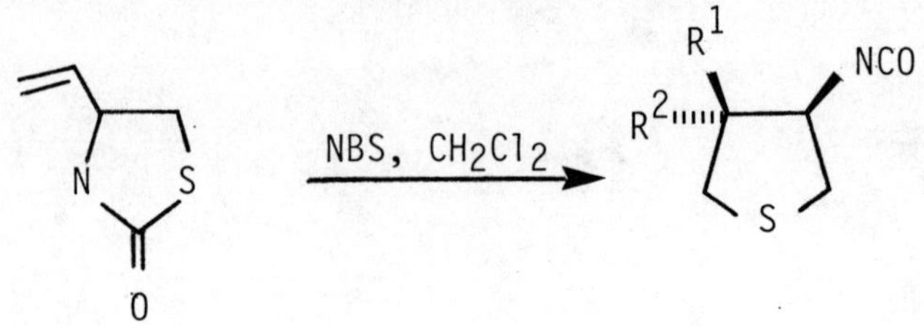

IV.E-21 J. Nakayama, Y. Nakamura, S. Murabayashi and M. Hoshino, Heterocycles, 26, 939 (1987).

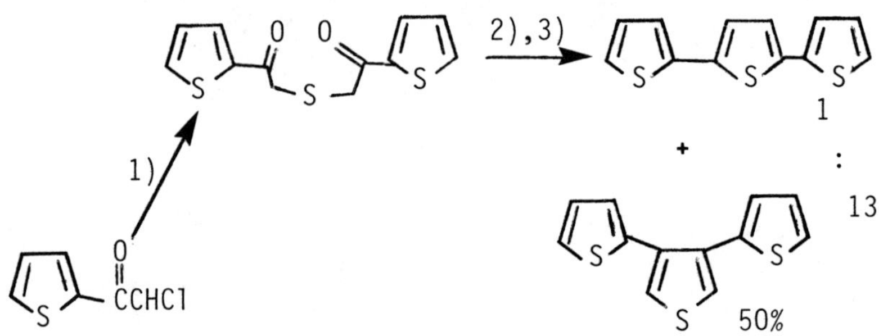

1) Na_2S, Me_2CO, H_2O; 2) p-$MeOC_6H_4$–P(=S)(μ-S)$_2$P(=S)C_6H_4OMe-p, PhH;

3) reflux, o-$C_6H_4Cl_2$

IV.E-22 R.A. Aitken and G. Burns, Tetrahedron Lett., 28, 3717 (1987).

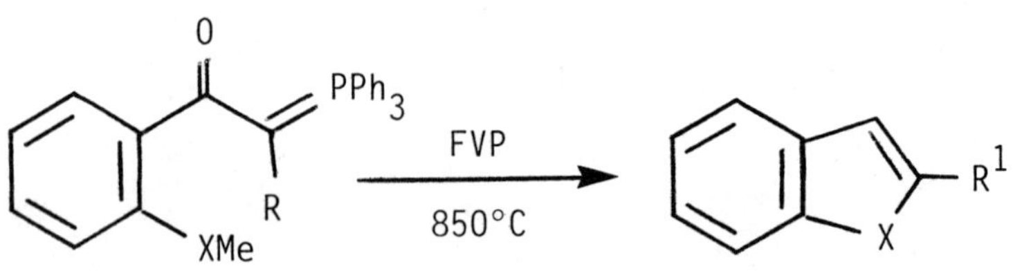

X = O or S

IV.F. Pyrroles, Indoles, Etc.

IV.F-1 Y. Watanabe et al., Tetrahedron Lett., 28, 3953 (1987); N.X. Hu, Y. Aso, T. Otsubo and F. Ogura, Chem. Lett., 1327 (1987); Y. Tsuji, Y. Yokoyama, K. Huh and Y. Watanabe, Bull. Chem. Soc. Jpn., 60, 3456 (1987).

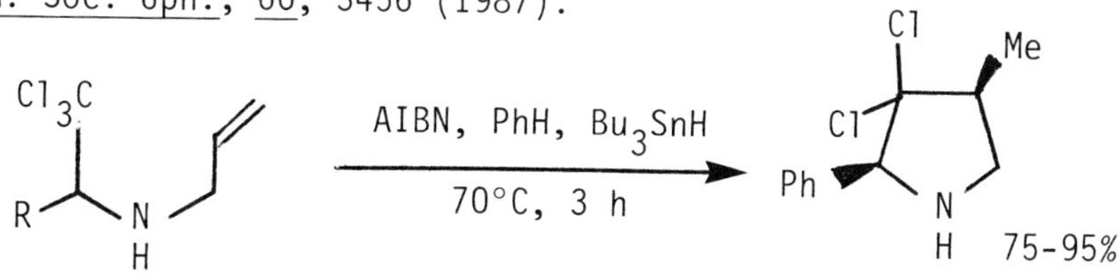

IV.F-2 J. Cooper, D.W. Knight and P.T. Gallagher, Tetrahedron Lett., 28, 3031 (1987).

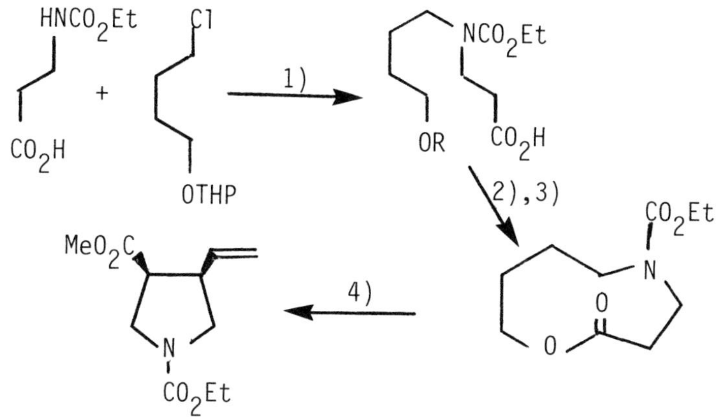

1) n-BuLi, THF, -78°C then 20°C; 2) PPTS, MeOH, 60°C, 8 h;

3) 2-chlor-1-methylpyridinium iodide, TEA, MeCN, 80°C;

4) LDA, TBDMSCl, THF.

IV.F-3 A. Padwa and W. Dent, J. Org.Chem., 52, 235 (1987).

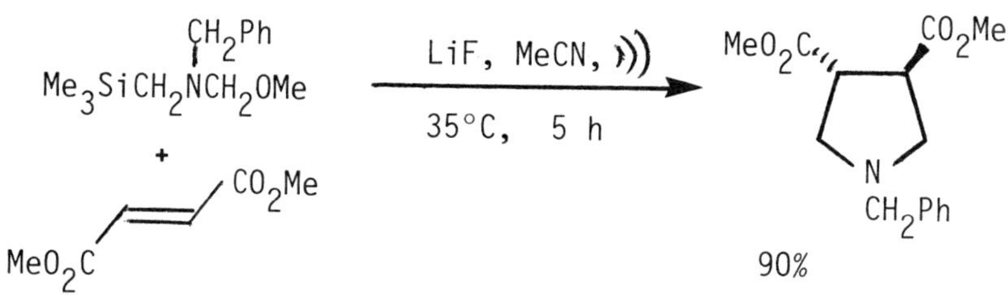

IV.F-4 N. Knouzi, M. Vaultier, L. Toupet and R. Carrie, Tetrahedron Lett., 28. 1757 (1987).

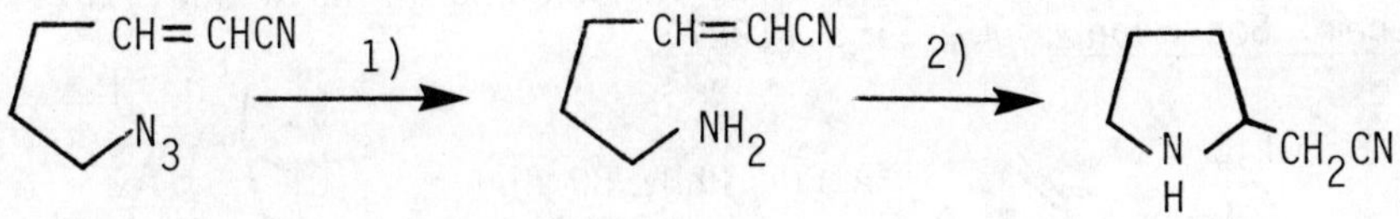

1) Ph_3P, THF, H_2O; 2) distillation.

IV.F-5 G.B. Itoua and J. Laronze, Synthesis, 353 (1987); H. Grützmacher and H.W. Roesky, Chem. Ber., 120, 995 (1987); O. Tsuge et al., Bull. Chem. Soc. Jpn., 60, 3347, 3359 (1987).

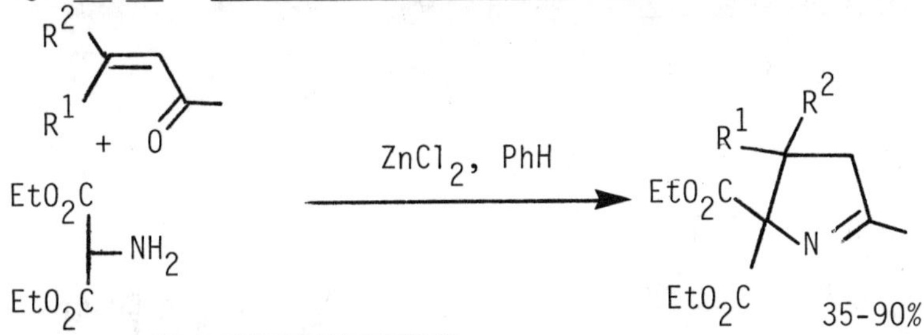

IV.F-6 J.P. Celerier, M. Haddad, D. Jacoby and G. Llommet, Tetrahedron Lett., 28, 6597 (1987).

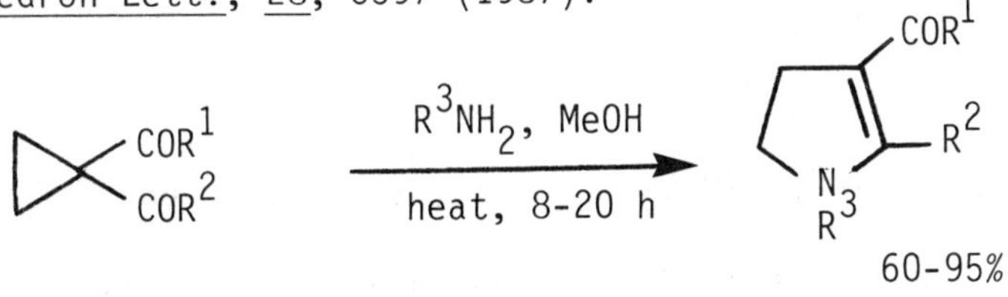

IV.F-7 A. Padwa, P. Eisenbarth, M.K. Venkatramanan and G.S.K. Wong J. Org. Chem., 52, 2427 (1987); O.Tsuge et al., J. Org. Chem., 52, 2523 (1987); P.D. Croce, P. Gariboldi and C. La Rosa, J. Heterocycl. Chem., 24, 1793 (1987).

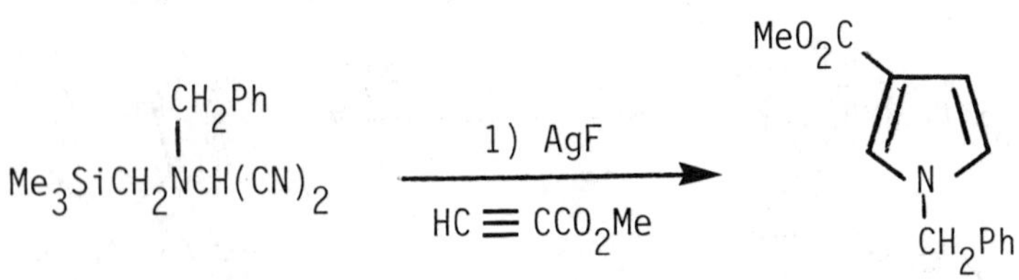

IV.F-8 R.J.P. Corriu, J.J.E. Moreau and C. Vernhet, Tetrahedron Lett., 28, 2963 (1987); S.A. Burns, R.J.P. Corriu, V. Huynh and J.J.E. Moreau, J. Organomet. Chem., 333, 281 (1987).

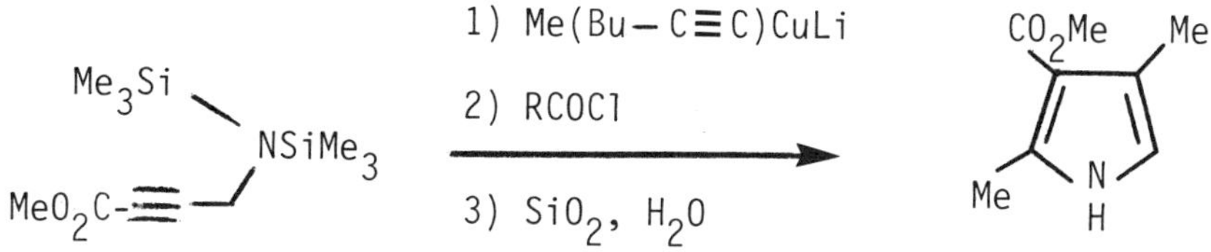

44-66%

IV.F-9 E. Cohnen and R. Dewald, Synthesis, 566 (1987); G. Baccolini and C. Sandali, J. Chem. Soc., Chem. Commun., 788 (1987); J.S. Kiely and S. Huang, J. Heterocycl. Chem., 24, 1137 (1987).

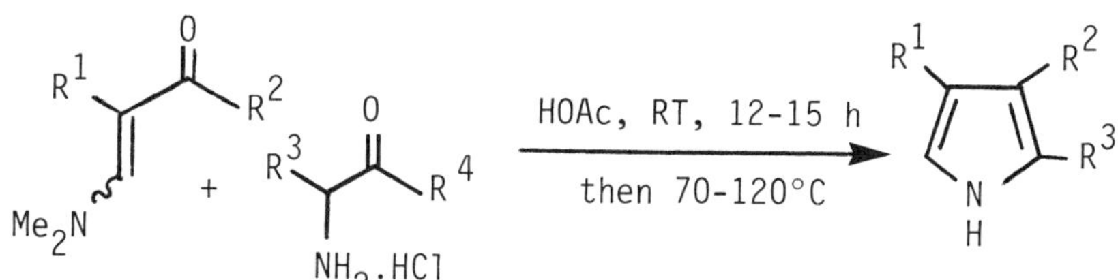

35-95%

IV.F-10 P. Chiu, K.Lui, P.N. Maini and M.P. Sammes, J. Chem. Soc., Chem. Commun., 109 (1987).

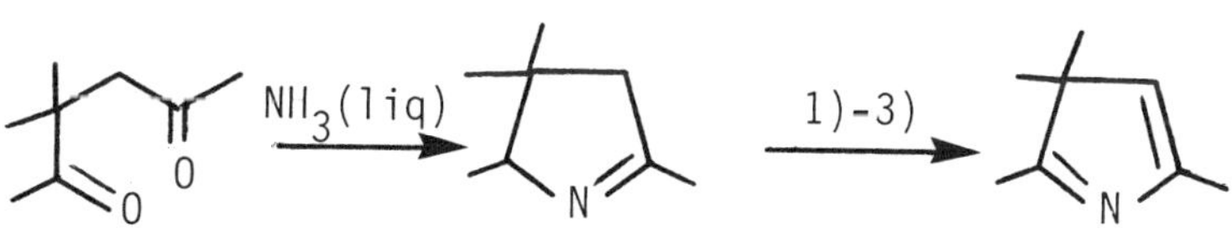

89% 46%

1) basic Al_2O_3, CH_2Cl_2, 4 Å mol. sieves, reflux;

2) HCl aq; 3) Na_2CO_3 aq, 0°C.

IV.F-11 P.A. Wender and C. B. Cooper, Tetrahedron Lett., 28, 6125 (1987).

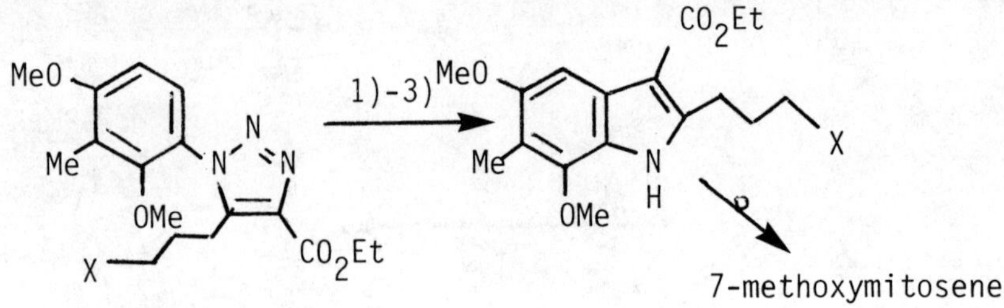

1) hν , Vycor, MeCN, (67%); 2) TBAF, THF, 0°C (99%);

3) p-TsCl, pyridine, 0°C (96%).

IV.F-12 M. Lim and D.G. Patil, Tetrahedron Lett., 28, 3775 (1987).

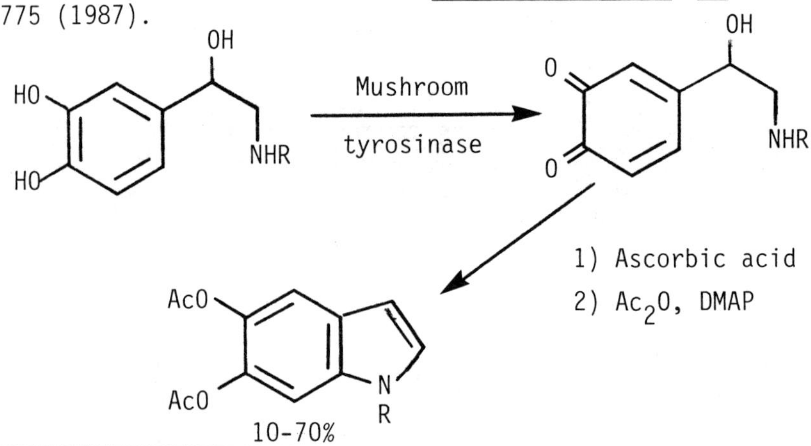

IV.F-13 R.C. Larock and S. Babu, Tetrahedron Lett., 28, 5291 (1987); E.O.M. Oriemans, A.H. Schreuder, P.G.M. Conti, W. Verboom and D.N. Reinhoudt, Tetrahedron, 43, 3817 (1987); T. Sakamoto, Y. Kondo, S. Iwashita and H. Yamanaka, Chem. Pharm. Bull., 35, 1823 (1987).

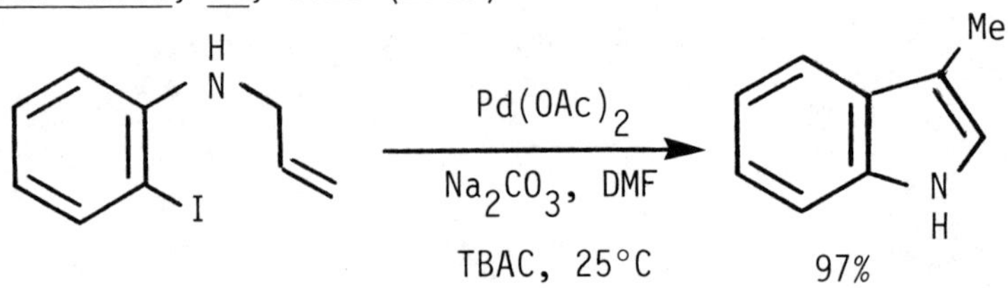

IV.F-14 A.P. Kozikowski, X. Cheng, C. Li and J.G. Scripko, Isr. J. Chem., 27, 61 (1986).

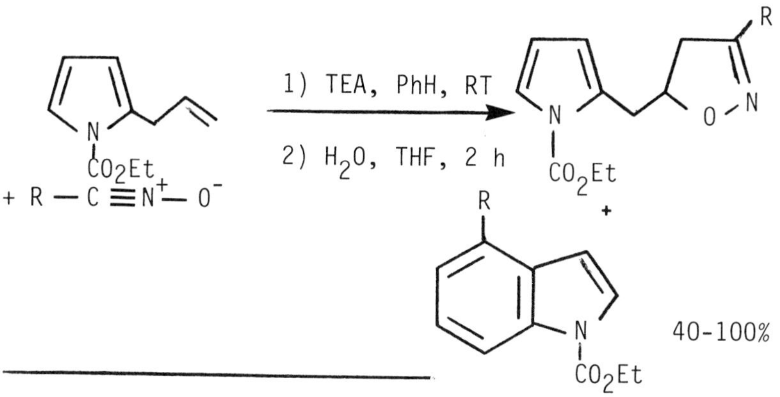

IV.F-15 C. Kashima, S. Hibi, T. Maruyama, K. Harada and Y. Omote, J. Heterocycl. Chem., 24, 913 (1987).

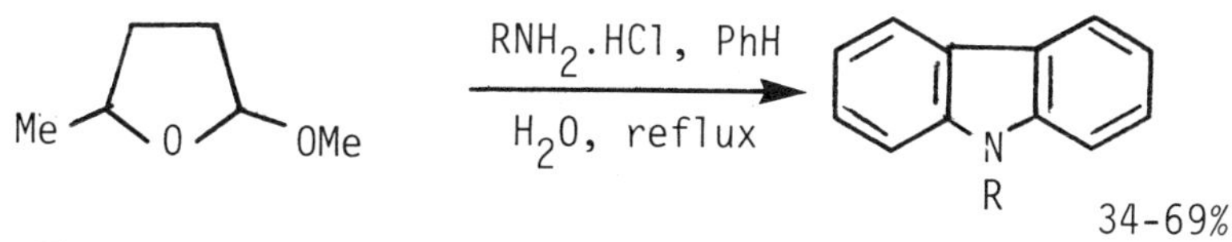

IV.G. Pyridines, Quinolines, Etc.

IV.G-1 E.N. Kayushina et al., Bull. Acad. Sci. USSR (Chem.) (Engl. Trans.), 35, 1474 (1986).

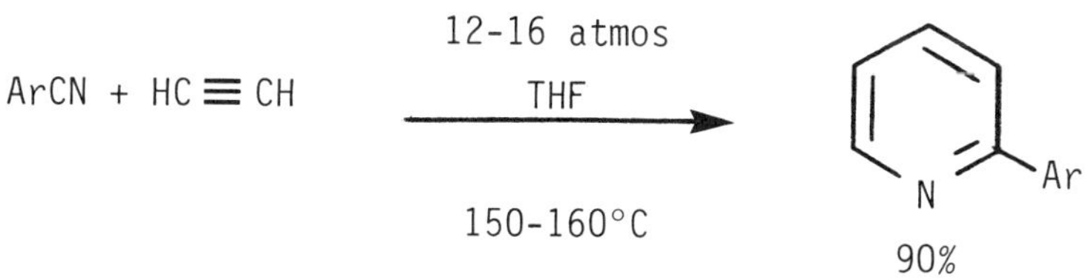

IV.G-2 T.L. Gilchrist, D. Hughes, W. Stretch and E.J.T. Chrystal, J. Chem. Soc., Perkin I, 2505 (1987).

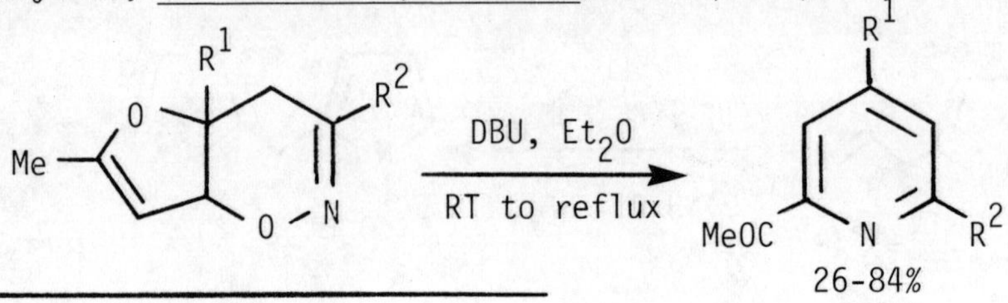

IV.G-3 H. Takahata, E. Wang, T. Nakajima and T. Yamazaki, Chem. Pharm. Bull., 35, 3139 (1987).

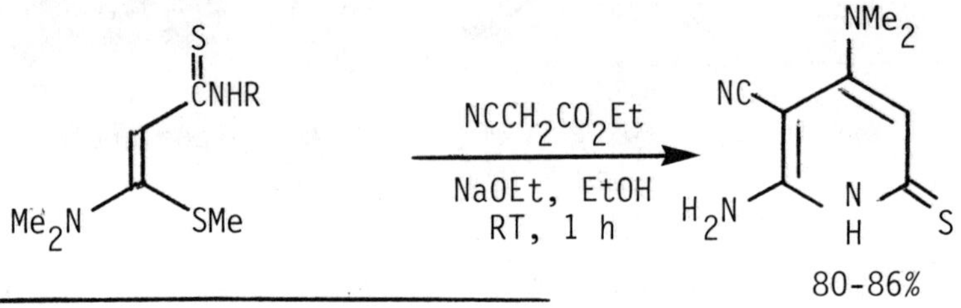

IV.G 4 A.G.M. Barrett and S.A. Lebold, Tetrahedron Lett., 28, 5791 (1987).

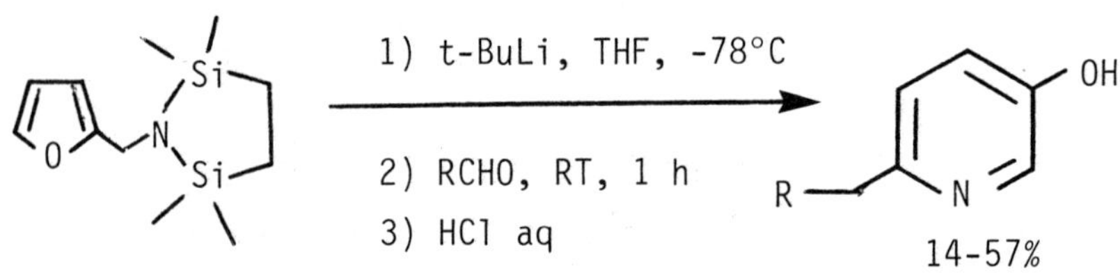

IV.G-5 W. Schroth, A. Hildebrandt and R. Spitzner, Z. Chem., 27, 256 (1987).

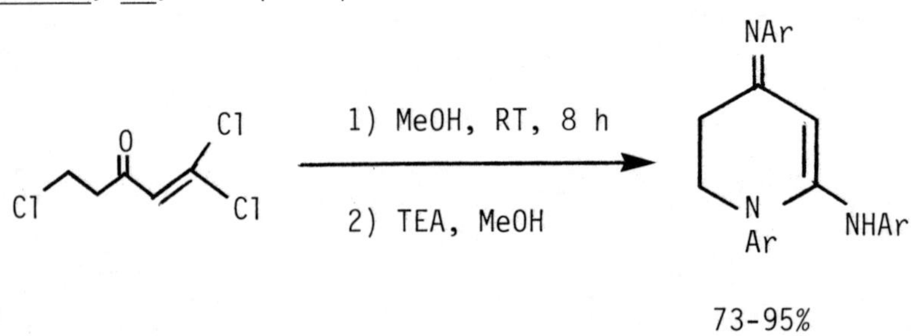

IV.G-6 K. Clinch, A. Vasella and R. Schauer, Tetrahedron Lett., 28, 6425 (1987).

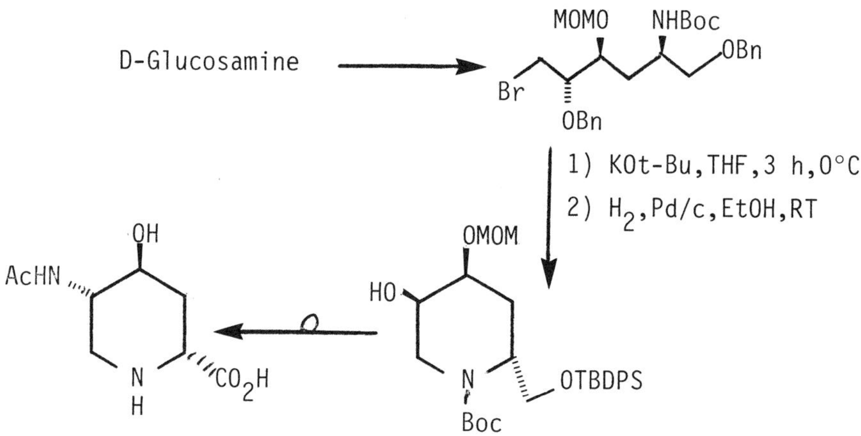

IV.G-7 M.D. Chordia, S.N. Karmarkar, S.L. Kelkar and M.S. Wadia, Synthesis, 810 (1987).

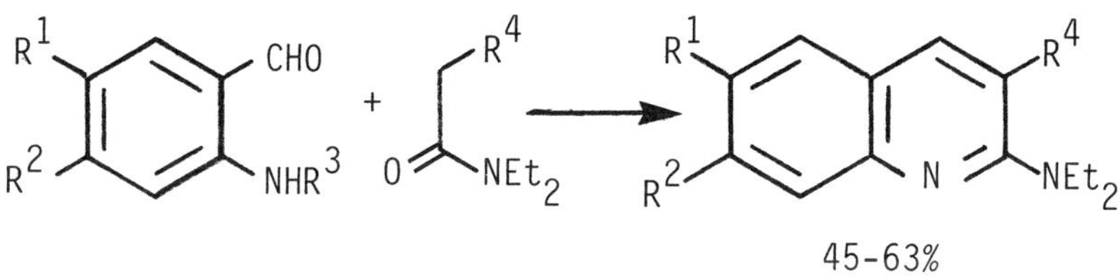

IV.G-8 T. Aubert, M. Farnier, B. Hanquet and R. Guilard, Synth. Commun., 17, 1831 (1987).

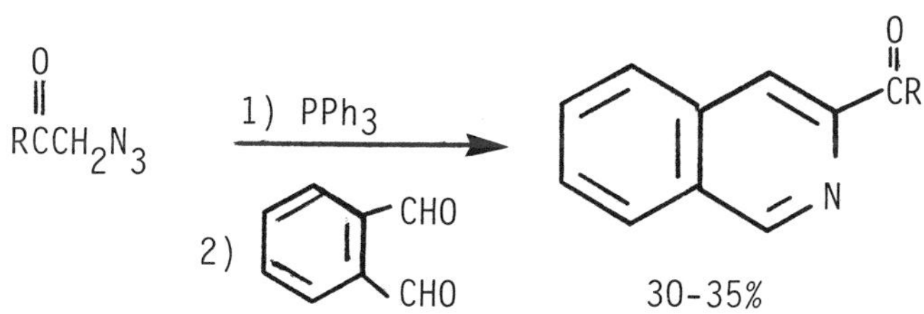

IV.H. Pyrans, Pyrones, Etc.

IV.H-1 M. Mitani, H. Takeuchi and K. Koyama, Chem. Lett., 333 (1987); R.R. Schmidt, W. Frick, B. Haag-Zeino and S. Apparao, Tetrahedron Lett., 28, 4045 (1987).

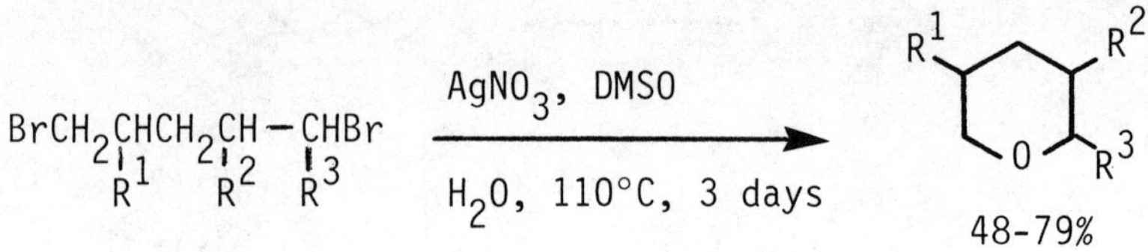

IV.H-2 M.L. Mihailović, S. Konstantinović and R. Vukićević, Tetrahedron Lett., 28, 4343 (1987); L. Coppi, A.Ricci and M. Taddei, Tetrahedron Lett., 28, 973 (1987); W. Kinzy and R.R. Schmidt, Tetrahedron Lett., 28, 1981 (1987).

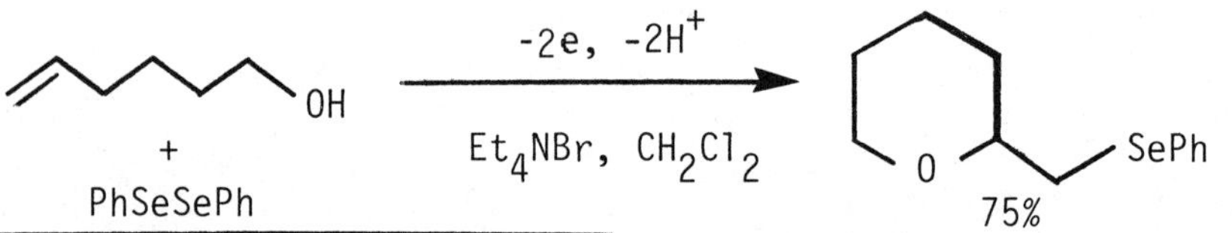

IV.H-3 Y. Huang and X. Lu, Tetrahedron Lett., 28, 6219 (1987); C. Iwata, et al., Tetrahedron Lett., 28, 3131 (1987).

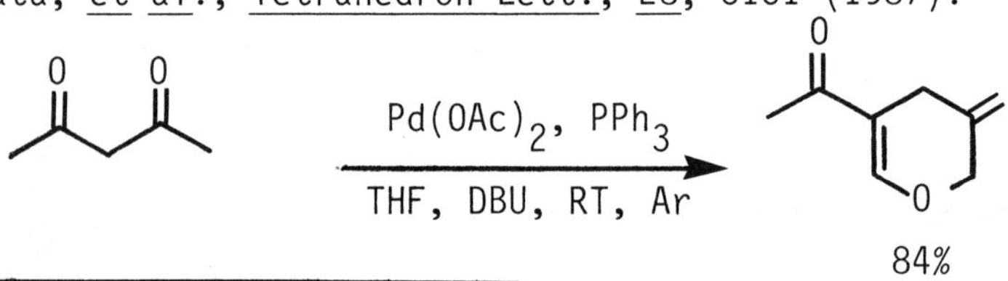

IV.H-4 J.R.I. Eubanks and J.G. Pacifici, Synth. Commun., 17, 829 (1987); F. Cottet, L. Cottier and G. Descotes, Synthesis, 497 (1987).

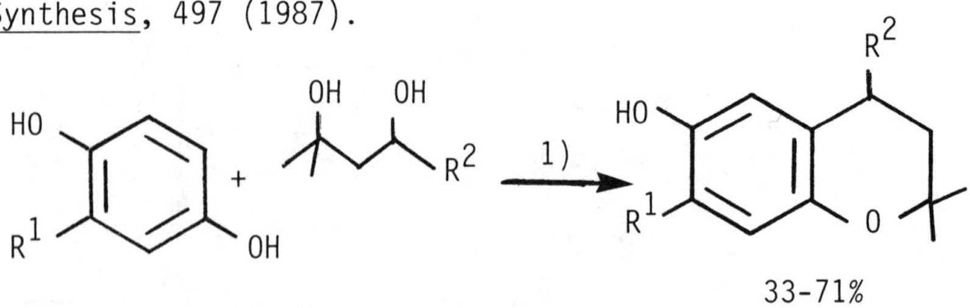

1) Amberlyst 15⁻

IV.H-5 J. Cervelló, J. Marquet and M. Moreno-Mañas, J. Chem. Soc., Chem. Commun., 644 (1987).

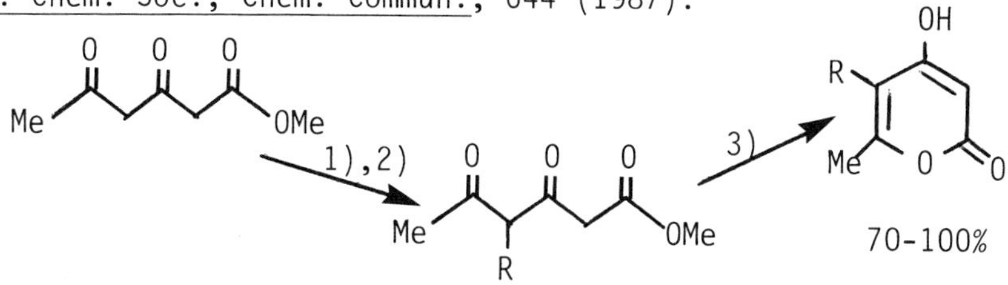

1) NaOH aq, $Co(OAc)_2$, MeOH, H_2O, RT; 2) RBr, $CHCl_3$, reflux, 24 h; 3) DBU, PhH, reflux.

IV.H-6 S.H. Cho and L.S. Liebeskind, J. Org. Chem., 52, 2631 (1987); T. Tsuda, R. Sumiya and T. Saegusa, Synth. Commun., 17, 147 (1987).

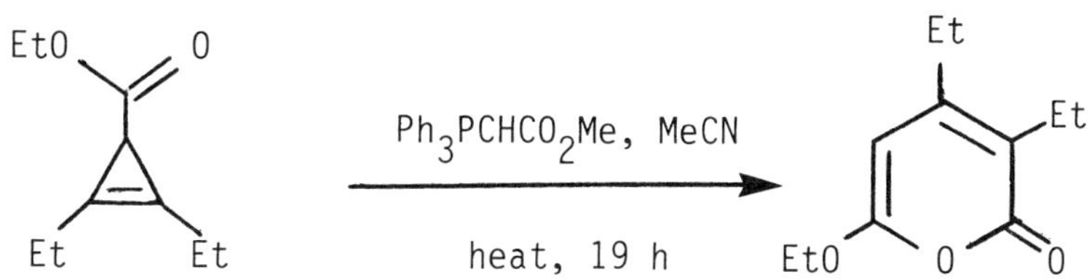

78%

IV.H-7 G. Himbert and L. Henn, Liebigs Ann. Chem., 381 (1987).

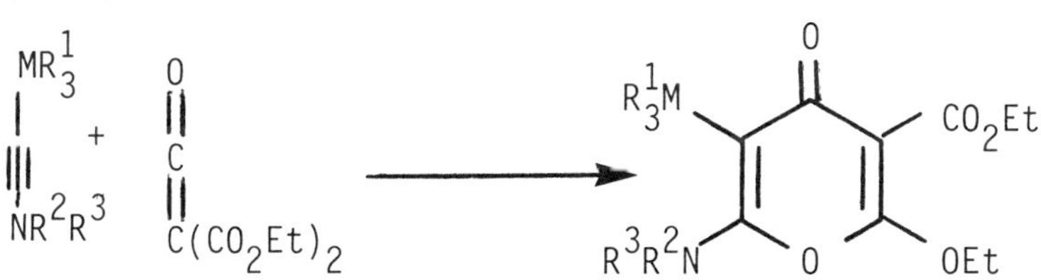

44-78%

IV.H-8 K. Sato _et al_., _J. Chem. Soc., Perkin I_, 1753 (1987); T. Ziegler, H. Möhler and F. Effenberger, _Chem. Ber._, _120_, 373 (1987); N. Oda, Y. Yoshida, S. Nagai, T. Ueda and J. Sakakibara, _Chem. Pharm. Bull._, _35_, 1796 (1987).

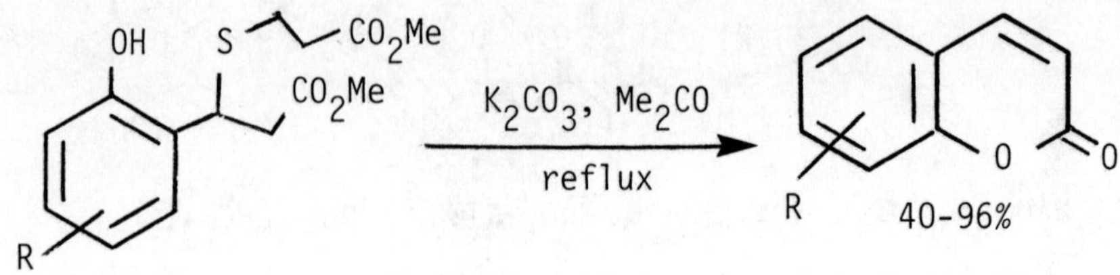

IV.H-9 E. Napolitano and A. Ramacciotti, _Gazz. Chim. Ital._, _117_, 443 (1987).

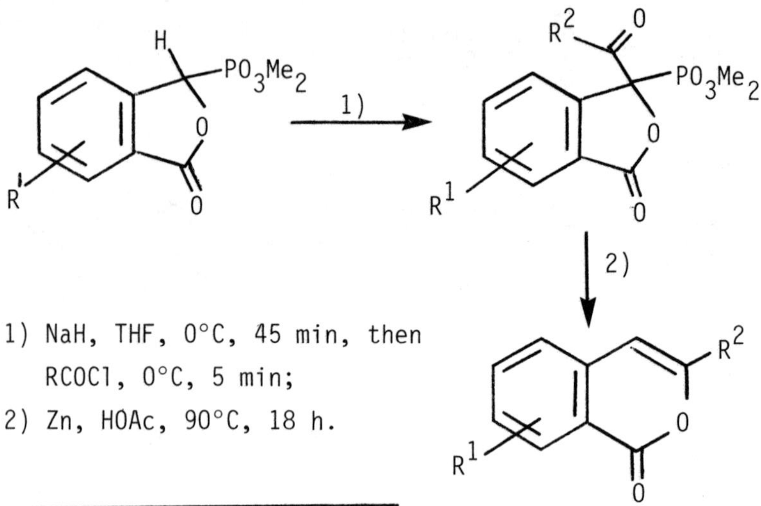

1) NaH, THF, 0°C, 45 min, then RCOCl, 0°C, 5 min;
2) Zn, HOAc, 90°C, 18 h.

IV.H-10 G. Shaw, _J. Chem. Soc., Chem. Commun._, 1735 (1987); S.M. Ali and M. Ilyas, _J. Chem. Res. Synop._, 52 (1987).

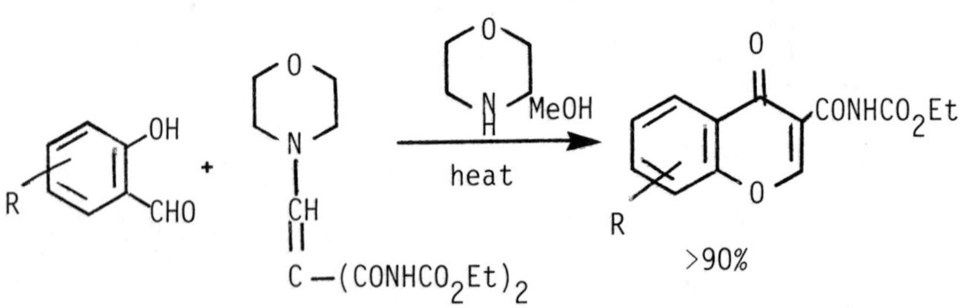

IV.H-11 P.J. Cremins, S.T. Saengchantra and T.W. Wallace, Tetrahedron, 43, 3075 (1987).

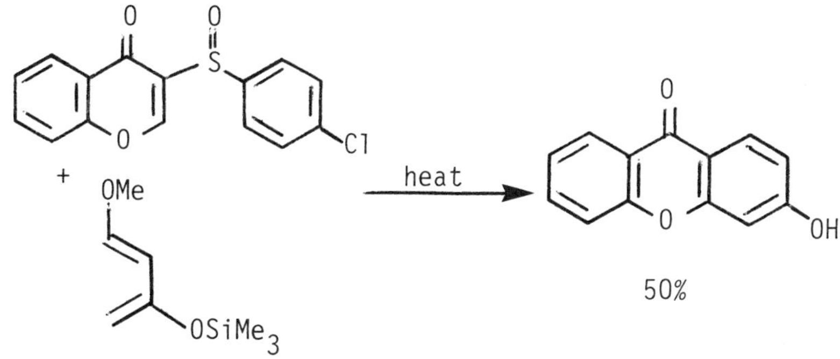

IV.H-12 D. Greif, M. Pulst and M. Weißenfels, Synthesis, 456 (1987); W.-D. Rudorf and R. Schwarz, Tetrahedron Lett., 28, 4267 (1987).

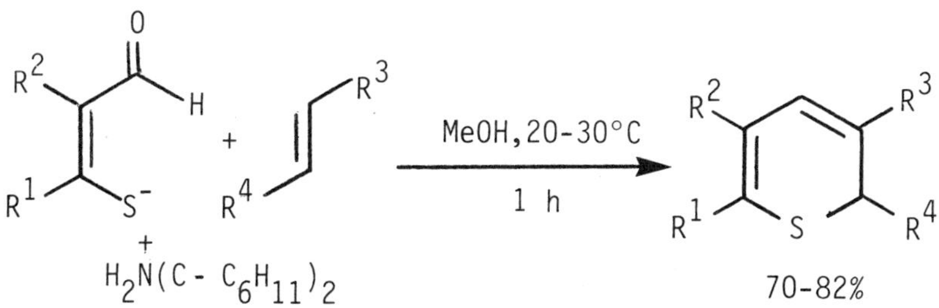

IV.H-13 J. Cossy, F. Henin and C. Leblanc, Tetrahedron Lett., 28, 1417 (1987); H. Ishibashi, H. Nakatani, H. Sakashita and M. Ikeda, J. Chem. Soc., Chem. Commun., 338 (1987); C.K. Lau, P.C. Bélanger, C. Dufresne and J. Scheigetz, J. Org. Chem., 52, 1670 (1987).

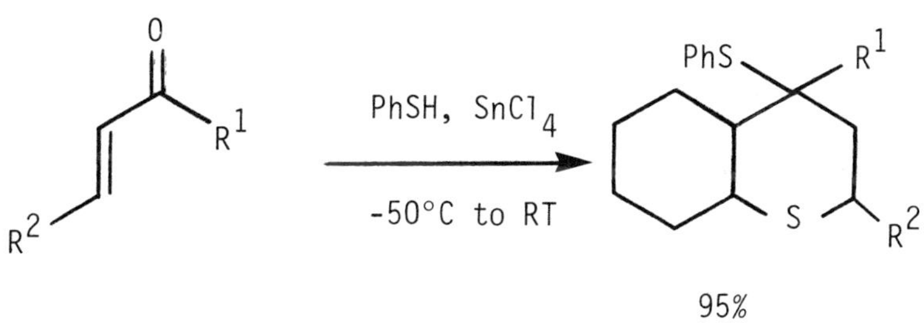

IV.H-14 G.A. Molander and D.C. Shubert, J. Am. Chem. Soc., 109, 6877 (1987).

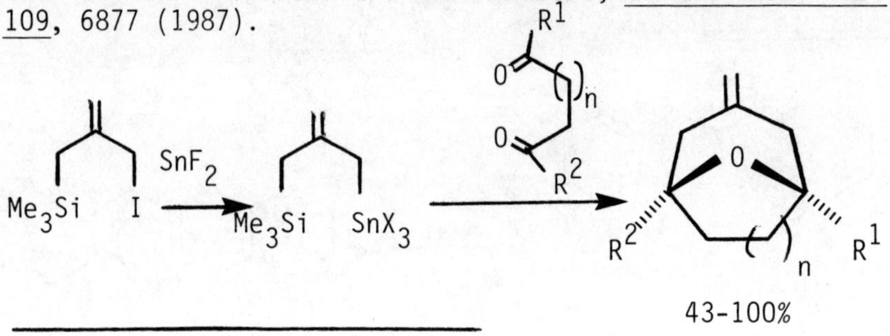

IV.H-15 K.C. Nicolaou, C. Hwang, M.E. Duggan and K. Balreddy, Tetrahedron Lett., 28, 1501 (1987).

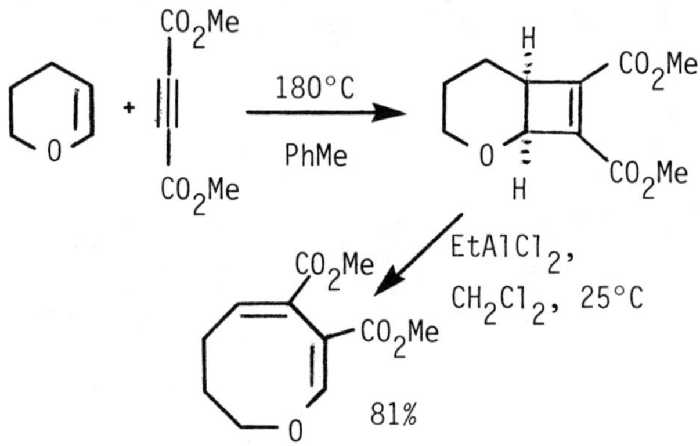

IV.I. Other Heterocycles with One Heteroatom

IV.I-1 O. Wagner, G. Maas and M. Regitz, Angew. Chem. Int. Ed. Engl., 26, 1257 (1987).

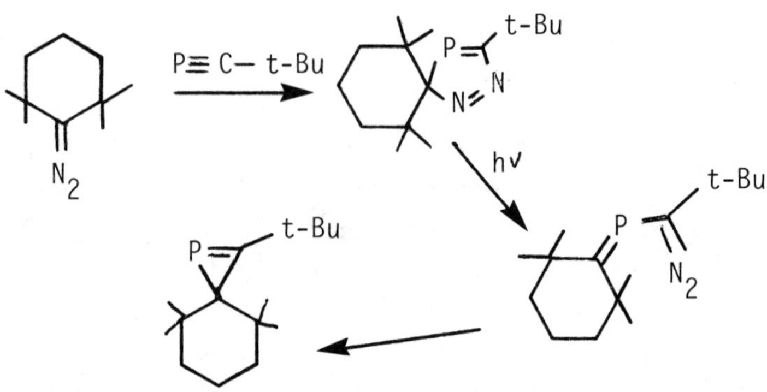

IV.J. Heterocycles with a Bridgehead Heteroatom

IV.J-1 R.A. Nugent and M. Murphy, J. Org. Chem., 52, 2206 (1987).

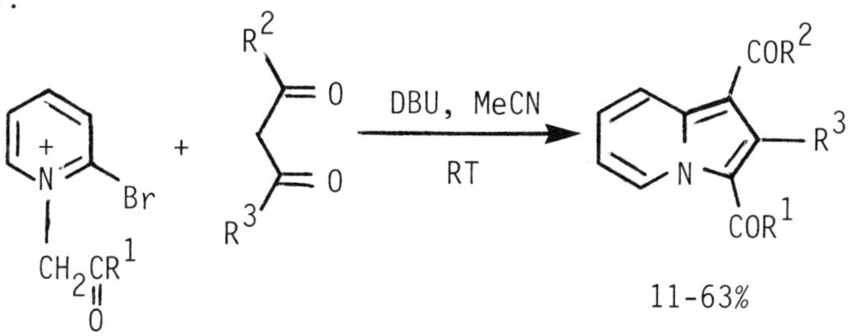

IV.J-2 K. Maruyama, N. Nagai and Y.Naruta, Chem. Lett., 97 (1987).

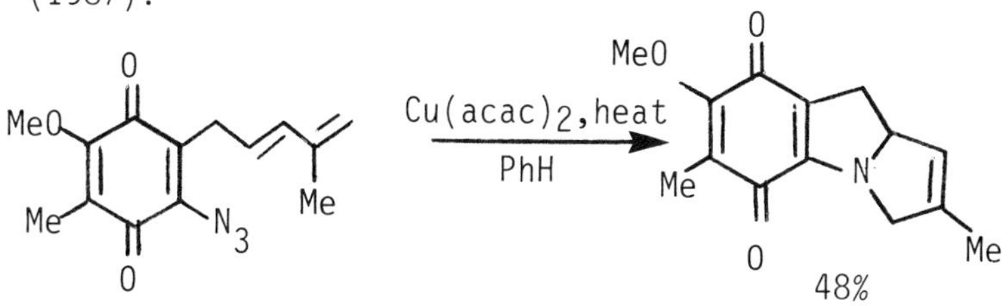

IV.J-3 R.C. Phadke and D.W. Rangnekar, Synthesis, 484 (1987).

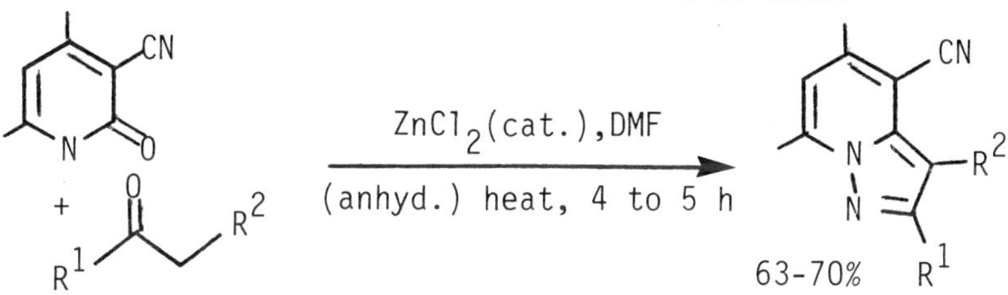

IV.J-4 A.M. Likhosherstov, V,P. Peresada, V.G. Vinokurov and A.P. Skoldinov, J. Org. Chem.(USSR), 22, 2341 (1986).

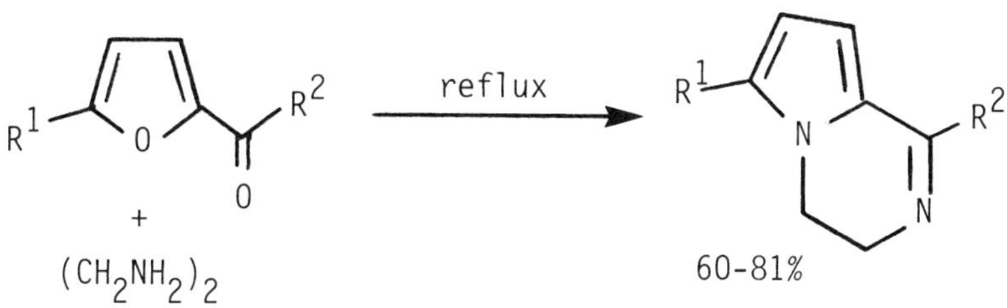

IV.J-5 J.E. Baldwin *et al.*, J. Chem. Soc., Chem. Commun., 374 (1987).

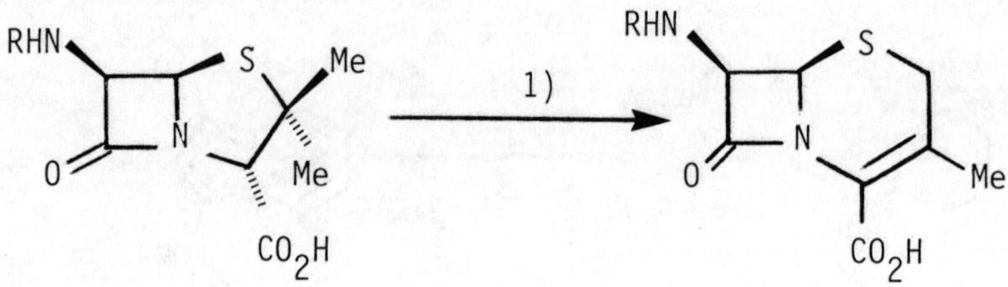

R= e.g. 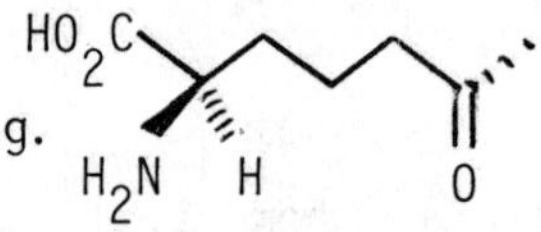

1) *C. acremonium* CO 728

IV.J-6 A. Kamal, N.A.V. Reddy and P.B. Sattur, Heterocycles, 26, 751 (1987).

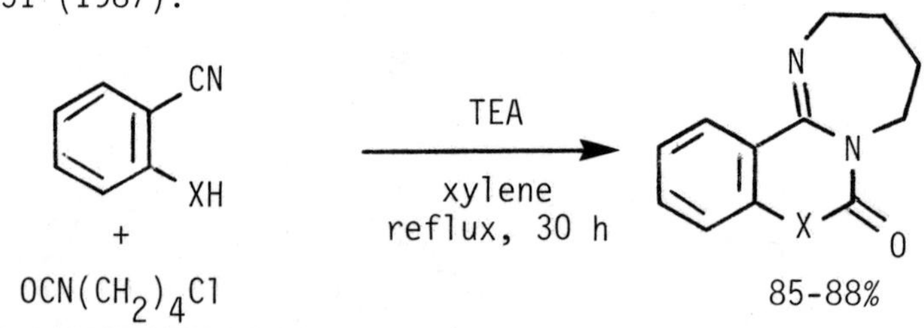

85-88%

IV.K. Heterocycles with Two or More Heteroatoms

1.a. 5-Membered Heterocycles with 2 N's

IV.K.1.a-1 A.W. Taylor and R.T. Cook, Tetrahedron, 43, 607 (1987); J.J. Vaquero, L. Fuentes, J.C. DelCastillo, M.L. Pérez, J.L. Garcia and J.L. Soto, Synthesis, 33 (1987); U. Klein, K. Mohrs, H. Wild and W. Steglich, Liebigs Ann. Chem., 485 (1987).

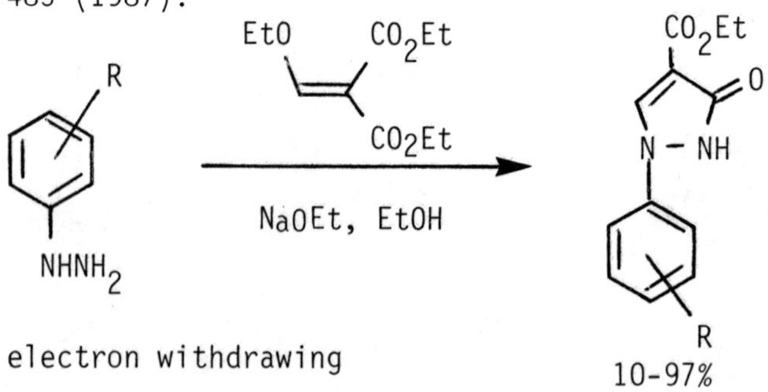

R = electron withdrawing

10-97%

IV.K.1.a-2 D. Pooranchand, H. Ila and H. Junjappa, <u>Synthesis</u>, 547 (1987).

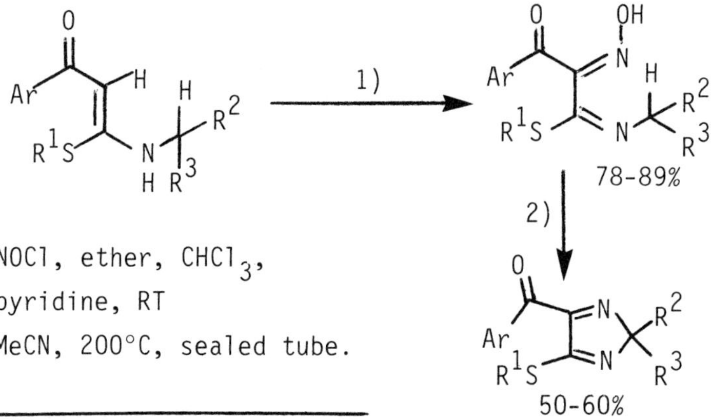

1) NOCl, ether, $CHCl_3$, pyridine, RT
2) MeCN, 200°C, sealed tube.

IV.K.1.a-3 O. Wong, N. Tsuzuki, M. Richardson, H. Rytting, R. Konishi and T. Higuchi, <u>Heterocycles</u>, <u>26</u>, 3153 (1987).

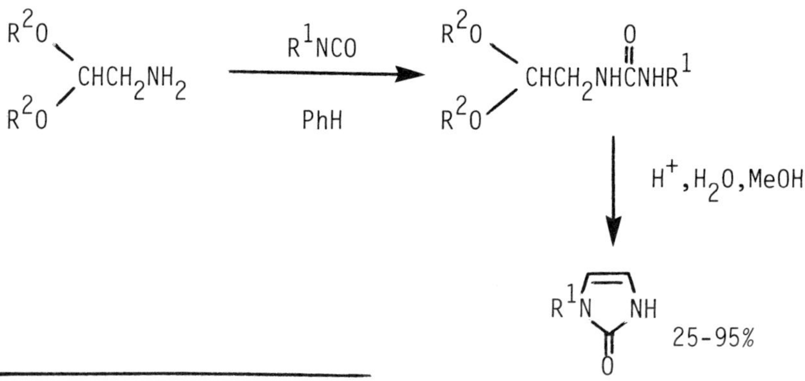

IV.K.1.a-4 J.B. Hendrickson and M.S. Hussoin, <u>J. Org. Chem.</u>, <u>52</u>, 4137 (1987); L.Y. Sheinberg, S.A. Konoratov, V.D. Boiko and S.M. Shein, <u>J. Org. Chem.(USSR)</u>, <u>22</u>, 2215 (1986); T. Yoshida, N. Kambe, S. Murai and N. Sonoda, <u>Bull. Chem. Soc. Jpn.</u>, <u>60</u>, 1793 (1987).

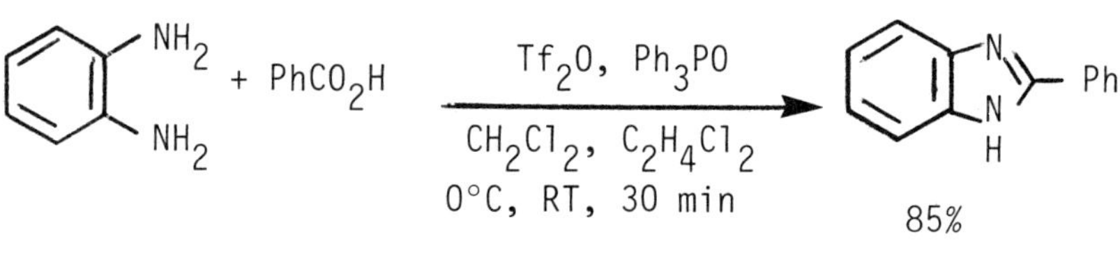

IV.K.1.b. 6-Membered with 2 N's

IV.K.1.b-1 J. Barluenga, M. Tomás, A. Ballesteros and V. Gotor, Synthesis, 489 (1987); H.W. Schmidt, G. Koitz and H. Juněk, J. Heterocycl. Chem., 24, 1305 (1987); A. Alberola, C. Andrés, A. González Ortega, R. Pedrose and M. Vincente, Synth. Commun., 17, 1309 (1987).

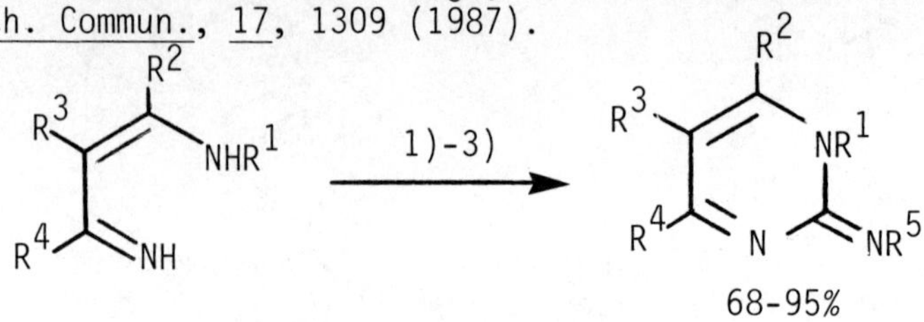

1) Cl_2SiPh_2, RT, PhH, TEA; 2) R^5NCO; 3) $2NH_2SO_4$

IV.K.1.b-2 A.G. Martinez et al., Tetrahedron Lett., 28, 1929 (1987); K.S. Atwal, B.C. O'Reilly, J.Z. Gougontas and M.F. Malley, Heterocycles, 26, 1189 (1987); E.A.A. Hafez, M.R.H. Elmoghayer and M.M.M. Ramiz, Liebigs Ann. Chem., 65 (1987).

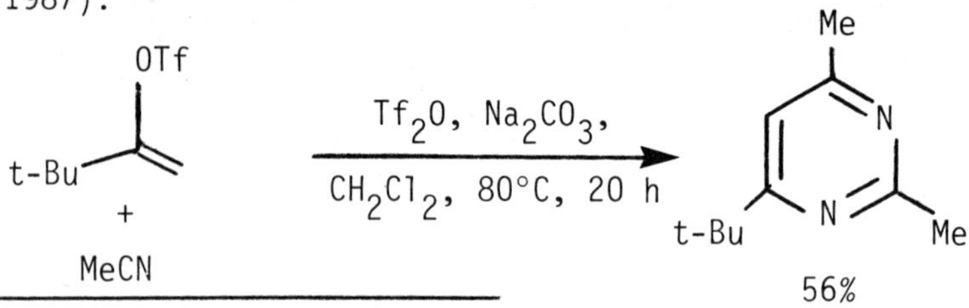

IV.K.1.b-3 P.J. Garrat et al., J. Chem. Soc., Chem. Commun., 568 (1987).

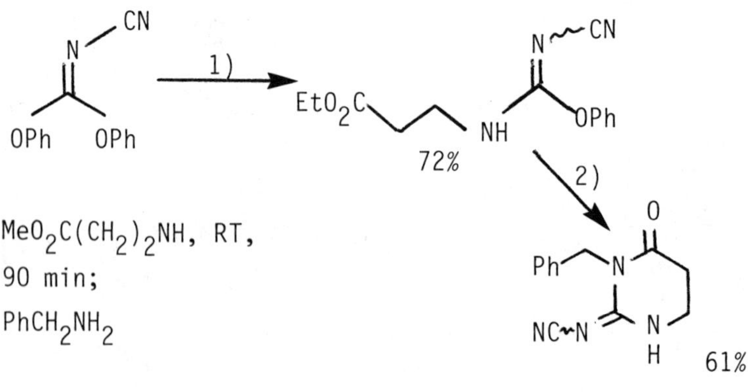

1) $MeO_2C(CH_2)_2NH$, RT, 90 min;

2) $PhCH_2NH_2$

IV.K.1.b-4 J.T. Gupton, J.F. Miller, R.D. Bryant, P.R. Maloney and B.S. Foster, Tetrahedron, 43, 1747 (1987); K.H.M. Hilmy, J. Mogensen and E.B. Pedersen, Acta Chem. Scand., 41B, 467 (1987).

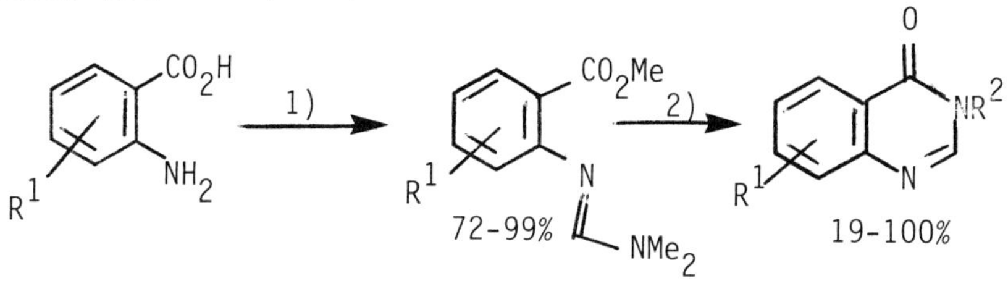

1) $HC(OMe)_2NMe_2$, reflux; 2) R^2NH_2, dioxane reflux.

IV.K.1.b-5 S. Murata, T. Sugimoto and S. Matsuura, Heterocycles, 26, 883 (1987).

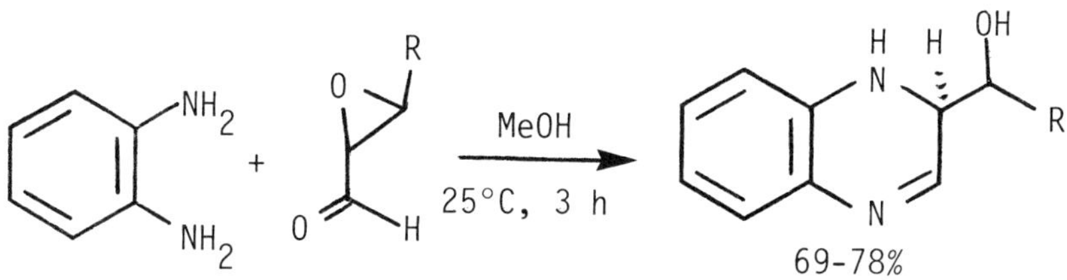

IV.K.1.b-6 G.W. Rewcastle and W.A. Denny, Synth. Commun., 17, 1171 (1987).

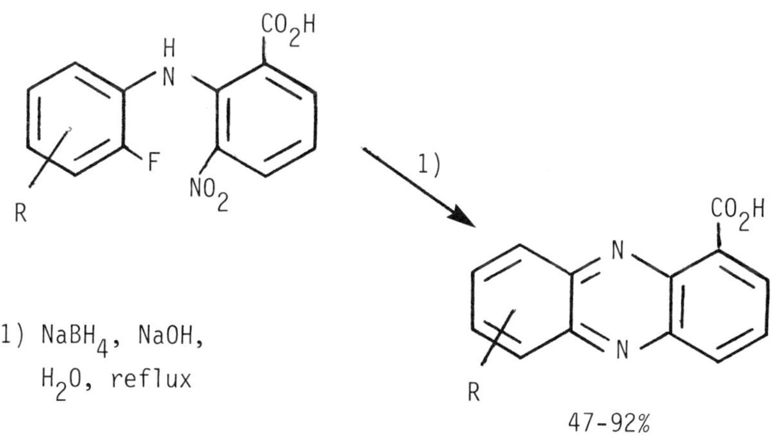

1) $NaBH_4$, NaOH, H_2O, reflux

IV.K.1.c. Other Heterocycles with 2 N's

IV.K.1.c-1 J. Barluenga, M.J. Iglesias and V. Gotor, J. Chem. Soc., Chem. Commun., 582 (1987).

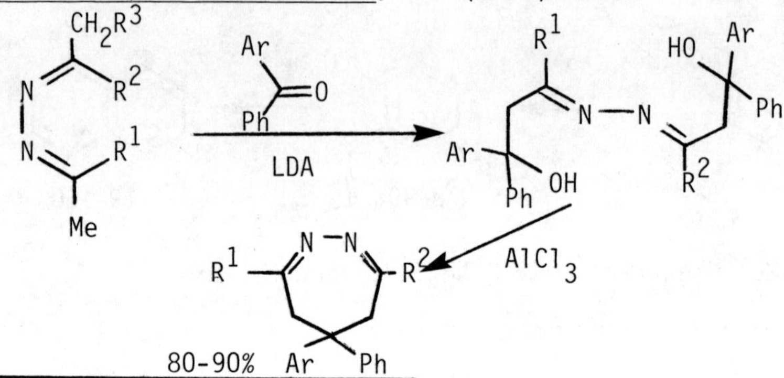

IV.K.1.c-2 K. Saito, Bull. Chem. Soc. Jpn., 60, 2105 (1987).

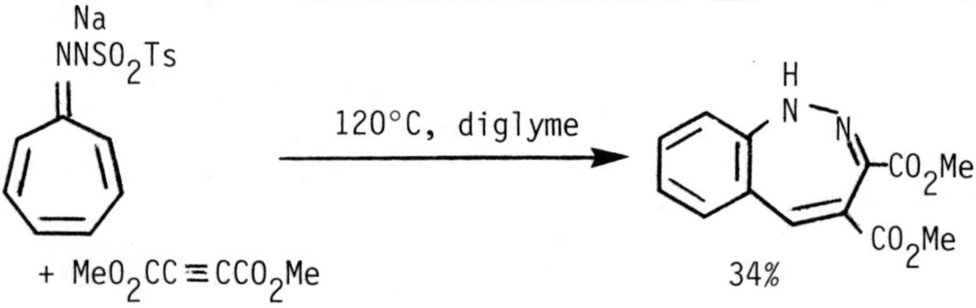

IV.K.1.c-3 H. Matsuyama, M. Kobayashi and H.H. Wasserman, Heterocycles, 26, 85 (1987).

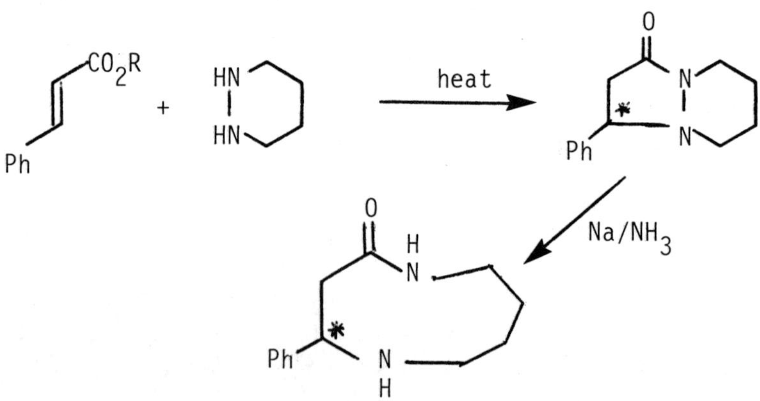

IV.K.1.c-4 P.J. Harrington, L.S. Hegedus and K.F. McDaniel, J. Am. Chem. Soc., 109, 4335 (1987).

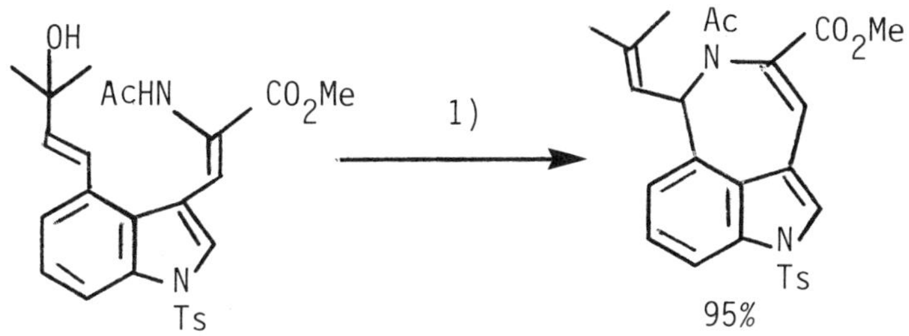

1) 15% $PdCl_2(MeCN)_2$, MeCN, Ar, heat, 2 h.

IV.K.2. Heterocycles with 2 O's or 2 S's

IV.K.2-1 L. Meier and H.-D. Scharf, Synthesis, 517 (1987).

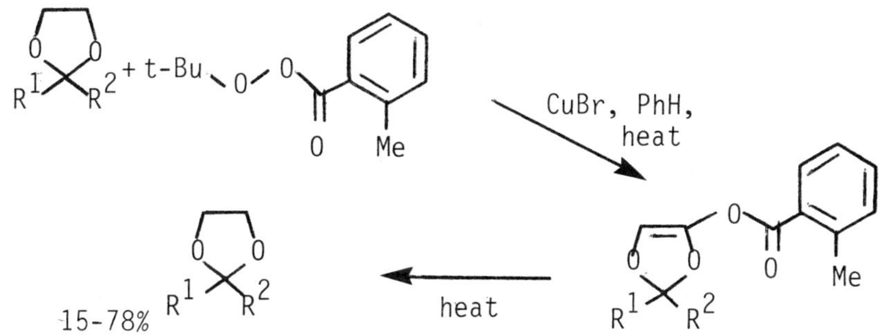

IV.K.2-2 Y. Inoue, J. Ishikawa, M. Taniguchi and H. Hashimoto, Bull. Chem. Soc. Jpn., 60, 1204 (1987); A. Baba, T. Nozaki and H. Matsuda, Bull. Chem. Soc. Jpn., 60, 1552 (1987).

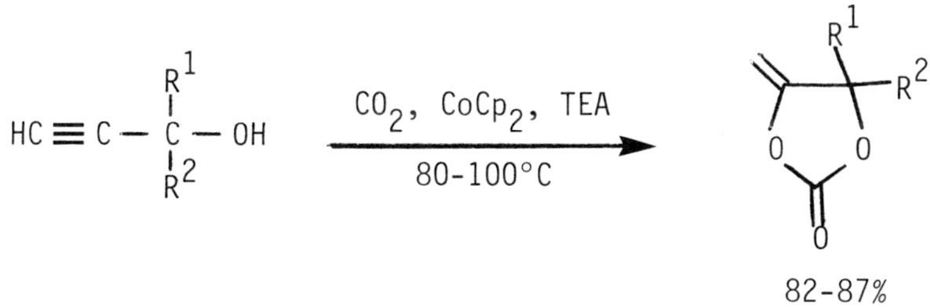

IV.K.2-3 M. Babero, S. Cadamuro, M. Ceruti, I. Degani, R. Fochi and V. Regondi, Gazz. Chim. Ital., 117, 227 (1987).

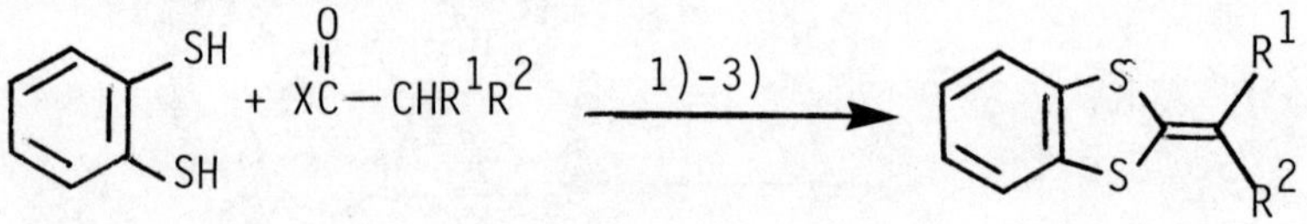

1) HBF_4, Et_2O, -105°C, then RT; 2) MeCN, 3) H_2O

IV.K.3. Heterocycles with 1 N and 1 O

IV.K.3-1 F. Toda and K. Tanaka, Chem. Lett., 2283 (1987)

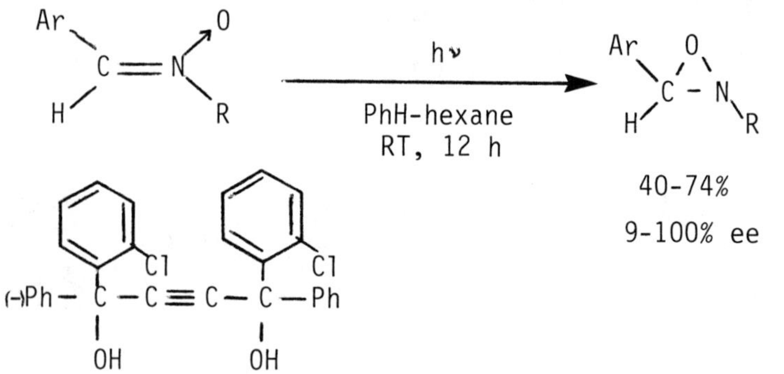

IV.K.3-2 D.P. Curran et al., J. Org. Chem., 52, 2137 (1987).

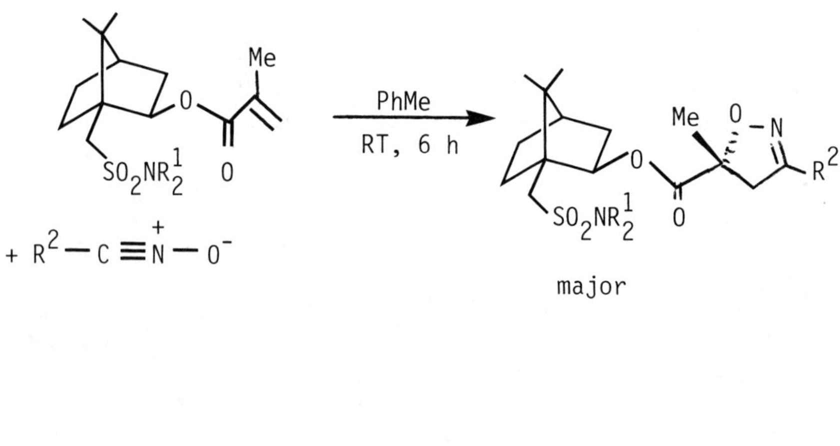

IV.K.3-3 A. Brandi, S. Carli, A. Guarna and F. DeSarlo, Tetrahedron Lett., 28, 3845 (1987).

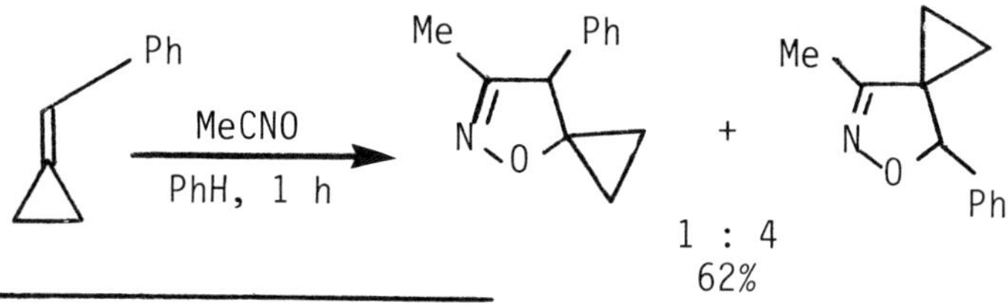

IV.K.3-4 H. Uno, N. Watanabe, S. Fujiki and H. Suzuki, Synthesis, 471 (1987); T. Shimizu, Y. Hayashi, H. Shibafuchi and K. Teramura, Bull. Chem. Soc. Jpn., 60, 1948 (1987); O. Tsuge, S. Kanemasa, H. Suga and N. Nakagawa, Bull. Chem. Soc. Jpn., 60, 2463 (1987); M.S. Akhtar, M. Seth and A. Bhaduri, Ind. J. Chem.(B), 26, 71 (1987).

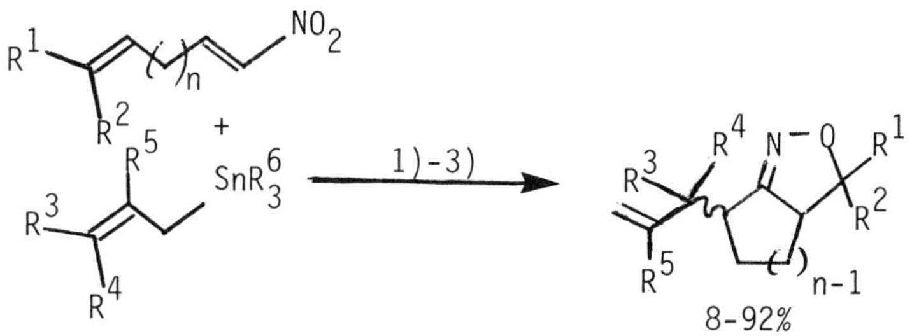

1) $TiCl_4$, CH_2Cl_2, -78°C to RT; 2) TEA, THF, RT; 3) $NaHCO_3$, KF, RT.

IV.K.3-5 R. Cassis, M. Fernandez, R. Tapia and J.A. Valderrama, Synth. Commun., 17, 1077 (1987).

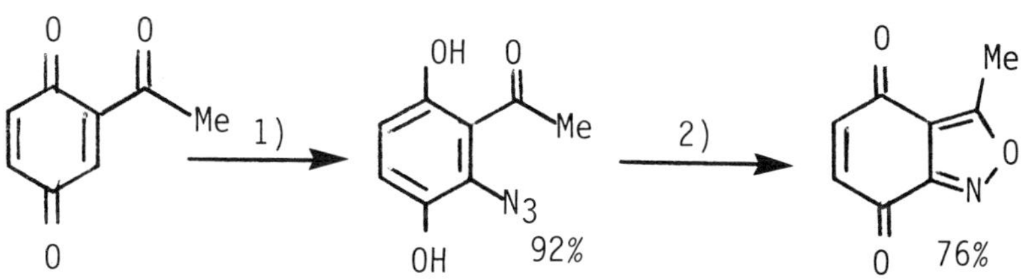

1) NaN_3, c.H_2SO_4, H_2O, CH_2Cl_2, 5-10°C;
2) MnO_2, CH_2Cl_2, RT, 15 min.

IV.K.3-6 B.M. Trost and A.R. Sudhakar, J. Am. Chem. Soc., 109, 3792 (1987); T. Toda and Y. Kitagawa, Angew. Chem. Int. Ed. Engl., 26, 334 (1987).

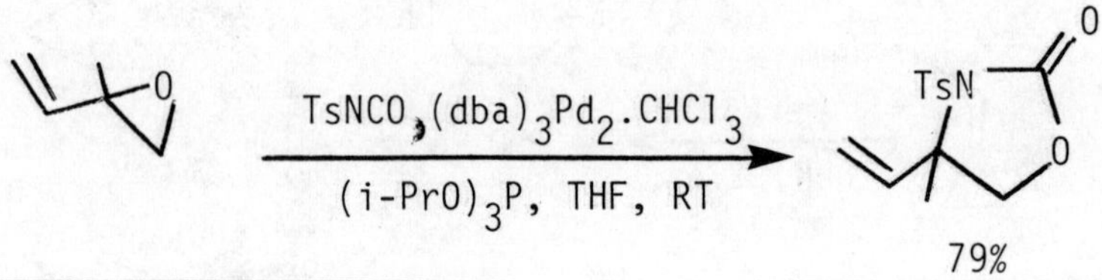

IV.K.3-7 T. Hayashi, A. Yamamoto and Y. Ito, Tetrahedron Lett., 28, 4837 (1987).

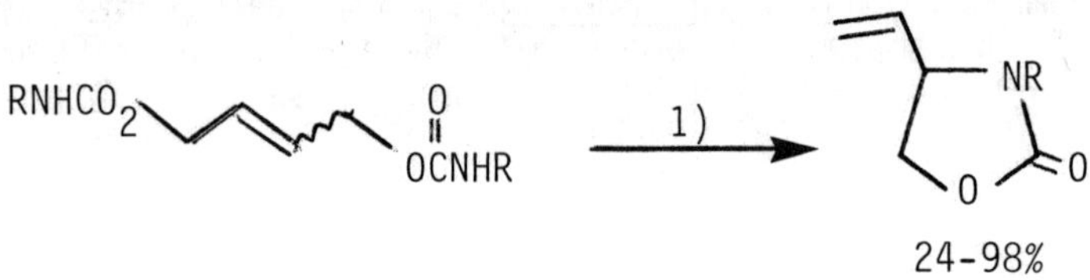

1) $Pd_2(dba)$, dppe, THF, reflux

IV.K.3-8 P. Wipf, R. Prewo, J.H. Bieri, H. Heimgartner, V. Nastopoulos and G. Germain, Helv. Chim. Acta, 70, 1380 (1987).

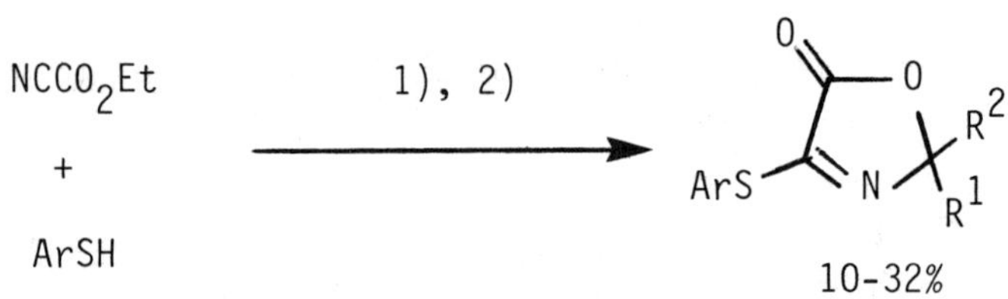

1) $TiCl_4$, TEA, Et_2O, 0°C, RT; 2) $R^1R^2C{=}O$, $Et_2O.BF_3$, RT

IV.K.3-9 C. Sund, J. Ylikoski and M. Kwiatkowski, Synthesis, 853 (1987).

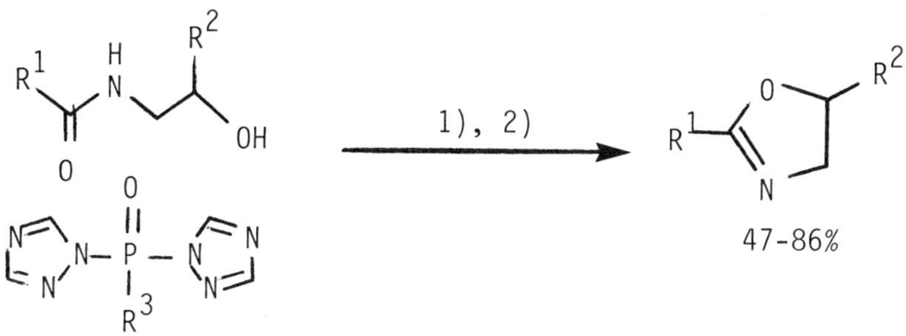

1) MeCN, pyridine, RT; 2) $NaHCO_3$ aq

IV.K.3-10 S.E. Denmark, C.J. Cramer and M.S. Dappen, J. Org. Chem., 52, 877 (1987).

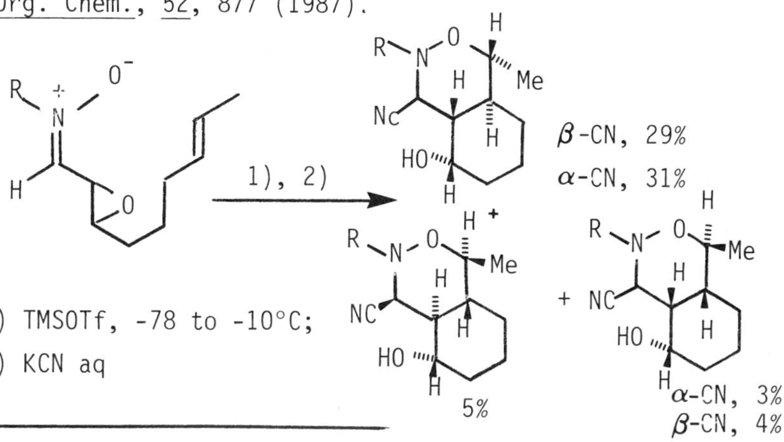

1) TMSOTf, -78 to -10°C;
2) KCN aq

IV.K.3-11 M. Sakaitani and Y. Ohfune, Tetrahedron Lett., 28, 3987 (1987); I. Shibata, T. Imoto, A. Baba and H. Matsuda, J. Heterocycl. Chem., 40, 361 (1987).

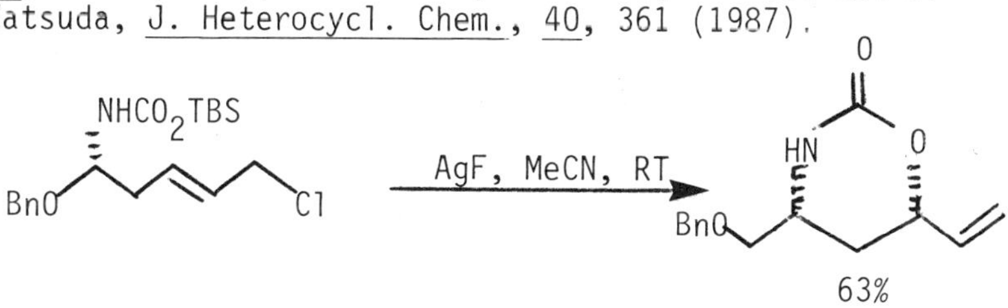

IV.K.3-12 W. Schroth and R. Spitzner, Monatsh. Chem., 118, 1263 (1987).

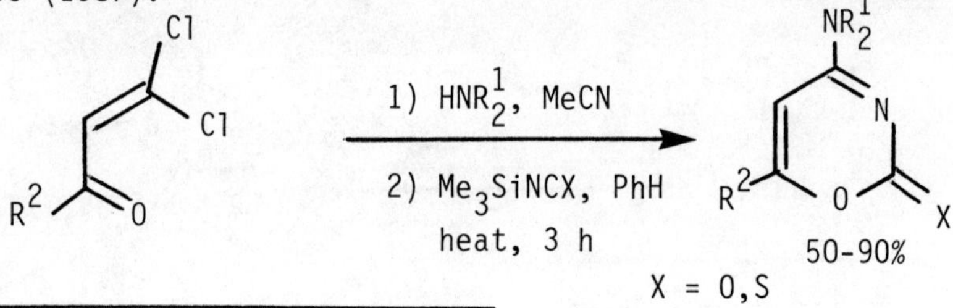

IV.K.3-13 A. Marchesini et al., J. Org. Chem., 52, 1666 (1987).

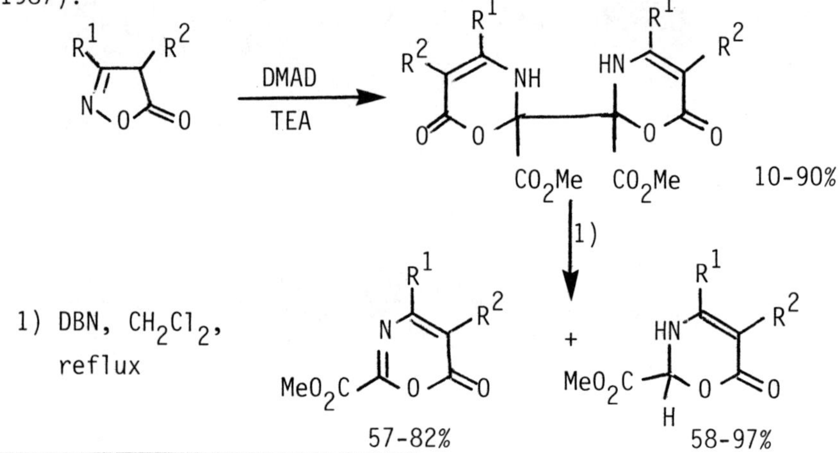

IV.K.4. Heterocycles with 1 N and 1 S

IV.K.4-1 M.C. McKie and R.M. Paton, J. Chem. Res.(S), 245 (1987); T. Ueda, Y. Shibata, Y. Kato and J. Sakakibara, Tetrahedron, 43, 3917 (1987).

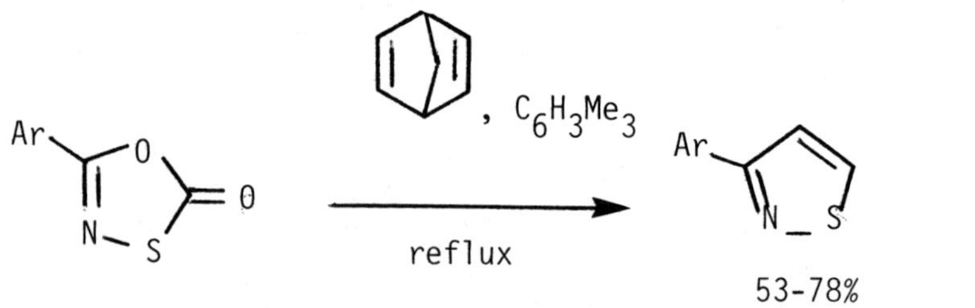

IV.K.4-2 D. Barbry and D. Couturier, Chem. Ber., 120, 1073 (1987).

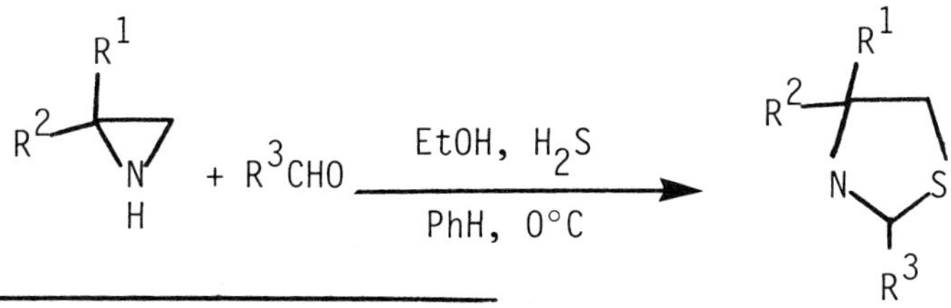

40-74%

IV.K.4-3 E.H.M. Abdelall, M.I. Alashmawy and J.M. Mellor, J. Chem. Soc., Perkin I, 2729 (1987); M. d.'Ischia, G. Prota, R.C. Rotteveel and W. Westerhof, Synth. Commun., 17, 1577 (1987).

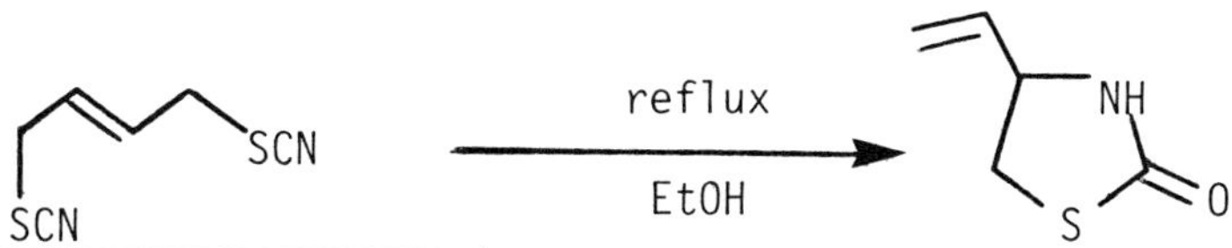

IV.K.4-4 W. Dolling, P. Kindt and M. Augustin, Z. Chem., 27, 212 (1987).

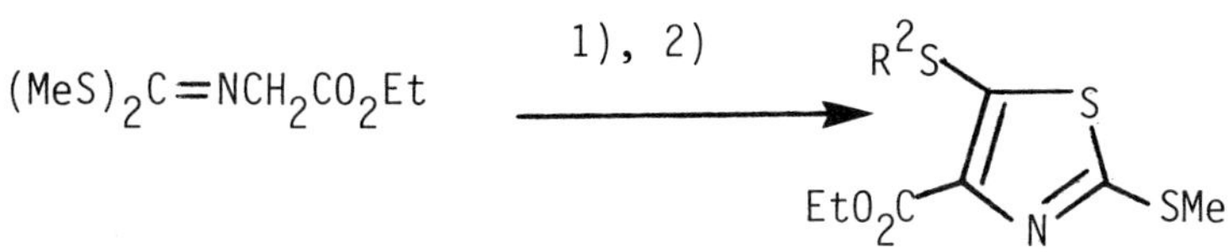

44-45%

1) CS_2; 2) R^2X, RT, 2 h

IV.K.4-5 P. Wipf, C. Jenny and H. Heimgartner, Helv. Chim. Acta, 70, 1001 (1987).

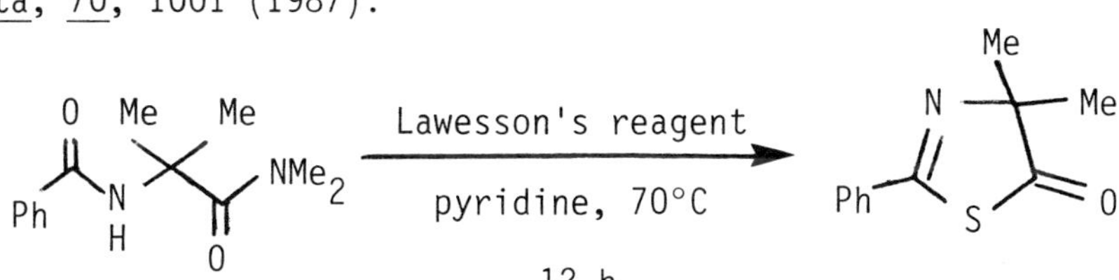

63%

IV.K.4-6 K. Takagi, T. Iwachido and N. Hayama, Chem. Lett., 839 (1987).

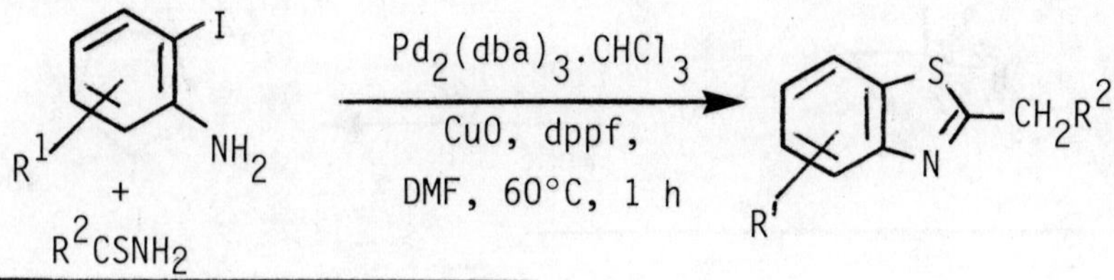

IV.K.4-7 M. Chehna, J.P. Pradere and A. Guingant, Synth. Commun., 17, 1971 (1987).

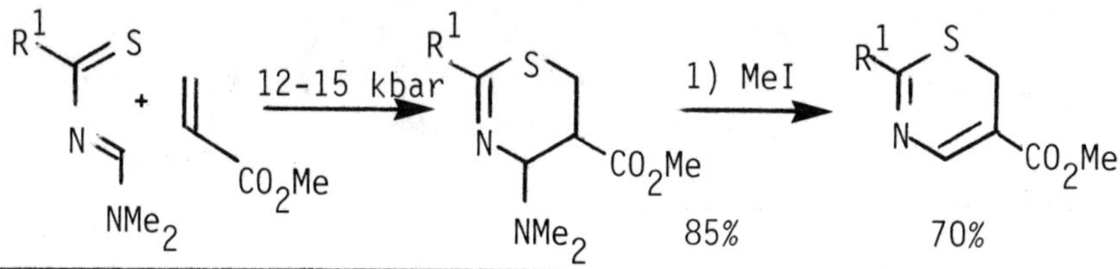

IV.K.4-8 Y. Kita, O. Tamura, T. Miki and Y. Tamura, Tetrahedron Lett., 28, 6479 (1987).

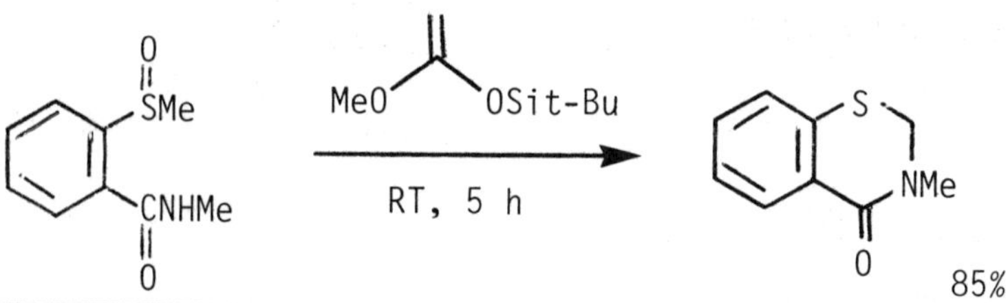

IV.K.4-9 V. Ambrogi and G. Grandolini, Synthesis, 724 (1987).

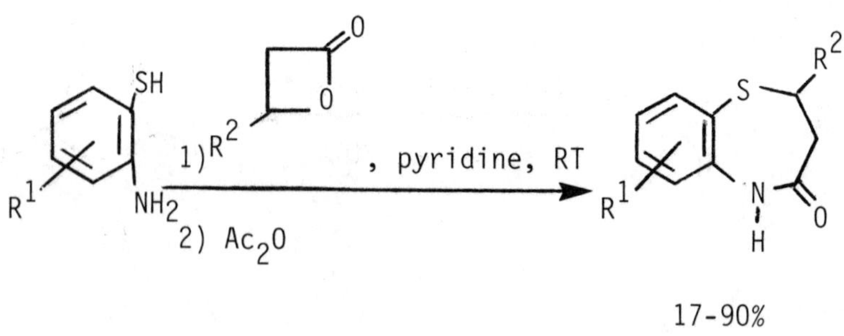

IV.K.5. Heterocycles with 1 O and 1 S

IV.K.5-1 W.S. Lee, O.S. Park, J.W. Choi and K.D. Nam, J. Org. Chem., 52, 5374 (1987).

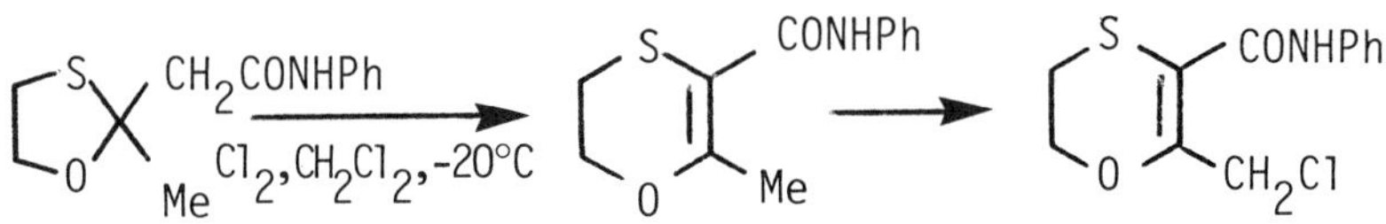

IV.K.5-2 A. Hosomi, S. Hayashi, K. Hoashi, S. Kohra and Y. Tominaga, J. Chem. Soc., Chem. Commun., 1442 (1987).

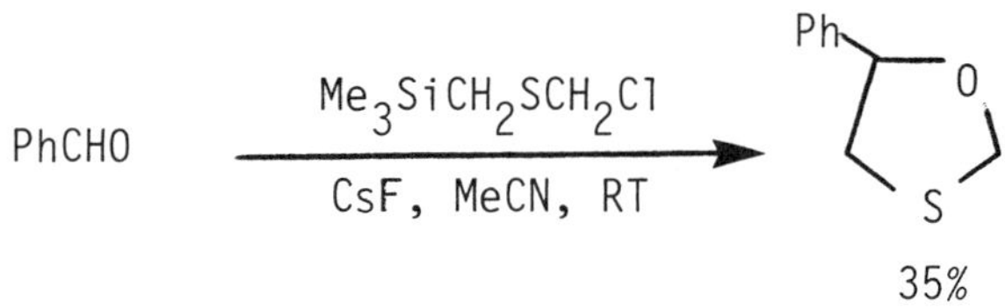

IV.K.6. Heterocycles with 3 or More N's

IV.K.6-1 J.E. Francis, L.A. Gorczyca, G. C. Mazzenga and H. Meckler, Tetrahedron Lett., 28, 5133 (1987).

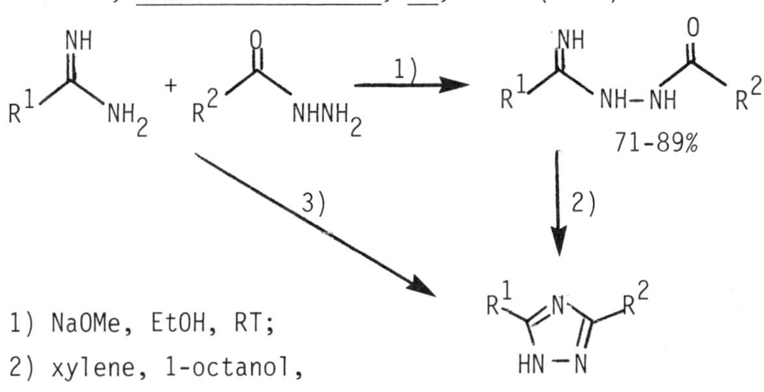

1) NaOMe, EtOH, RT;
2) xylene, 1-octanol, reflux;
3) NaOMe, EtOH, reflux, 48 h.

66-97%

IV.K.6-2 M. Veverka and M. Marchalin, Coll. Czech. Chem. Commun., 52, 113 (1987).

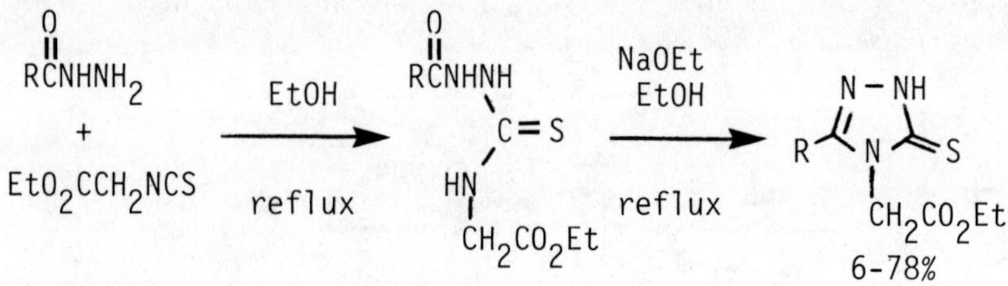

IV.K.6-3 J.G. Schantl, H. Gstach and N. Lanznaster, Synthesis, 986 (1987).

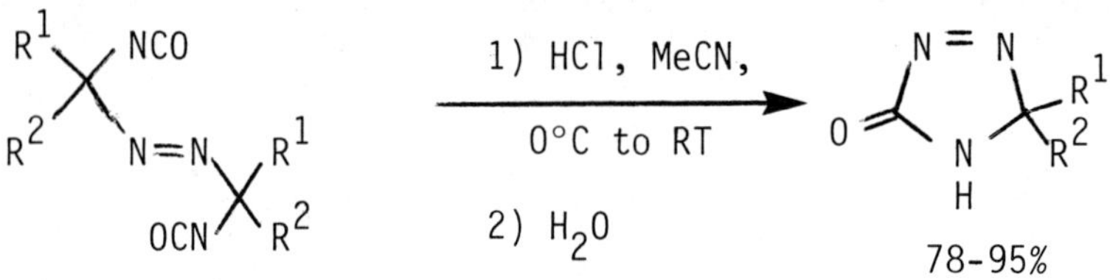

IV.K.6-4 G.R. Weisman, D.J. Vachon, V.B. Johnson and D.A. Gronbeck, J. Chem. Soc., Chem. Commun., 886 (1987).

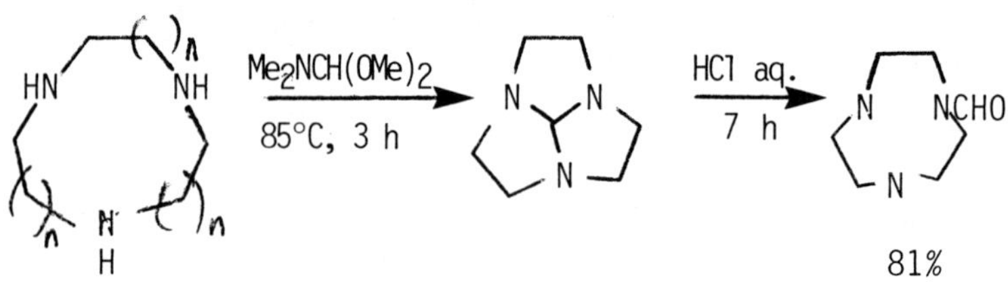

IV.K.6-5 P.R. Bernstein and E. P. Vacek, Synthesis, 1133 (1987).

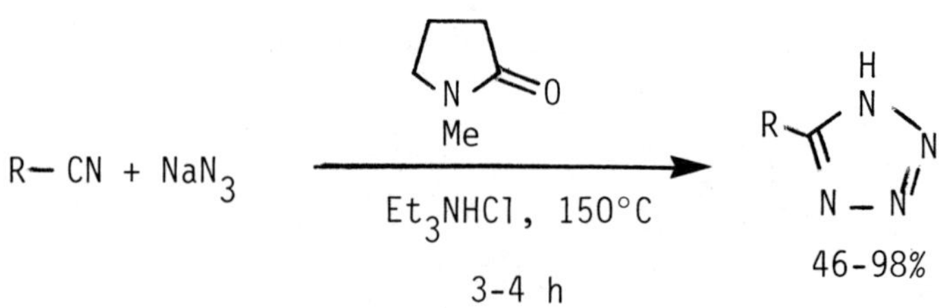

IV.K.7. Heterocycles with 2 N's and 1 O

IV.K.7-1 N.R. Ayyangar, S.M. Kumar and K.V. Srinivasan, Synthesis, 616 (1987).

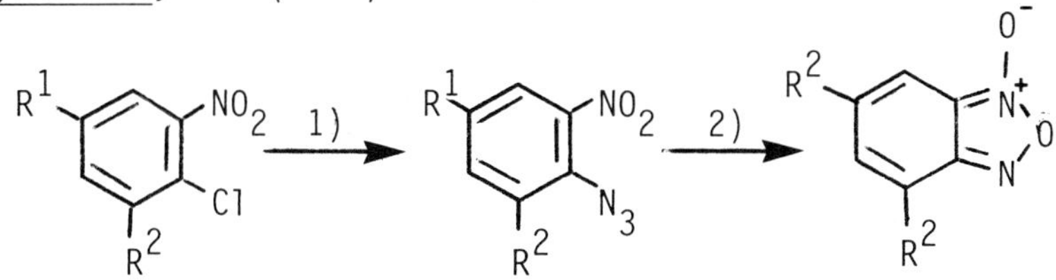

91-97% 42-97%

1) NaN_3, $PhCH_2\overset{+}{N}Bu_3Br^-$, $ClCH_2CH_2Cl$, 30-55°C; 2) 60°C.

IV.L. Other Heterocycles

IV.L.1-1 E.C. Taylor and J.L. Pont, Tetrahedron Lett., 28, 379 (1987).

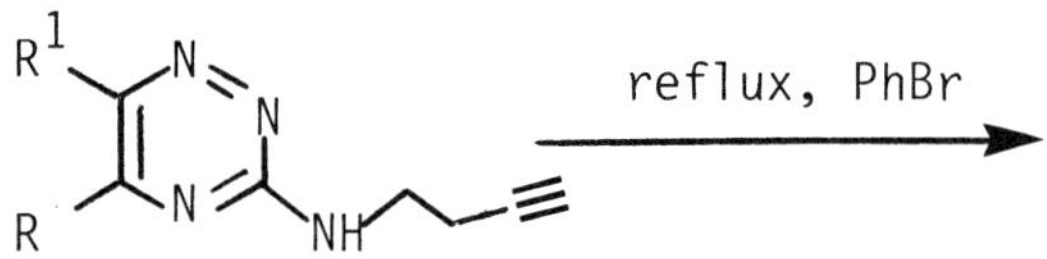

20-73%

IV.L.1-2 K. Eger, J.G. Pfahl, G. Folkers and H.J. Roth, J. Heterocycl. Chem., 40, 425 (1987).

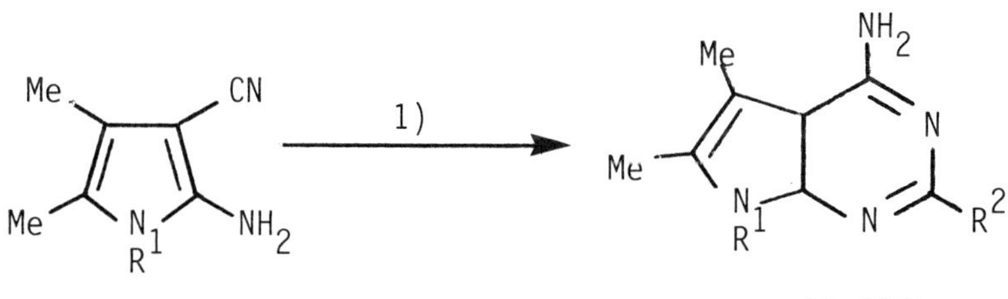

21-98%

1) NaOMe, i-PrOH, reflux.

IV.L.1-3 R. Madroñero and S. Vega, Synthesis, 628 (1987).

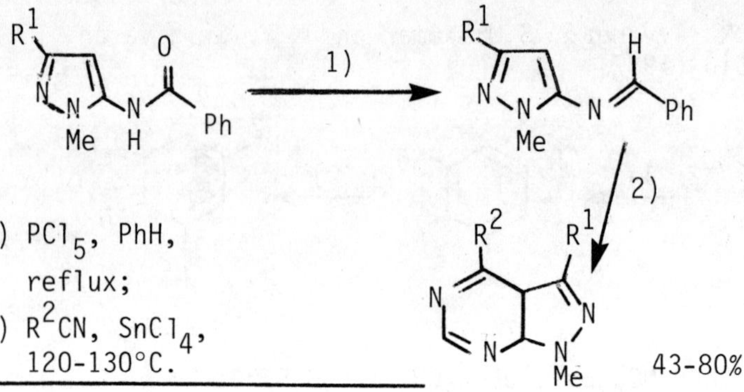

1) PCl_5, PhH, reflux;
2) R^2CN, $SnCl_4$, 120-130°C.

IV.L.1-4 J. Garin et al., Heterocycles, 26, 1303 (1988).

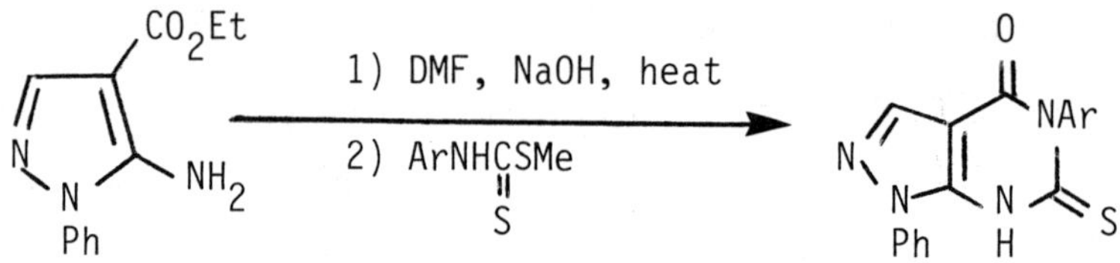

1) DMF, NaOH, heat
2) ArNHC(=S)SMe
3) H^+, DMF, 150°C

IV.L.1-5 J. Garin et al., Heterocycles, 26, 2371 (1987).

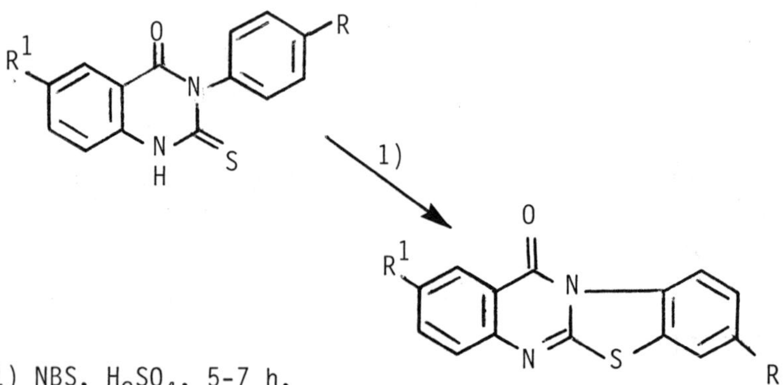

1) NBS, H_2SO_4, 5-7 h, 70-100°C.

94-96%

IV.L.1-6 Y.L. Chen and T.J. Barton, Organometallics, 6, 2590 (1987).

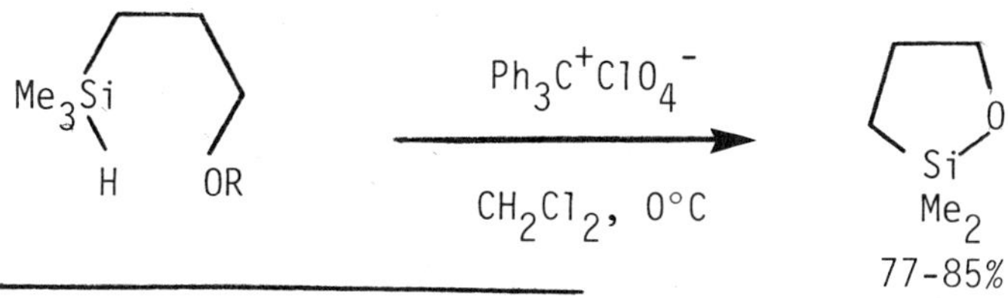

IV.L.1-7 V. Bertini, A. De Munno, N. Picci, F. Lucchesini and M. Pocci, Heterocycles, 26, 2153 (1987).

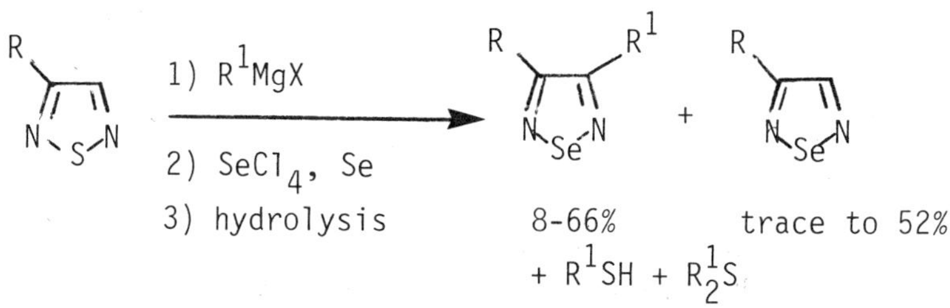

IV.L.1-8 P.J. Dunn and C.W. Rees, J. Chem. Soc., Chem. Commun., 59 (1987).

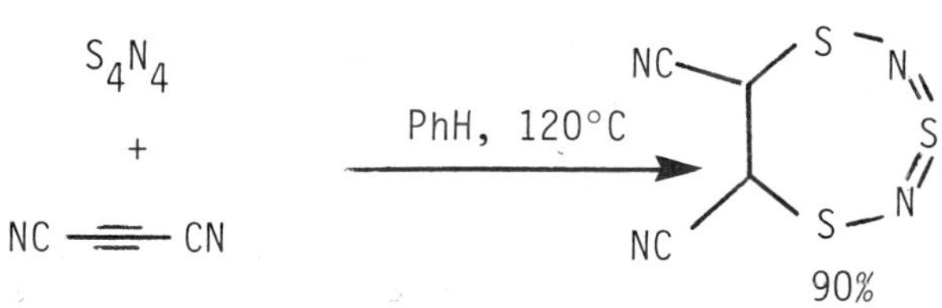

IV.M. Reviews

IV.M-1 J. Tsuji, Heterocycles, 25, 1 (1987).

Review: "Contributions of Professor Gilbert Stork in Heterocyclic Chemistry"

IV.M-2 H. Takahata and T. Yamazaki, J. Synth. Org. Chem. Jpn., 45, 682 (1987).

Review: "Synthesis of Heterocycles Using Thioamide Groups"

IV.M-3 E. Lukevits and I.D. Segal, Chem. Heterocycl. Compds. (Engl. Transl.), 23, 1 (1987).

Review: "Pyridylsilanes in the Synthesis of Heterocycles"

IV.M-4 Y. Tamaru and Z. Yoshida, J. Organomet. Chem., 334, 213 (1987).

Review: "Heterocyclic Synthesis by the Use of the Oxidizing Potential of Palladium (II)"

IV.M-5 K. Tanaka and K. Mitsuhashi, J. Synth. Org. Chem. Jpn., 45, 269 (1987).

Review: "Synthesis of Trifluoromethyl Azoles Using Building Blocks"

IV.M-6 J. Chanet-Ray and R. Vessiere, Org.Prep. Proced. Int., 18, 157 (1986).

Review: "Synthesis and Reactions of β-Sultams"

IV.M-7 L.D. Freedman and H.S. Freeman, Chem. Rev., 87, 289 (1987).

Review: "Preparation, Reactions, Structures and Possible Uses of 5,10-Dihydrophenophosphazine Derivatives"

IV.M-8 R. Grigg, Chem. Soc. Rev., 16, 89 (1987).

Review: "Prototropic Routes to 1,3- and 1,5-Dipoles and 1,2-Ylides: Applications to the Synthesis of Heterocyclic Compounds"

IV.M-9 S. Eguchi, T. Okano and H. Takeuchi, Heterocycles, 26, 3265 (1987).

Review: "Synthesis of Azamodified Adamantane Derivatives via Bridgehead and Bridge Imines"

IV.M-10 M.H. Elnagdi, S.M. Sherif and R.M. Mohareb, Heterocycles, 26, 497 (1987).

Review: "The Synthetic Potentialities of Nitriles in Heterocyclic Synthesis"

IV.M-11 B.Y. Riad, A.M. Negm, S.E. Abdou and H.A. Daboun, Hetrocycles, 26, 205 (1987).

Review: "The Chemistry of α-Cyanothioacetamide"

IV.M-12 A. Kamal and P.B. Sattur, Heterocycles, 26, 1051 (1987).

Review: "Chlorosulfonyl Isocyanate: A Novel Reagent for the Synthesis of Heterocycles"

IC.M-13 J.M. Bobbitt and A.J. Bourque, Heterocycles, 25, 601 (1987).

Review: "Synthesis of Heterocycles Using Aminoacetals"

IV.M-14 L.Y. Brezhnev, M.M. Vartanyan, T.Y. Solov'eva, V.A. Zetirova, N.P. Karzhavina and R.A. Karakhanov, Chem. Heterocycl. Compds., (Engl. Transl.), 22, 931 (1986).

Review: "Synthesis and Properties of 1,6-Dioxaspiro[4.4]nonane and its Derivatives"

IV.M-15 A.R. Katritzky, D.L. Ostercamp and T.I. Yousaf, Tetrahedron, 43, 5171 (1987).

Tetrahedron Report 225: "The Mechanisms of Heterocyclic Ring Closures"

IV.M-16 G.A. Shvekhgeimer, V.I. Zvolinskii and K.I. Kobrakov, Chem. Heterocycl. Compds., 22, 353 (1986).

Review: "Synthesis of Heterocycles on the Basis of Aliphatic Nitro Compounds"

IV.M-17 A.R. Katritzky, H. Faid-Allah and C.M. Marson, Heterocycles, 26, 1333 (1987).

Review: "The N-Carboxylic Acids of Nitrogen Heterocycles"

IV.M-18 A.R. Katritzky, C.M. Marson and H. Faid-Allah, Heterocycles, 26, 1657 (1987).

Review: "Heterocyclic N-Thiodicarboxylic Acids"

IV.M-19 E. Breuer, Isr. J. Chem., 27 (1986).

Review: "Heterocycles in Organic Synthesis"

IV.M-20 H.C. van der Plas, M. Simonyi, F.C. Alderweireldt and J.A. Lepoiure, Elsevier Science Publishers, 1986.

Book: "Bio-organic Heterocycles, 1986: Synthesis, Mechanisms and Bioactivity"

IV.M-21 L.I. Belen'kii, Chem. Heterocycl. Compd., 22 (1986).

Review: "Effect of the Acid-base Properties of Heteroaromatic Compounds on Their Electrophilic Substitution Reactions"

IV.M-22 C. Kashima and T. Tajima, J. Synth. Org. Chem., Jpn., 45, 863 (1987).

Review: "The Synthesis and Reactions of β-Imidazolyl Enones"

IV.M-23 H. Heimgartner, Isr. J. Chem., 27, 3 (1986).

Review: "The Application of 3-Amino-2H-Azirines as Amino Acid Equivalents"

IV.M-24 A.K. Mukerjee, Heterocycles, 26, 1077 (1987).

Review: "Azlactones: Retrospect and Prospect"

IV.M-25 S. Gronowitz, J. Wiley (1987).

Book: "The Chemistry of Heterocyclic Compounds. Volume 55, Part 3: Thiophene and its Derivatives"

IV.M-26 B.L. Feringa, Rec. Trav. Chim., 106, 469 (1987).

Review: "Photo-oxidation of Furans"

IV.M-27 V.A. Bakul'ev and V.S. Mokrushin, Chem. Heterocycl. Compds., (Engl. Transl.), 22, 811 (1986).

Review: "Structures, Synthesis and Properties of 1,2,3-Thiadiazoles"

IV.M-28 N. Furakawa, J. Synth. Org. Chem. Jpn., 45, 624 (1987).

Review: "Reactions of Pyridyl and Related Hetero-aromatics Activated by Sulphur Functional Groups"

IV.M-29 A.A. Tolmachev and E.S. Kozlov, Chem. Heterocycl. Compds., (Engl. Trans.), 22, 1277 (1986).

Review: "Phosphorylated, Pyridines and Their Benzo Homologues"

IV.M-30 L.A. Summers, J. Heterocycl. Chem., 24, 533 (1987).

Review: "Oxybispyridines, Thiobispyridines and Selénobispyridines"

IV.M-31 D.M. Flanagan and M.M. Joullié, Heterocycles, 26, 2247 (1987).

Review: "Synthetic Strategies for the Construction of 3-Pyrrolidinol, A Versatile Nitrogen Heterocycle"

IV.M-32 M.I. Knyazhanskii, Y.R. Tymyanskii, V.M. Feigelman and A.R. Katritzky, Heterocycles, 26, 2963 (1987).

Review: "Pyridinium Salts: Luminescent Spectroscopy and Photochemistry"

IV.M-33 R. Gawinecki and D. Rasala, Heterocycles, 26, 2727 (1987).

Review: "Synthesis of Vicinally Substituted Nitropyridine Derivatives and Their N-Oxides"

IV.M-34 G. Hall, J.K. Sugden and M.B. Waghela, Synthesis, 10 (1987).

Review: "Pyrrazoline Chemistry"

IV.M-35 G. Heinisch, Heterocycles, 26, 481 (1987).

Review: "Advances in the Synthesis of Substituted Pyridazines via Introduction of Carbon Functional Groups into the Parent Heterocycle"

IV.M-36 A. Katoh, T. Nishio and C. Kashima, Heterocycles, 26, 2223 (1987).

Review: "Synthesis and Reactions of N-Substituted 2(1H)-Pyrimidones and Pyrimidinethiones"

IV.M-37 K. Senga, J. Synth. Org. Chem. Jpn., 45, 318 (1987).

Review: "Synthesis of Fused Pyrimidines From Hydrazinopyrimidines and Their Related Compounds"

IV.M-38 M.J. Hearn and F. Levy, Org. Prep. Proced. Intl., 19, 215 (1987).

Review: "Recent Progress in the Synthesis and Reactions of Tetrazines"

IV.M-39 S. Saraf, F. Al-Omran and B. Al-Saleh, Heterocycles, 26, 239 (1987).

Review: "Recent Advances in the Synthesis of Phenothiazines"

IV.M-40 T.M. Romney-Alexander, Heterocycles, 26, 1899 (1987).

Review: "Benzopyran-4-ones Containing a 2,3-Fused Heterocyclic Ring"

IV.M-41 C.K. Ghosh, C. Bandyopadhyay and J. Maiti, Heterocycles, 26, 1623 (1987).

Review: "Heterocycles Fused with the 2,3-Bond of [1]Benzopyran"

IV.M-42 A. Leval, Chem. Heterocycl. Compds., (Engl. Trans.), 22, 1161 (1986).

Review: "Synthesis of Benzothiazepines"

IV.M-43 I.I. Grandberg, L.K. Denisov and O.A. Popova, Chem. Heterocycl. Compd. (Engl. Trans.), 23, 117 (1987).

Review: "7-Aminocoumarins"

IV.M-44 Y. Okamoto and K. Takagi, J. Hetrocycl. Chem., 24, 885 (1987).

Review: "Chemistry of 4-Amino-1H-1,5-Benzodiazepine-3-Carbonitrile"

IV.M-45 G. Mohiuddin, P.S. Reddy, K. Ahmed and C.V. Ratnam, Heterocycles, 24, 3489 (1986).

Review: "Recent Advances in the Synthesis of Annulated 1,4-Benzodiazepines"

IV.M-46 S. Grasso, M. Zappalà and A. Chimirri, Heterocycles, 26, 2477 (1987).

Review: "Benzodiazocines"

IV.M-47 Y. Matsuda and H. Goton, Heterocycles, 26, 2757 (1987).

Review: "The Chemistry of Antiaromatic Azacycl[3.3.3]azines"

IV.M-48 K.M. Smith and J.A.S. Cavaleiro, Heterocycles, 26, 1947 (1987).

Review: "Protoporphyrin IX: Some Useful Substituent Manipulations"

V.A. Hydroxyl Protecting Groups

V.A-1 M.S. Shashidhar and M.V. Bhatt, J. Chem. Soc., Chem. Commun., 654 (1987).

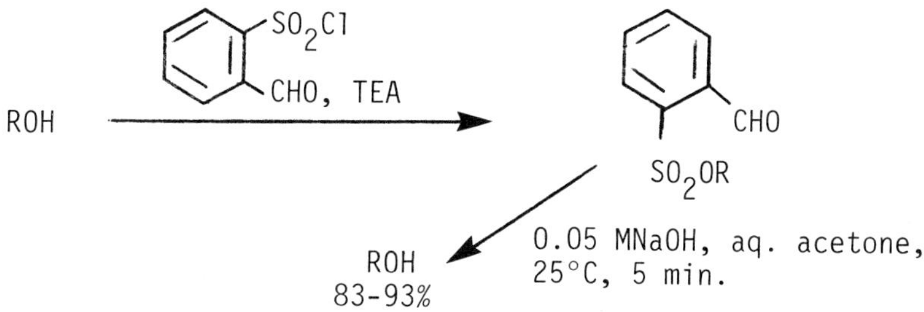

V.A-2 C. Malanga, Chem. Ind (London), 856 (1987).

R-OTr —— Amberlyst 15-H / MeOH, RT, 5-10 min ——→ ROH

69-90%

V.A-3 S. Kim and J.H. Park, Tetrahedron Lett., 28, 439 (1987).

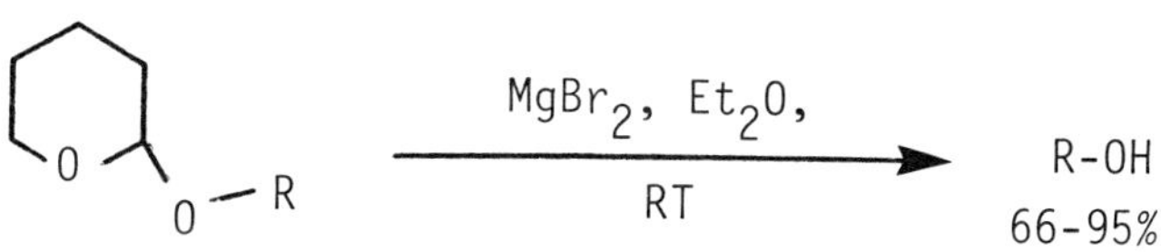

t-Butyldimethylsilyl ethers are unaffected under the above conditions.

V.A-4 Y. Guindon, M.A. Bernstein and P.C. Anderson, Tetrahedron Lett., 28, 2225 (1987).

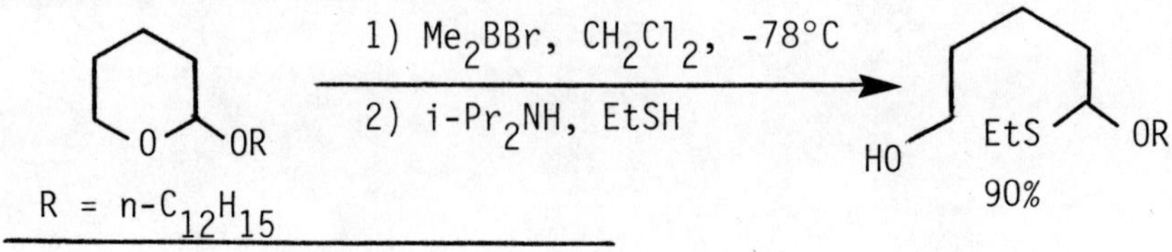

V.A-5 A.P. Kozikowski and J. Wu, Tetrahedron Lett., 28, 5125 (1987).

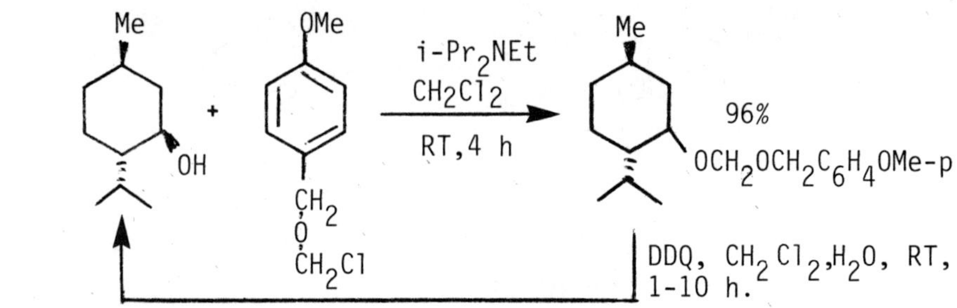

V.A-6 R.A. Moss, P. Scrimin, S. Bhattacharya and S. Chatterjee, Tetrahedron Lett., 28, 5005 (1987).

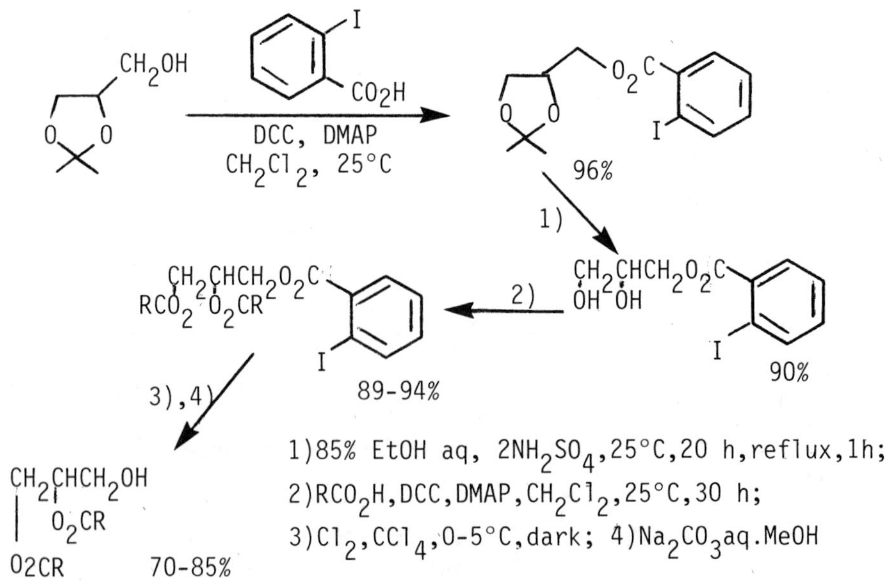

V.A-7 A.K. Sinhababu, M. Kawase and R.T. Borchardt, Tetrahedron Lett., 28, 4139 (1987).

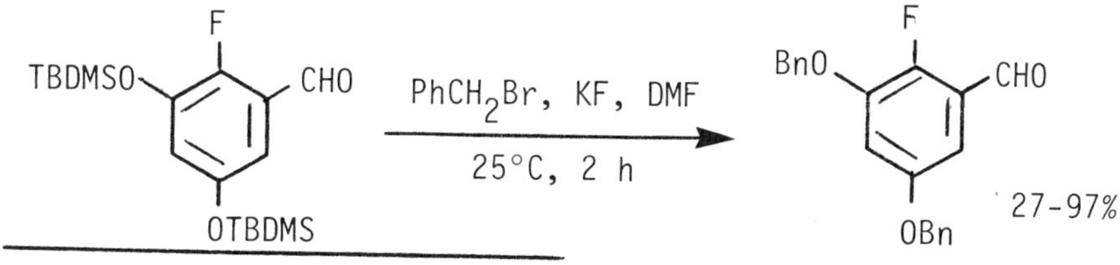

V.A-8 J.L. Gras, R. Nouguier and M. Mchich, Tetrahedron Lett., 28, 6601 (1987).

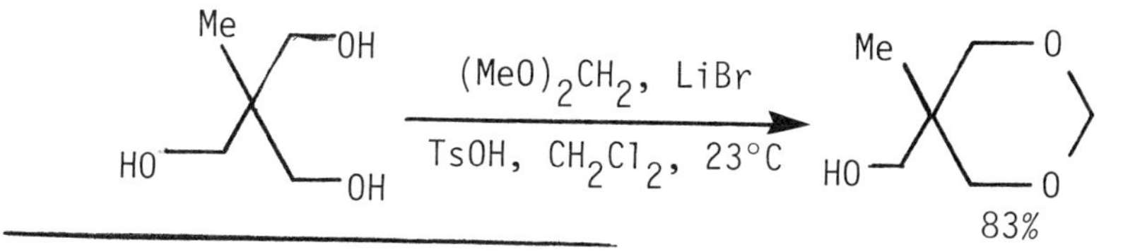

V.A-9 K. Tamao, T. Yamauchi and Y. Ito, Chem. Lett., 171 (1987).

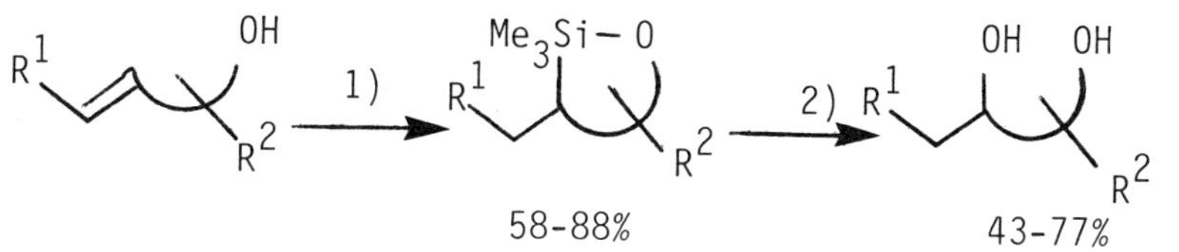

1) $RCOCl, ZnCl_2$, RT, 1-3 h; 2) $KF, KHCO_3$, MeOH, THF.

V.A-10 T.R. Hoye, B.H. Peterson and J.D. Miller, J. Org. Chem., 52, 1351 (1987).

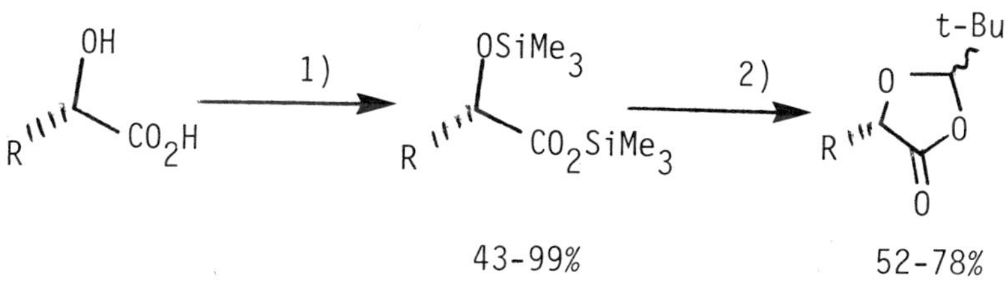

1) $(Me_3Si)_2NH, Me_3SiCl$; 2) Me_3SiOTf, Me_3CCHO.

V.A-11 E. de Vroom, G.A. van der Marel and J.H. van Boom, Rec. Trav. Chim., 106, 577 (1987).

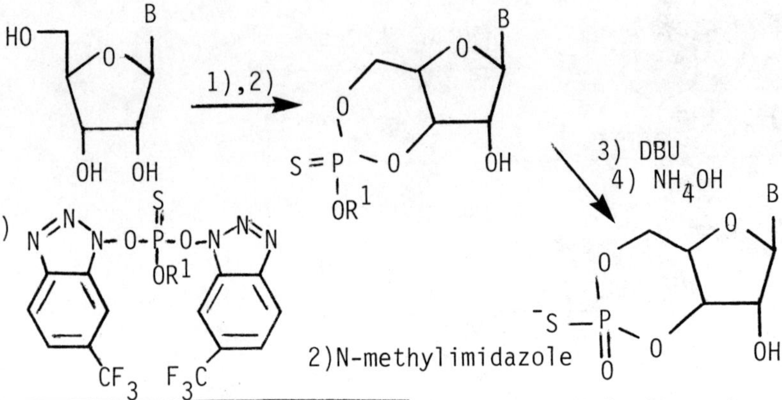

V.A-12 M. Kloosterman, E.W.J. Mosmuller, H.E. Shoemaker and E.M. Meijer, Tetrahedron Lett., 28, 2989 (1987); M.H. Park, R. Takeda and K. Nakanishi, Tetrahedron Lett., 28, 3823 (1987).

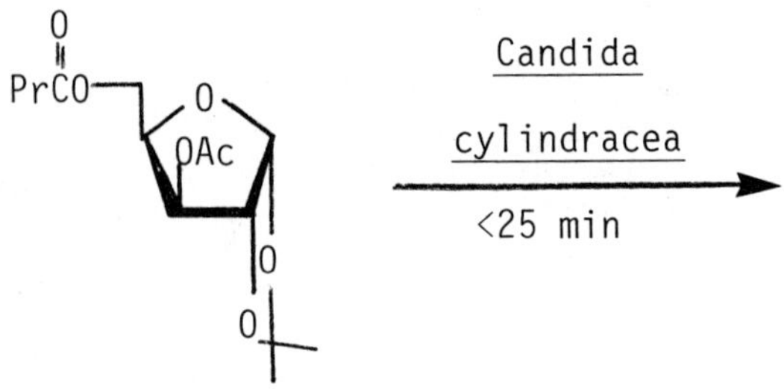

V.A-13 D. Plusquellec and M. Lefeuvre, Tetrahedron Lett., 28, 4165 (1987).

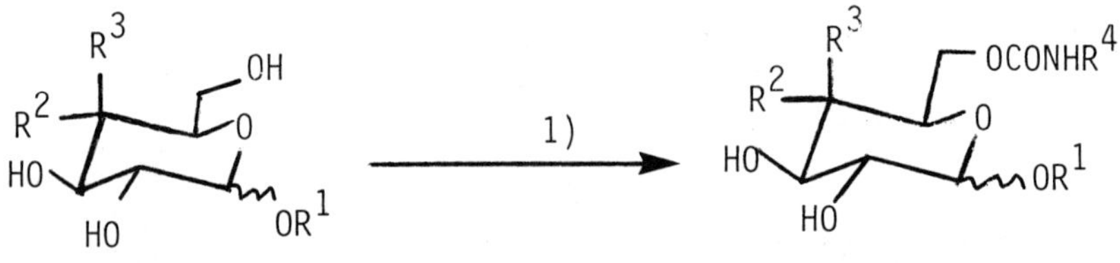

34-92%

1) R^4NCO, pyridine, RT, 10 to 15 h.

V.A-14 K. Narasaka, Y. Ichikawa and H. Kubota, Chem. Lett., 2139 (1987).

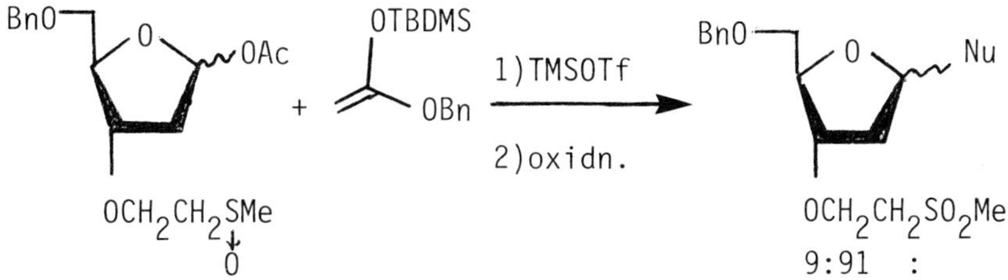

9:91 :
86%

V.A-15 J.C. Schulhof, D. Molko and R. Teoule, Tetrahedron Lett., 28, 51 (1987).

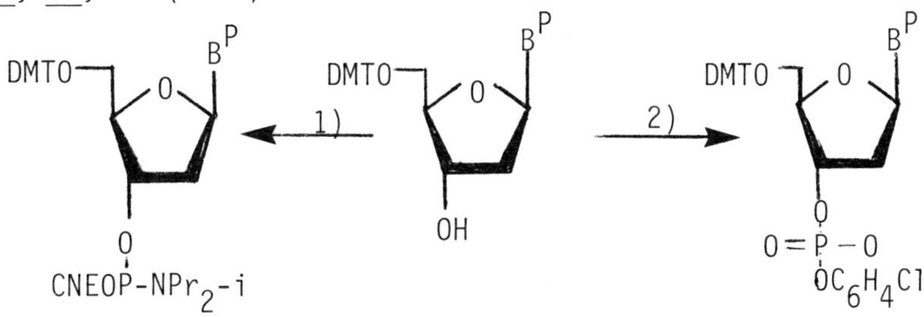

1) Bis(diisopropylamino)cyanoethyl phosphine, diisopropyl ammonium tetrazolidate, CH_2Cl_2; 2) chlorophenylbis triazolidate, MeCN, pyridine.

V.A-16 F. Himmelsbach, R. Charubala and W. Pfleiderer, Helv. Chim. Acta, 70, 1286 (1987).

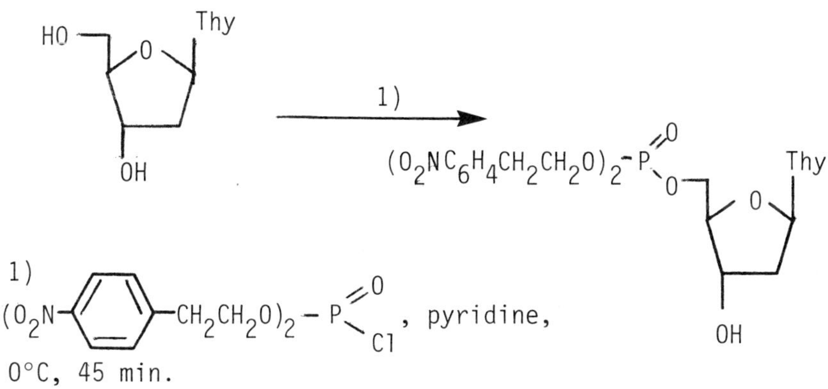

V.A-17 N. Kunesch, C. Miet and J. Poisson, Tetrahedron Lett., 28, 3569 (1987).

$$\text{pentaacetyl glucose} \xrightarrow[\text{RT}]{\text{guanidine, EtOH, } CH_2Cl_2} \text{glucose quant.}$$

Acetamides, benzoates and pivalates are stable under these conditions.

V.A-18 J.L. Fourrey, J. Varenne, C. Blonski, P. Dousset and D. Shire, Tetrahedron Lett., 28, 5157 (1987).

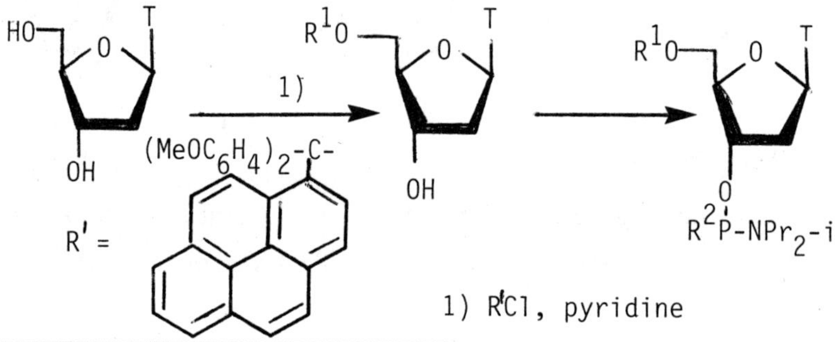

1) R'Cl, pyridine

V.A-19 M. Fujii, S. Yamakage, H. Takaku and T. Hata, Tetrahedron Lett., 28, 5713 (1987).

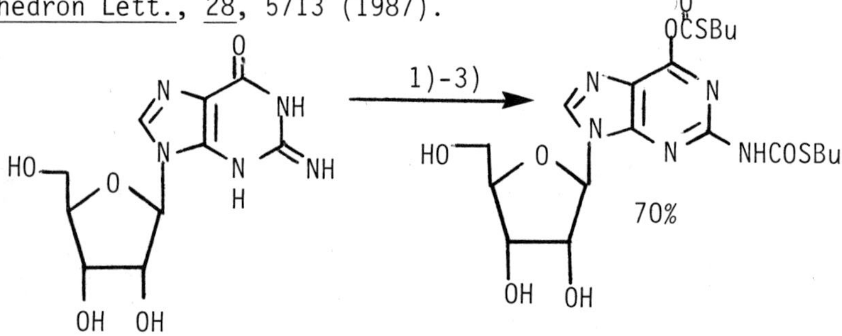

1) Me_3SiCl, pyridine, 1 h; 2) BuSCOCl, i-Pr_2NEt, RT, 24 h; 3) H_2O

V.B. Amine Protecting Groups

V.B-1 M.A. Casadei and D. Pletcher, Synthesis, 1118 (1987).

$$R^1R^2NCO_2CH_2Ph \xrightarrow{1)} R^1R^2NH + CO_2$$

90-99% PHMe

1) Pd/C cathode, MeOH, HOAc, $NaClO_4$

V.B-2 D.P. Kudav, S.P. Samant and B.D. Hosangadi, Synth. Commun., 17, 1185 (1987).

$$R^1R^2NTs \xrightarrow[100°C,\ 1H]{HClO_4,\ HOAc} R^1R^2NH$$

30-75%

R^1R^2 = H, alkyl or aryl

V.B-3 S. Ram and L.D. Spicer, Synth. Commun., 17, 415 (1987).

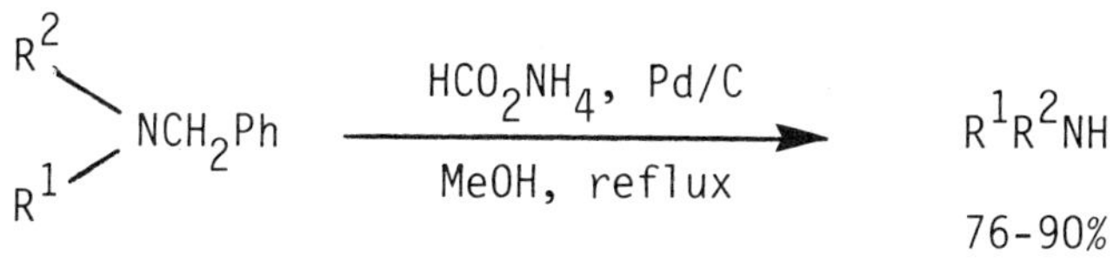

V.B-4 P.I. Suirskaya, C.C. Leznoff and M. Steinman, J. Org. Chem., 52, 1362 (1987).

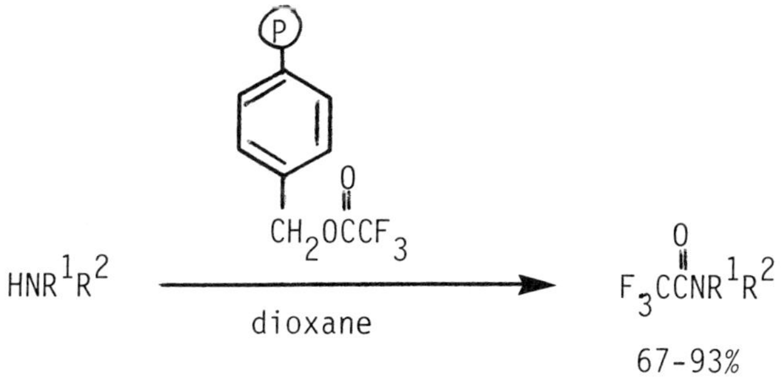

V.B-5 V.K. Haridasan, A. Ajayaghosh and V.N.R. Pillai, J. Org. Chem., 52, 2662 (1987).

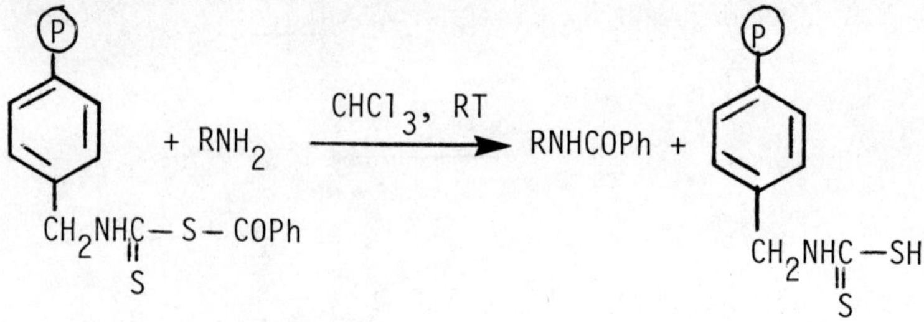

V.B-6 O. Dangles, F. Guibé, G. Balavoine, S. Lavielle and A. Marquet, J. Org. Chem., 52, 4984 (1987).

$$RNHCO_2CH_2CH = CH_2 \xrightarrow{1)} RNH_2$$

70-100%

1) Bu_3SnH, $Pd(PPh_3)_4$, HOAc

V.B-7 A.R. Katritzky, G.R. Khan and C.M. Marson, J. Heterocycl. Chem., 24, 641 (1987).

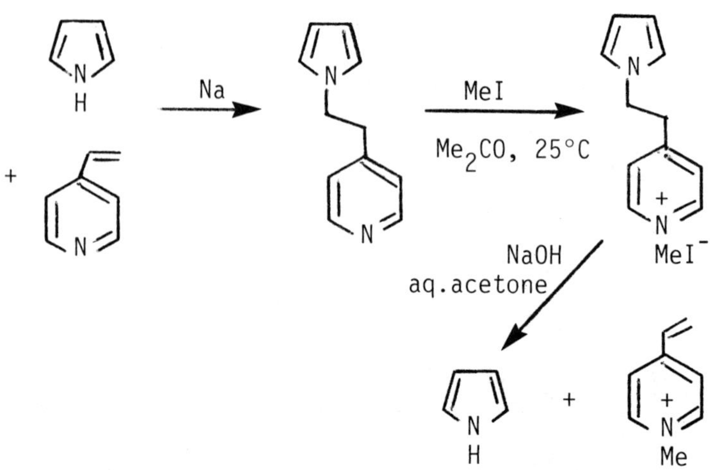

V.B-8 M. Lurdes, S. Almeida, L. Grehn and U. Ragnarsson, J. Chem. Soc., Chem. Commun., 1250 (1987).

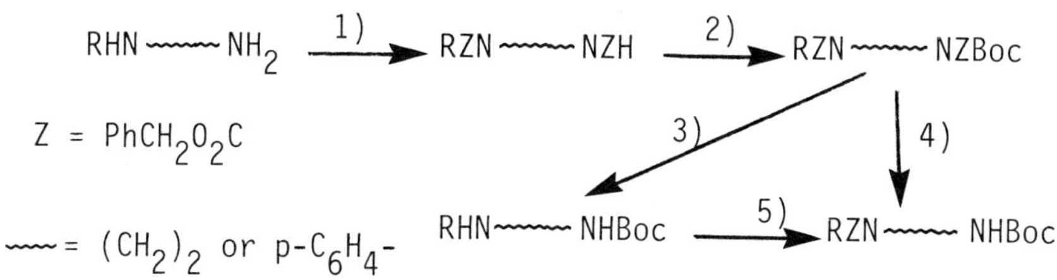

1) Z-Cl, pyridine; 2) Boc_2O, DMAP, MeCN, RT, 15 min; 3) HCO_2NH_4, Pd/C, HOAc, RT, 2 h; 4) 1,1,3,3-tetramethyl-guanidine, MeOH; 5) Z_2O, CH_2Cl_2.

V.B-9 A.L. Campbell et al., Tetrahedron Lett., 28, 2331 (1987).

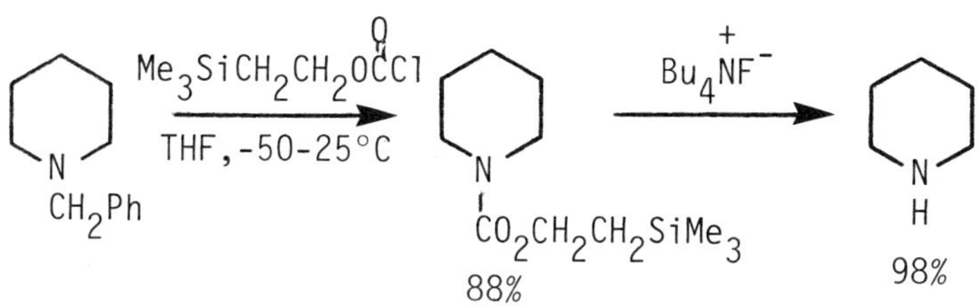

V.B-10 A. Dicko, M. Montury and M. Baboulene, Tetrahedron Lett., 28, 6041 (1987).

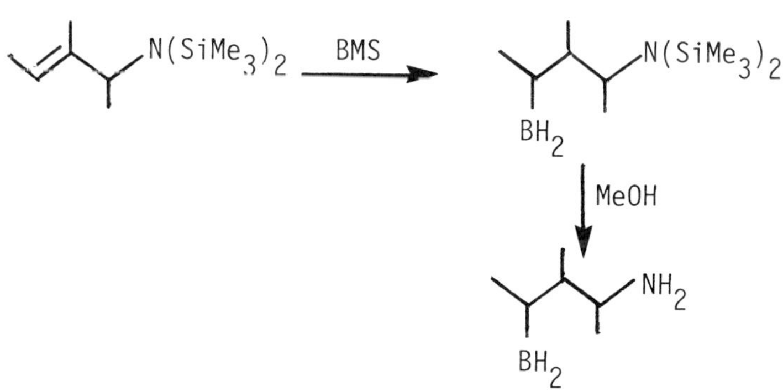

V.B-11 I. Minami, M. Yuhara and J. Tsuji, Tetrahedron Lett., 28, 2737 (1987).

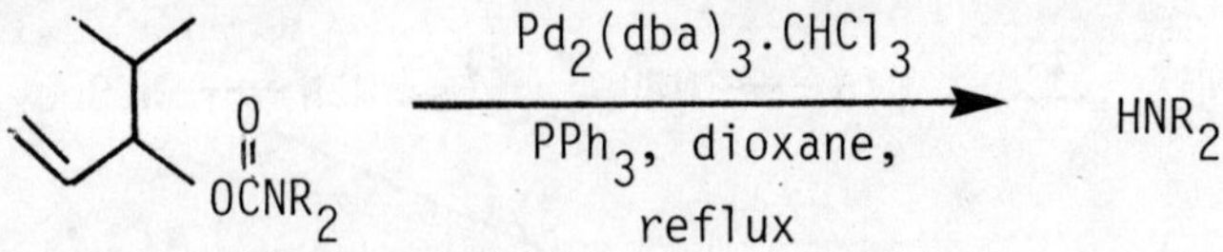

V.C. Carboxyl Protecting Groups

V.C-1 J. Chen and X. Zhou, Synth. Commun., 161 (1987).

$$RCO_2CH_2CH_2X \xrightarrow[\text{RT}]{\text{NaHTe, EtOH}} RCO_2H$$

84-92%

V.C-2 M. Iwamura, T. Ishikawa, Y. Koyama, K. Sakuma and H. Iwamura, Tetrahedron Lett., 28, 679 (1987).

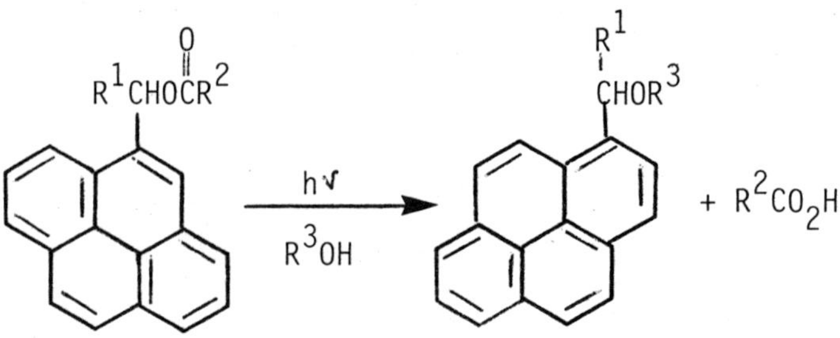

V.C-3 H. Tsutsui, M. Muto, K. Motoyoshi and O. Mitsubono, Chem. Lett., 1595 (1987).

$$RC(=O)OCH_2CH_2Ts \xrightarrow{\text{1) - 3)}} RCO_2H + CH_2{=}CHTs$$

52-95%

1) TBAF, THF, 0°C, 1 h; 2) $NaHCO_3$ aq; 3) H^+

V.D. Protecting Groups for Aldehydes and Ketones

V.D-1 J.R. Hwu *et al.*, *J. Org. Chem.*, 52, 188 (1987).

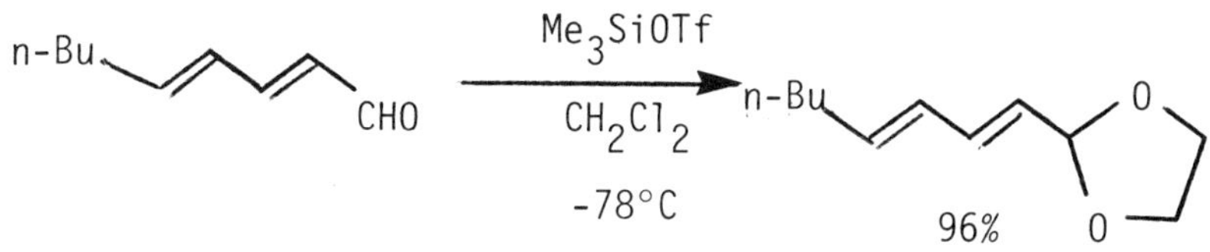

V.D-2 K. Sukata, *Bull. Chem. Soc. Jpn.*, 60, 3820 (1987).

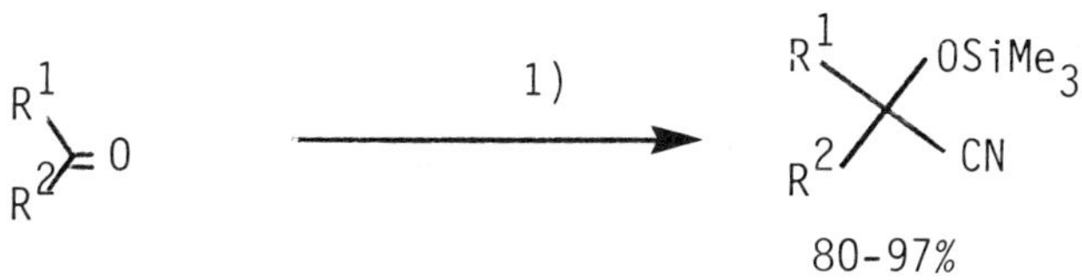

1) Me_3SiCl, KCN-Amberlite XAD resin, MeCN, 60°C, 4-25 h.

V.D-3 J. Cossy, *Synthesis*, 1113 (1987).

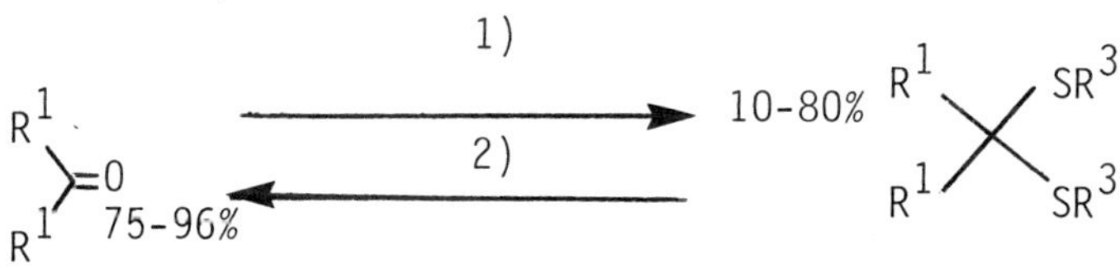

1) $Et_2O.BF_3$, RT, R^3SH

2) m-$ClC_6H_4CO_3H$, CF_3CO_2H, CH_2Cl_2, 0°C.

V.D-4 R.V. Hoffman and G.A. Buntain, Synthesis, 831 (1987).

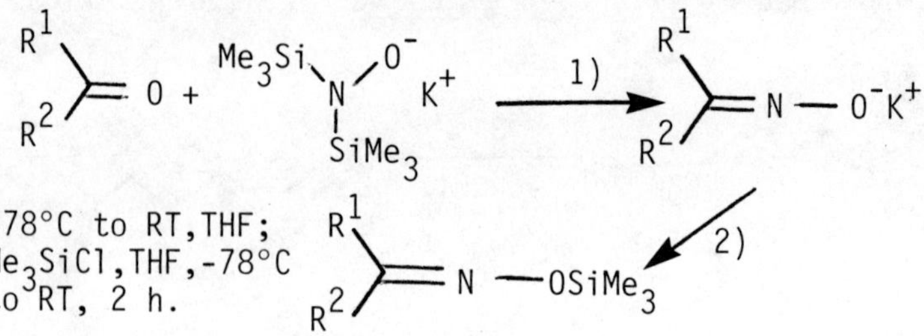

1) -78°C to RT, THF;
2) Me_3SiCl, THF, -78°C to RT, 2 h.

V.D-5 W.H. Pearson and M. Cheng, J. Org. Chem., 52, 1353 (1987).

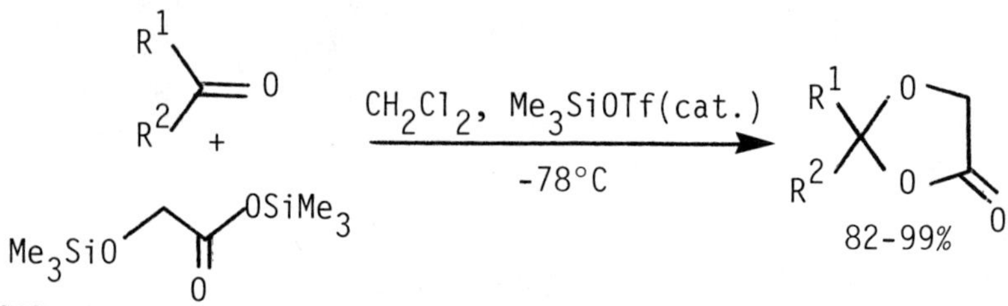

V.D-6 A. Fadel, R. Yefsah and J. Salaün, Synthesis, 37 (1987).

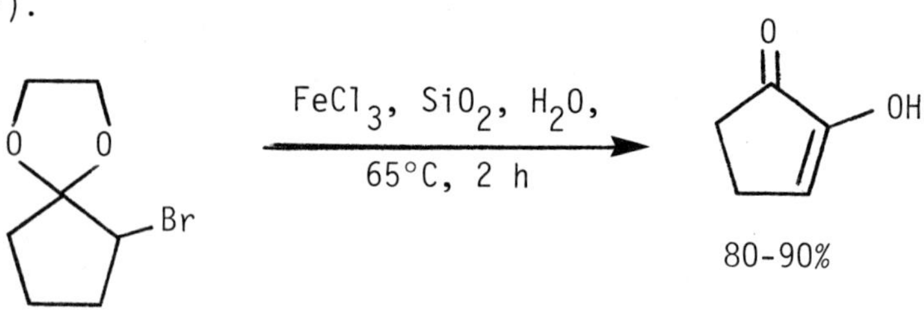

V.D-7 S.E. Denmark, R.P. Hammer, E.J. Weber and K.L. Habermas, J. Org. Chem., 52, 165 (1987).

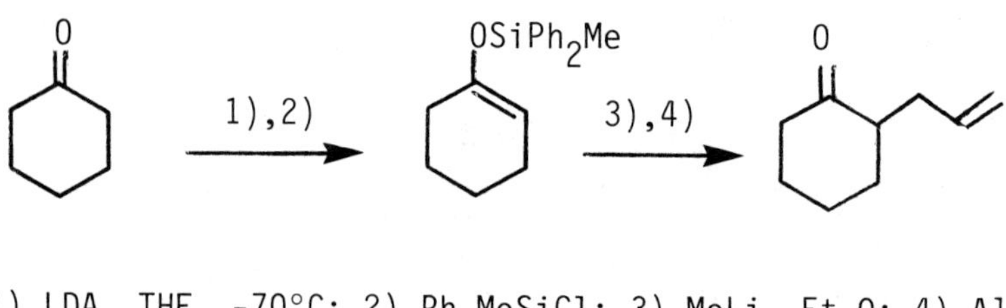

1) LDA, THF, -70°C; 2) $Ph_2MeSiCl$; 3) MeLi, Et_2O; 4) AllBr

V.D-8 M.E. Wright and S.R. Pulley, J. Org. Chem., 52, 5036 (1987).

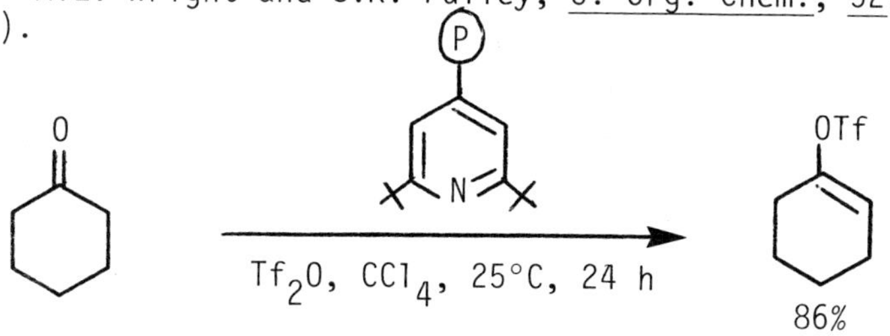

V.D-9 A.R. Katritzky, W. Fan and O. Li, Tetrahedron Lett., 28, 1195 (1987).

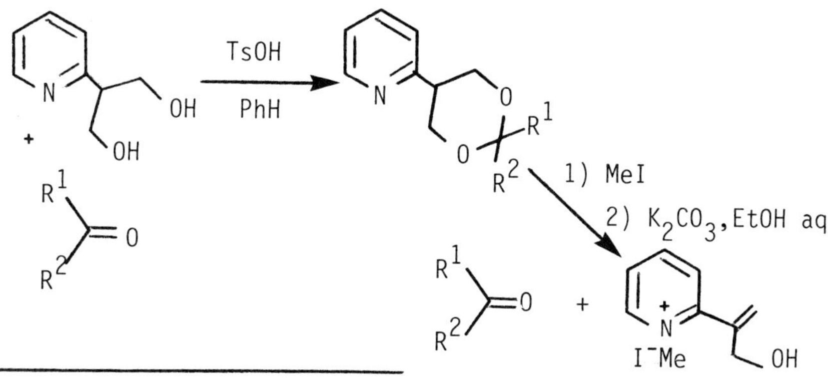

V.E. Amino Acid Protection

V.E-1 C. Froussios and M. Kolovos, Synthesis, 1106 (1987).

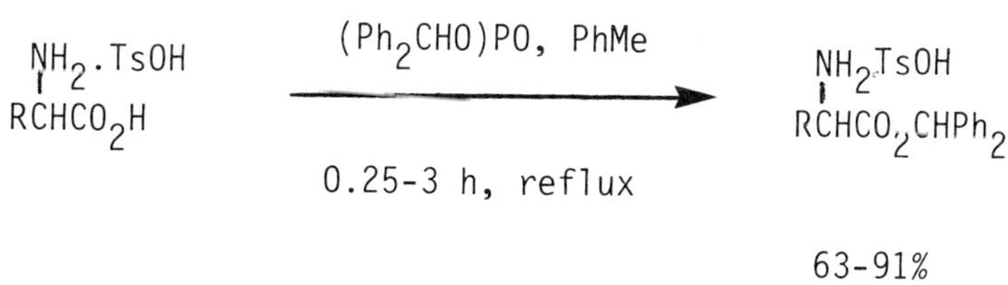

V.E-2 G. Sennyey, G. Barcelo and J.-P. Senet, <u>Tetrahedron Lett.</u>, <u>28</u>, 5809 (1987).

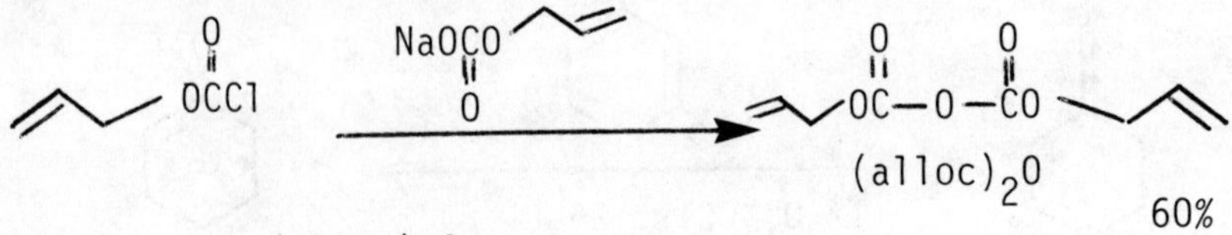

$$\text{L-Phe} \xrightarrow[\text{H}_2\text{O, dioxane, 18 h, reflux}]{(\text{alloc})_2\text{O}} \text{L-PheNH alloc}$$

96%

V.E-3 S. Ram and L.D. Spicer, <u>Tetrahedron Lett.</u>, <u>28</u>, 515 (1987).

$$PhCH_2NH(CHR)_nCO_2R^1 \xrightarrow[\text{10\% Pd-C, MeOH}]{HCO_2NH_4} H_2N(CHR)_nCO_2R^1$$

V.E-4 P. Henklein, H. Heyne, W. Halatsch and H. Niedrich, <u>Synthesis</u>, 166 (1987).

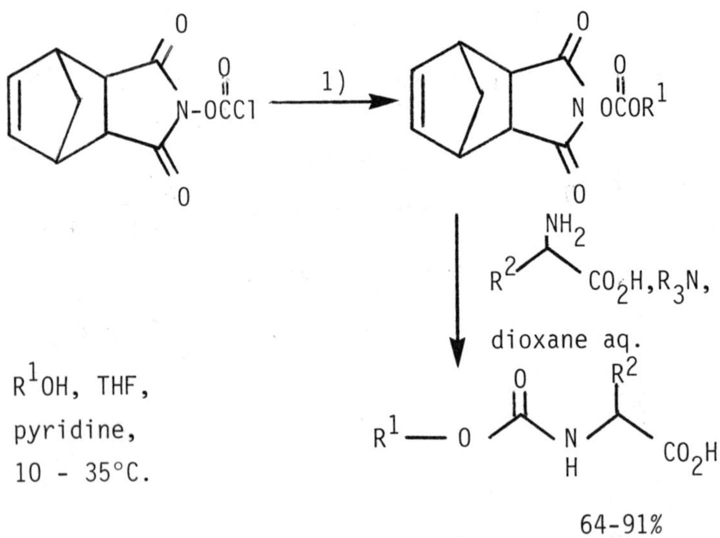

1) R^1OH, THF, pyridine, 10 - 35°C.

64-91%

V.E-5 R.E. Shute and D.H. Rich, Synthesis, 346 (1987).

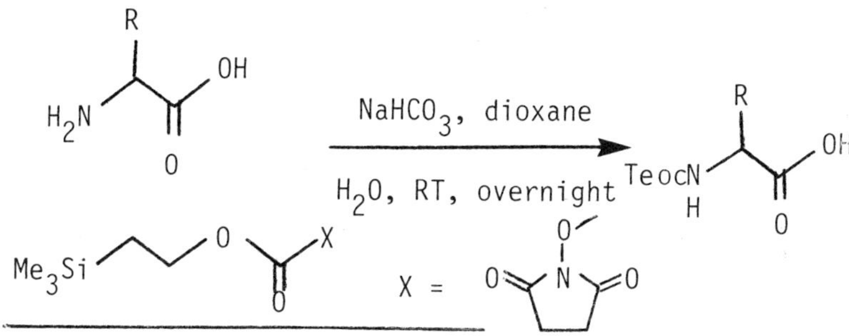

V.E-6 D.G. Mullen and G. Barany, Tetrahedron Lett., 28, 491 (1987).

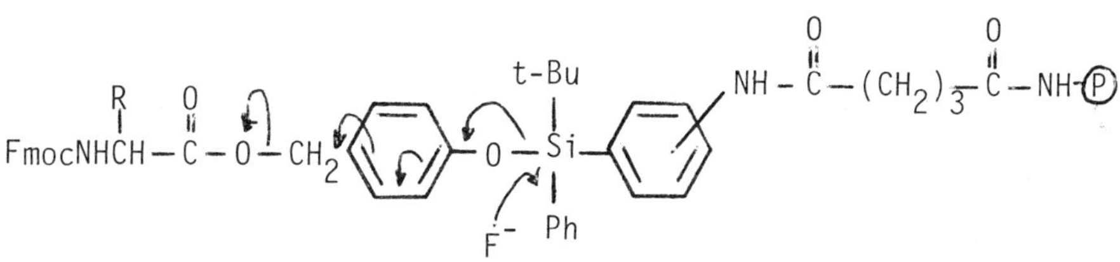

Deprotection with TBAF.

Ⓟ = polystyrene

V.E-7 R. Ramage and J. Green, Tetrahedron Lett., 28, 2287 (1987).

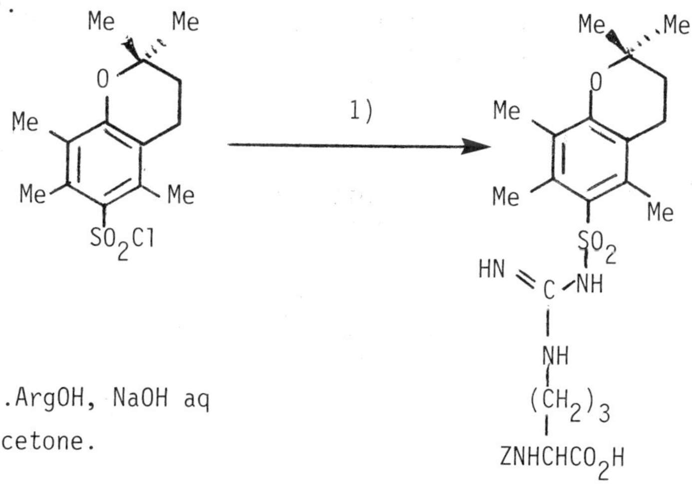

1) Z.ArgOH, NaOH aq acetone.

V.E-8 S. Plaue and D. Heissler, Tetrahedron Lett., 28, 1401 (1987).

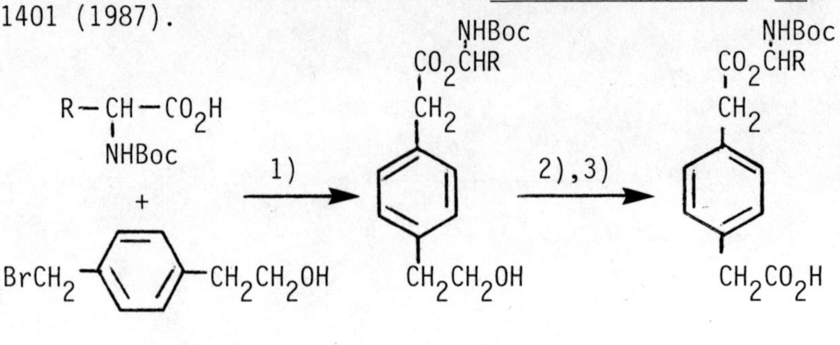

1) i-Pr_2NEt, EtOAc, reflux; 2) CrO_3.py, CH_2Cl_2; 3) $NaClO_2$, $Me_2C{=}CHMe$, NaH_2PO_4, t-BuOH, H_2O.

V.E-9 P. Sieber and B. Riniker, Tetrahedron Lett., 28, 6035 (1987).

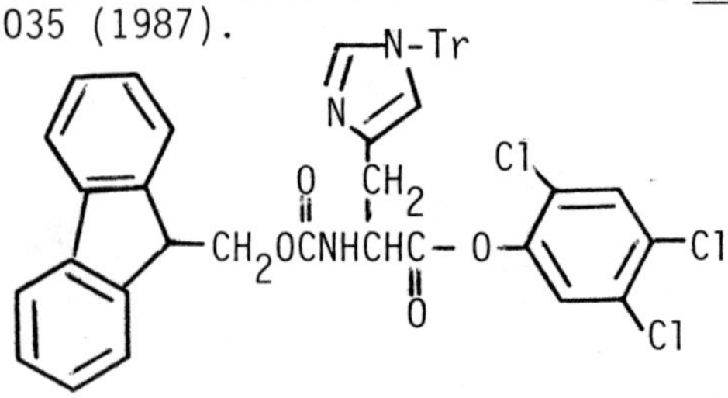

Fmoc-His(Tr)-OTcp is a suitable derivative for the Fmoc method of peptide synthesis.

V.F. Oxime Protection

V.F-1 T. Yamada, K. Goto, Y. Mitsuda and J. Tsuji, Tetrahedron Lett., 28, 4557 (1987).

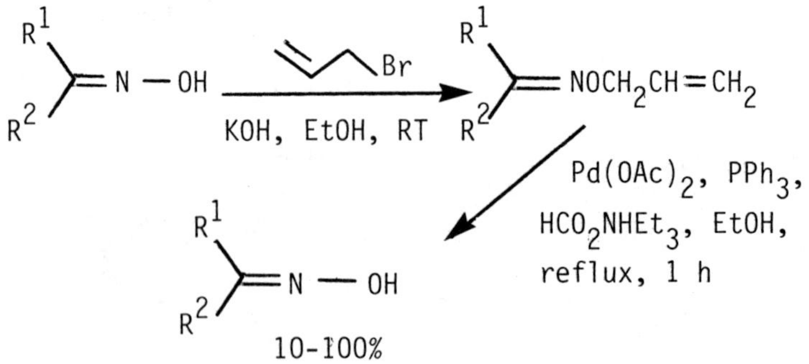

10-100%

V.G. Reviews

V.G-1 U. Hacksell and T. Hogberg, Acta Pharm. Suecica, 23, 6 (1986).

Review: "Protective Groups in Organic Synthesis"

V.G-2 K. Troev and G. Borisov, Phosphorus and Sulphur, 29, 129 (1987).

Review: "Characteristic Reactions of the Alkoxy Groups of the Acid Diesters of Phosphorus Acid"

V.G-3 L.A. Carpino, Acc. Chem. Res., 20, 401 (1987).

Review: "The 9-Fluoroenylmethoyloxycarbonyl Family of Base-Sensitive Amino-Protecting Groups"

VI.A. Functional Group Preparations

1. Acids and Anhydrides

VI.A.1-1 X. He and E.L. Eliel, Tetrahedron, 43, 4979 (1987).

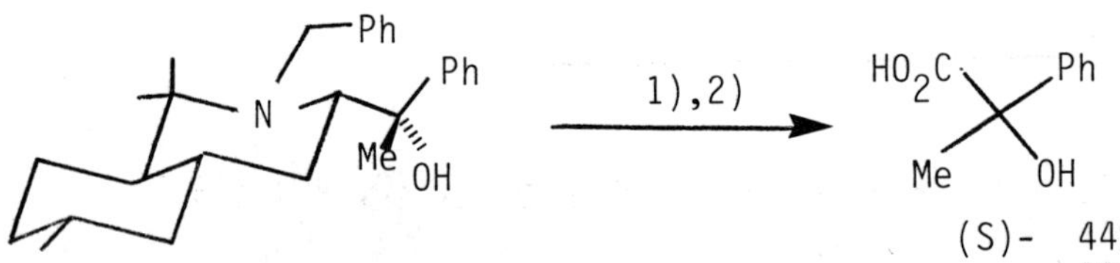

(S)- 44%
98% ee

1) HCl aq, EtOH, reflux; 2) $NaClO_4$, $Me_2C{=}CHMe$, KH_2PO_4, H_2O, Me_2CO, RT

VI.A.1-2 S. Calet, H. Alper, J.F. Petrignani and H. Arzoumanian, Organometallics, 6, 1625 (1987).

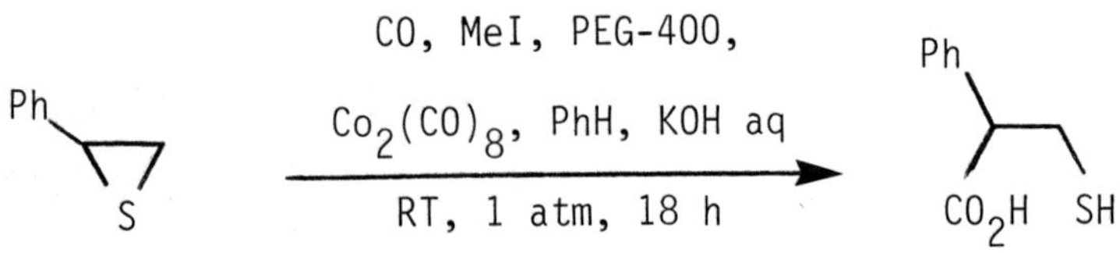

78%

VI.A.1-3 M.C. Luz and W.P. Daly, Org. Prep. Proced. Int., 19, 468 (1987).

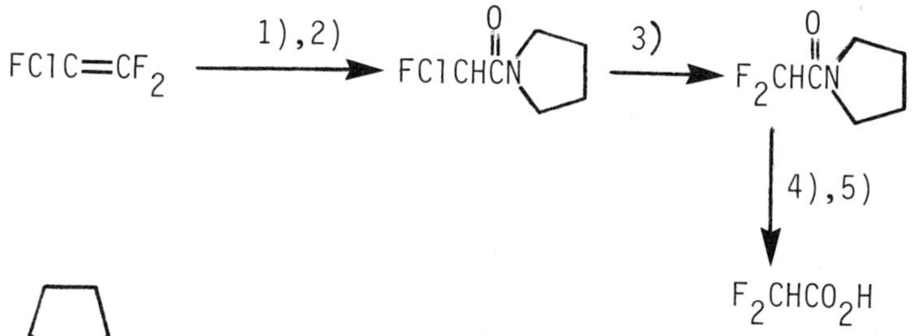

1) 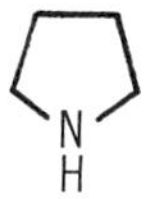, 10°C, 2 h; 2) H_2O, RT, 2 h; 3) KF, diethylene-glycol to 180°C; 4) NaOH aq; 5) H_2SO_4

VI.A.1-4 S. Rao, U.T. Bhalerao and B.D. Tilak, Ind. J. Chem., 26B, 208 (1987).

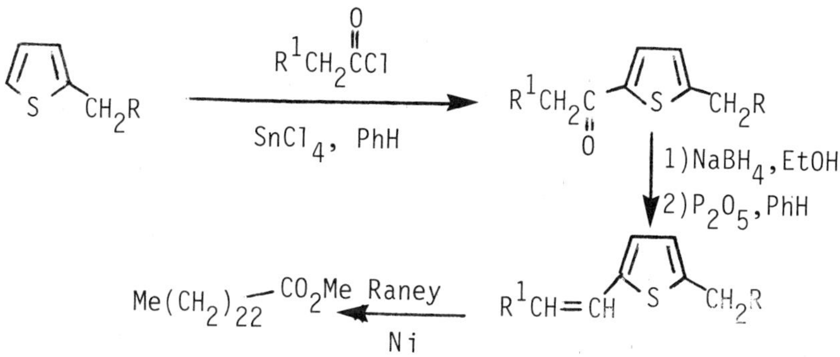

VI.A.1-5 T. Kitagawa, H. Kuroda and H. Sasaki, Chem. Pharm. Bull., 35, 1262 (1987).

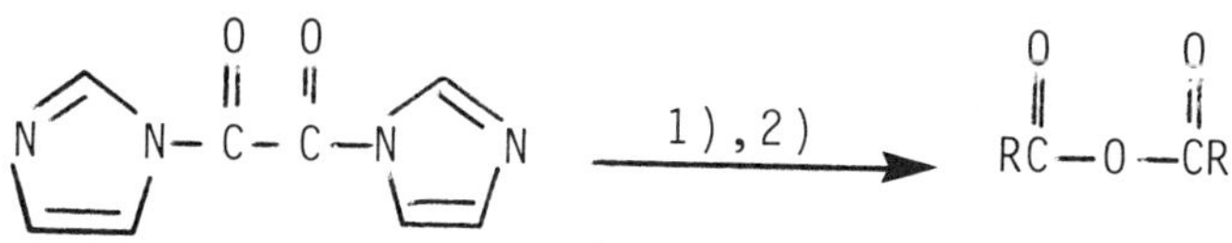

30-98%

1) MeCN, 40°C, 1-2 h; 2) RCO_2H, $MeSO_3H$, MeCN, heat.

VI.A.1-6 B.Y. Dakova, M.J. Evers, L.R. Christiaens and M.R. Guillaume, Bull. Soc. Chim. Belg., 96, 219 (1987).

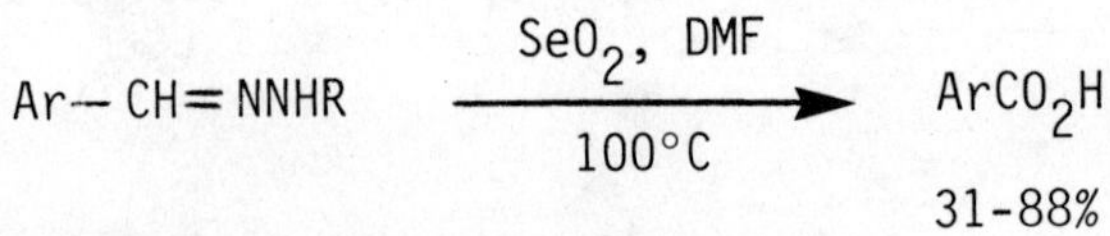

VI.A.1-7 M. Alpegiani and S. Hanessian, J. Org. Chem., 52, 278 (1987).

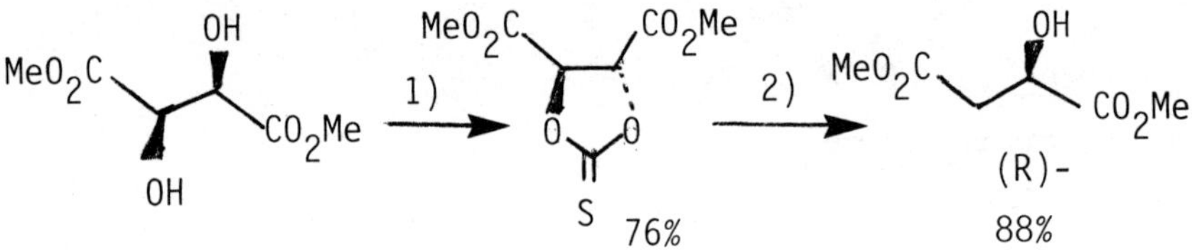

1) $Im_2C{=}S$, THF, Ar, RT, 2 h; 2) n-Bn_3SnH, reflux, Ar.

VI.A.1-8 A.G.M. Barrett and D. Dhanak, Tetrahedron Lett., 28, 3327 (1987).

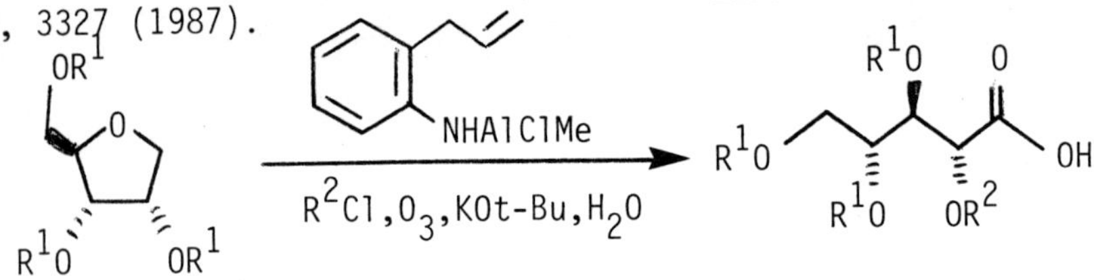

R^1 = TBDMS R^2 = $CH_2OCH_2CH_2OMe$

VI.A.1-9 R.M. Moriarty, R.K. Vaid, M.P. Duncan and B.K. Vaid, Tetrahedron Lett., 28, 2845 (1987).

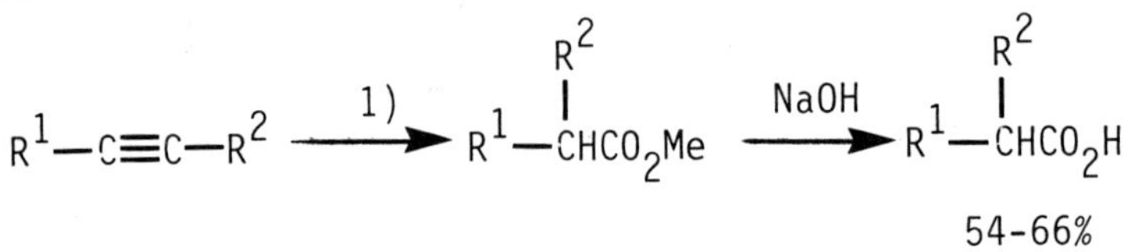

1) PhI(OH)OTs, MeOH, reflux.

VI.A.1-10 T. Kobayashi, T. Sakakura and M. Tanaka, Tetrahedron Lett., 28, 2721 (1987).

$$RX \xrightarrow[\text{i-PrOH, } H_2O]{\text{CO, CaOH, } PdCl_2(PMe_3)_2} RCH(OH)CO_2H$$

53-80%

VI.A.2 Alcohols

VI.A.2-1 J. Nakayama, S. Yamaoka and M. Hoshino, Tetrahedron Lett., 28, 1799 (1987).

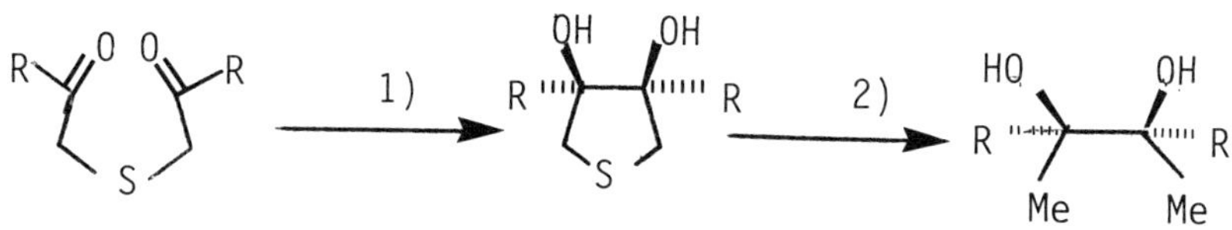

75-82%

1) $TiCl_4$-Zn, THF, 0°C; 2) Raney-Ni, EtOH, reflux.

VI.A.2-2 K. Tamao, E. Nakajo and Y. Ito, J. Org. Chem., 52, 4412 (1987).

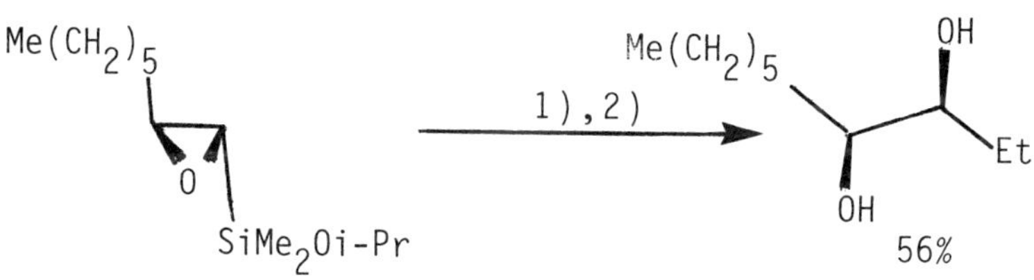

56%

1) EtMgBr, CuCN, Et_2O, -20 to -30°C;

2) H_2O_2, KF, $KHCO_3$, MeOH, THF, RT.

VI.A.2-3 G.A. Molander and D.C. Shubert, J. Am. Chem. Soc., 109, 576 (1987).

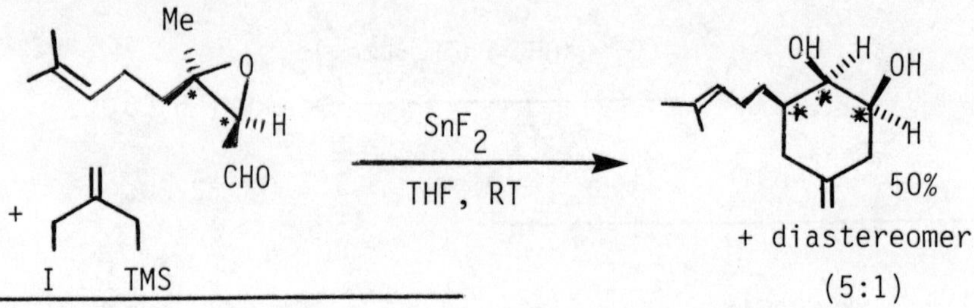

VI.A.2-4 C.W. Jefford, J.-C. Rossier and J. Boukouvalas, J. Chem. Soc. Chem. Commun., 1593 (1987); S.Y. Ko, H.Masamune and K.B. Sharpless, J. Org. Chem., 52, 667 (1987).

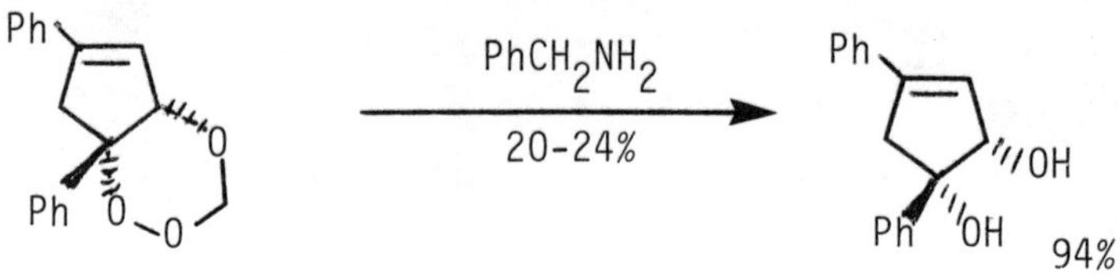

VI.A.2-5 K. Tomioka, M. Nakajima and K. Koga, J. Am. Chem. Soc., 109, 6213 (1987).

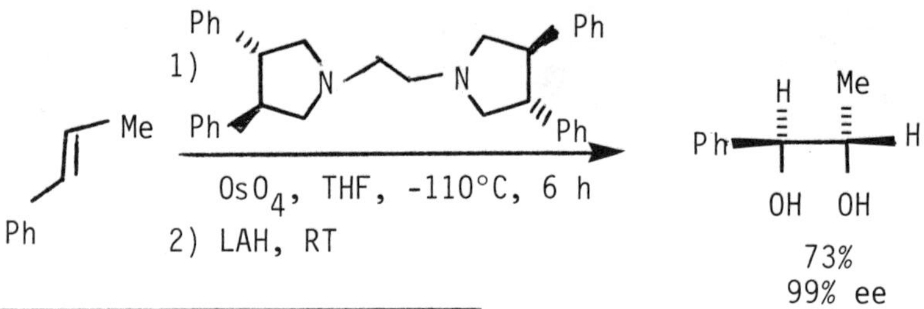

VI.A.2-6 N. Kambe et al., Chem. Lett., 269 (1987).

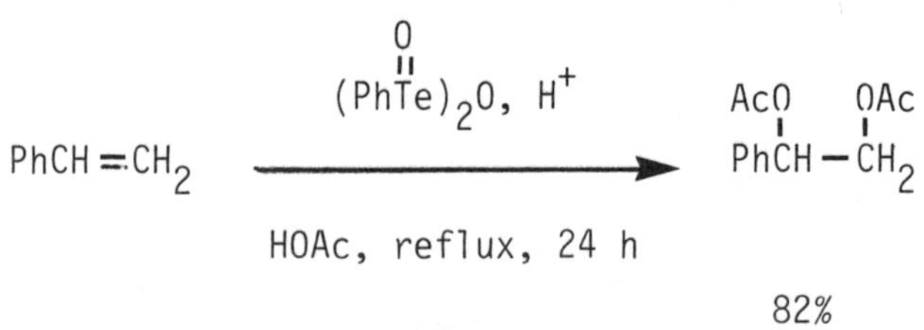

VI.A.2-7 M. Cinouini et al., Tetrahedron Lett., 28, 3139 (1987).

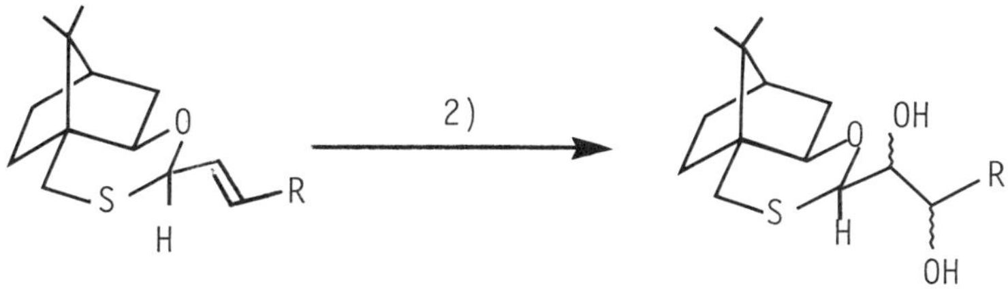

1) dihydroquinidine acetate
OsO_4, Me_3NO., THF aq., -20°C.

63%
diast. ratio 40:1

VI.A.2-8 M. Koreeda and Y. Tanaka, Tetrahedron Lett., 28, 143 (1987).

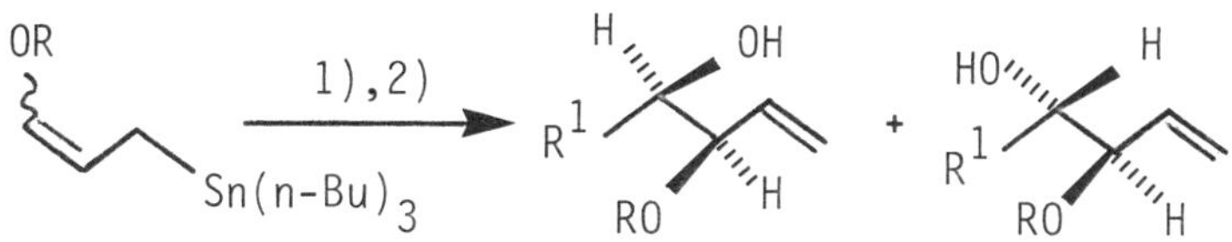

10-14:1
86%

1) R^1-CHO, $BF_3.OEt_2$, -78°C, CH_2Cl_2;
2) MeOH, H_2O.

VI.A.2-9 D. Seebach, R. Imwinkelried and G. Stucky, Helv. Chim. Acta, 70, 448 (1987).

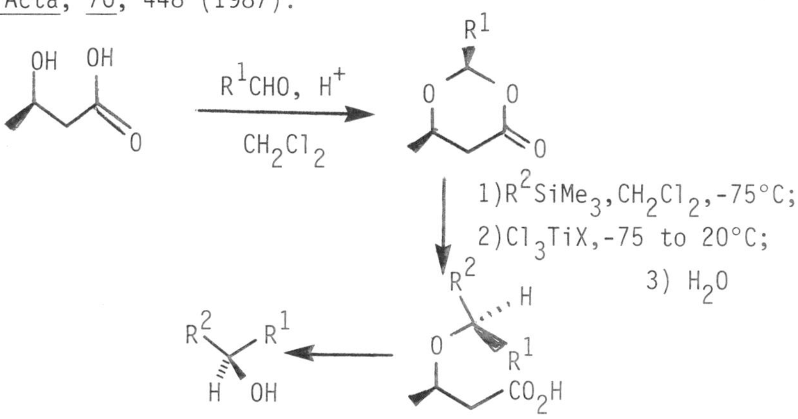

67-86%

VI.A.2-10 K. Soai, A. Ookawa, T. Kaba and K. Ogawa, J. Am. Chem. Soc., 109, 7111 (1987).

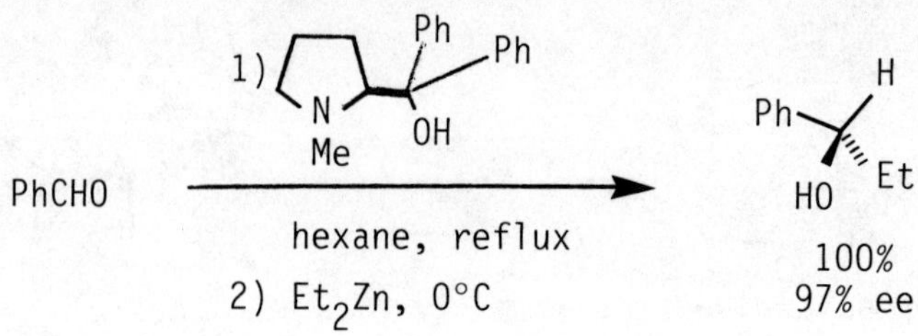

VI.A.2-11 K. Yamamoto, H. Ando and H. Chikamatsu, J. Chem. Soc., Chem. Commun., 334 (1987).

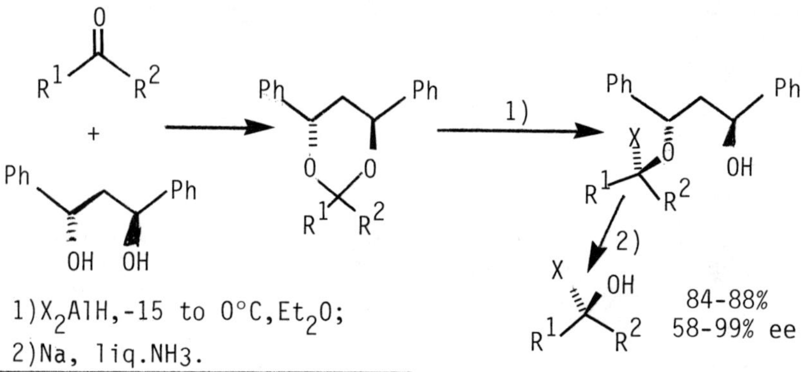

VI.A.2-12 M. Asami and H. Kirihara, Chem. Lett., 389 (1987); C. Narayana and M. Periasamy, J. Chem. Soc., Chem. Commun., 1857 (1987).

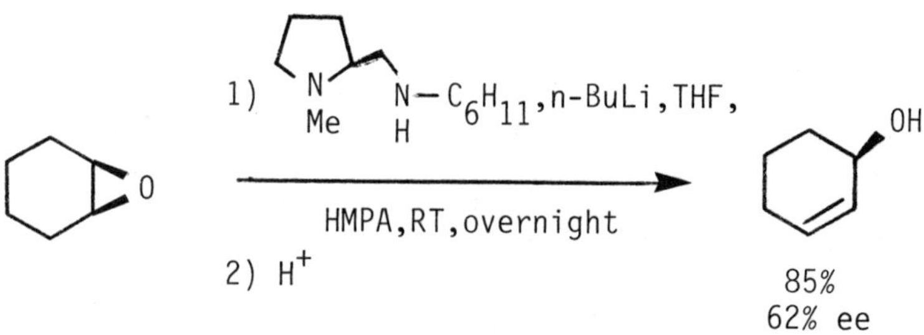

VI.A.2-13 A.R. Katritzky, W. Fan and K. Akutagawa, Synthesis, 415 (1987).

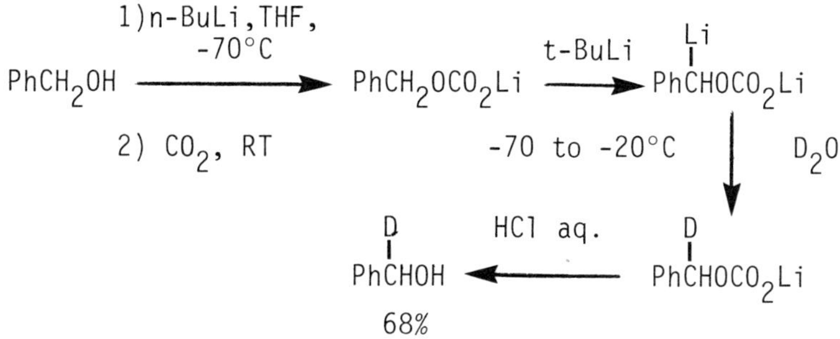

VI.A.2-14 J.T. Lin, T. Yamazaki and T. Kitazume, J. Org. Chem., 52, 3211 (1987).

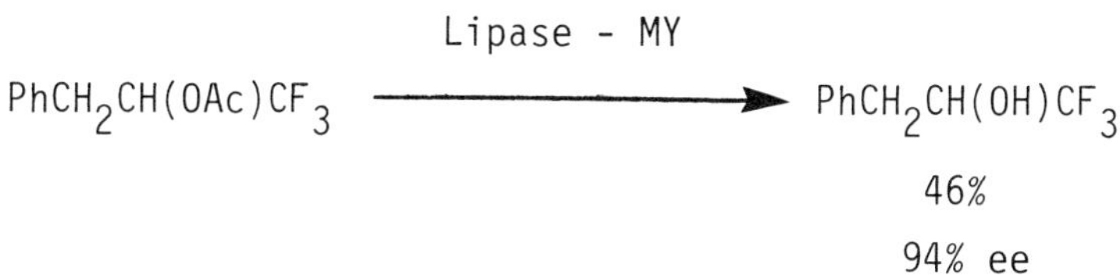

VI.A.2-15 H.A. Zahalka and H. Alper, Tetrahedron Lett., 28, 2215 (1987).

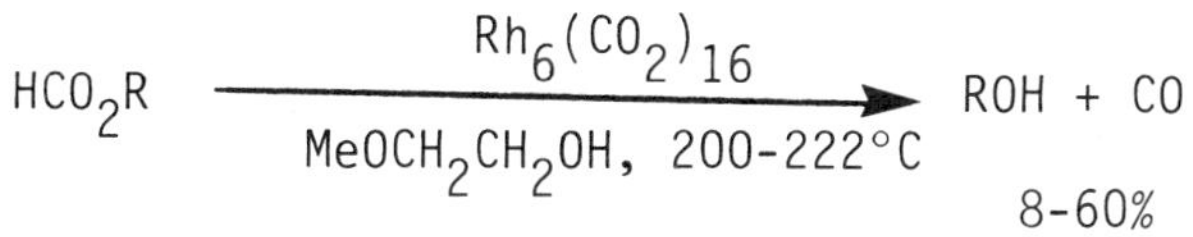

VI.A.2-16 T. Ishizuka and T. Kuieda, Tetrahedron Lett., 28, 4185 (1987).

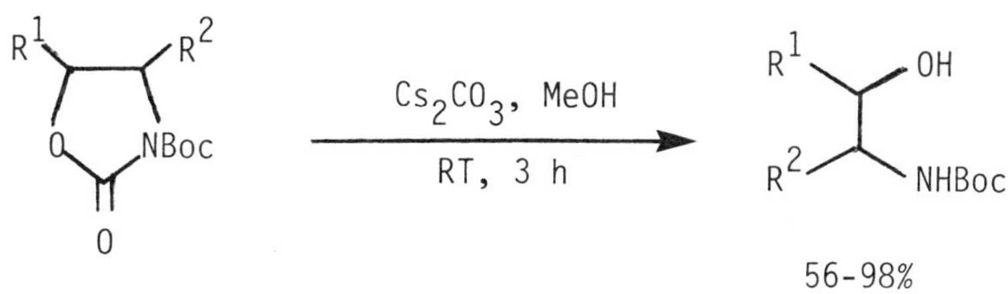

VI.A.2-17 I. Fleming and P.E.J. Sanderson, Tetrahedron Lett., 28, 4229 (1987); M. Larchevêque and S. Henrot, Tetrahedron Lett., 28, 1781 (1987).

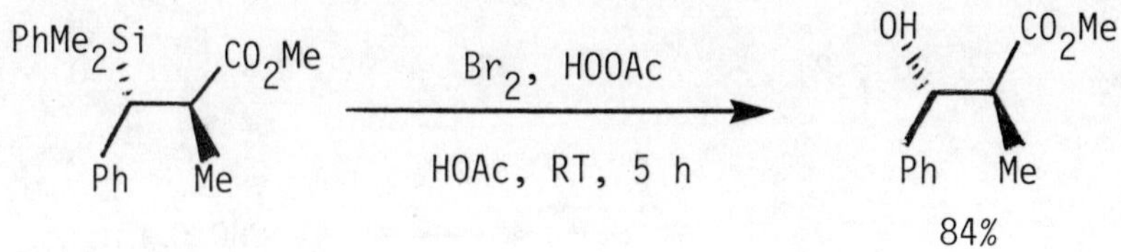

VI.A.2-18 S.L. Schreiber and M.T. Goulet, Tetrahedron Lett., 28, 1043 (1987); P. Mohr and C. Tamm, Tetrahedron Lett., 28, 391 (1987); K. Chen, K.G. Gunderson, G.E. Hardtmann, K. Prasad, O. Repic and M.J. Shapiro, Chem. Lett., 1923 (1987); A.V.Rama Rao, E.R. Reddy, B.V. Joshi and J.S. Yadav, Tetrahedron Lett., 28, 6497 (1987); J.M. Chong, D.R. Cyr and E.K. Mar, Tetrahedron Lett., 28, 5009 (1987).

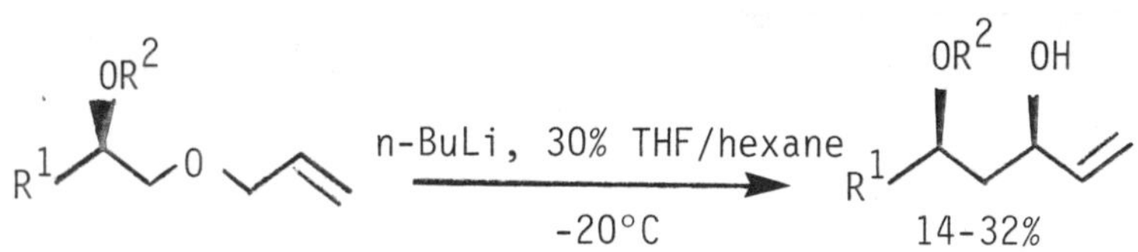

VI.A.2-19 T. Harada, H. Kurokawa and A. Oku, Tetrahedron Lett., 28, 4843, 4847 (1987).

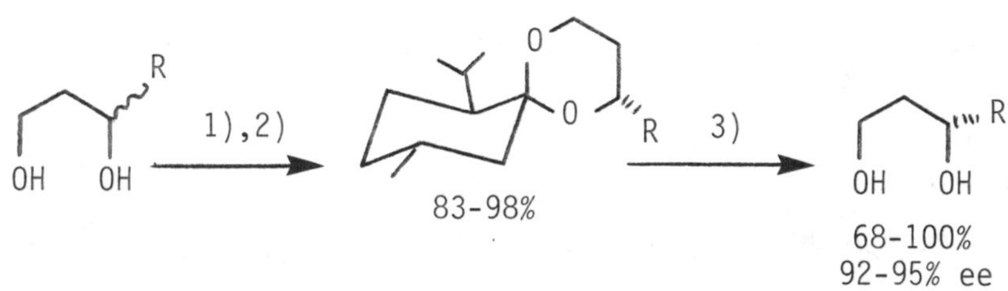

1) $(TMS)_2NH$, TMSOTf, THF, 0°C;

2) l-menthone, TMSOTf, CH_2Cl_2, -40°C; 3) conc HCl, MeOH, RT.

VI.A.2-20 K. Tomooka, T. Okinaga, K. Suzuki and G. Tsuchihashi, Tetrahedron Lett., 28. 6335 (1987).

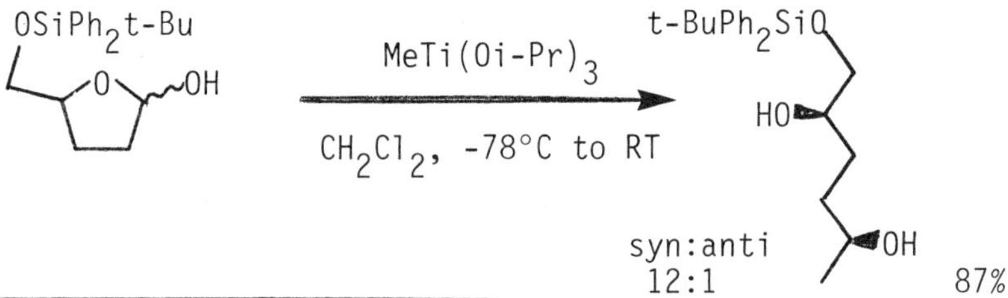

VI.A.3. Alkyl, Aryl Halides

VI.A.3-1 S.T. Purrington, C.L. Bumgardner, N.V. Lazaridis and P. Singh, J. Org. Chem., 52, 4307 (1987).

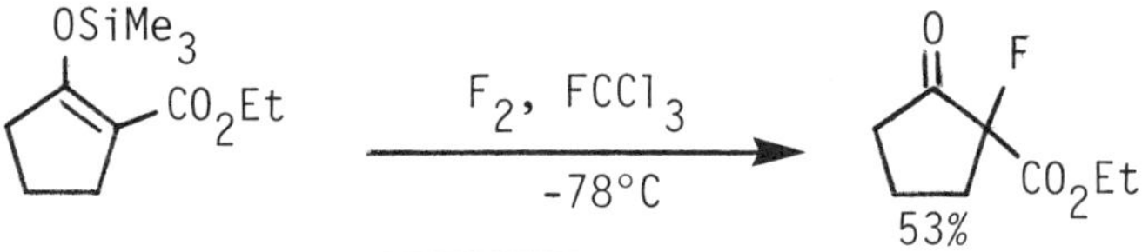

VI.A.3-2 A. Gregorcic and M. Zupan, Bull. Chem. Soc. Jpn., 60, 3083 (1987).

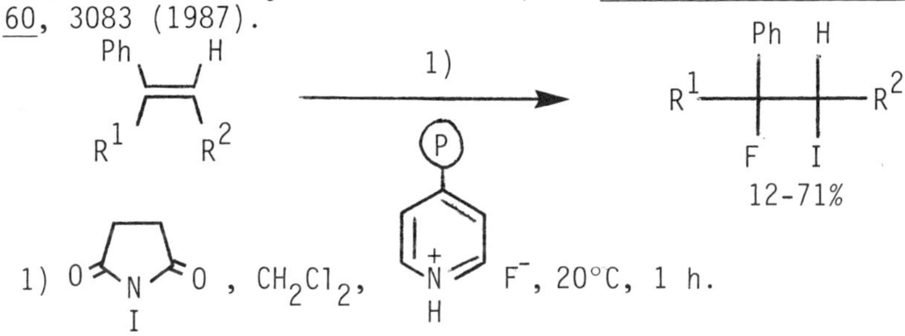

VI.A.3-3 S. Stauber and M. Zupan, J. Org. Chem., 52, 5022 (1987).

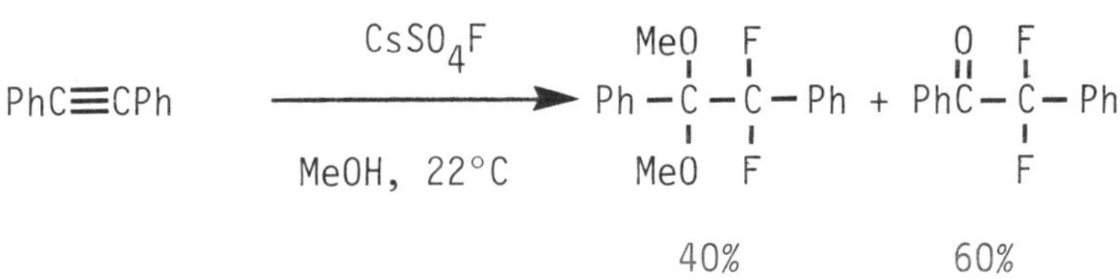

VI.A.3-4 S. Kanemoto, M. Shimizu and H. Yoshioka, Tetrahedron Lett., 28, 663 (1987).

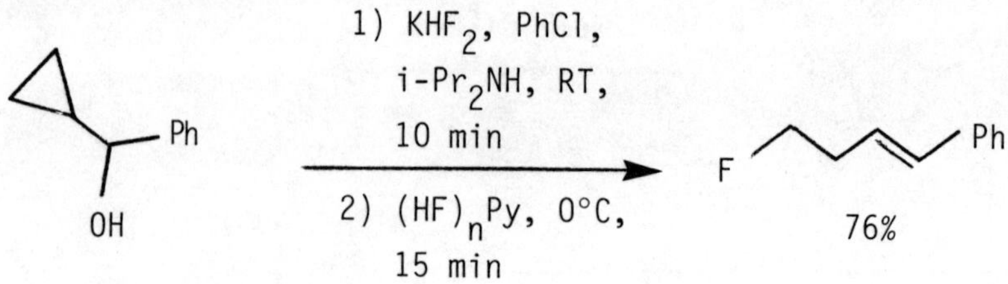

VI.A.3-5 M.R. Bryce, R.D. Chambers, S.T. Mullins and A. Parkin, Bull. Soc. Chim. Fr., 930 (1986).

$$\text{ArM} \xrightarrow{CF_3OF,\ CH_2Cl_2} \text{ArF}$$

11-87%

VI.A.3-6 P. Bosch, F. Camps, E. Chamorro, V. Gasol and A. Guerrero, Tetrahedron Lett., 28, 4733 (1987).

$$\text{RX} \xrightarrow[\text{THF, HMPT}]{Bu_4N^+HF_2^-} \text{RF}$$

51-100%

VI.A.3-7 J. Ichikawa, K. Sugimoto, T. Sonada and H. Kobayashi, Chem. Lett., 1985 (1987).

$$R^1SR^2 \xrightarrow[CH_2Cl_2]{FSO_3Me} R^2\text{—}\overset{R^1}{\underset{R^3}{S^+}}FSO_3^- \xrightarrow[\text{reflux}]{CsF} R^1F + R^2SR^3$$

VI.A.3-8 T. Fukuhara, N. Yoneda, T. Sawada and A. Suzuki
Synth. Commun., 17, 685 (1987).

$$ArNH_2 \xrightarrow[\text{2) }NaNO_2\text{, 0°C, then 55-130°C}]{\text{1) HF-pyridine, 0°C, RT}} ArF \quad 15\text{-}99\%$$

VI.A.3-9 A.J. Bloodworth, K.J. Bowyer and J.C. Mitchell,
Tetrahedron Lett., 28, 5347 (1987).

$$PhCH_2CCl_2Me \xrightarrow[\text{1 h, RT}]{AgBF_4\text{, }CH_2Cl_2} PhCH_2CF_2Me \quad 50\%$$

VI.A.3-10 S. Rozen, M. Brand, D. Zamir and D. Hebel,
J. Am. Chem. Soc., 109, 896 (1987).

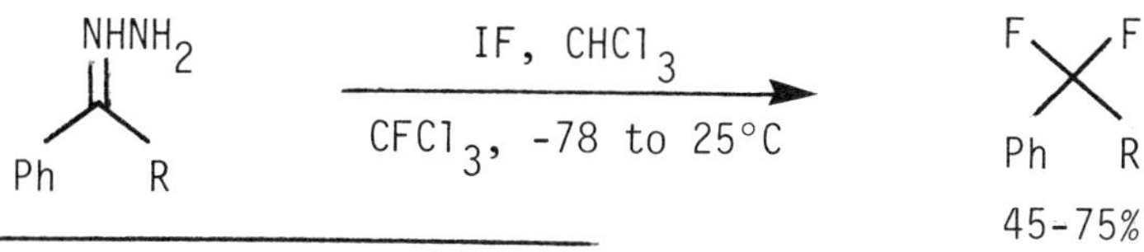

VI.A.3-11 D.H.R. Barton, B. Lacher and S.Z. Zard,
Tetrahedron, 43, 4321 (1987); D. Crich and S.M. Fortt,
Synthesis, 35 (1987).

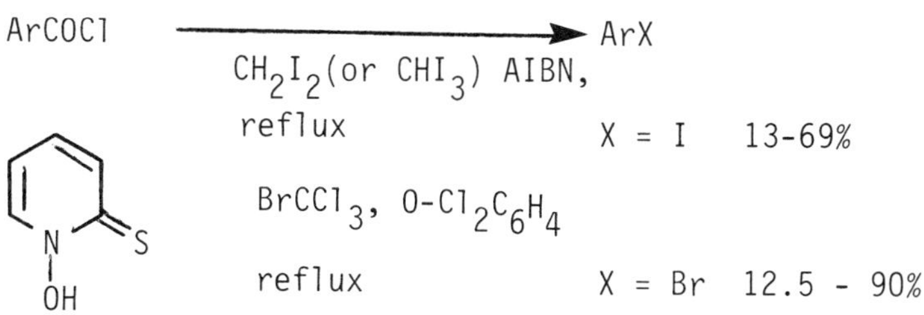

VI.A.3-12 L. Engman, J. Org. Chem., 52, 4086 (1987); M. Kameyama and N. Kamigata, Bull. Chem. Soc. Jpn., 60, 3687 (1987).

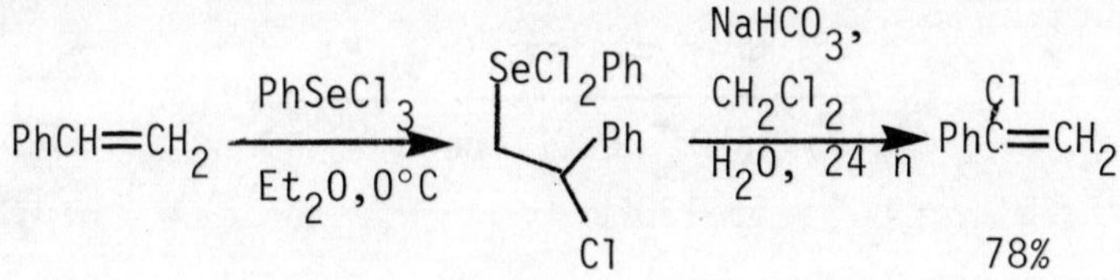

VI.A.3-13 P. Tundo, P. Venturello and E. Angeletti, J. Chem. Soc., Perkin I, 2157 (1987); X. Gu, I. Ikeda and M. Okahara, Bull. Chem. Soc. Jpn., 60, 397 (1987).

$$\text{BuOH} \xrightarrow[\text{HCl, } H_2O\text{, 170°C}]{Bu_4P^+Cl^-\text{-silica gel}} \text{BuCl} \quad 87\%$$

VI.A.3-14 K.B. Yoon and J.K. Kochi, J. Chem. Soc., Chem. Commun., 1013 (1987).

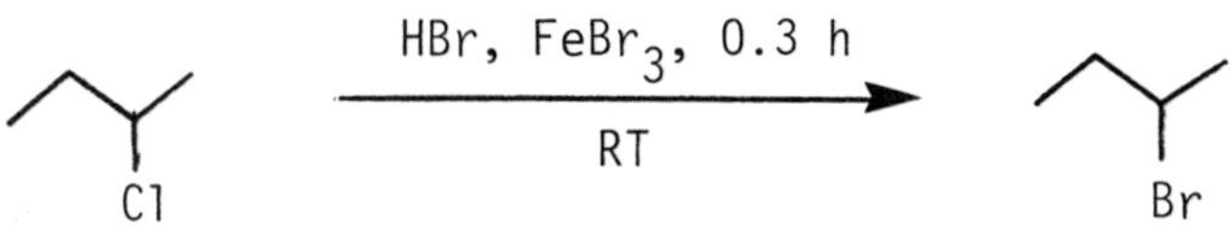

VI.A.3-15 S.P. Schmidt and D.W. Brooks, Tetrahedron Lett., 28, 767 (1987); H. Mattes and C. Benezra, Tetrahedron Lett., 28, 1697 (1987).

$$R^1OR^2 \xrightarrow[\text{0 to 23°C}]{(Ph_2PCX_2)_2\text{, } CH_2Cl_2} R^1X \quad 42\text{-}98\%$$

R^2 = H or THP

X = Br or I

VI.A.3-16 A. Hassner and K. Murthy, Tetrahedron Lett., 28, 683 (1987).

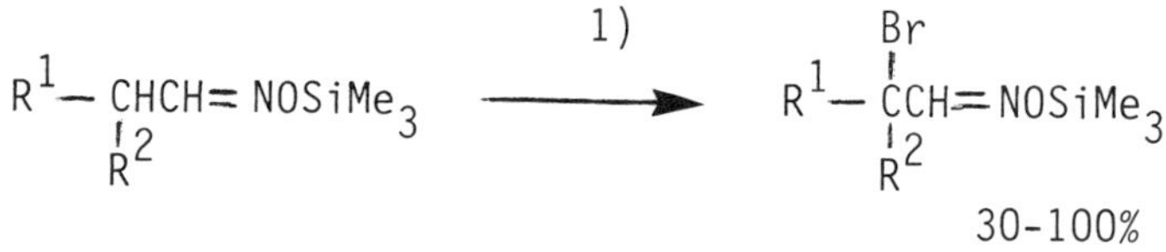

1) hν, NBS, CCl_4, RT.

VI.A.3-17 F. Camps, V. Gasol and A. Guerro, Synthesis, 511 (1987); S.P. Collingwood, A.P. Davies and B.T. Golding, Tetrahedron Lett., 28, 4445 (1987).

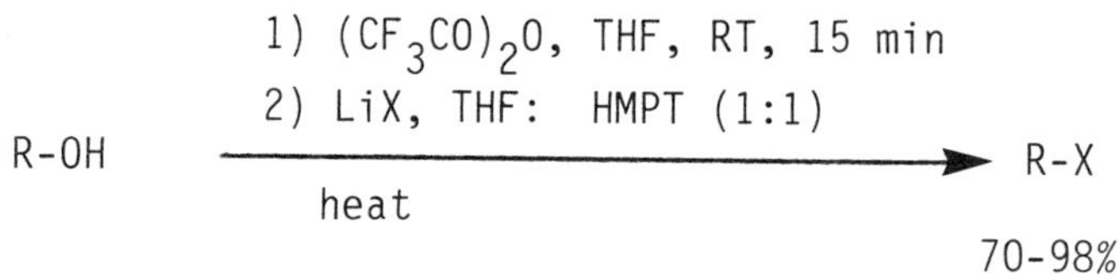

VI.A.3-18 A. Oliva and A. Molinari, Synth. Commun., 17, 837 (1987).

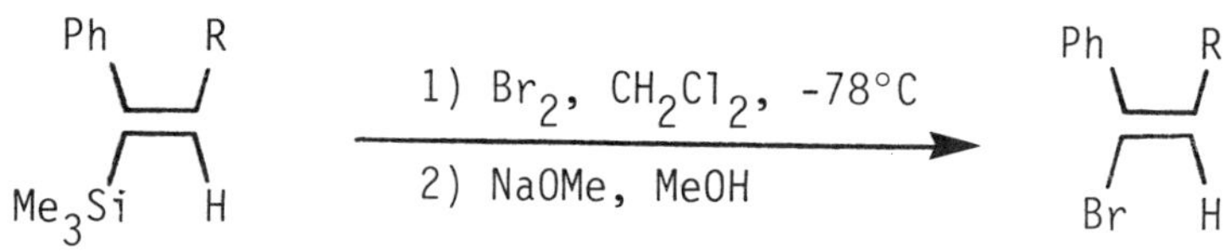

VI.A.3-19 J. Barluenga et al., Synthesis, 661 (1987).

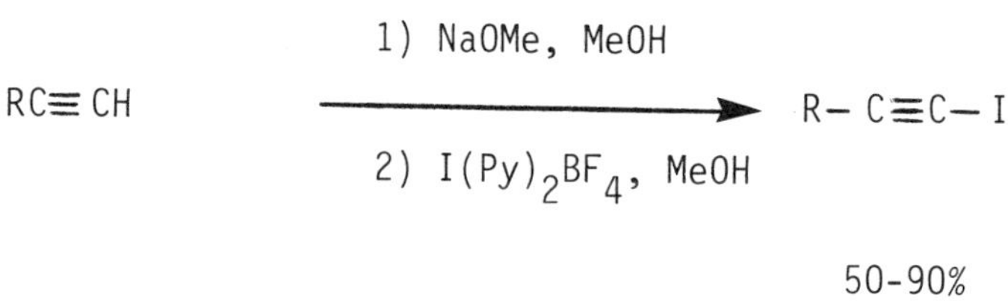

VI.A.3-20 L.J. Stewart, D. Gray, R.M. Pagni and G.W. Kabalka, Tetrahedron Lett., 28, 4497 (1987).

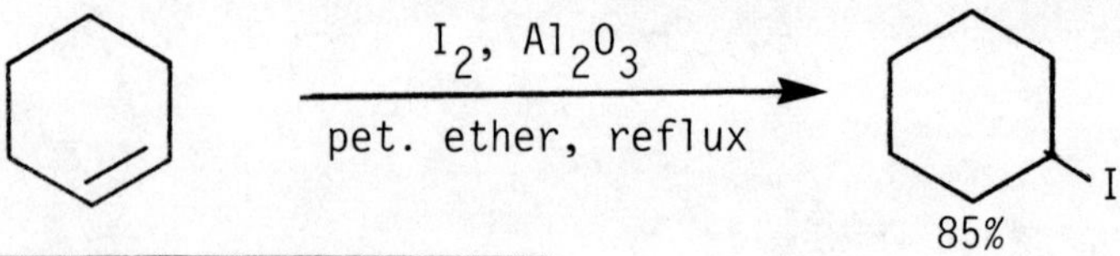

VI.A.3-21 S.H. Yang, C.S. Li and C.H. Cheng, J. Org. Chem., 52, 691 (1987).

$$\text{ArBr} \xrightarrow[\text{150°C}]{\text{KI, Ni, DMF}} \text{ArI}$$

73-95%

VI.A.3-22 E. Keinan and D. Perez, J. Org. Chem., 52, 4846 (1987); A. Garcia et al., Tetrahedron Lett., 28, 6441 (1987).

$$\text{MeCH(OH)C}_6\text{H}_{13} \xrightarrow[\text{RT, 2 h}]{\text{I}_2\text{SiH}_2\text{, CH}_2\text{Cl}_2} \text{MeCHIC}_6\text{H}_{13}$$

93%

VI.A.3-23 E. Laurent, B. Marquet, R. Tardivel and H. Thiebault, Bull. Soc. Chim. Fr., 955 (1986).

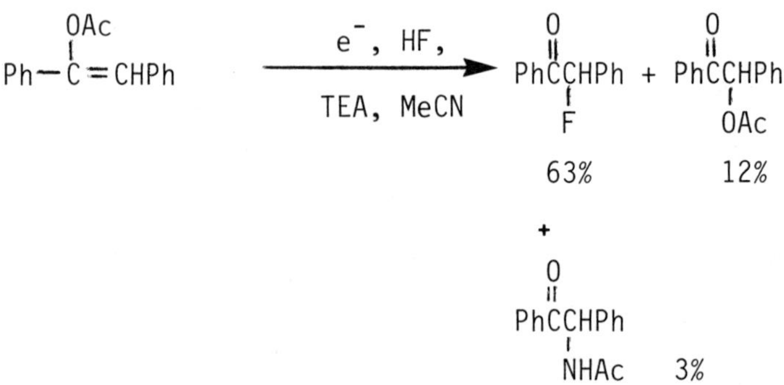

VI.A.4. Amides

VI.A.4-1 S. Rajeswari, R.J. Jones and M.P. Cava, Tetrahedron Lett., 28, 5099 (1987).

$$R^1\overset{O}{\overset{\|}{C}}F + Me_3SiNR^2R^3 \xrightarrow[\text{MeCN, RT, 6 h}]{Bu_4NF} R^1\overset{O}{\overset{\|}{C}}NR^2R^3 \quad 73\text{-}98\%$$

VI.A.4-2 T. Högberg, P. Ström, M. Ebner and S. Rämsby, J. Org. Chem., 52, 2033 (1987); K. Arai, C. Shaw, K. Nozawa, K. Kawai and S. Nakajima, Tetrahedron Lett., 28, 441 (1987).

$$R^1CO_2Et \xrightarrow[\text{MeOH, 50°C}]{R^1R^2NH,\ NaCN,} R^1CONR^2R^3 \quad 47\text{-}87\%$$

VI.A.4-3 S.V. Ley and P.R. Woodward, Tetrahedron Lett., 28, 3019 (1987).

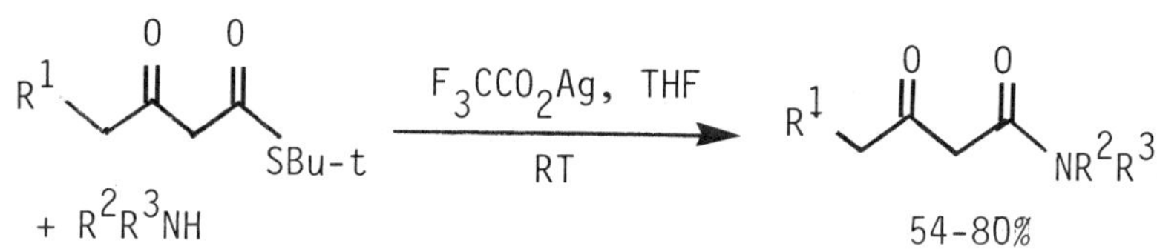

VI.A.4-4 W. Mormann and E. Hissmann, Tetrahedron Lett., 28, 3087 (1987); Y. Zhang, J. Jiang and Y. Chen, Tetrahedron Lett., 28, 3815 (1987).

$$OCN(CH_2)_nCOCl + Me_3SiXR \xrightarrow{0\text{-}120°C,\ N_2,\ PhMe} OCN(CH_2)_n\overset{O}{\overset{\|}{C}}XR \ (70\text{-}92\%) + TMSCl$$

X = O or NH

VI.A.4-5 S. Murahashi, T. Naota and N. Nakajima, Chem. Lett., 879 (1987).

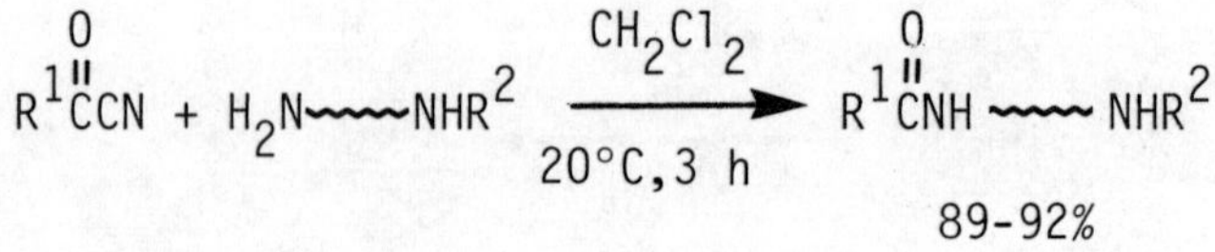

VI.A.4-6 K. Matsumoto, S. Hashimoto, T. Okamoto, S. Otani and J. Hayami, Chem. Lett., 803 (1987).

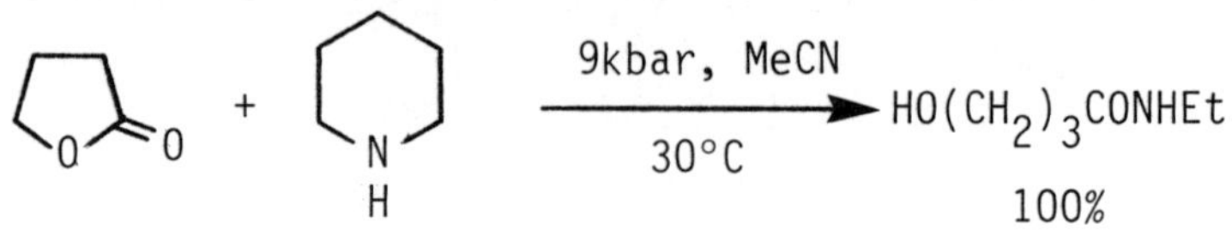

VI.A.4-7 L. Nakonieczna and A. Chimiak, Synthesis, 418 (1987).

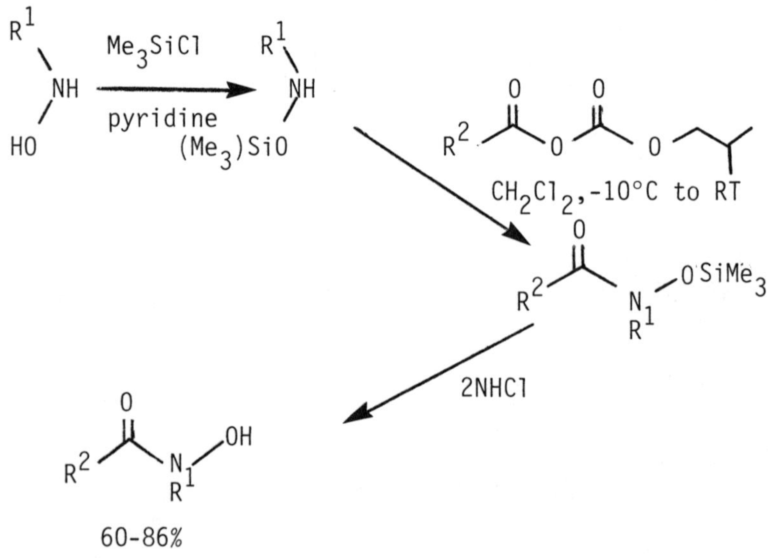

VI.A.5. Amines and Carbamates

VI.A.5-1 S.E. Denmark, T. Weber and D.W. Piotrowski, J. Am. Chem. Soc., 109, 2224 (1987).

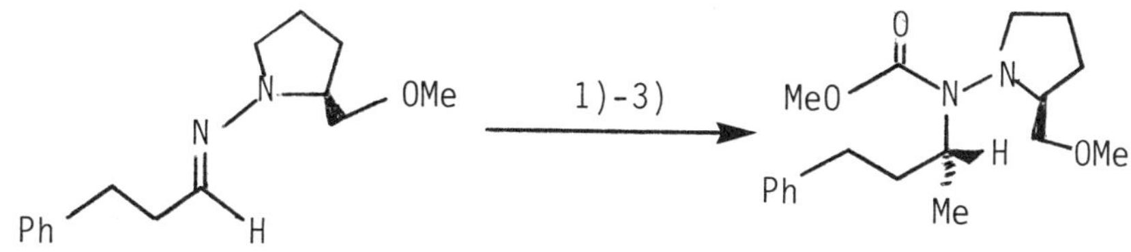

1) MeLi, $CeCl_3$, THF, -78°C, 1 h;
2) -78°C to RT, 2-3 h; 3) $ClCO_2Me$, 20°C, 12 h.

81%
98% d.s.

VI.A.5-2 J.L. Marco, J. Royer and H. Husson, Synth. Commun., 17, 669 (1987); J. Yaozhong, L. Guilan, L. Jinchu and Z. Changyou, Synth. Commun., 17, 1545 (1987).

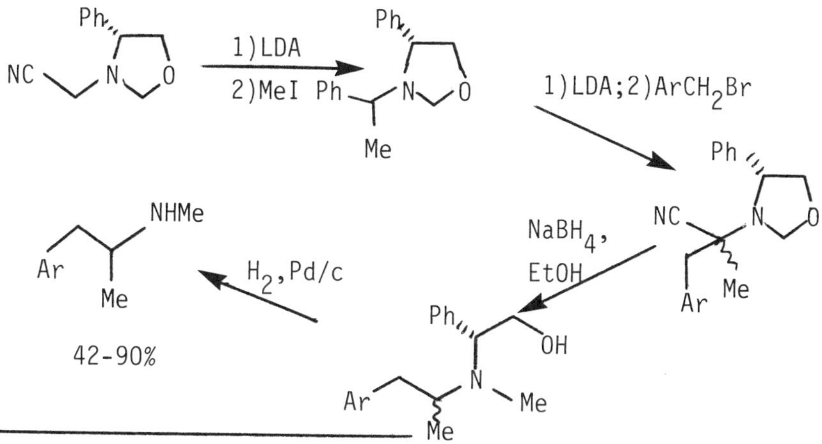

VI.A.5-3 H.C. Brown, K. Kim, M. Srebnik and B. Singaram, Tetrahedron, 43, 4071 (1987); G.W. Kabalka, D.A. Henderson and R.S. Varma, Organometallics, 6, 1369 (1987); A. Koziara, K. Osowska-Pacewicka, S. Zawadzki and A. Zwierzak, Synthesis, 487 (1987).

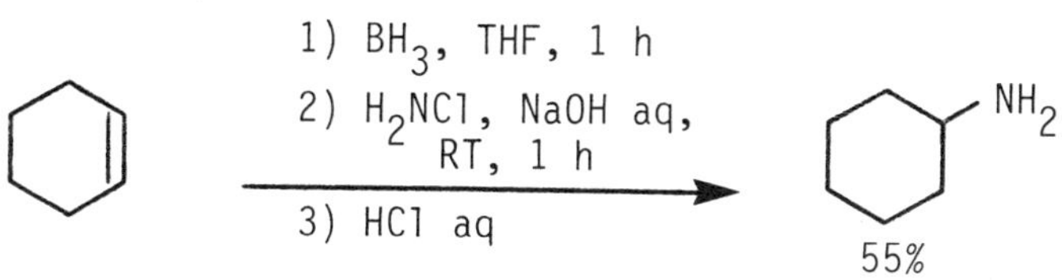

VI.A.5-4 A.R. Katritzky, S. Rachwal and B. Rachwal, <u>J. Chem. Soc., Perkin I</u>, 799, 805 (1987).

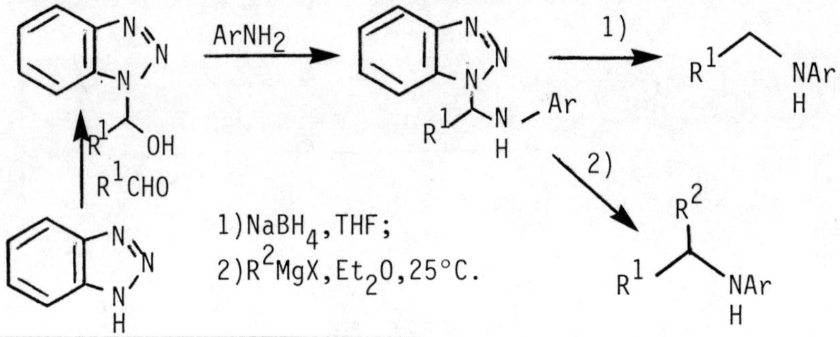

VI.A.5-5 A.R. Katritzky, M. Drewniak and J.M. Aurrecoechea, <u>J. Chem. Soc., Perkin I</u>, 2539 (1987).

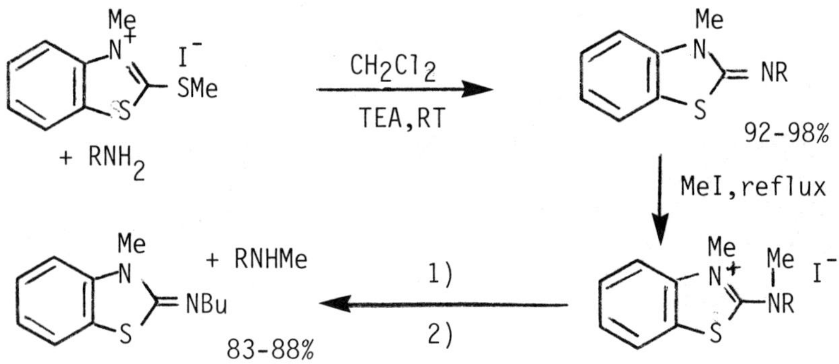

1)$BuNH_2$,CH_2Cl_2,25°C,3 h; 2)KOH,25°C,3 h.

VI.A.5-6 H.C. Brown, M.M. Midland and A.B. Levy, <u>Tetrahedron Lett.</u>, <u>43</u>, 4079 (1987).

$$R^1_3B + R^2N_3 \xrightarrow[\text{2) HCl aq}]{\text{1) reflux, PhH}} R^1NHR^2$$

14-80%

VI.A.5-7 T. Pienemann and H. Schäfer, Synthesis, 1005 (1987); N. Kambe, T. Inagaki, N. Miyoshi, A. Ogawa and N. Sonoda, Chem. Lett., 1275 (1987); R.O. Hutchins, A. Abdel-Magid, Y.P. Stercho and A. Wambsgans, J. Org. Chem., 52, 704 (1987).

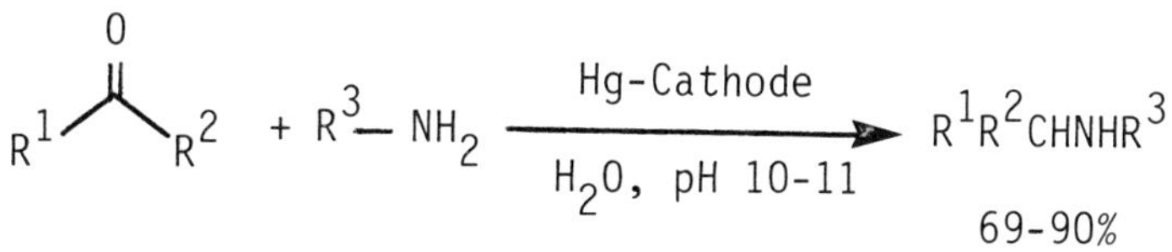

VI.A.5-8 D.H.R. Barton, J.-P. Finet and J. Khamsi, Tetrahedron Lett., 28, 887 (1987); D.H.R. Barton, N. Yadav-Bhatnagar, J.P. Finet and J. Khamsi, Tetrahedron Lett., 28, 3111 (1987).

$$\text{p-MeOC}_6\text{H}_4\text{NH}_2 \xrightarrow[\text{RT, CH}_2\text{Cl}_2\text{, Ar, 24 h}]{\text{Ph}_3\text{Bi, Cu(OAc)}_2} \text{p-MeOC}_6\text{H}_4\text{NHPh}$$

82%

VI.A.5-9 J. Kang and K.S. Kim, J. Chem. Soc., Chem. Commun., 897 (1987).

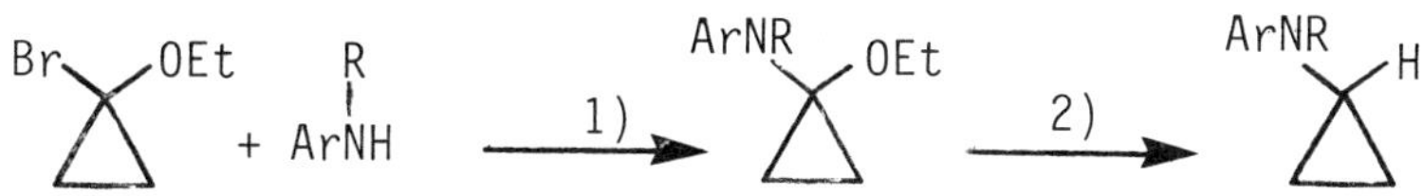

1) CH_2Cl_2, reflux; 2) $NaBH_4$, $Et_2O.BF_3$, THF, 0°C

VI.A.5-10 S. Gauthier and J.M.J. Fréchet, Synthesis, 383 (1987).

$$\text{Ph}_2\text{NH} + \text{ArI} \xrightarrow[\text{18-crown-6, o-dichlorobenzene, reflux}]{\text{Cu, K}_2\text{CO}_3} \text{Ph}_2\text{NAr}$$

78-93%

VI.A.5-11 G.W. Gribble and C.F. Nutaitis, <u>Synthesis</u>, 709 (1987); A. Basha and D.W. Brooks, <u>J. Chem. Soc., Chem. Commun.</u>, 305 (1987).

$$R^1NH_2 \xrightarrow[\text{THF, RT}]{(HCHO)_n,\ NaBH_4,\ CF_3CO_2H,} R^1NMe_2$$

59-74%

VI.A.5-12 S. Knapp and A.T. Lavorse, <u>Tetrahedron Lett.</u>, <u>28</u>, 3213 (1987).

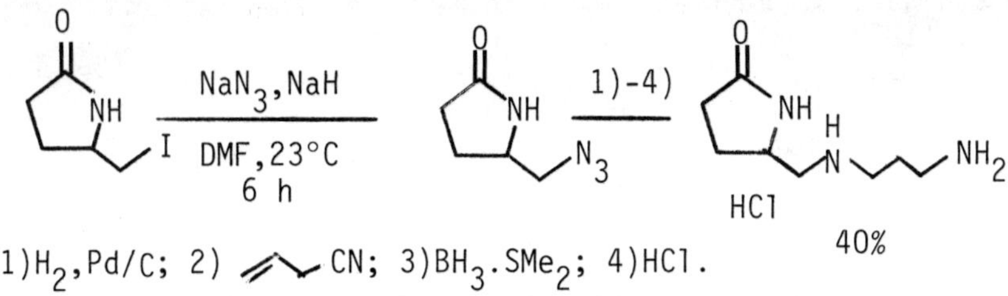

VI.A.5-13 E.J. Roskamp and S.F. Pedersen, <u>J. Am. Chem. Soc.</u>, <u>109</u>, 3152 (1987).

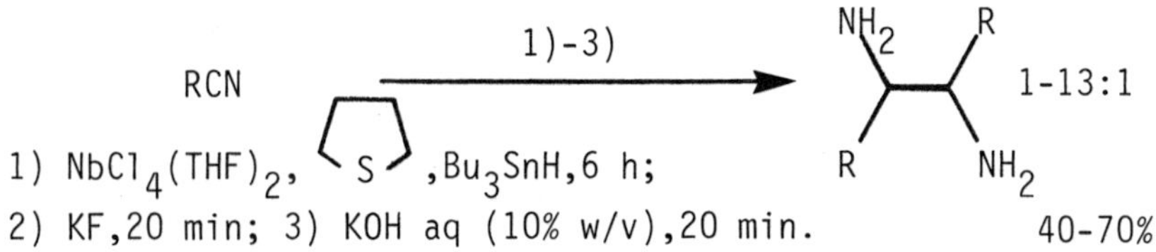

VI.A.5-14 C. Betschart and D. Seebach, <u>Helv. Chim. Acta</u>, <u>70</u>, 2215 (1987).

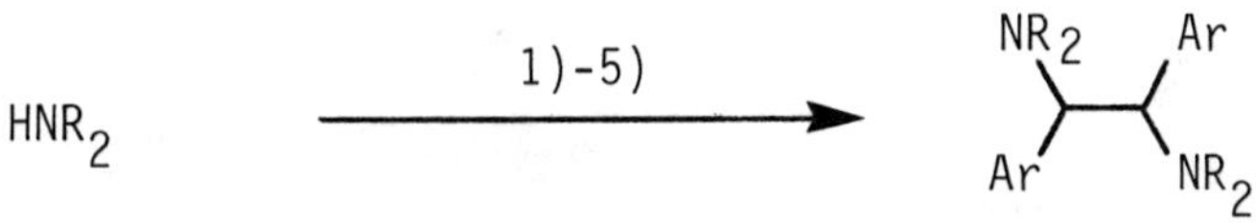

1) BuLi, hexane; 2) ArCHO; 3) $TiCl_4$, -70°C to -10°C, RT; 4) $TiCl_4$/Mg, -70°C to RT; 5) hydrolysis

VI.A.5-15 E.J. Roskamp and S.F. Pedersen, J. Am. Chem. Soc., 109, 6551 (1987); J.A. Marsella, J. Org. Chem., 52, 467 (1987).

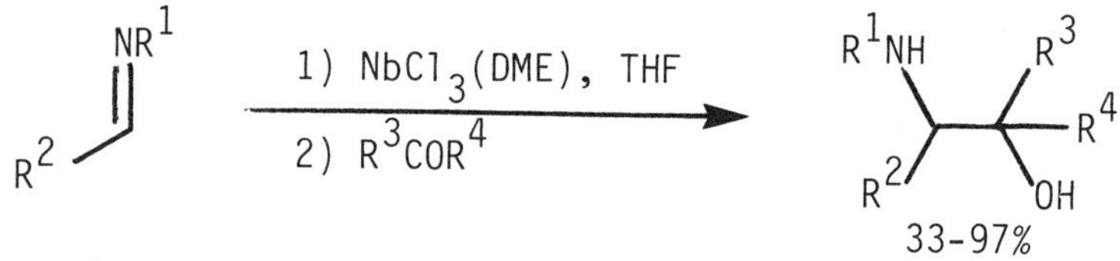

VI.A.5-16 T. Gajda, Synthesis, 1108 (1987).

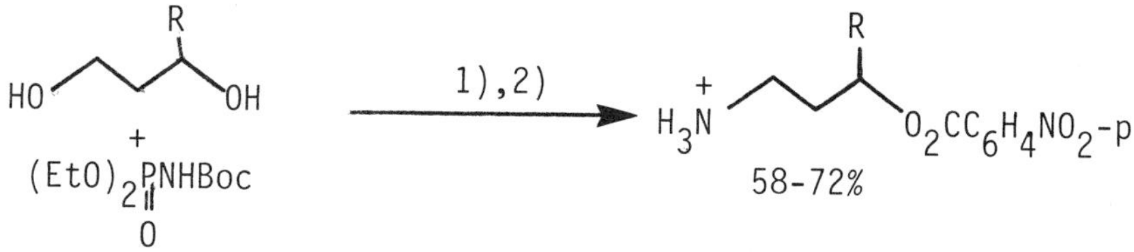

1) $Ph_3P, EtO_2CN{=}NCO_2Et, Et_2O, 0°C$;
2) $4\text{-}O_2NC_6H_4COCl$, Py, 0°C to RT; 3) HCl, EtOAc.

VI.A.5-17 M. Kira, S. Nagai, M. Nishimura and H. Sakurai, Chem. Lett., 153 (1987).

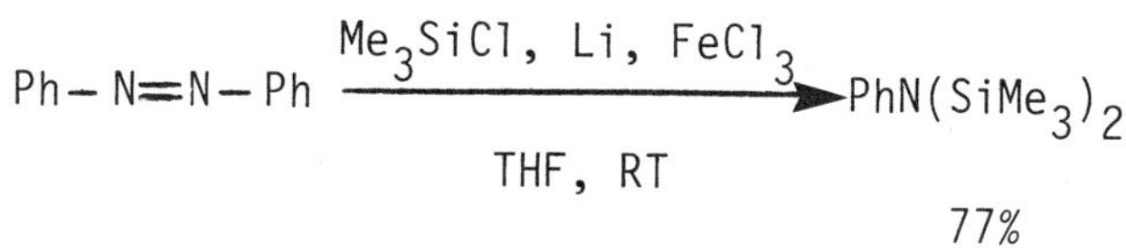

VI.A.5-18 G. Vasapollo and H. Alper, Tetrahedron Lett., 28, 6411 (1987); G. Sennyey et al., Tetrahedron Lett., 28, 5823, 5827 (1987); T. Mitsudo, Y. Hori, Y. Yamakawa and Y. Watanabe, Tetrahedron Lett., 28, 4417 (1987); H. Ogura et al., Synthesis, 557 (1987); C. Bruneau and P.H. Dixneuf, Tetrahedron Lett., 28, 2005 (1987); U.R. Nayak, V.B. Deodhar and V.S. Dalavoy, Chem. Ind. (London), 627 (1987); H. Alper et al., Organometallics, 6, 2391 (1987); L.D. Arnold, J.C.G. Drover and J.C. Vederas, J. Am. Chem. Soc., 109, 4649 (1987).

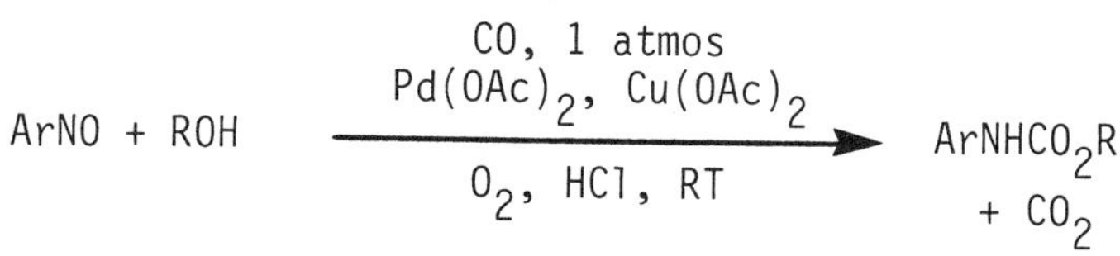

VI.A.6. Amino Acids and Derivatives

VI.A.6-1 J.R. McCarthy, C.L. Barney, M.J. O'Donnell and J.C. Huffman, J. Chem. Soc., Chem. Commun., 469 (1987); T. Kolasa, S. Sharma and M.J. Miller, Tetrahedron Lett., 28, 4973 (1987).

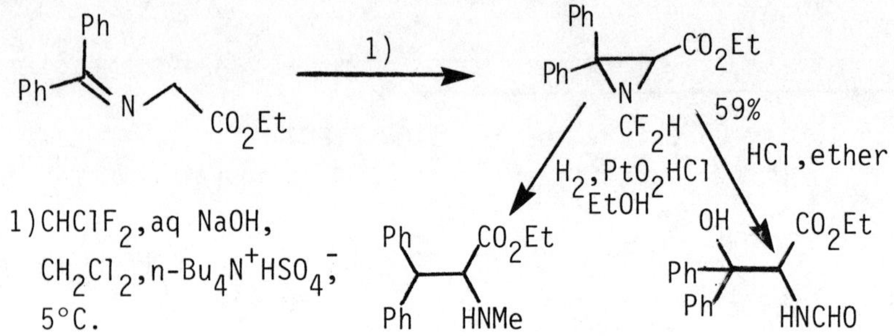

VI.A.6-2 J.E. Baldwin, R.M. Adlington and N.G. Robinson, J. Chem. Soc., Chem. Commun., 153 (1987).

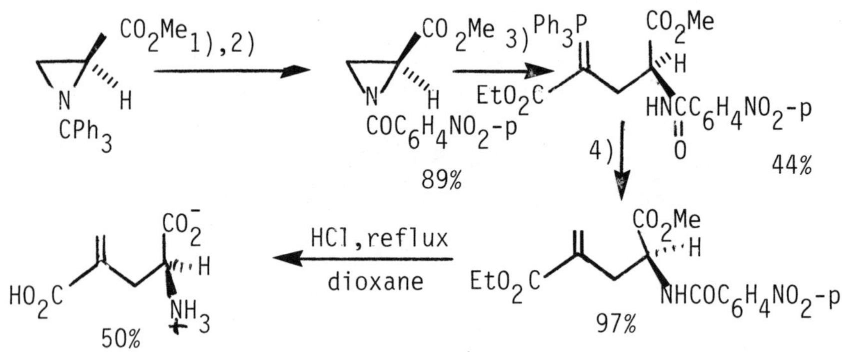

1)CF_3CO_2H;2)p-$NO_2C_6H_4COCl$;3)Ph_3P=CO_2Et;4)paraformaldehyde.

VI.A.6-3 D.A. Evans, E.B. Sjogren, A.E. Weber and R.E. Conn, Tetrahedron Lett., 28, 39 (1987); D.A. Evans, J.A. Ellman and R.L. Dorow, Tetrahedron Lett., 28, 1123 (1987).

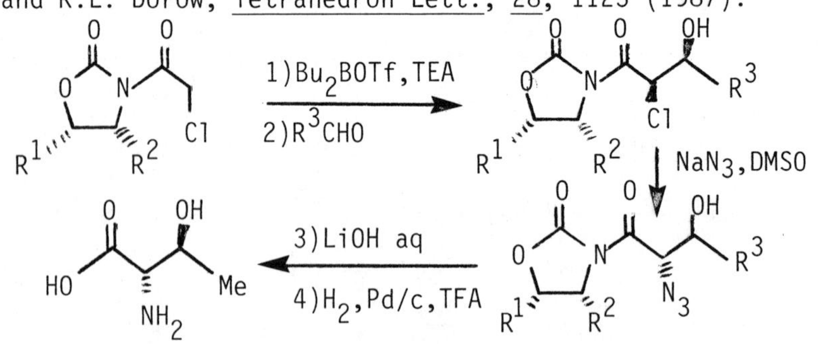

VI.A.6-4 D.S. Matteson and E.C. Beedle, Tetrahedron Lett., 28, 4499 (1987); S. Takano, M. Yanase, Y. Sekiguchi and K. Ogasawara, Tetrahedron Lett., 28, 1783 (1987); M. Utaka, S. Konishi, T. Okubo, S. Tsuboi and A. Takeda, Tetrahedron Lett., 28, 1447 (1987).

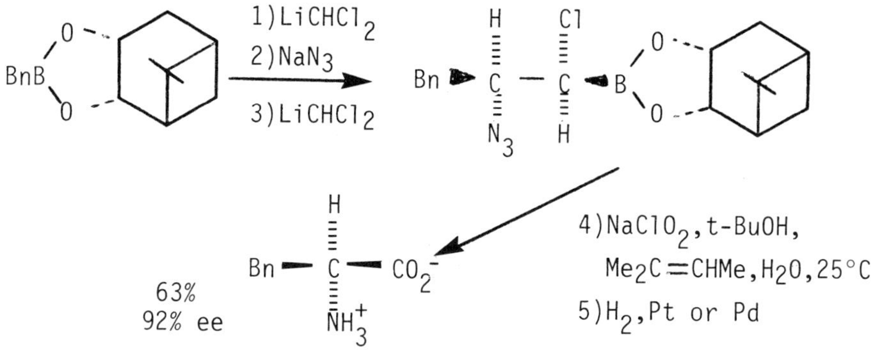

VI.A.6-5 J.E. Baldwin, R.L. Dyer, S.C. Ng, A.J. Pratt and M.A. Russell, Tetrahedron Lett., 28, 3745 (1987).

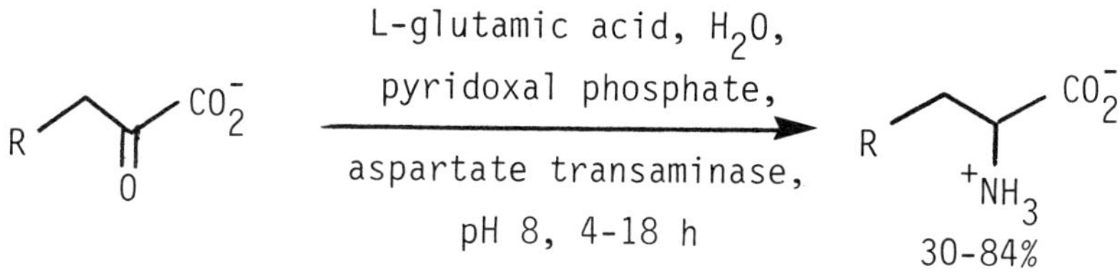

VI.A.6-6 B.I. Glänzer, K. Faber and H. Griengl, Tetrahedron, 43, 771 (1987).

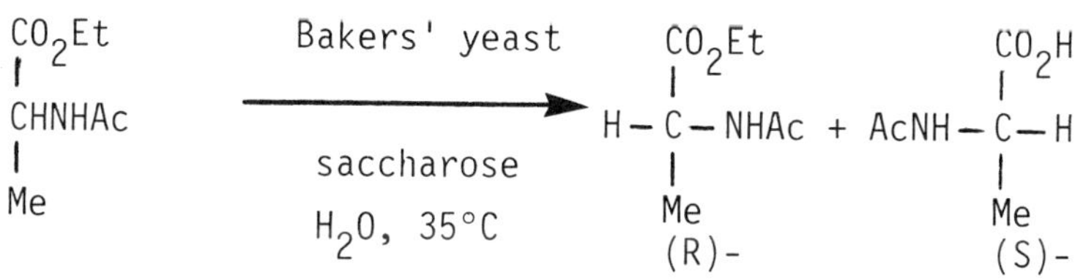

47%
100% ee

VI.A.6-7 R.W. Feenstra, E.H.M. Stokkingreef, R.J.F. Nivard and H.C.J. Ottenheijm, Tetrahedron Lett., 28, 1215 (1987); S.M. Heilmann, K.M. Jensen, L.R. Krepski, D.M. Moren, J.K. Rasmussen and H.K. Smith, II, Synth. Commun., 17, 843 (1987).

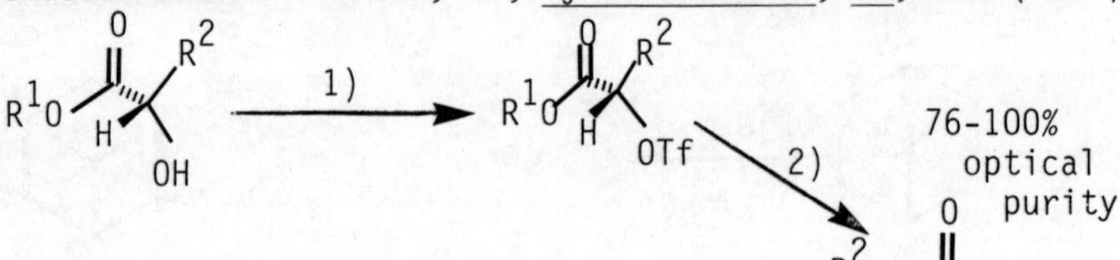

1) Tf_2O, CH_2Cl_2;
2) $PhCH_2ONH_2, CH_2Cl_2$, RT

VI.A.6-8 S. Murahashi and T. Shiota, Tetrahedron Lett., 28, 6469 (1987); M. Mitani, O. Tachizawa, H. Takeuchi and K. Koyama, Chem. Lett., 1029 (1987).

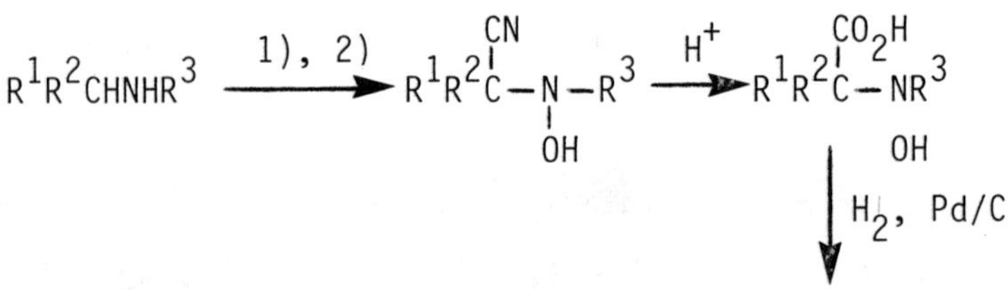

1) H_2O_2, Na_2WO_4;
2) KCN, HCl

VI.A.6-9 S.V. Pansare and J.C. Vederas, J. Org. Chem., 52, 4804 (1987); R. Ramage, C.A. Barron, S. Bielecki and D.W. Thomas, Tetrahedron Lett., 28, 4105 (1987); P. Jouin et al., Tetrahedron Lett., 28, 1661, 1665 (1987); J. Viret, J. Gabard and A. Collet, Tetrahedron, 43, 891 (1987); P. Wipf and H. Heimgartner, Helv. Chim. Acta, 70, 354 (1987); R. Albert, J. Danklmaier, H. Hönig and H. Kandolf, Synthesis, 635 (1987); Y. Vo-Quang et al., Tetrahedron Lett., 28, 4057 (1987).

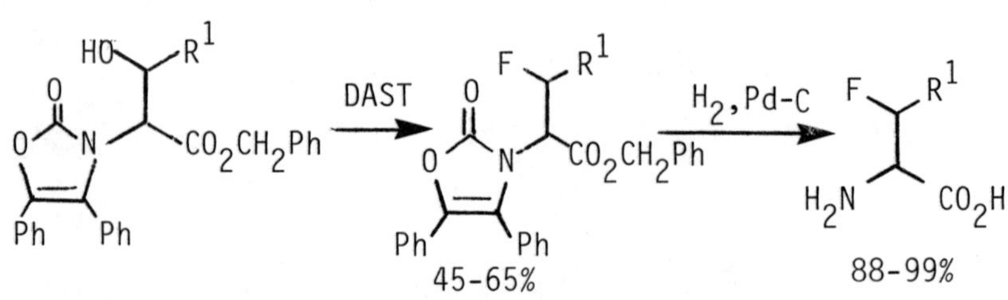

VI.A.6-10 A.L. Margolin and A.M. Klibanov, J. Am. Chem. Soc., 109, 3802 (1987); J.B. West and C. Wong, Tetrahedron Lett., 28, 1629 (1987); P.A. Grieco and A. Bahsas, J. Org. Chem., 52, 5746 (1988).

$$\text{N-Ac-L-Phe-OCH}_2\text{CH}_2\text{Cl} + \text{L-Leu-NH}_2 \xrightarrow{1)} \text{N-Ac-L-Phe-L-Leu-NH}_2$$

1) Porcine pancreatic lipase, PhMe, 45°C

VI.A.7. Esters

VI.A.7-1 H. Ogura et al., Synthesis, 560 (1987); S. Ahmad and J. Iqbal, Chem Lett., 953 (1987); S. Fukuoka, S. Takimoto, T. Katsuki and M. Yamaguchi, Tetrahedron Lett., 28, 4711 (1987).

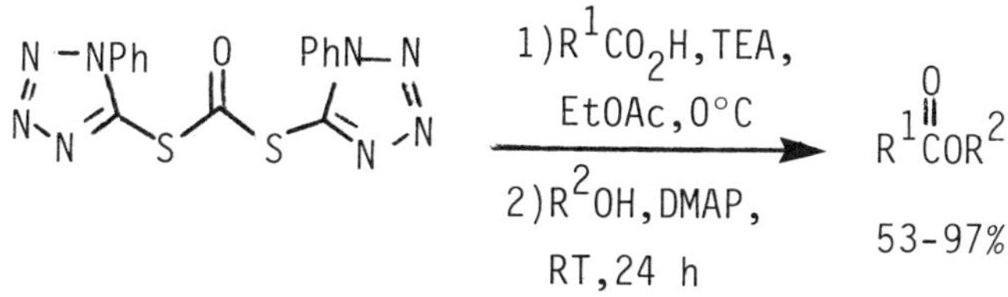

VI.A.7-2 M. Degueil-Castaing, B. de Jeso, S. Drouillard and B. Maillard, Tetrahedron Lett., 28, 953 (1987); M. Mlekuz, F. Joo and H. Alper, Organometallics, 6, 1591 (1987).

$$n\text{-PrCO}_2\text{CH=CH}_2 + \text{EtOH} \xrightarrow[\text{4 Å mol sieves, reflux, THF}]{\text{PPL}} n\text{-PrCO}_2\text{Et} + \text{CH}_3\text{CHO} \quad 83\%$$

VI.A.7-3 B. O'Connor and G.Just, Tetrahedron Lett., 28, 3235 (1987); M. Kuo and K. Liu, J. Org. Chem., 52, 2927 (1987).

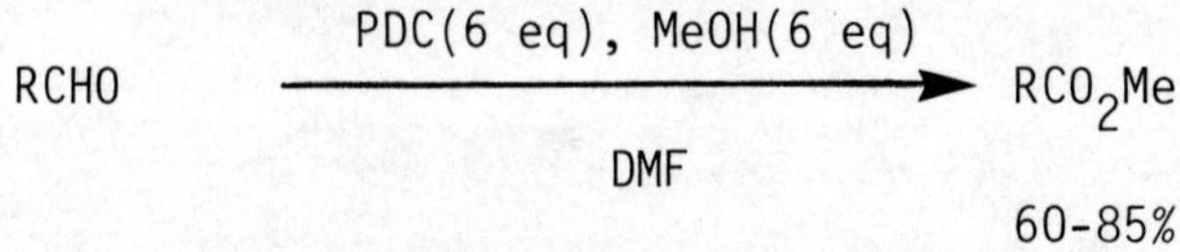

VI.A.7-4 D.G. Salunkhe, M.H. Jagdale, M.M. Salunkhe and P.P. Wadgaonkar, Bull. soc. Chim. Belg., 96, 247 (1987); M. Ohshima and T. Mukaiyama, Chem. Lett., 377 (1987); E. Guibé-Jampel, G. Rousseau and J. Salaün, J. Chem. Soc., Chem. Commun., 1080 (1987).

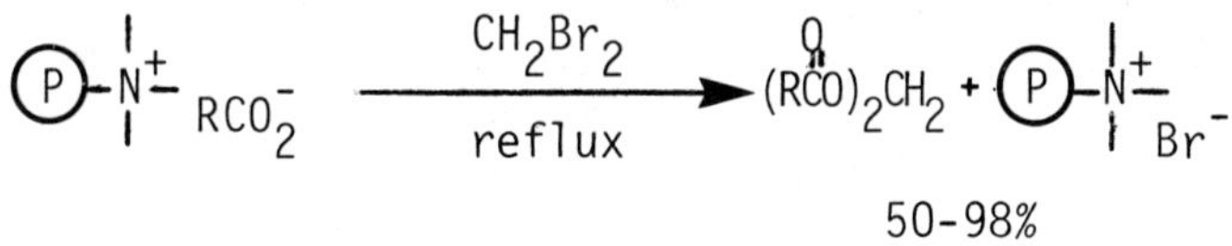

VI.A.7-5 D. Plusquellec and K. Baczko, Tetrahedron Lett., 28, 3809 (1987).

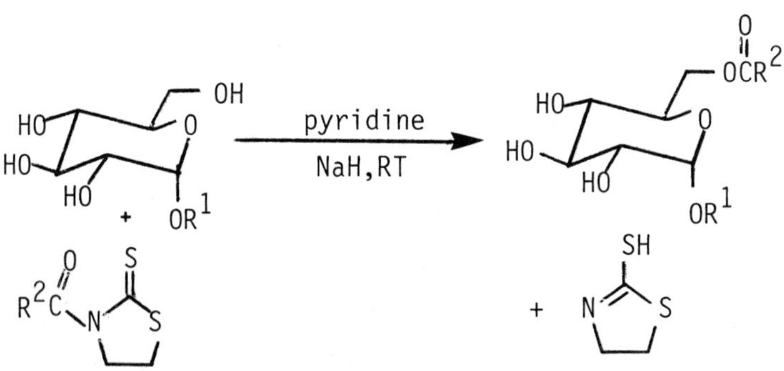

VI.A.7-6 H. Feuer, H.S. Bevinakatti and X. Luo, J. Heterocycl. Chem., 24, 9 (1987).

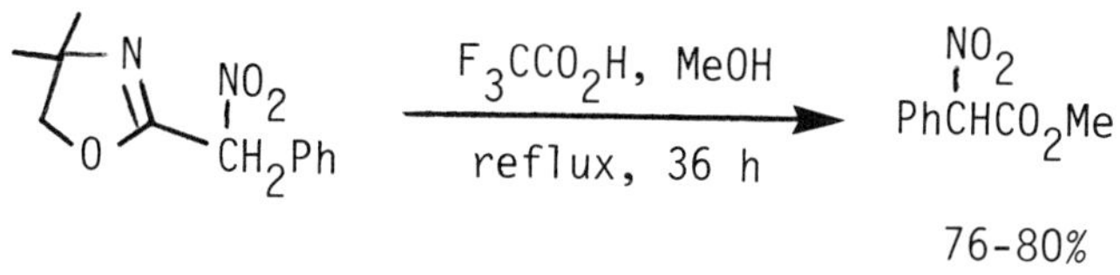

VI.A.7-7 W.H. Pearson and M. Cheng, J. Org. Chem., 52, 3176 (1987); N. Nagashima and M. Ohno, Chem. Lett., 141 (1987); M. Larchevêque and Y. Petit, Tetrahedron Lett., 28, 1993(1987).

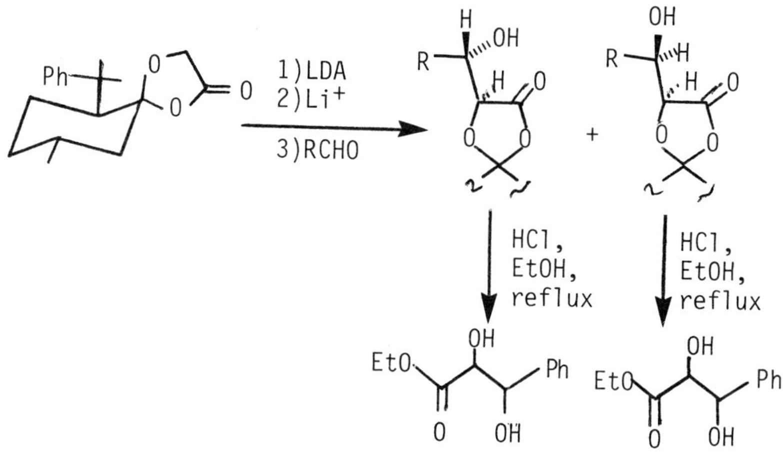

VI.A.7-8 G.R. Newkome and G.R. Baker, Org. Prep. Proced. Int., 18, 117 (1986).

Review: "The Chemistry of Methanetricarboxylic Esters"

VI.A.7-9 T. Ohmaye, J. Synth. Org. Chem. Jpn., 45, 691 (1987).

Review: "Review on Synthetic Methods for Vinyl Acetate"

VI.A.8. Ethers

VI.A.8-1 S.E. Denmark, M.A. Harmata and K.S. White, J. Org. Chem., 52, 4031 (1987); A. Moyano, F. Charbonnier and A.E. Green, J. Org. Chem., 52, 2919 (1987); H. Ogawa, T. Hagiwara, T. Chihara, S. Teratani and K. Taya, Bull. Chem. Soc. Jpn., 60, 627 (1987).

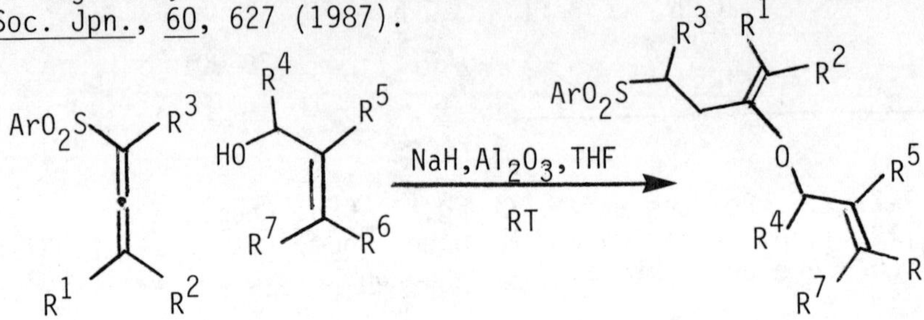

VI.A.8-2 S. Kim, K.N. Chung and S. Yang, J. Org. Chem., 52, 3917 (1987); P.G. Sammes, D. Thetford and M. Voyle, J. Chem. Soc., Chem. Commun., 1373 (1987).

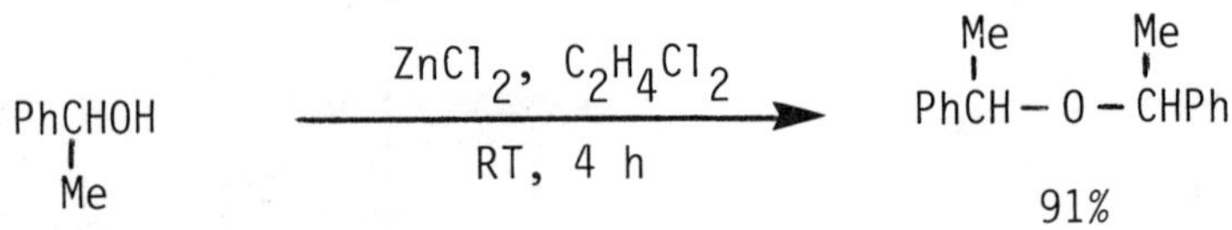

VI.A.8-3 R.D. Walkup, Tetrahedron Lett., 28, 511 (1987); T. Koizumi, T. Fuchigami and T. Nonaka, Chem. Lett., 1095 (1987).

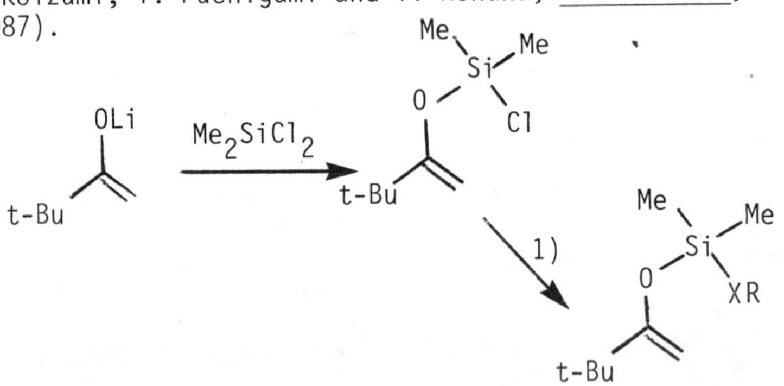

1) RXH,TEA,CH_2Cl_2

VI.A.9. Aldehydes and Ketones

VI.A.9-1 J. Lou and W. Lou, *Synthesis*, 179 (1987); C.L. Willis, *Tetrahedron Lett.*, 28, 2175 (1987); J.L. Graff and M.G. Romanelli, *J. Chem. Soc., Chem. Commun.*, 337 (1987); B.C. Ranu and D.C. Sarkar, *Synth. Commun.*, 17, 155 (1987).

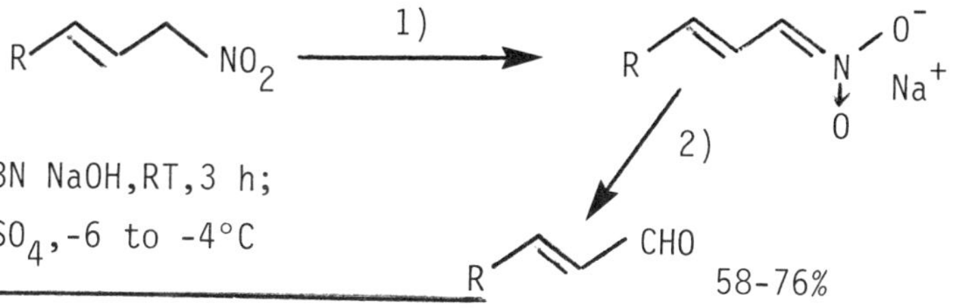

1) 1.3N NaOH, RT, 3 h;
2) H_2SO_4, -6 to -4°C

58-76%

VI.A.9-2 M. Srebnik, T.E. Cole and H.C. Brown, *Tetrahedron Lett.*, 28, 3771 (1987); C.J. Sunley, F.P. Fanizzi, I.M. Saez and P.M. Maitlis, *J. Organomet. Chem.*, 330, C27 (1987); S. Collins and Y. Hong, *Tetrahedron Lett.*, 28, 4391 (1987); F. Clerici, M.L. Gelmi and L.M. Rossi, *Synthesis*, 1025 (1987); J.M. Aizpurua, M. Oiarbide and C. Palomo, *Tetrahedron Lett.*, 28, 5361 (1987); A. Thurkauf, P. Hillery, A.E. Jacobson and K.C. Rice, *J. Org. Chem.*, 52, 5466 (1987); T. Satoh, M. Itoh, T. Ohara and K. Yamakawa, *Bull. Chem. Soc. Jpn.*, 60, 1839 (1987).

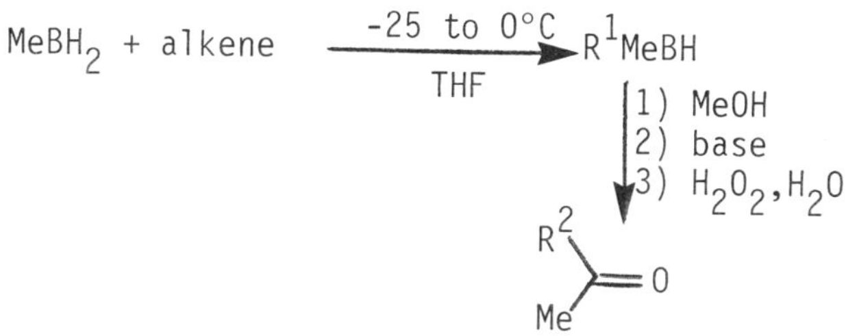

VI.A.9-3 A.R. Katritzky and W. Kuzmierkiewicz, *J. Chem. Soc., Perkin I*, 819 (1987).

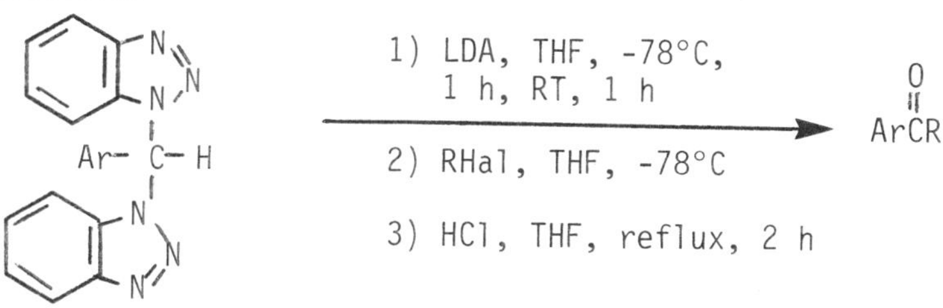

VI.A.9-4 D. Enders and B.B. Lohray, Angew. Chem. Int. Ed. Engl., 26, 351 (1987); G. Guanti, E. Narisano and L. Banfi, Tetrahedron Lett., 28, 4331 (1987).

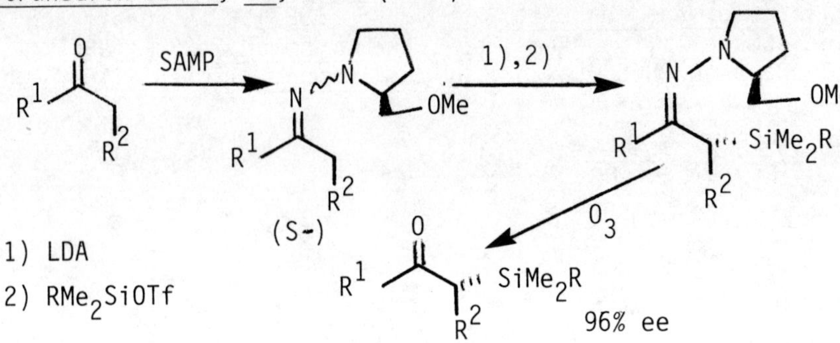

VI.A.9-5 I.R. Silverman, C. Edington, J.D. Elliott and W.S. Johnson, J. Org. Chem., 52, 180 (1987); R.W. Hoffmann, D.K. Ditrich and D.S. Froch, Liebigs Ann. Chem., 977 (1987); M. Shimazaki, H. Hara, K. Suzuki and G. Tsuchihashi, Tetrahedron Lett., 28, 5891 (1987).

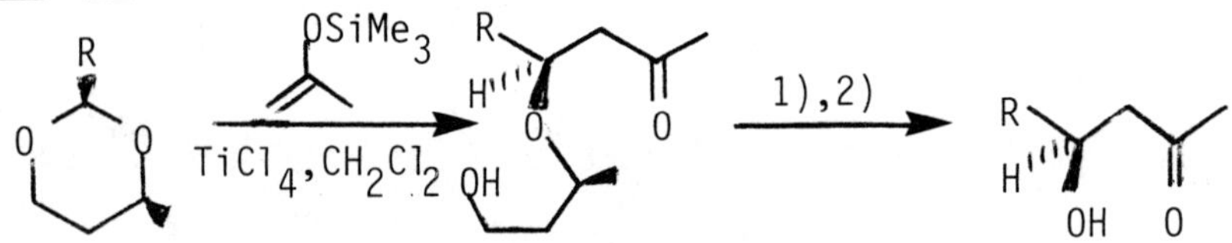

1)pyridinium chloroformate; 2)PhH,$(PhCH_2)_2NH$

VI.A.9-6 A.R. Katritzky and S. Sengupta, Tetrahedron Lett., 28, 1847 (1987); F.A. Davis and A.C. Sheppard, J. Org. Chem., 52, 955 (1987); P.G. Baraldi, A. Barco, S. Benetti, S. Manfredini and D. Simoni, Synthesis, 276 (1987).

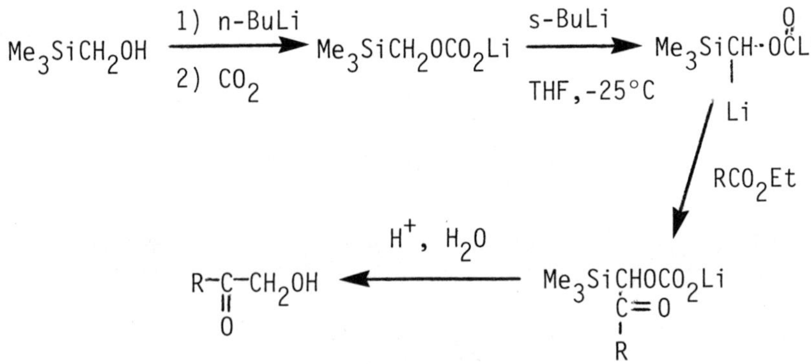

VI.A.9-7 R.M. Moriarty et al., J. Org. Chem., 52, 150 (1987); R.M. Moriarty, I. Prakash and R. Penmasta, Synth. Commun., 17, 709 (1987); R.M. Moriarty, R. Penmasta and I. Prakash, Tetrahedron Lett., 28, 873 (1987).

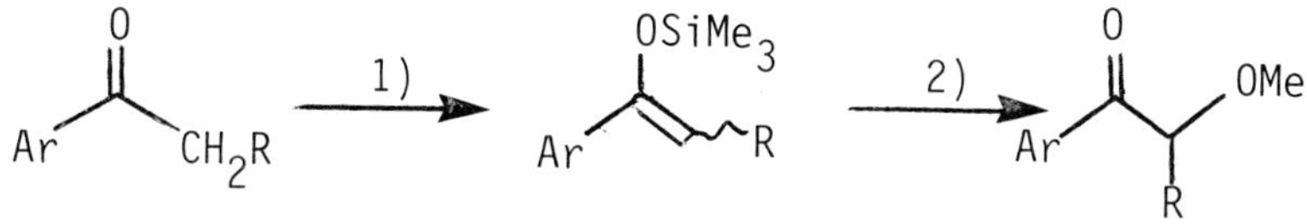

1)Me_3SiCl,TEA,DMF; 2)$(PhIO)_n$,$BF_3.Et_2O$,MeOH. 68-78%

VI.A.9-8 A. Padwa, D.N. Kline and J. Perumattam, Tetrahedron Lett., 28, 913 (1987); W. Bretsch and H.-U. Reissig, Liebigs Ann. Chem., 175 (1987); K. Van Sant and M.S. South, Tetrahedron Lett., 28, 6019 (1987)

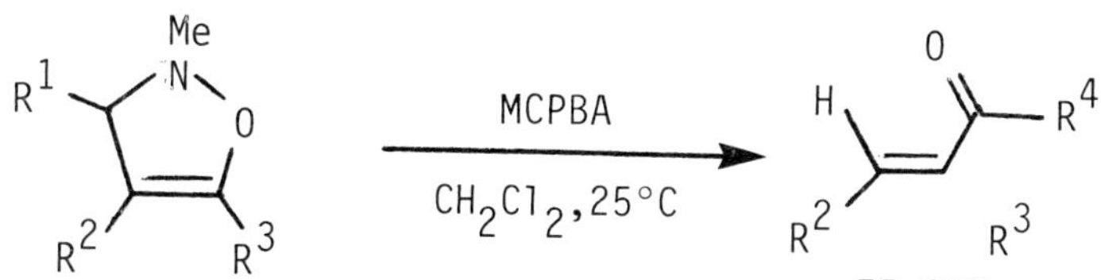

75-98%

VI.A.9-9 D. Bernard, A. Doutheau and J. Gore, Tetrahedron, 43, 2721 (1987)

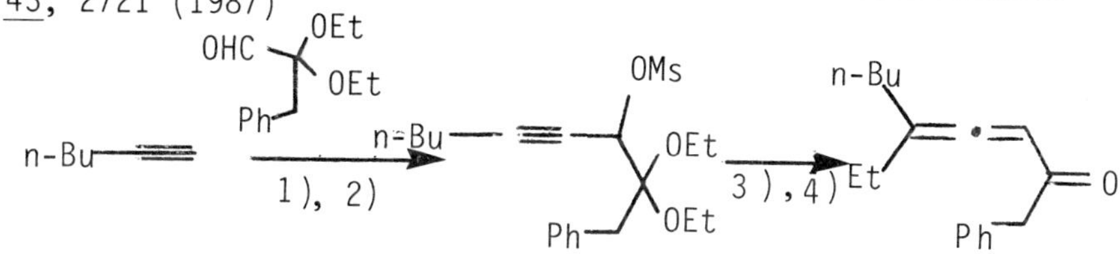

57%

1)n-BuLi;2)MsCl;3)EtMgBr,CuBr,THF;3)SiO_2.

VI.A.9-10 C. Portella and M. Iznaden, Tetrahedron Lett., 28, 1655 (1987); C. Sha, J. Young and T. Jean, J. Org. Chem., 52, 3919 (1987); S. Martin, R. Sauvetre and J.-F. Normant, Bull. Soc. Chim. Fr., 900 (1986).

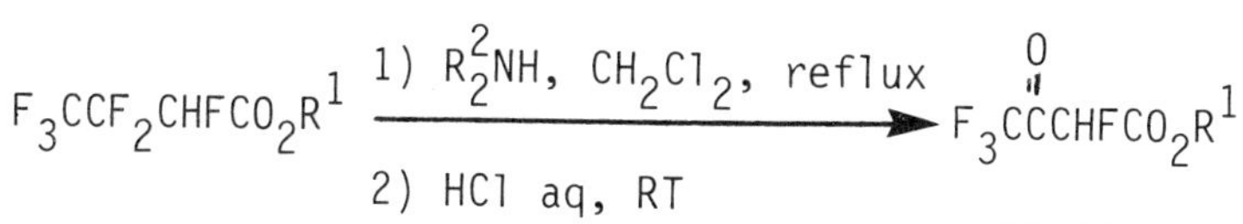

80-90%

VI.A.9-11 J. Hooz, J. Oudenes, J.L. Roberts and A. Benderly, J. Org. Chem., 52, 1347 (1987); S. B. Hendi, M.S. Hendi and J.F. Wolfe, Synth. Commun., 17, 13 (1987).

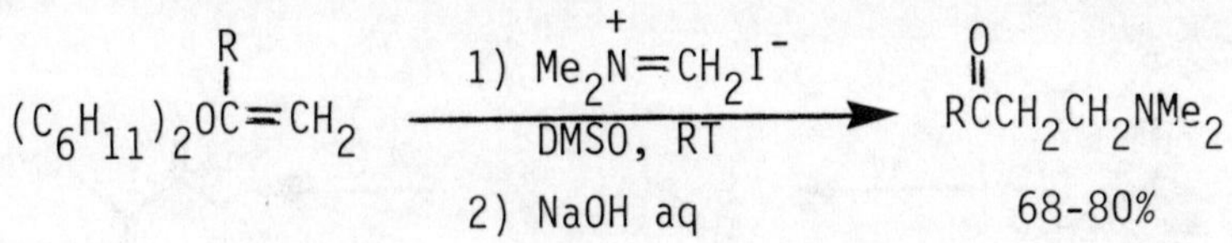

VI.A.9-12 C.K. Bradsher, Chem. Rev., 87, 1277 (1987).

Review: "Formation of Six-Membered Aromatic Rings by Cyclialkylation of Some Aldehydes and Ketones"

VI.A.10. Nitriles and Imines

VI.A.10-1 G. Petrillo, M. Novi, G. Garbarino and C. Dell'Erba, Tetrahedron, 43, 4625 (1987); M. Novi, G. Petrillo and C. Dell'Erba, Tetrahedron Lett., 28, 1345 (1987); A. Liguori, G. Sindona, G. Romeo and N. Uccella, Synthesis, 168 (1987); M. Meier and C. Rüchardt, Chem. Ber., 120, 1 (1987); E.M. Campi, P.S. Elmes, W.R. Jackson, C.G. Lovel and M.K.S. Probert, Aust. J. Chem., 40, 1053 (1987).

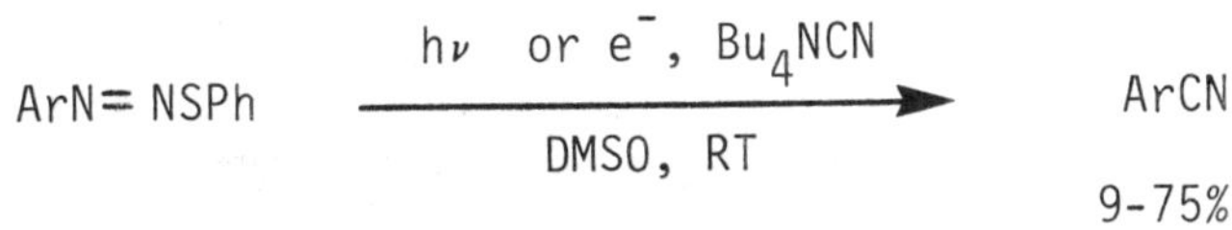

VI.A.10-2 F. Effenberger, T. Ziegler and S. Förster, Angew. Chem. Int. Ed. Engl., 26, 458 (1987); T. Hanafusa, J. Ichihara and T. Ashida, Chem. Lett., 687 (1987); H. Kunz and W. Sager, Angew. Chem. Int. Ed. Engl., 26, 557 (1987); T. Takeda, Y. Kaneko, H. Nakagawa and T. Fujiwara, Chem. Lett., 1963 (1987).

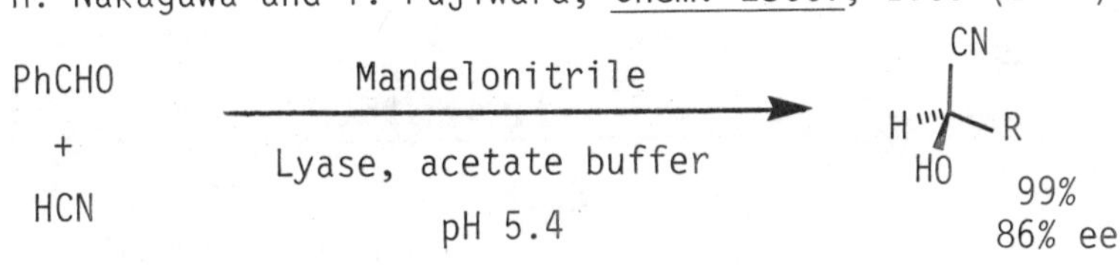

VI.A.10-3 P.A. Wade and J.F. Bereznak, J. Org. Chem., 52, 2973 (1987).

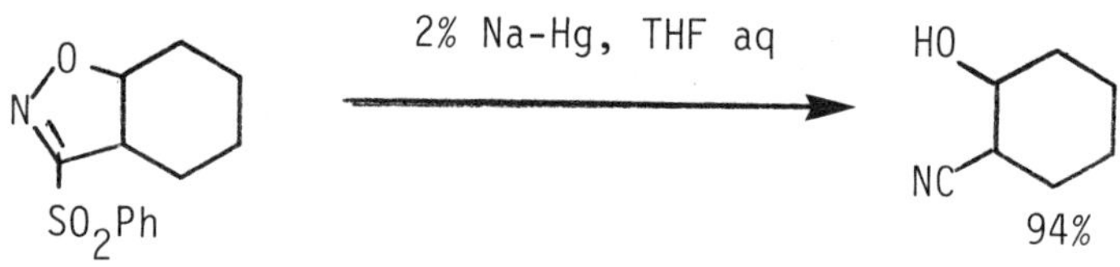

VI.A.10-4 J.R. Hwu and N. Wang, J. Chem. Soc., Chem. Commun., 427 (1987).

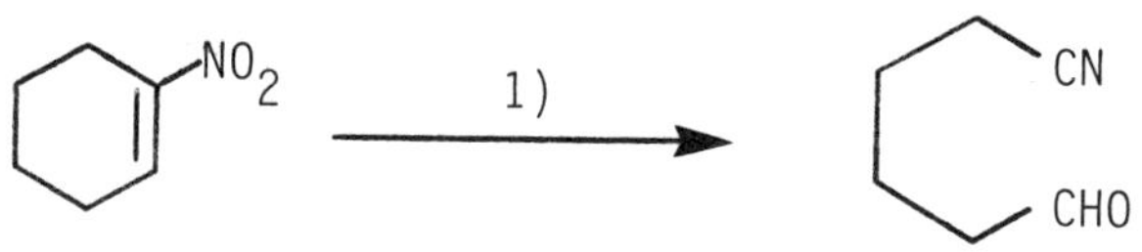

1) KOMe, DME, RT, 2 h; 2) Et_3OBF_3, heat;
3) $SOCl_2$, 0°C, 30 min; 4) H_2O.

VI.A.10-5 D. Bacos, J.P. Celerier and G. Lhommet, Tetrahedron Lett., 28, 2353 (1987).

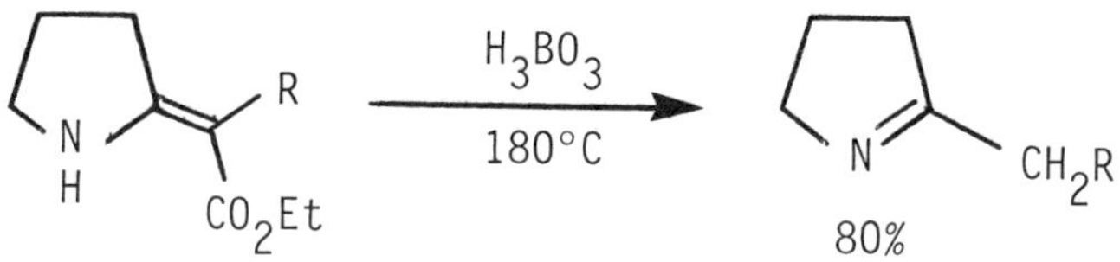

VI.A.11. Azides

VI.A.11-1 A. Hassner and A.S. Amarasekara, Tetrahedron Lett., 28, 5185 (1987); V. Pozsgay and H.J. Jennings, Tetrahedron Lett., 28, 5091 (1987); J. Garcia and J. Vilarrasa, Tetrahedron Lett., 28, 341 (1987); G. Janairo, W. Kowollik and W. Voelter, Liebigs Ann. Chem., 165 (1987); A. Koziara and A. Zwierzak, Tetrahedron Lett., 28, 6513 (1987); S. M. Kumar, Synth. Commun., 17, 1015 (1987).

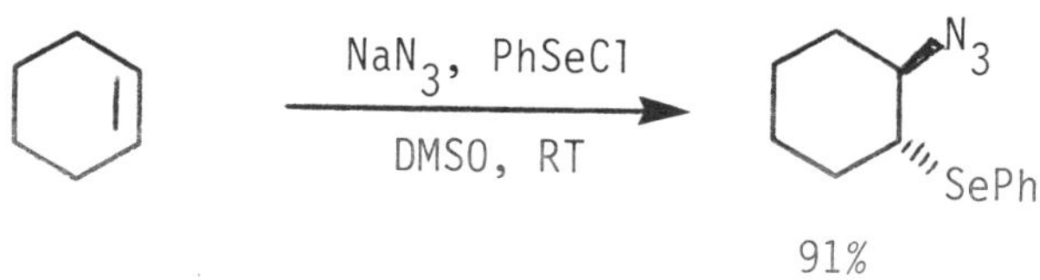

VI.A.11-2 D.A. Evans and T.C. Britton, J. Am. Chem. Soc., 109, 6881 (1987).

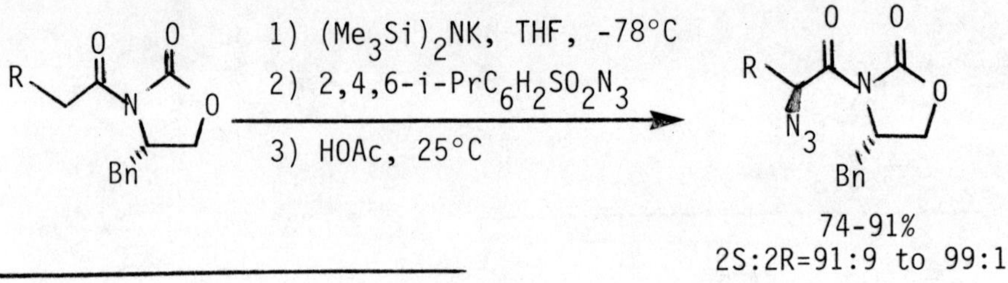

VI.A.11-3 E.R. Koft, J. Org. Chem., 52, 3466 (1987).

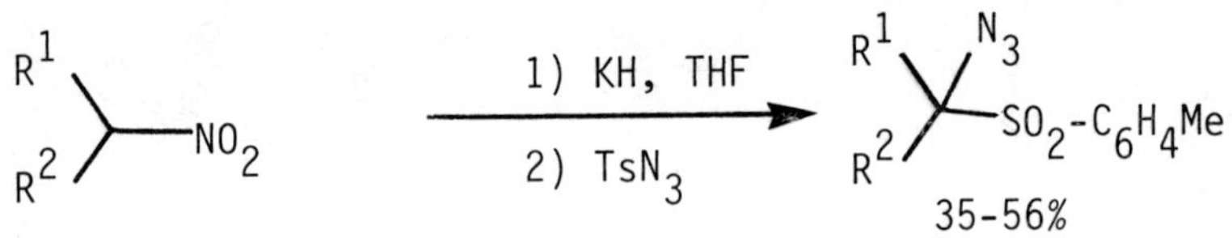

VI.A.11-4 M. Arimoto, H. Yamaguchi, E. Fujita, M. Ochiai and Y. Nagao, Tetrahedron Lett., 28, 6289 (1987).

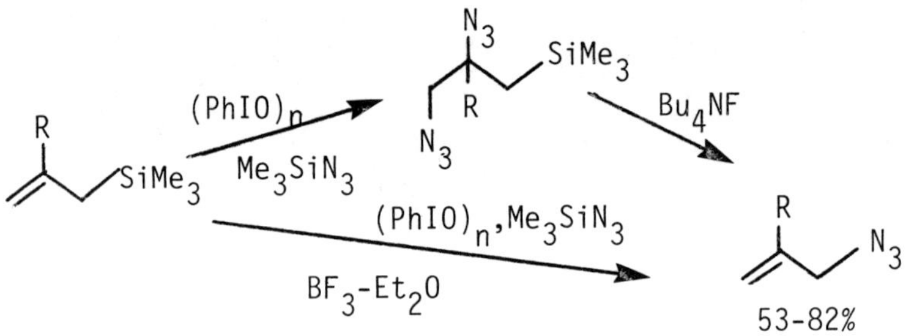

VI.A.11-5 K. Nishiyama, M. Oba and A. Watanabe, Tetrahedron, 43, 693 (1987).

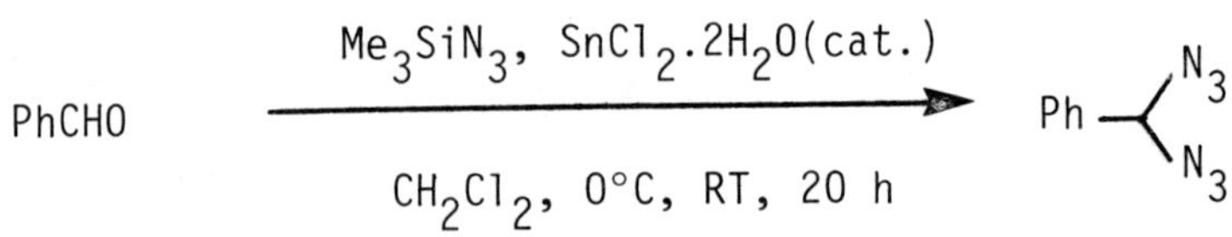

VI.A.12. Other N-Containing Functional Groups

VI.A.12-1 S. Zawadzki, K. Osowska-Pacewicka and A. Zwierzak, *Synthesis*, 485 (1987); D. Konwar, R.C. Boruah and J.S. Sandhu, *Tetrahedron Lett.*, *28*. 955 (1987); J.P. Demers and D.H. Klaubert, *Tetrahedron Lett.*, *28*, 4933 (1987); T. Okawara, Y. Kanazawa, T. Yamasaki and M. Furkawa, *Synthesis*, 183 (1987).

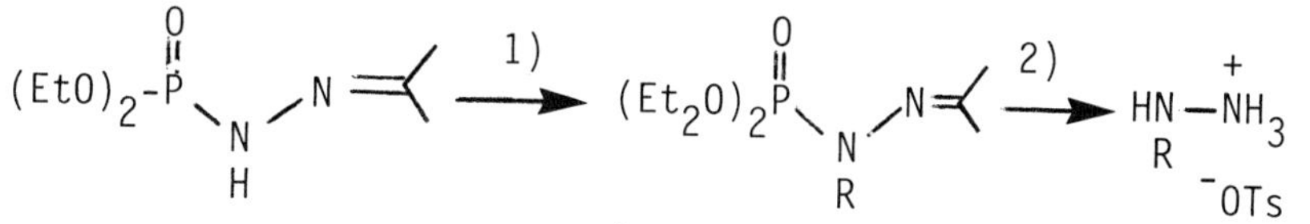

47-77%

1) RBr, NaOH, K_2CO_3, PhH, $n\text{-}Bu_4\overset{+}{N}HSO_4^-$, 80°C, 3 h;
2) TsOH.H_2O, EtOH, 78°C, 8 h.

VI.A.12-2 R.T. Boere, R.T. Oakley and R.W. Reed, *J. Organomet. Chem.*, *331*, 161 (1987).

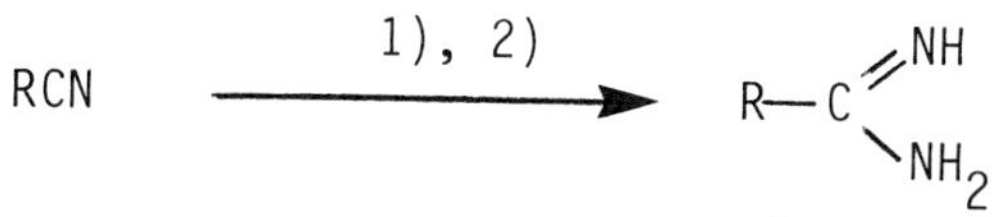

75-88%

1) $(Me_3Si)_2NLiOEt_2$, Et_2O, 24 h;
2) HCl aq, Et_2O, 0°C

VI.A.12-3 R. Hernandez, E.I. Leon, J.A. Salazar and E. Suarez, *J. Chem. Soc., Chem. Commun.*, 312 (1987).

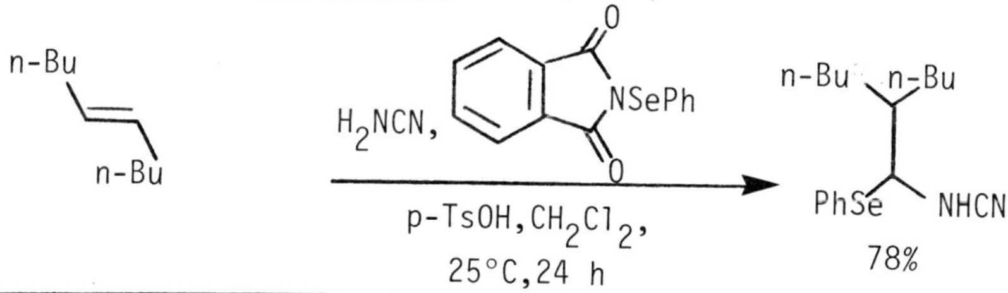

VI.A.12-4 K. Tanaka *et al.*, *Chem. Pharm. Bull.*, *35*, 364 (1987).

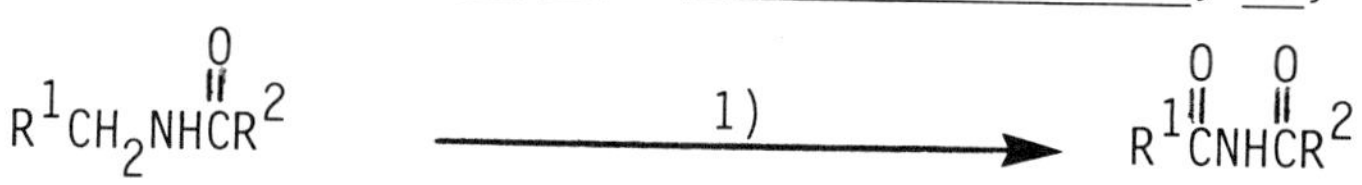

25-96%

1) RuO_4, EtOAc, H_2O, 0° to RT

VI.A.12-5 P. Giannoccaro, <u>J. Organomet. Chem.</u>, <u>336</u>, 271 (1987); G. Barcelo, J.-P. Senet and G. Sennyey, <u>Synthesis</u>, 1027 (1987).

$$RNH_2 \xrightarrow[\text{EtOH, 80°C, 6-12 h}]{CO, O_2,\ PdCl_2,\ PhNH_2} (RNH)_2CO \quad 62\text{-}92\%$$

VI.A.12-6 R.W. Hoffmann and A. Endesfelder, <u>Liebigs Ann.Chem.</u>, 215 (1987); G. Doleschall, <u>Tetrahedron Lett.</u>, <u>28</u>, 2993 (1987).

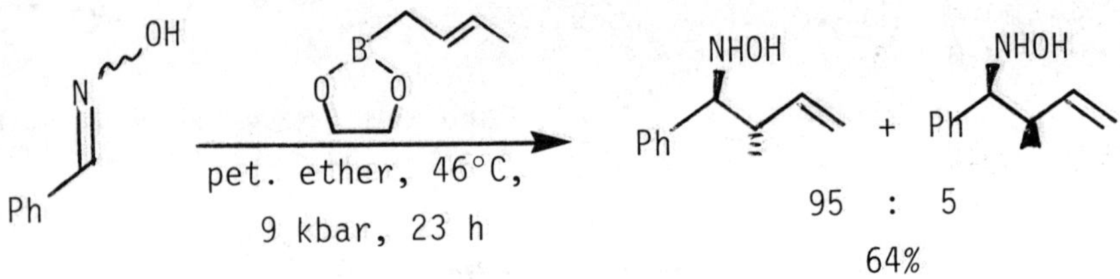

VI.A.12-7 K. Mai and G. Patil, <u>J. Org. Chem.</u>, <u>52</u>, 275 (1987).

$$NCNH_2 + Me_3SiCN \xrightarrow[\text{1 min}]{\text{RT}} Me_3SiN{=}C{=}NSiMe_3$$

VI.A.12-8 W. Mormann, S. Hoffmann and W. Hoffmann, <u>Chem. Ber.</u>, <u>120</u>, 285 (1987).

$$OCN{-}R^1{-}\overset{R^2}{\overset{|}{C}}HCOCl \xrightarrow[\text{RT, 1.5 h}]{\text{1) TEA, PhMe}} OCN{-}R^1{-}\overset{R^2}{\overset{|}{C}}{=}C{=}O$$

2) 50°C

VI.A.12-9 D. Armesto et al., Org. Prep. Proced. Intl., 19, 181 (1987).

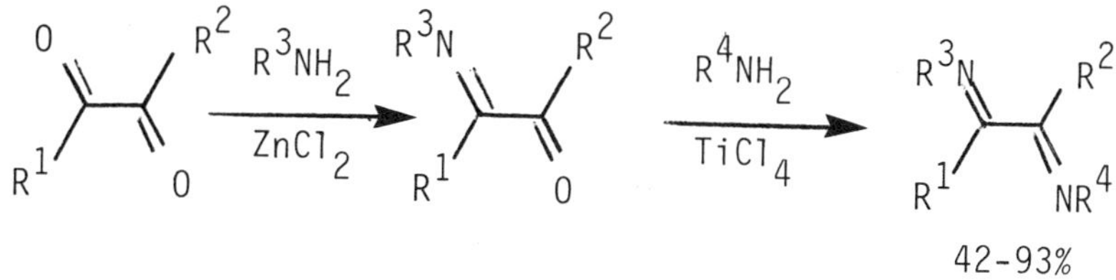

VI.A.12-10 S.K. Danek, D.P. Kelly, A.K. Serelis and P.J. Steel, J. Org. Chem., 52, 2911 (1987).

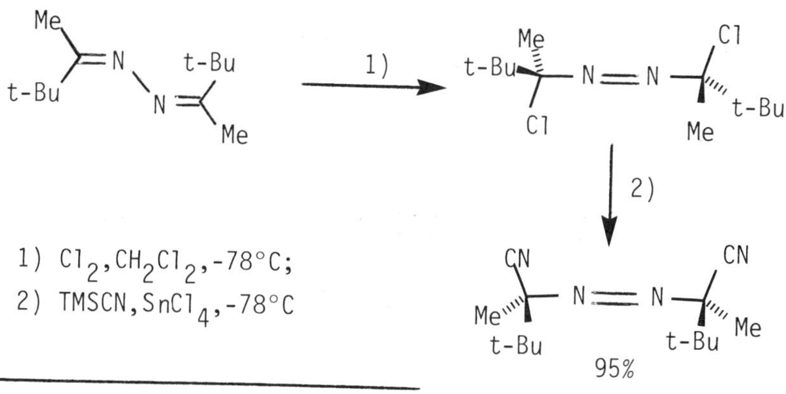

VI.A.12-11 J.S. Baum, D.A. Shook, H.M.L. Davies and H.D. Smith, Synth. Commun., 17, 1709 (1987); H.J. Monteiro, Synth.Commun., 17, 983 (1987).

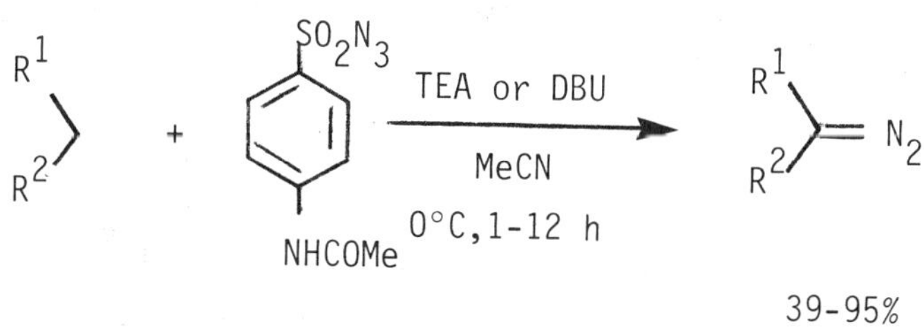

VI.A.12-12 G.A. Olah and C. Rochin, J. Org. Chem., 52, 701 (1987); R.J. Schmitt et al., J. Org. Chem., 52, 2294 (1987); A.J. Bloom and J.M. Mellor, J. Chem. Soc., Perkin I, 2737 (1987); J.C. Bottaro, R.J. Schmitt and C.D. Bedford, J. Org. Chem., 52, 2292 (1987).

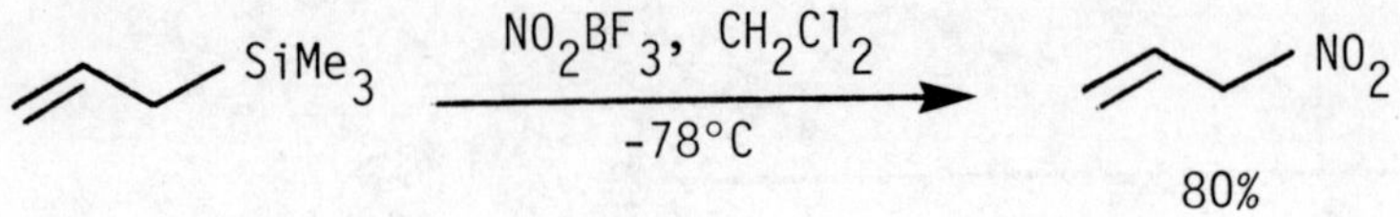

VI.A.12-13 J.E. Saavedra, Org. Prep. Proced. Intl., 19, 83 (1987).

Review: "Recent Synthetic Applications of N-Nitrosamines and Related Compounds"

VI.A.12-14 G. Kresze, B. Ascherl, H. Braun and H. Felber, Org. Prep. Proced. Intl., 19, 329 (1987).

Review: "α-Halonitroso Compounds"

VI.B. Sulfur Compounds

VI.B.1-1 D.H.R. Barton, D. Bridon and S.Z. Zard, Tetrahedron, 43, 5307 (1987); D.R. Graber, R.A. Morge and J.C. Sih, J. Org. Chem., 52, 4620 (1987); T.L. Guggenheim, Tetrahedron Lett., 28, 6139 (1987).

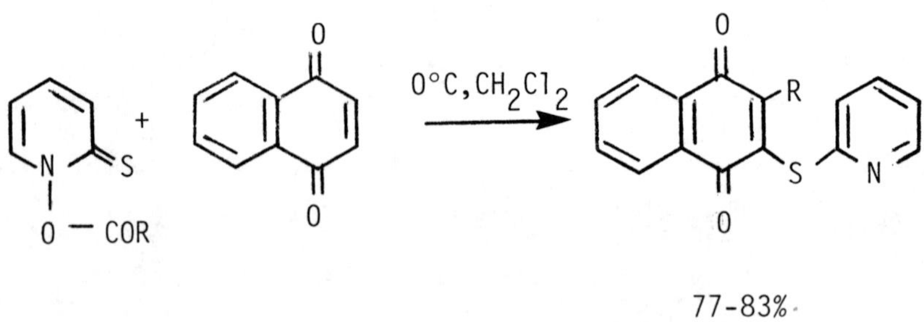

VI.B.1-2 Y. Ichinose, K. Wakamatsu, K. Nozaki, J. Birbaum, K. Oshima and K. Utimoto, Chem. Lett., 1647 (1987); A.E. Vougioukas and H.B. Kagan, Tetrahedron Lett., 28, 6065 (1987); V. Pozsgay and H.J. Jennings, Tetrahedron Lett., 28, 1375 (1987).

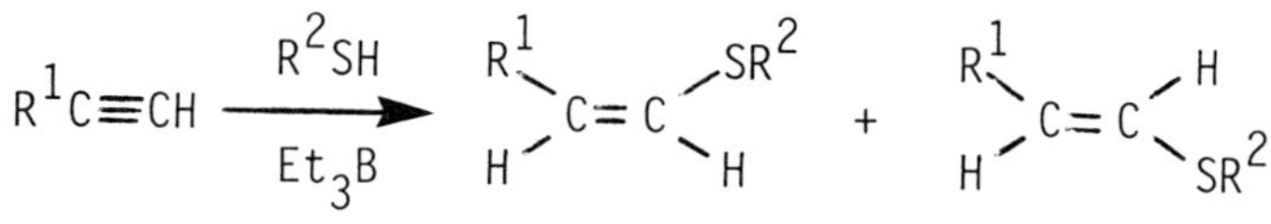

82:15 to 0:100

60 - 95%

VI.B.1-3 D.N. Harpp and M. Gingras, Tetrahedron Lett., 28, 4373 (1987); K. Takagi, Chem. Lett., 2221 (1987); D.N. Harpp, M. Gingras, T. Aida and T.H. Chan, Synthesis, 1122 (1987); X. Lu and Z. Ni, Synthesis, 66 (1987).

$$2RX + R^1_3SnSSnR^1_3 \xrightarrow[\text{MeCN}]{Bu_4NF,\ CsF} RSR + 2R^1_3SnF + 2X^-$$

13-99%

VI.B.1-4 Z. Chen. Y. Jin and P.J. Stang, J. Org. Chem., 52, 4117 (1987); M. Paventi and J.T. Edward, Can. J. Chem., 65, 282 (1987).

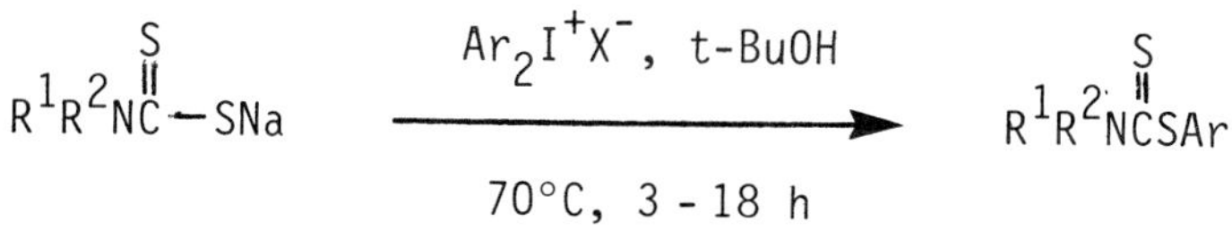

56-86%

VI.B.1-5 S. Kato et al., Synthesis, 304 (1987); A. Sugawara et al., Bull. Chem. Soc. Jpn., 60, 435 (1987).

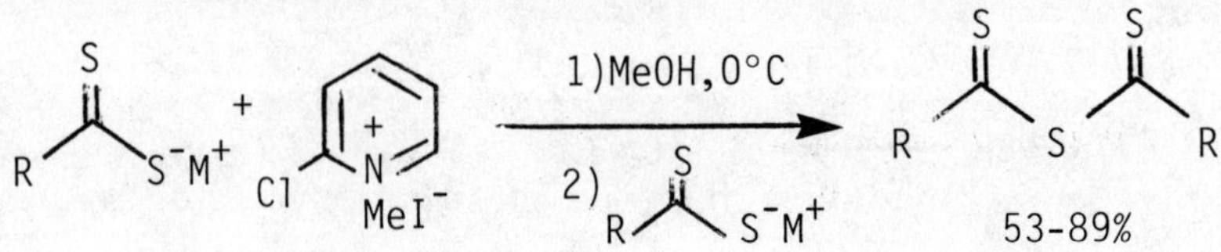

VI.B.1-6 H.D. Verkruijsse and L. Brandsma, J. Organomet. Chem., 332, 95 (1987).

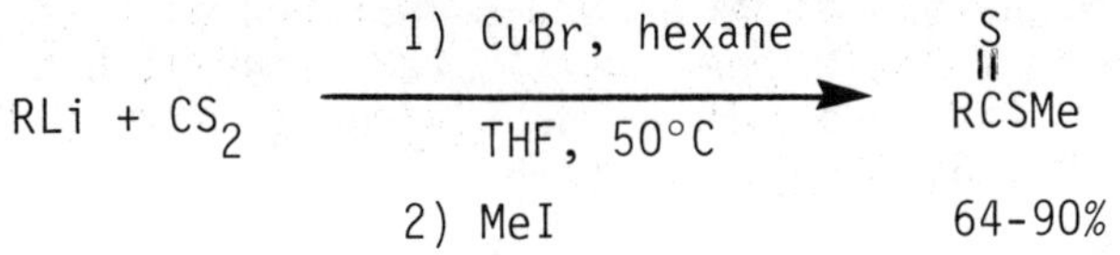

VI.B.1-7 S.V. Ley and P.R. Woodward, Tetrahedron Lett., 28, 345 (1987).

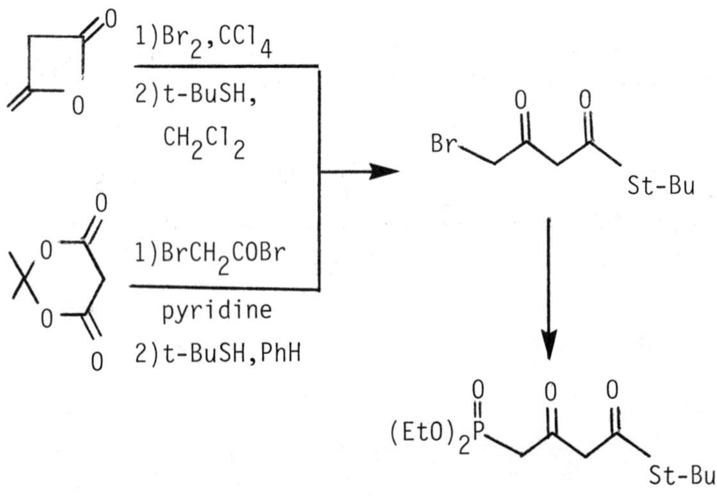

VI.B.1-8 H. Gotthardt, T. Löhr and D.J. Brauer, Chem. Ber., 120, 747 (1987).

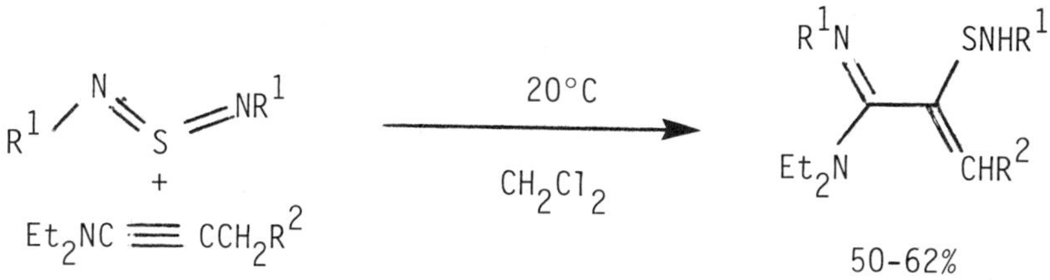

VI.B.1-9 T. Ando et al., J. Org. Chem., 52, 681 (1987); T. Ando, J.H. Clark, D.G. Cork, M. Fujita and T. Kimura, J. Chem. Soc., Chem. Commun., 1301 (1987); K. Nishiyama and M. Oba, Bull. Chem. Soc. Jpn., 60, 2692 (1987).

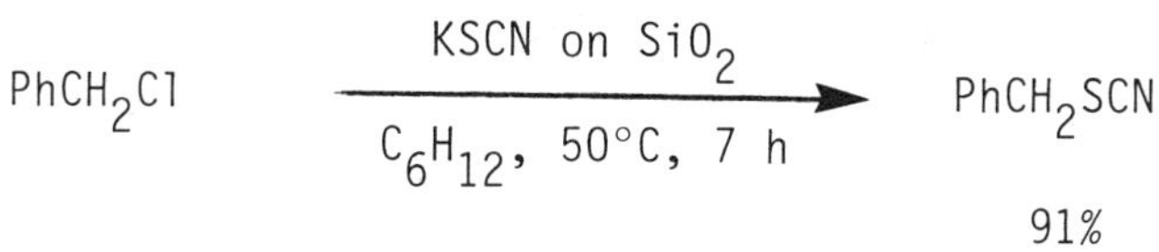

VI.B.1-10 S.T. Purrington adn J.H. Pittman, Tetrahedron Lett., 28, 3901 (1987); G.F. Koser, P.B. Kokil and M. Shah, Tetrahedron Lett., 28, 5431 (1987); S. Colonna, G. Germinario and C.J.M. Stirling, Gazz. Chim. Ital., 117, 67 (1987).

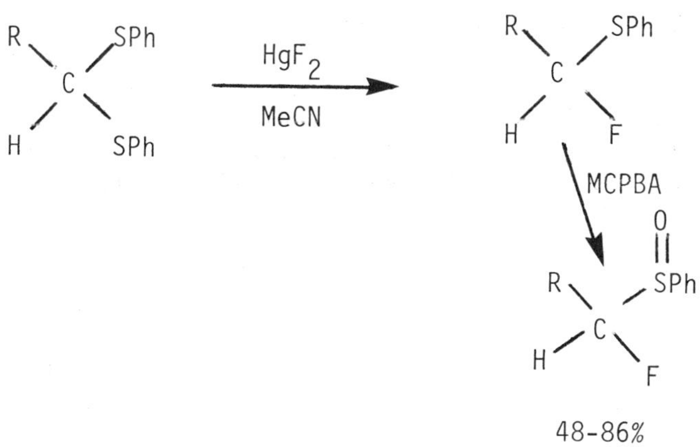

VI.B.1-11 C. Wakselman and M. Tordeux, Bull. Soc. Chim. Fr., 868 (1986); R. Karaman, H. Leader, A. Goldblum and E. Breuer, Chem. Ind. (London), 857 (1987); J.C. Carretero, M. Demillequand and L. Ghosez, Tetrahedron, 43, 5125 (1987).

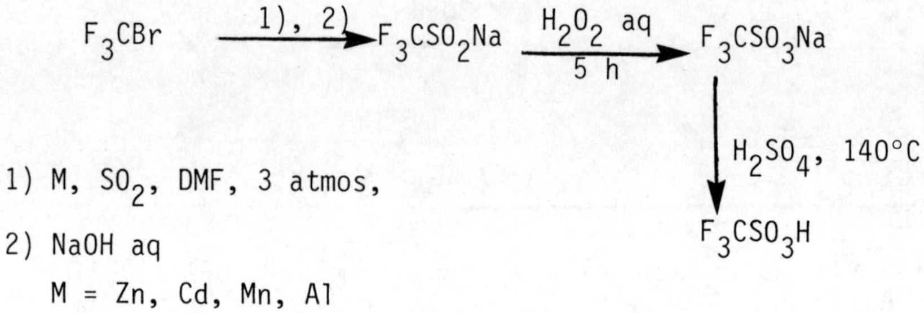

VI.B.1-12 G. Bram et al., Synthesis, 56 (1987); S.M. Andrews, C. Konstantinou and W.A. Feld, Synth. Commun., 17, 1041 (1987); N.P. Peet, S. Sunder, R.J. Barbuch, E.W. Huber and E.M. Bargar, J. Heterocycl. Chem., 24, 1531 (1987); J.M. Klunder and K.B. Sharpless, J. Org. Chem., 52, 2598 (1987); T.G. Back, S. Collins and M.V. Krishna, Can. J. Chem., 65, 38 (1987); T.G. Back, M.V. Krishna and K.R. Muralidharan, Tetrahedron Lett., 28, 1737 (1987).

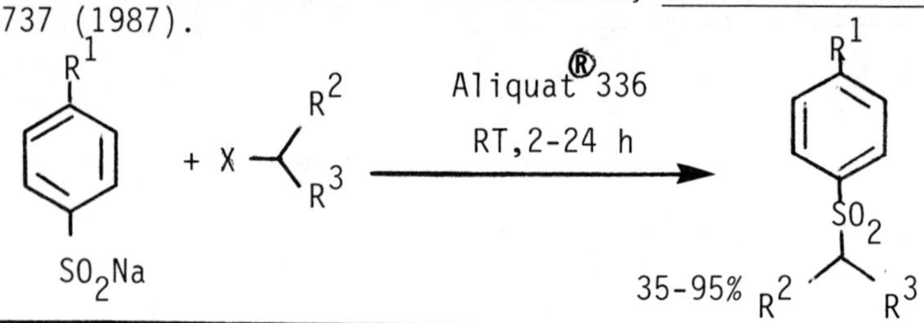

VI.B.1-13 S. Kim and S.S. Kim, Tetrahedron Lett., 28, 1913 (1987); S. Kim et al., J. Org. Chem., 42, 2114 (1987); A.G. Martinez et al., Synthesis, 49 (1987); R. Caputo et al., Synthesis, 386 (1987); M. Kakimoto et al., Synthesis, 164 (1987); K. Wahala et al., Synth. Commun., 17, 137 (1987); B. Costisella and I. Keitel, Synthesis, 44 (1987); T. Sato et al., Chem Lett., 1661 (1987); G. Drosten et al., Chem.Ber., 120, 1757 (1987).

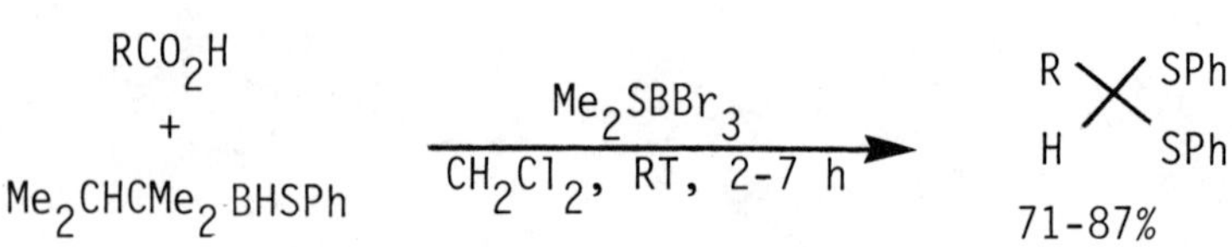

VI.B.1-14 J. Liebscher, A. Knoll and B. Abegas, Z. Chem., 27, 8 (1987).

Review: "Chemistry of the 3-Aminothioacrylamides -- Syntheses and Properties"

VI.B.1-15 M. Evers, Chemica Scripta, 26, 585 (1986).

Review: "Alkane Thiolate and Selenolate Anions: Useful Reagents for the Cleavage of the Alkyl-Chalcogen Bond of Alkyl Aryl Chalcogenides"

VI.B.1-16 A.A. Martin and G. Barnikow, Z. Chem., 27, 90 (1987).

Review: "Thiocarboxylic Anhydrides -- A New Class of Thioacylating Reagents"

VI.B.1-17 D.W. Roberts and D.L. Williams, Tetrahedron, 43, 1027 (1987).

Review: "Sultone Chemistry"

VI.B.1-18 G. L'abbé, Synthesis, 525 (1987).

Review: "Six-Membered Heterocyclic Isocyanates and Isothiocyanates: Synthesis and Reactions"

VI.C. Phosphorus and Tellurium Compounds

VI.C.1-1 S.E. Tunney and J.K. Stille, J. Org. Chem., 52, 748 (1987); J.C. Folest, J.Y. Nedelec and J. Perichon, Tetrahedron Lett., 28, 1885 (1987); P. Majewski, Synthesis, 554 (1987).

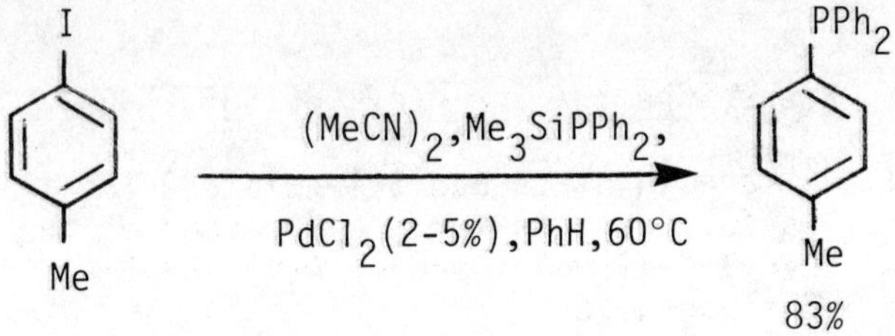

VI.C.1-2 T. Takanami, K. Suda, H. Ohmori and M. Masui, Chem. Lett., 1335 (1987).

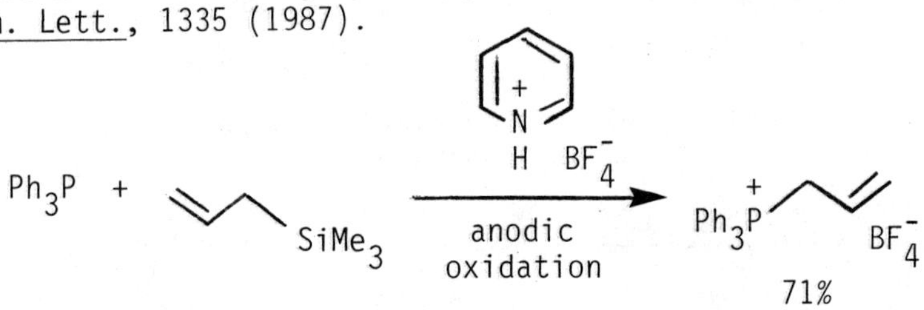

VI.C.1-3 H. Kottmann, J. Skarzewski and F. Effenberger, Synthesis, 797 (1987); T. Calogeropoulou, G.B. Hammond and D.F. Wiemer, J. Org. Chem., 52, 4185 (1987); N.N. Bhongle, R.H. Notter and J.G. Turcotte, Synth. Commun., 17, 1071 (1987); Y. W tanabe, H. Nakahira, M. Bunya and S. Ozaki, Tetrahedron Lett., 28, 4179 (1987).

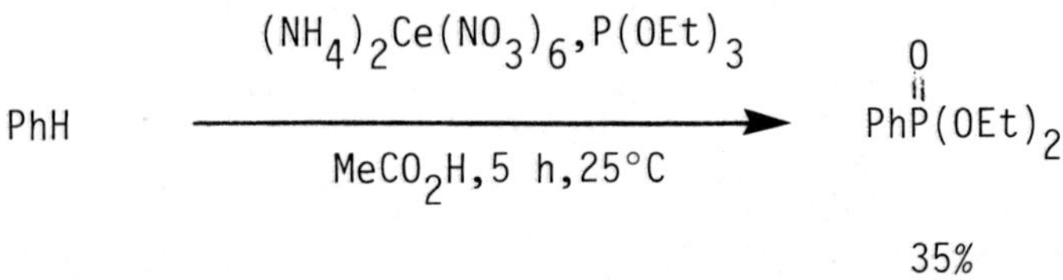

VI.C.1-4 V. Schöllkopf and R. Schütze, Liebigs Ann. Chem., 45 (1987).

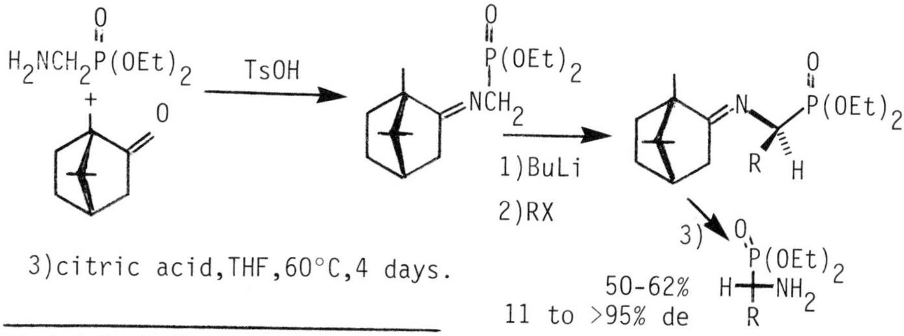

3)citric acid,THF,60°C,4 days.

VI.C.1-5 P.K. Chakravarty, P. Combs, A. Roth and W.J. Greenlee, Tetrahedron Lett., 28, 611 (1987); B.R. Shadid, H.C. van der Plas, E. de Vroom, G.A. van der Marel and J.H. van Boom, Rec.Trav. Chim., 106, 509 (1987).

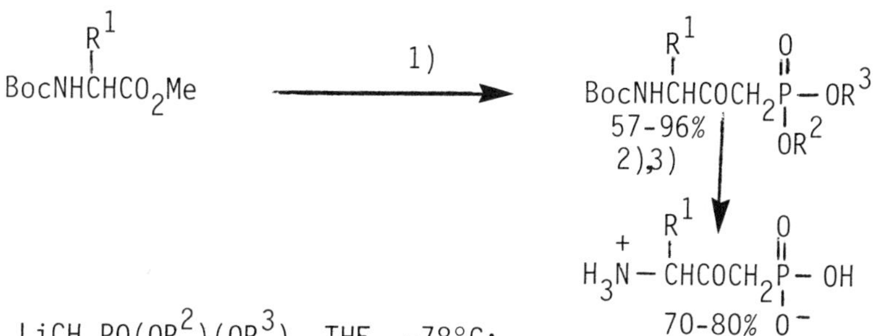

1) $LiCH_2PO(OR^2)(OR^3)$, THF, -78°C;
2) Me_3SiCl, CH_2Cl_2, 25°C, 18 h; 3) MeOH

VI.C.1-6 P. Majewski, Synthesis, 555 (1987).

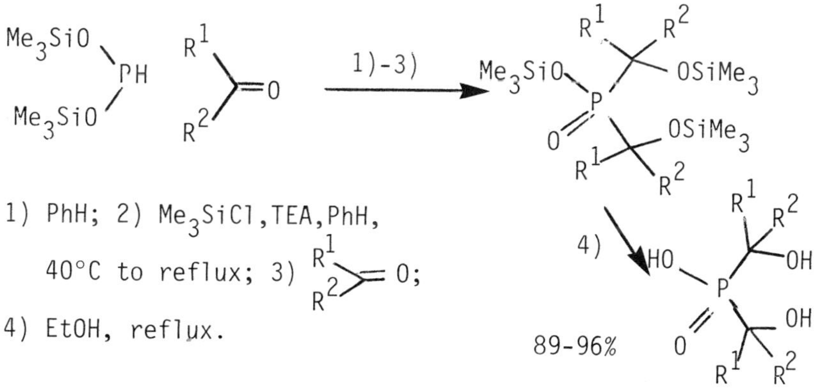

1) PhH; 2) Me_3SiCl,TEA,PhH, 40°C to reflux; 3) $R^1R^2C=O$;
4) EtOH, reflux.

VI.C.1-7 J.W. Perich and R.B. Johns, Tetrahedron Lett., 28, 101 (1987).

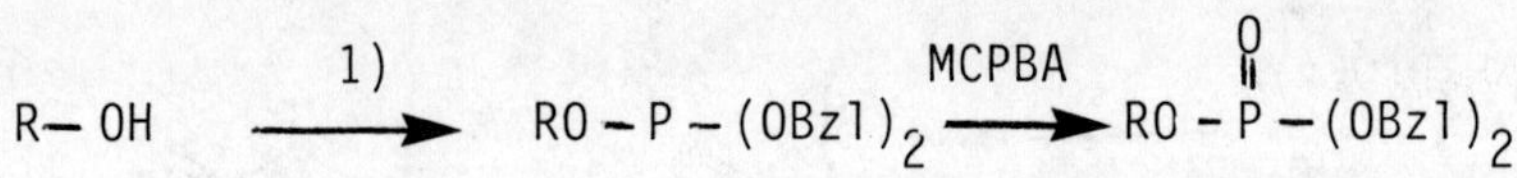

1) $(BzlO)_2PNEt_2$, methyltetrazole, THF.

VI.C.1-8 T. Kakigano, T. Kanda, M. Ishida and S. Kato, Chem. Lett., 475 (1987).

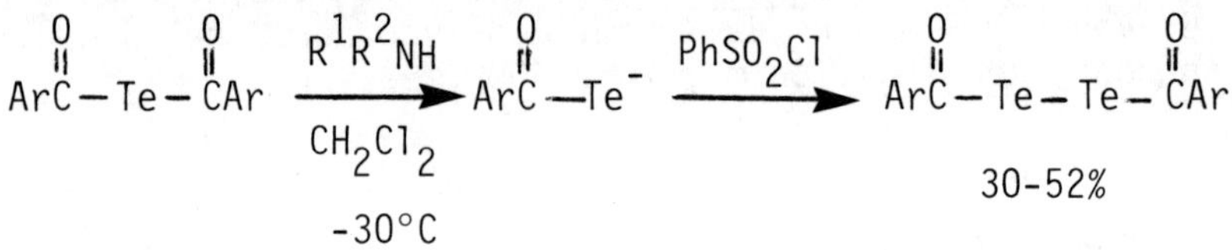

Ar = m-methoxyphenyl

VI.C.1-9 R. Appel, F. Knoll and I. Ruppert, Phosphorus Sulphur, 30 (1987).

Tenth International Conference on Phosphorus Chemistry. Part I.

VI.C.1-10 R. Appel, F. Knoll and I. Ruppert, Phosphorus Sulphur, 30, (1987).

Tenth International Conference on Phosphorus Chemistry. Part II.

VI.D. Nucleotides, Etc.

VI.D.1-1 J.H. van Boom, *et al.*, *Rec. Trav. Chim.*, *106*, 65 (1987); J.H. van Boom *et al.*, *Rec. Trav. Chim.*, *106*, 72 (1987).

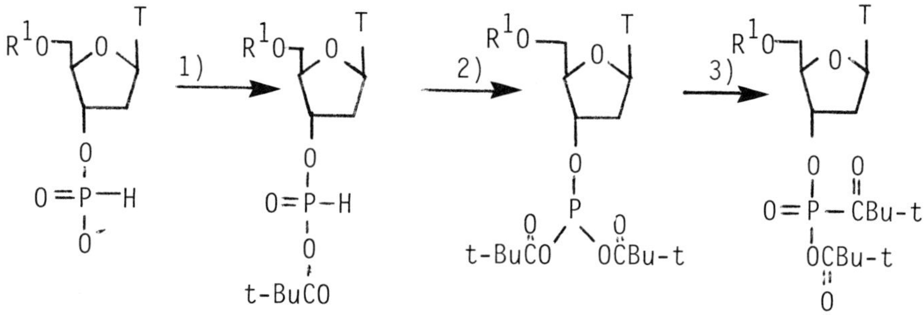

1),2) and 3) t-BuCOCl, pyridine.

VI.D.1-2 H. Hotoda, T. Wada, M. Sekine and T. Hata, *Tetrahedron Lett.*, *28*, 1681 (1987).

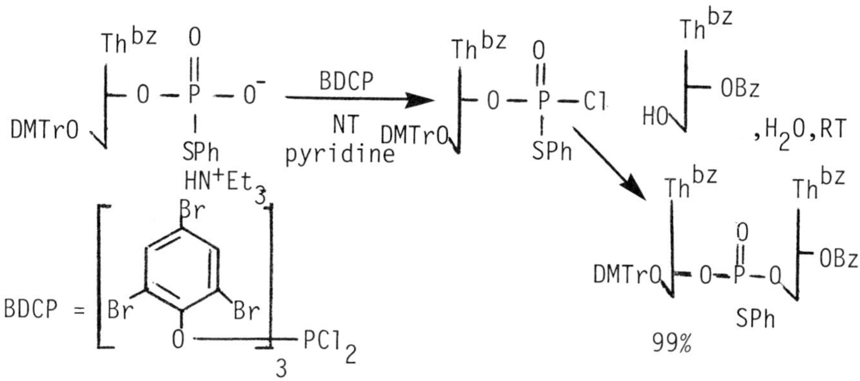

NT=3-nitro-1,2,4-triazole

VI.D.1-3 M. Sekine and T. Hata, *J. Org. Chem.*, *52*, 946 (1987).

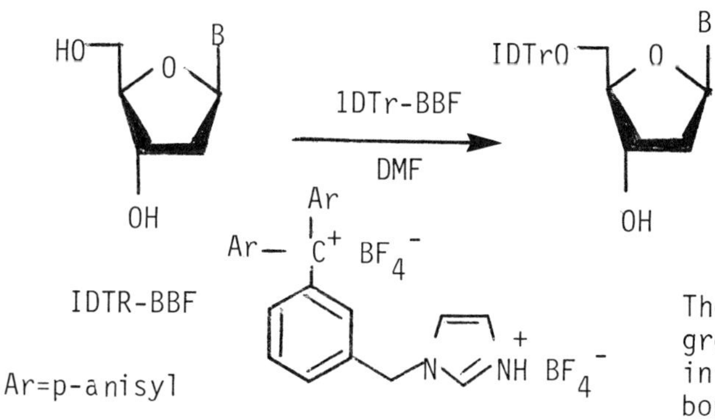

The IDTr group greatly enhances internucleotide bond formation.

VI.D.1-4 T. Nagamatsu and T. Kunieda, Tetrahedron Lett., 28, 2375 (1987).

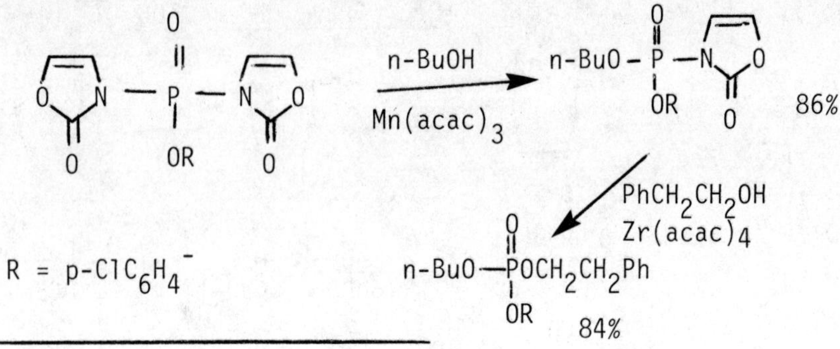

VI.D.1-5 W. Dabkowski, F. Cramer and J. Michalski, Tetrahedron Lett., 28, 3561 (1987); W. Dabkowski, F. Cramer and J. Michalski, Tetrahedron Lett., 28, 3559 (1987).

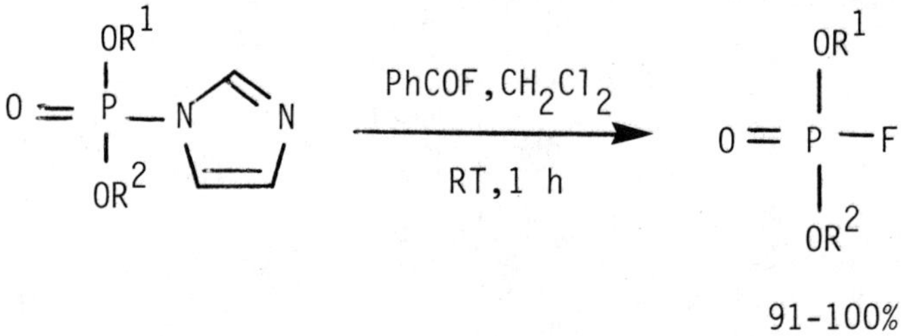

VI.D.1-6 M.P. Reddy, J.B. Rampal and S.L. Beaucage, Tetrahedron Lett., 28, 23 (1987).

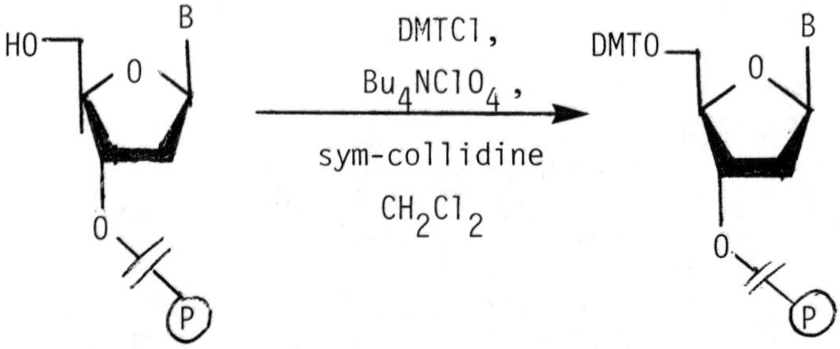

DMT = dimethoxytritylchloride

VI.D.1-7 T.Tanaka, S. Tamatsukuri and M. Ikehara, Tetrahedron Lett., 28, 2611 (1987).

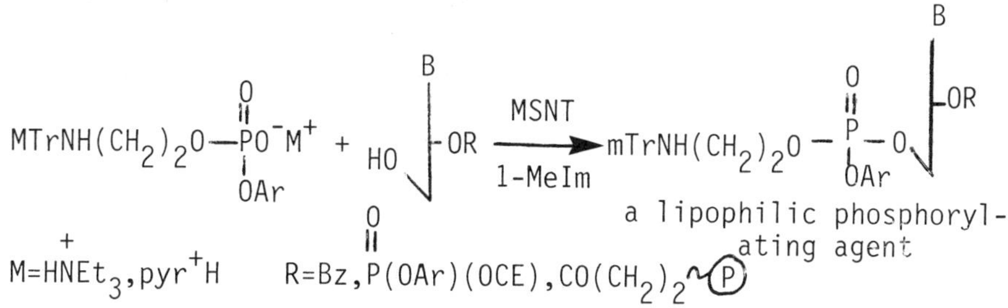

VI.D.1-8 H. Seki and O. Mitsunobu, Chem. Lett., 775 (1987).

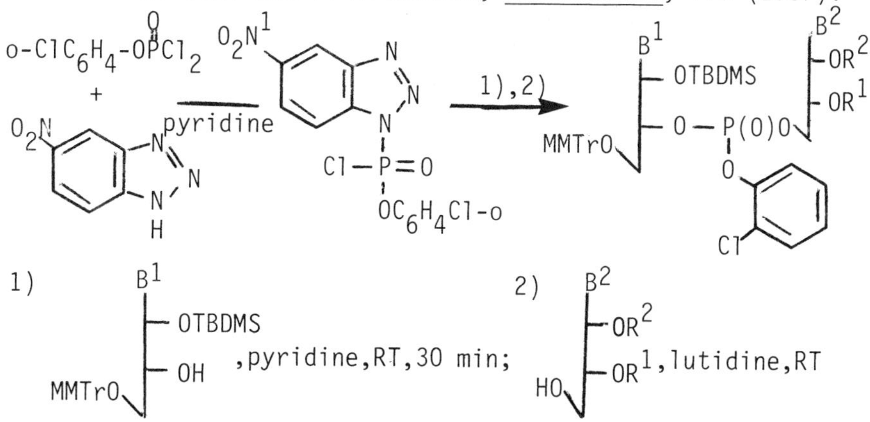

VI.D.1-9 H. Takaku, J. Synth. Org. Chem. Jpn., 45, 196 (1987).

Review: "Recent Developments in Phosphorylation of Nucleosides"

VI.D.1-10 S.J. Blunden, P.A. Cusack and P.J. Smith, J. Organomet. Chem., 325, 141 (1987).

Review: "The Use of Tin Compounds in Carbohydrate and Nucleoside Chemistry"

VI.D.1-11 H. Seliger _et_ _al_., _Chemica Scripta_, _26_, 561 (1986).

Review: "New Preparative Methods in Oligonucleotide Chemistry and Their Application to Gene Synthesis. I. Improvements in the Chemistry of Solid-Phase Nucleotide Synthesis"

VI.D.1-12 H. Seliger _et_ _al_., _Chemica Scripta_, _26_, 569 (1986).

Review: "New Preparative Methods in Oligonucleotide Chemistry and Their Application to Gene Synthesis. II. Recent Developments of Apparatus"

VII.A. Techniques

VII.A-1 S. Butt and S.M. Roberts, Chem. Br., 23, 127 (1987).

Review: "Opportunities for Using Enzymes in Organic Synthesis"

VII.A-2 C.-H. Wong and G.M. Whitesides, Organic chemistry series, 9.

Review: "Enzymes in Synthetic Organic Chemistry"

VII.A-3 J.P. Lorimer and T.J. Mason, Chem. Soc. Rev., 16, 239 (1987).

Review: "Sonochemistry. Part 1 -- The Physical Aspects"

VII.A-4 J. Lindley and T.J. Mason, Chem.Soc. Rev., 16, 275 (1987).

Review: "Sonochemistry. Part 2 -- Synthetic Applications"

VII.A-5 R.P. Sheridan and R. Venkataraghavan, Acc. Chem. Res., 20, 322 (1987).

Review: "New Methods in Computer-Aided Drug Design"

VII.A-6 A.J. Kresge, Acc. Chem. Res., 20, 364 (1987).

Review: "Unusual Reactivity of Prostacyclin: Rational Drug Design through Physical Organic Chemistry"

VII.A-7 T.J. Simpson, Chem. Soc. Rev., 16, 123 (1987).

Review: "Applications of Multinuclear NMR to Structural and Biosynthetic Studies of Polyketide Microbial Metabolites"

VII.A-8 D.D. Perrin and W.L.F. Armarego, Pergamon Press.

Book: "Purification of Laboratory Chemicals" (3rd ed.)

VII.A-9 N.S. Isaacs and A.V. Goerge, Chem. Br., 23, 47 (1987).

Review: "Chemical Synthesis at High Pressures"

VII.A-10 S. Suzuki and Y. Ono, J. Synth. Org. Chem. Jpn., 45, 672 (1987).

Review: "Novel Solid Acid Catalyst -- Polyorganosiloxane Bearing SO_3H Groups"

VII.B. Asymmetric Synthesis

VII.B-1 S. Itsuno, J. Synth. Org. Chem. Jpn., 45, 101 (1987).

Review: "Catalytic Asymmetric Reduction Using Optically Active Amino Alcohol-Borane Complex"

VII.B-2 M. Srebnik and P.V. Ramachandran, Aldrichimica Acta, 20, 9 (1987).

Review: "The Utility of Chiral Organoboranes in the Preparation of Optically Active Compounds"

VII.B-3 G.H. Posner, Acc.Chem. Res., 20, 72 (1987).

Review: "Asymmetric Synthesis of Carbon-Carbon Bonds Using Sulfinyl Cycloalkenones, Alkenolides and Pyrones"

VII.B-4 W. Oppolzer, Tetrahedron, 43, 1969 (1987).

Tetrahedron Report 216: "Camphor Derivatives as Chiral Auxiliaries in Asymmetric Synthesis"

VII.B-5 S.G. Davies et al., Bull Soc. Chim. Fr., 608 (1987).

Review: "Synthesis and Stereoselective Reactions of α,β-Unsaturated Acyl Ligands Bound to the Chiral Auxiliary $(\eta^5\text{-}C_5H_5)Fe(CO)(PPh_3)$ "

VII.B-6 Y. Yamamoto, Acc. Chem. Res., 20, 243 (1987).

Review: "Acyclic Stereocontrol via Allylic Organometallic Compounds"

VII.V-7 H.E. Zimmerman, Acc. Chem. Res., 20, 263 (1987).

Review: "Kinetic Protonation of Enols, Enolates and Analogs. The Stereochemistry of Ketonization"

VII.B-8 T. Nagahara and T. Kametani, Heterocycles, 25, 729 (1987).

Review: "Enantioselective Syntheses of Carbapenam Antibiotics"

VII.B-9 R.M. Williams, Pergamon Press, Organic Chemistry Vol.8.

Book: "Synthesis of Optically Active α-Amino Acids"

VII.B-10 H. Kosugi, J. Synth. Org. Chem., Jpn., 45, 472 (1987).

Review: "Asymmetric Synthesis of Natural Products Using Chiral Sulphoxides"

VII.B-11 H. Hashimoto and N.Kawaguchi, J. Synth. Org. Chem., Jpn., 45, 408 (1987).

Review: "New Chiral Synthons Derived From D-Glucose"

VII.B-12 F. Toda, J. Synth. Org. Chem., Jpn., 45, 745 (1987).

Review: "Optical Resolution by Host-Guest Complexation"

VII.B-13 H. Brunner and H. Leyerer, Bull. Soc. Chim. Belg., 96, 353 (1987).

Review: "Methylbutylphosphines and Their Rhodium Complexes -- Enantioselective Catalysts"

VII.B-14 T. Katsuki, J. Synth. Org. Chem., Jpn., 45, 90 (1987).

Review: "Recent Developments in Asymmetric Epoxidation"

VII.B-15 K. Furuhashi, J. Synth. Org. Chem., Jpn., 45, 162 (1987).

Review: "Production of Optically Active Epoxides by Microbial Oxidation of Olefins"

VII.B-16 H.N.C. Wong, C.C.M. Fok and T. Wong, Heterocycles, 26, 1345 (1987).

Review: "Stereospecific Deoxygenation of Epoxides to Olefins"

VII.B-17 M. Braun, Angew. Chem. Int. Ed. Engl., 26, 24 (1987).

Review: "Stereoselective Aldol Reactions with α-Unsubstituted Chiral Enolates"

VII.B-18 H. Kunz, Angew. Chem. Int. Ed. Engl., 26, 294 (1987).

Review: "Synthesis of Glycopeptides, Partial Structures of Biological Recognition Components"

VII.B-19 R.W. Hoffmann, Angew. Chem. Int. Ed. Engl., 26, 489 (1987).

Review: "Stereoselective Syntheses of Building Blocks with Three Consecutive Stereogenic Centers: Important Precursors of Polyketide Natural Products"

VII.B-20 V. Ramamurthy, J.R. Scheffer and N.J. Turro, Tetrahedron, 43, 1197 (1987).

Symposium in Print 29: "Organic Chemistry in Anisotropic Media"

VII.C. Reactions

VII.C-1 W. Carruthers, Pergamon Press, Organic Chemistry Vol.7.

Book: "Cycloaddition Reactions in Organic Synthsis"

VII.C-2 J.S. Sandhu and B. Sain, Heterocycles, 26, 777 (1987).

Review: "Some Recent Advances in the Chemistry of Imines, in Particular Cycloaddition Reactions"

VII.C-3 A.J. Fatiadi, Synthesis, 749 (1987).

Review: "Addition and Cycloaddition Reactions of Tetracyanoethylene (TCNE) in Organic Chemistry"

VII.C-4 A. Ichihara, Synthesis, 207 (1987).

Review: "Retro-Diels-Alder Strategy in Natural Product Synthesis"

VII.C-5 D. Craig, Chem. Soc. REv., 16, 187 (1987).

Review: "Stereochemical Aspects of the Intramolecular Diels-Alder Reaction"

VII.C-6 J.L. Charlton and M.M. Alauddin, Tetrahedron, 43, 2873 (1987).

Tetrahedron Report 220: "Ortho-Quinodimethanes"

VII.C-7 M.L. Petrov and A.A. Petrov, Russ. Chem. Rev., 56, 152 (1987).

Review: "The 1,3-Anionic Cycloaddition Reactions of α,β-Unsaturated Thiolates and Their Analogs"

VII.C-8 M. Makosza and J. Winiarski, Acc. Chem. Res., 20, 282 (1987).

Review: "Vicarious Nucleophilic Substitution of Hydrogen"

VII.C-9 A. Kaji, R. Tanikaga, K. Tanaka and N. Ono, J. Synth. Org. Chem., Jpn., 45, 421 (1987).

Review: "Studies on the Development of New Reactions for Organic Synthesis. Synthetic Reactions Using Organic Sulphur and Nitro Compounds"

VII.C-10 N.S. Zefirov, Acc. Chem. Res., 20, 237 (1987).

Review: "An Approach to Systematization and Design of Organic Reactions"

VII.C-11 M. Chanon, Acc. Chem. Res., 20, 214 (1987).

Review: "Electron-Transfer-Induced Chain Reactions and Catalysis"

VII.C-12 T. Mukaiyama and M. Murakami, Synthesis, 1043 (1987).

Review: "Cross-Coupling Reactions Based on Acetals"

VII.C-13 R. Noack and K. Schwetlick, Z. Chem., 27, 77 (1987).

Review: "Insertion Reactions of Isocyanates"

VII.C-14 A.W. Frank, Phosphorus and Sulphur, 29, 297 (1987).

Review: "Non-Enzymic Methods for the Phosphorylation of Proteins"

VII.C-15 S. Chandrasekhar, Chem. Soc. Rev., 16, 313 (1987).

Review: "Product Stability in Kinetically Controlled Organic Reactions"

VII.D. Reactive Intermediates

VII.D-1 M. Ramaiah, Tetrahedron, 43, 3541 (1987).

Tetrahedron Report 223: "Radical Reactions in Organic Synthesis"

VII.D-2 D. Crich, Aldrichimica Acta, 20, 35 (1987).

Review: "O-Acyl Thiohydroxamates: New and Versatile Sources of Alkyl Radicals for Use in Organic Synthesis"

VII.D-3 J. Mattay, Angew. Chem. Int. Ed. Engl., 26, 825 (1987).

Review: "Charge Transfer and Radical Ions in Photochemistry"

VII.D-4 E.T. Denisov and I.V. Khudyakov, Chem. Rev., 87, 1313 (1987).

Review: "Mechanisms of Action and Reactivities of the Free Radicals of Inhibitors"

VII.D-5 H.D. Roth, Acc. Chem. Res., 20, 343 (1987).

Review: "Organic Radical Cations in Fluid Solution: Unusual Structures and Rearrangements"

VII.D-6 J.L. Courtneidge and A.G. Davies, Acc. Chem. Res., 20, 90 (1987).

Review: "Hydrocarbon Radical Cations"

VII.D-7 N.L. Bauld *et al*., Acc. Chem. Res., 20, 371 (1987).

Review: "Cation Radical Pericyclic Reactions"

VII.D-8 A.L. Campbell and G.R. Lenz, Synthesis, 421 (1987).

Review: "The Ground State and Excited State Reactions of Enamides"

VII.D-9 L. Eberson and F. Radner, Acc. Chem. Res., 20, 53 (1987).

Review: "Electron-Transfer Mechanisms in Electrophilic Aromatic Nitration"

VII.D-10 D. Lloyd, I. Gosney and R.A. Ormiston, Chem. Soc. Rev., 16, 45 (1987).

Review: "Arsonium Ylides"(with some mention also of Arsinimines, Stibonium and Bismuthonium Ylides)

VII.D-11 M. Brookhart and W.B. Studabaker, Chem. Rev., 87, 411 (1987).

Review: "Cyclopropanes from Reactions of Transition-Metal-Carbene Complexes with Olefins"

VII.D-12 O.A. Attanasi and L. Caglioti, Org. Prep. Proced. Int., 18, 299 (1986).

Review: "Conjugated Azoalkenes: Attractive Products and Versatile Intermediates"

VII.D-13 F. Mathey, Angew. Chem. Int. Ed. Engl., 26, 275 (1987).

Review: "The Development of a Carbene-like Chemistry with Terminal Phosphinidene Complexes"

VII.D-14 Yu.G. Gololobov et al., Tetrahedron, 43, 2609 (1987).

Review: "Twenty-five Years of Dimethylsulfoxonium Methylide (Corey's Reagent)"

VII.D-15 O. Meth-Cohn, Acc. Chem. Res., 20, 18 (1987).

Review: "New Synthetic Applications of Oxycarbonylnitrenes"

VII.D-16 H. Bock and R. Dammel, Angew. Chem. Int. Ed. Engl., 26, 504 (1987).

Review: "The Pyrolysis of Azides in the Gas Phase"

VII.D-17 J.A. Chudek and R. Foster, Gazz. Chim. Ital., 117, 629 (1987).

Review: "Formation and Reactions of Sigma-anionic Compounds (Jackson-Meisenheimer Adducts) in Liquid Ammonia Studies by ^{1}H and ^{13}C NMR"

VII.D-18 K.M. Nicholas, Acc. Chem. Res., 20, 207 (1987).

Review: "Chemistry and Synthetic Utility of Cobalt-Complexed Propargyl Cations"

VII.D-19 F. Tureček, Coll Czech. Chem. Commun., 1928 (1987).

Review: "Stereochemistry of Organic Ions in the Gas Phase"

VII.D-20 T.A. O'Donnell, Chem. Soc. Rev., 16, 1 (1987).

Review: "Stabilization of Unusual Cationic Species in Protonic Superacids and Acidic Melts"

VII.D-21 J. Dunogues, Bull. Soc. Chim. Fr., 659 (1987).

Colloquium Seminar: "Organosilicons, Selective Intermediates for the Synthesis of Fine Organics"

VII.E. Organo-metallics and Metalloids

VII.E-1 J.-I. Setsune and D. Dolphin, Can. J. Chem., 65, 459 (1987).

Review: "Organometallic Aspects of Cytochrome P-450 Metabolism"

VII.E-2 J.M. O'Connor and C.P. Casey, Chem. Rev., 87, 307(1987).

Review: "Ring-Slippage Chemistry of Transition-Metal Cyclopentadienyl and Indenyl Complexes"

VII.E-3 J. Dubac and A. Laporterie, Chem. Rev., 87, 319 (1987).

Review: "Ene and Retro-Ene Reactions of Group 14 Organometallic Chemistry"

VII.E-4 A.J. Fatiadi, Synthesis, 959 (1987).

Review: "New Applications of Tetracyanoethylene in Organometallic Chemistry"

VII.E-5 B.H. Lipshutz, Synthesis, 325 (1987).

Review: "Applications of Higher-Order Mixed Organocuprates to Organic Synthesis"

VII.E-6 M. Veith, Angew. Chem. Int. Ed. Engl., 26, 1 (1987).

Review: "Unsaturated Molecules Containing Main Group Metals"

VII.E-7 H. Kagan, Bull. Soc. Chim. Fr., 4, 607 (1987).

Review: "Chemistry of Organometallics and Their Applications in Organic Synthesis"

VII.E-8 E. Negishi, Acc. Chem. Res., 20, 65 (1987).

Review: "Controlled Carbometallation as a New Tool for Carbon-Carbon Bond Formation and Its Application to Cyclization"

VII.E-9 E. Erdik, Tetrahedron, 43, 2203 (1987).

Tetrahedron Report 217: "Use of Activation Methods for Organozinc Reagents"

VII.E-10 P.A. Sutton and D.A. Buckingham, Acc. Chem. Res., 20, 357 (1987).

Review: "Cobalt(III)-Promoted Hydrolysis of Amino Acid Esters and Peptides and the Synthesis of Small Peptides"

VII.E-11 Y. Yamamoto, Aldrichimica Acta, 20, 45 (1987).

Review: "Allylic Tin Compounds in Organic Synthesis"

VII.E-12 H.C. Colquhoun, Chem. Ind.(London), 612 (1987).

Review: "Homogeneous, Palladium-Catalyzed, Carbon-Carbon Bond Formation"

VII.E-13 K. Utimoto, J. Synth. Org. Chem. Jpn., 45, 112 (1987).

Review: "Organic Synthesis with Palladium Catalysts. Intramolecular Addition of OH, NH, and CO_2H to an Acetylene Moiety"

VII.E-14 P.A. Chaloner, J. Organomet. Chem., 324, 283 (1987).

Review: "Nickel, Palladium and Platinum: Annual Survey Covering the Year 1980"

VII.E-15 P.A. Chaloner, J. Organomet. Chem., 337, 431 (1987).

Review: "Nickel, Palladium and Platinum: Annual Survey Covering the Year 1981"

VII.E-16 J. Tsuji and I. Minami, Acc. Chem. Res., 20, 140 (1987).

Review: "New Synthetic Reactions of Allyl, Alkyl Carbonates, Allyl β-Keto Carboxylates and Allyl Vinylic Carbonates Catalyzed by Palladium Complexes"

VII.E-17 S. Kato, T. Murai and M. Ishida, Org. Prep. Proced. Int., 18, 369 (1986).

Review: "Selenium and Tellurium Isologues of Carboxylic Acid Derivatives"

VII.E-18 M.H. Chisholm, New. J. Chem., 11, 459 (1987).

Review: "Cleavage and Formation of Carbon-Carbon and Carbon-Oxygen Bonds in Reactions Involving Dimolybdenum and Ditungsten Hexaalkoxides"

VII.E-19 M. Hirobe, J. Synth. Org. Chem., Jpn., 45, 755 (1987).

Review: "Catalytic Reactions by Iron-Sulphur Complexes"

VII.E-20 B.M. Rockett and G. Marr, J. Organomet. Chem., 318, 231 (1987).

Review: "Ferrocene: Annual Survey Covering the Year 1985"

VII.E-21 R.C. Kerber, J. Organomet. Chem., 318, 157 (1987).

Review: "Organoiron Chemistry: Annual Survey Covering the Year 1985"

VII.E-22 T. Mitsudo, Y. Hori and Y. Watanabe, J. Organomet. Chem., 334, 157 (1987).

Review: "Novel Organic Syntheses Catalyzed by (Cyclooctadiene)(Cyclooctatriene) Ruthenium and its Derivatives"

VII.E-23 J.B. Keister, J. Organomet. Chem., 318, 297 (1987).

Review: "Ruthenium and Osmium: Annual Survey for the Year 1983"

VII.E-24 P.A. Shapley, J. Organomet. Chem., 318, 409 (1987).

Review: "Annual Survey of Ruthenium and Osmium for the Year 1984"

VII.E-25 W.H. Ilsley, J. Organomet. Chem., 337, 1 (1987).

Review: "Mercury: Annual Survey Covering the Year 1980"

VII.E-26 H.B. Kagan and J.L. Namy, Tetrahedron, 42, 6573 (1987).

Tetrahedron Report 213: "Lanthanides in Organic Synthesis"

VII.E-27 L.D. Freedman and G. Doak, J. Organomet. Chem., 324, 1 (1987).

Review: "Antimony: Annual Survey Covering the Year 1985"

VII.E-28 G.O. Doak and L.D. Freedman, J. Organomet.Chem., 324, 39 (1987).

Review: "Bismuth: Annual Survey Covering Year 1985"

VII.E-29 J.T. Mague, J. Organomet.Chem., 324, 57 (1987).

Review: "Cobalt, Rhodium and Iridium: Annual Survey Covering the Year 1985"

VII.E-30 G.L. Larson, J. Organomet. Chem., 337, 195 (1987).

Review: "Organosilicon Survey 1985. The Silicon-Carbon Bond"

VII.E-31 H. Suzuki, J. Synth. Org. Chem., Jpn., 45, 603 (1987).

Review: "Organic Transformations Based on Tellurium Compounds"

VII.E-32 D.H.R. Barton et al., Pure Appl. Chem., 59, 937 (1987).

Review: "Bismuth(v) Reagents in Organic Synthesis"

VII.E-33 A. Maercker and M. Theis, Top. Curr. Chem., 138, 1 (1987).

Review: "Some Aspects of the Chemsitry of Polylithiated Aliphatic Hydrocarbons"

VII.E-34 A. Maercker, Angew. Chem., Int. Ed. Engl., 26, 972 (1987).

Review: "Ether Cleavage with Organo-Alkali-Metal Compounds and Alkali Metals"

VII.F. Halogen-Compounds and Halogenation

VII.F-1 J. Fluorine Chem., 35, 1 (1987).

Symposium: "Proceedings of the International Symposium to Mark the Centenary of the Isolation of Fluorine"

VII.F-2 J.-F. Normant and C. Wakselman, Bull. Soc. Chem. Fr., 6, 858 (1986).

Review: "New Trends in Fluorinated Organic Compounds Chemistry"

VII.F-3 L.M. Yagupolskii, J. Fluor. Chem., 36, 1 (1987).

Review: "Aromatic Compounds with new Fluorine-Containing Substituents"

VII.F-4 D.W. Ribbons et al., J. Fluorine Chem., 37, 299 (1987).

Review: "Biotransformations of Fluoroaromatic Compounds"

VII.F-5 T. Kitazume and T. Yamazaki, J. Synth. Org. Chem.,Jpn., 45, 888 (1987).

Review: "New Catalytic Reactions of Hydrolytic Enzymes in Fluorine Chemistry"

VII.F-6 J.T. Welch, Tetrahedron, 43, 3123 (1987).

Tetrahedron Report 221: "Advances in the Preparation of Biologically Active Organofluorine Compounds"

VII.F-7 J.E. Jackson et al., Tetrahedron, 43, 653 (1987).

Review: "Reaction of Difluorocarbene with Small Bicyclic Molecules"

VII.F-8 B.P. Cho and R.G. Harvey, J. Org. Chem., 52, 5668 (1987).

Review: "Polycyclic Fluoranthene Hydrocarbons. 2. A New General Synthesis"

VII.G. Natural Products

VII.G-1 S.J. Danishefsky, Aldrichimica Acta, 19, 59 (1986).

Review: "Reflections on Organic Synthesis: The Evolution of a General Strategy for the Stereoselective Construction of Polyoxygenated Natural Products"

VII.G-2 J.P. Kutney, Heterocycles, 25, 617 (1987).

Review: "Studies in Plant Tissue Culture. The Synthesis and Biosynthesis of Indole Alkaloids"

VII.G-3 S. Rajeswari et al., Heterocycles, 25, 659 (1987).

Review: "Syntheses of Indolizidine Alkaloids"

VII.G-4 V.H. Rawal, R.J. Jones and M.P. Cava, Heterocycles, 25, 701 (1987).

Review: "Synthetic Studies Toward CC-1065 and PDE-11"

VII.G-5 F. Piozzi, B. Rodriguez and G. Savona, Heterocycles, 25, 807 (1987).

Review: "Advances in the Chemistry of the Furanoditerpenoids from Teucrium Species"

VII.G-6 B.S. Joshi and S.W. Pelletier, Heterocycles, 26, 2503 (1987).

Review: "X-Ray Crystal Structures of C-19 Diterpenoid Alkaloids"

VII.G-7 J.W. Huffman, <u>Tetrahedron</u>, <u>23</u>, 5467 (1987).

Symposia-in-Print 30: "Current Topics in Sesquitenpene Synthesis"

VII.G-8 N.K. Kochetkov, <u>Tetrahedron</u>, <u>43</u>, 2389 (1987).

Tetrahedron Report 218: "Synthesis of Polysaccharides with a Regular Structure"

VII.G-9 S.J. Danishefsky and M.P. DeNinno, <u>Angew. Chem. Int. Ed. Engl.</u>, <u>26</u>, 15 (1987).

Review: "Totally Synthetic Routes to the Higher Monosaccharides"

VII.G-10 D.M. Goodall and I.T. Norton, <u>Acc. Chem. Res.</u>, <u>20</u>, 59 (1987).

Review: "Polysaccharide Conformations and Kinetics"

VII.G-11 T.L.B. Boivin, <u>Tetrahedron</u>, <u>43</u>, 3309 (1987).

Tetrahedron Report 222: "Synthetic Routes to Tetrahydrofuran, Tetrahydropyran and Spiroketal Units of Polyether Antibiotics and a Survey of Spiroketals of Other Natural Products"

VII.G-12 P.G. Baraldi <u>et al</u>., <u>Synthesis</u>, <u>52</u>, 857 (1987).

Review: "Synthesis of Natural Products <u>via</u> Isoxazoles"

VII.G-13 M.Fujita and T. Hiyama, J. Synth. Org. Chem. Jpn., 45, 664 (1987).

Review: "New Approach to Synthetic Pyrethroids Having a Trifluoromethyl Group"

VII.G-14 I. Nitta and H. Veno, J. Synth. Org. Chem., Jpn., 45, 445 (1987).

Review: "New Syntheses of Corticoids"

VII.G-15 H.H. Wasserman, Aldrichimica Acta, 20, 63 (1987).

Review: "New Methods in the Formation of Monocyclic Lactams and Lactones of Biological Interest"

VII.G-16 Ae. de Groot and T.A. van Beek, Recl. Trav. Chim., 106, 1 (1987).

Review: "Terpenoid Antifeedants (Part II). The Synthesis of Drimane and Clerodane Insect Antifeedants"

VII.G-17 E.D. Matveeva et al., Russ. Chem. Rev., 55, 672(1987).

Review: "The Synthesis of Lepidoptera Pheromones"

VII.G-18 V.A. Dombrovskii et al., Russ. Chem. Rev., 55, 978 (1986).

REview: "The Synthesis of Prostacyclin and its Analogues"

VII.G-19 G. Adam and V. Marquardt, Z. Chem., 41 (1987).

Review: "Synthesis of Brassinosteroids"

VII.G-20 T. Suami, Pure Appl. Chem., 59, 1509 (1987).

Review: "Synthetic Ventures in Pseudo-Sugar Chemistry"

VII.H. Others

VII.H-1 E. Breitmaier et al., Synthesis, 1 (1987).

Review: "3-Alkoxyacroleins in Organic Synthesis"

VII.H-2 B. Alcermark, Chem. Scr., 27 (1987).

Symposium: "Synthetic Selectivity: A Time of Opportunity"

VII.H-3 V. Ramamurthy and K. Venkatesan, Chem. Rev., 87, 433 (1987).

Review: "Photochemical Reactions of Organic Cystals"

VII.H-5 J.J. McCullough, Chem. Rev., 87, 811 (1987).

Review: "Photoadditions of Aromatic Compounds"

VII.H-6 C.O. Buchecker and J.-P. Sauvage, Chem. Rev., 87, 795 (1987).

Review: "Interlocking of Molecular Threads: From the Statistical Approach to the Templated Synthesis of Catenands"

VII.H-7 A.S. Kertes and C.J. King, Chem. Rev., 87, 687 (1987).

Review: "Extraction Chemistry of Low Molecular Weight Aliphatic Alcohols"

VII.H-8 G.P. Ellis and T.M. Romney-Alexander, Chem. Rev., 87, 779 (1987).

Review: "Cyanation of Aromatic Halides"

VII.H-9 G.A. Olah et al., Chem. Rev., 87, 671 (1987).

Review: "Formylating Agents"

VII.H-10 T.W. Bastock, Spec. Chem., 7, 382 (1987).

Review: "New Synthetic Routes to Old Molecules"

VII.H-11 Z.V. Todres, Tetrahedron, 43, 3839 (1987).

Tetrahedron Report 224: "Ethylenic Compounds as Probes for the Study of Donor-Acceptor Interaction"

VII.H-12 G.A. Olah, Acc. Chem. Res., 20, 422 (1987).

Review: "Electrophilic Methane Conversion"

VII.H-13 R. West, Angew. Chem. Int. Ed. Engl., 26, 1201 (1987).

Review: "Chemistry of the Silicon-Silicon Double Bond"

VII.H-14 A. Collet, Tetrahedron, 43, 5725 (1987).

Tetrahedron Report 226: "Cyclotriveratrylenes and Cryptophanes"

VII.H-15 B. Halton and P.J. Stang, Acc. Chem. Res., 20, 443 (1987).

Review: "Alkylidenecycloproparenes and Related Compounds"

VII.H-16 G.R. Krow, Tetrahedron, 43, 1 (1987).

Review: "One Carbon Ring Expansions of Bridged Bicyclic Ketones"

VII.H-17 M. Tanaka, J. Synth. Org. Chem., Jpn., 45, 716 (1987).

Review: "A Perspective View of Carbon Monoxide Chemistry"

VII.H-18 E. Breitmaier *et al.*, *Synthesis*, 1 (1987).

Review: "3-Alkoxyacroleins in Organic Synthesis"

VII.H-19 D.J.R. Massy, *Synthesis*, 589 (1987).

Review: "α-Thioalkylation *via* Aldehydes and Thiols"

VII.H-20 H. Viola, *Z.Chem.*, *27*, 15 (1987).

Review: "Reagents for Thiolation in Organic Chemistry"

VII.H-21 J.L. Charlton and M.M. Alauddin, *Tetrahedron*, *43*, 2873 (1987).

Tetrahedron Report: "Orthoquinodimethanes"

VII.H-22 M. Tisler, *Org. Prep. Proced. Int.*, *18*, 17 (1986).

Review: "Synthetic Approaches to Bisnaphthalenes"

VII.H-23 B.R. Venepalli and W.C. Agosta, *Chem. Rev.*, *87*, 399 (1987).

Review: "Fenestranes and the Flattening of Tetrahedral Carbon"

VII.H-24 M.A. Parish, *Spec. Chem.*, *7*, 366 (1987).

Review: "Cyclodextrins, A Review"

VII.H-25 N. Celebi, FABAD Farm. Bil. Der., 12, 5 (1987).

Review: "Cyclodextrins. I. Properties, Methods of Preparation and Their Clathrate Compounds"

VII.H-26 H. Hartmann and D. Klemm, Z. Chem., 27, 1 (1987).

Review: "Biologically Active Functionalization of Macromolecular Organic Compounds"

VII.H-27 G.R.Krow, Tetrahedron, 43, 3 (1987).

Review: "One Carbon Ring Expansions of Bridged Bicyclic Ketones"

VII.H-28 U. Pindur et al., chem. Soc. Rev., 16, 75 (1987).

Review: "Ortho Esters and Dialkoxycarbenium Ions: Reactivity, Stability, Structure and New Synthetic Applications"

VII.H-29 M.G. Vinogradov et al., Russ. Chem. Rev., 55, 1130 (1986).

Review: "Oxallyl Complexes in Organic Synthesis"

VII.H-30 L.A. Pavlova et al., Russ. Chem. Rev., 55, 1026 (1986).

Review: "Alkyl Orthoesters and Their Applications in Organic Synthesis"

VII.H-31 D. Ginsburg, Top. Curr.Chem., 137, 1 (1987).

Review: "Of Propellanes -- And of Spirans"

VII.H-32 V.M. Dzhemilev et al., Russ. Chem. Rev., 56, 36 (1987).

Review: "Norbornadienes in the Synthesis of Polycyclic Strained Hydrocarbons with Participation of Metal Complex Catalysts"